AutoCAD 14 Companion

James A. Leach
University of Louisville

I R W I N
GRAPHICS
S E R I E S

WCB
McGraw-Hill

Boston   Burr Ridge, IL   Dubuque, IA   Madison, WI   New York   San Francisco   St. Louis
Bangkok   Bogotá   Caracas   Lisbon   London   Madrid
Mexico City   Milan   New Delhi   Seoul   Singapore   Sydney   Taipei   Toronto

The Irwin Graphics Series

Providing you with the highest quality textbooks that meet your changing needs requires feedback, improvement, and revision. The team of authors and McGraw-Hill are committed to this effort. We invite you to become part of our team by offering your wishes, suggestions, and comments for future editions and new products and texts.

Please mail or fax your comments to: Jim Leach
c/o McGraw-Hill
1333 Burr Ridge Parkway
Burr Ridge, IL 60521
fax 630-789-6946

TITLES IN THE IRWIN GRAPHICS SERIES INCLUDE:

Technical Graphics Communication, 2e
Bertoline, Wiebe, Miller, and Mohler, 1997

Fundamentals of Graphics Communication, 2e
Bertoline, Wiebe, and Miller, 1998

Engineering Graphics Communication
Bertoline, Wiebe, Miller, and Nasman, 1995

Problems for Engineering Graphics Communication and Technical Graphics Communication Workbook #1, *1995*

Problems for Engineering Graphics Communication and Technical Graphics Communication Workbook #2, *1995*

AutoCAD Instructor Release 12
James A. Leach, 1995

AutoCAD Companion Release 12
James A. Leach, 1995

AutoCAD 13 Instructor
James A. Leach, 1996

AutoCAD 13 Companion
James A. Leach, 1996

AutoCAD 14 Instructor
James A. Leach, 1998

AutoCAD 14 Companion
James A. Leach, 1998

The Companion for CADKEY 97
John Cherng, 1998

Hands-On CADKEY: A Guide to Versions 5, 6, and 7
Timothy J. Sexton, 1996

Modeling with AutoCAD Designer
Dobek and Ranschaert, 1996

Engineering Design and Visualization Workbook
Dennis Stevenson, 1995

Graphics Interactive CD-ROM
Dennis Lieu, 1997

To John Damron Collins

"You can do anything if you
just believe in yourself."

WCB/McGraw-Hill

A Division of The **McGraw·Hill** *Companies*

AutoCAD 14 Companion

This book is printed on acid-free paper.

1   2   3   4   5   6   7   8   9   0   CSI/CSI   1   0   9   8   7

ISBN 0256266921

Vice president and editorial director: *Kevin T. Kane*
Publisher: *Tom Casson*
Executive editor: *Elizabeth A. Jones*
Senior developmental editor: *Kelley Butcher*
Marketing manager: *John T. Wannemacher*
Senior project manager: *Beth Cigler*
Production supervisor: *Heather Burbridge*
Designer: *Jennifer McQueen Hollingsworth*
Compositor: *Karen Collins and Bryan Chamberlain*
Printer: *Courier Stoughton, Inc.*

http://www.mhhe.com

PREFACE

ABOUT THIS BOOK

AutoCAD in One Semester
AutoCAD 14 Companion is designed to provide you with the material typically covered in a one-semester AutoCAD course. *AutoCAD 14 Companion* covers the essentials of 2D design and drafting as well as solid modeling.

Your AutoCAD 14 Companion
Because AutoCAD instruction is typically one component of your curriculum, *AutoCAD 14 Companion* is designed to be used in conjunction with other discipline-specific graphics books, such as Technical Graphics Communication or Fundamentals of Graphics Communication by Bertoline, et al.

Graphically Oriented
Because *AutoCAD 14 Companion* discusses concepts that are graphical by nature, many illustrations (approximately **1000**) are used to communicate the concepts, commands, and applications.

Pedagogical Progression
AutoCAD 14 Companion begins with small pieces of information explained in a simple form and then builds on that experience to deliver more complex ideas, requiring a synthesis of earlier concepts. The chapter exercises follow the same progression, beginning with a simple tutorial approach and ending with more challenging problems requiring a synthesis of earlier exercises.

Easy Upgrade from Release 12 or 13
AutoCAD 14 Companion is helpful if you are already an AutoCAD user but upgrading from Release 12 or 13. Release 13 and Release 14 commands, concepts, features, and variables are denoted by a "R13" or "R14" vertical bar on the edges of the pages. In this way, the book provides a useful reference to AutoCAD topics, but allows you to easily locate the Release 13 and Release 14 features.

Valuable Reference Guide
AutoCAD 14 Companion is structured to be used as a reference guide to AutoCAD. Every command throughout the book is given with a "command table" listing the possible methods of invoking the command. A complete index gives an alphabetical listing of all AutoCAD commands, command options, system variables, and concepts discussed.

For Students in Diverse Areas
AutoCAD 14 Companion is written for students in the fields of engineering, architecture, design, construction, manufacturing, and any other field that has a use for AutoCAD. Applications and examples from many fields are given throughout the text. The applications and examples are not intended to have an inclination toward a particular field. Instead, applications to a particular field are used when they best explain an idea or use of a command.

Additional Topics Available
For instruction in the full range of AutoCAD Release 14's commands and features, you may want to purchase *AutoCAD 14 Instructor* by James A. Leach, WCB/McGraw-Hill. *AutoCAD 14 Instructor* covers all of the topics discussed in this book as well as advanced selection sets, block attributes, external references, object linking and embedding, raster images and other file formats, advanced paper space, wireframe modeling, surface modeling, rendering, creating 2D drawings from 3D models, Internet tools, script and slide files, basic customization, menu customization, bonus features, and a variety of additional reference material. *AutoCAD 14 Instructor* includes 1550 illustrations in 1200 pages.

www.mhhe.com/leach
Please visit our web page at the above address. Beginning early 1998, ancillary material is available for reading or download. Questions (true-false, multiple choice, and written answer) are available for review and testing. Additional drawing problems specifically for architectural, mechanical engineering, and other engineering applications are available. Solutions for drawing problems and questions can be downloaded by requesting a password at the web site.

Have Fun
I predict you will have a positive experience learning AutoCAD. Although learning AutoCAD is not a trivial endeavor, you will have fun learning this exciting technology. In fact, I predict that more than once in your learning experience you will say to yourself, "Cool!" (or something to that effect).

James A. Leach

ABOUT THE AUTHOR

James A. Leach (B.I.D., M.Ed.) is an associate professor of engineering graphics at the University of Louisville. He began teaching AutoCAD at Auburn University early in 1984 using Version 1.4, the first version of AutoCAD to operate on IBM personal computers. Jim is currently the director and primary instructor at the AutoCAD Training Center (ATC) at the University of Louisville, one of the first fifteen centers to be authorized by Autodesk, having been established in 1985.

In his 21 years of teaching Engineering Graphics and AutoCAD courses, Jim has published numerous journal and magazine articles, drawing workbooks, and textbooks about AutoCAD and engineering graphics instruction. He has designed CAD facilities and written AutoCAD-related course materials for Auburn University, University of Louisville, the ATC at the University of Louisville, and several two-year and community colleges. Jim is the author of six AutoCAD textbooks published by Richard D. Irwin and McGraw-Hill.

CONTRIBUTING AUTHORS

Steven H. Baldock is an engineer at a consulting firm in Louisville and operates a CAD consulting firm, Infinity Computer Enterprises (ICE). Steve is an Autodesk Certified Instructor and teaches several courses at the University of Louisville AutoCAD Training Center. He has ten years experience using AutoCAD in architectural, civil, and structural design applications. Steve has degrees in engineering, computer science, and mathematics. Steve Baldock prepared material for several sections of *AutoCAD 14 Companion*, such as, text, and dimensioning. Steve also created several hundred figures used in *AutoCAD 14 Companion*.

Michael E. Beall is the owner of Computer Aided Management and Planning in Shelbyville, Kentucky. Michael offers contract services and professional training on AutoCAD as well as CAP products from Sweets Group, a division of McGraw-Hill. He has recently co-authored *AutoCAD 14 Fundamentals,* and was a contributing author to *Inside AutoCAD 14* from New Riders Publishing. Other efforts include co-authoring *AutoCAD Release 13 for Beginners* and *Inside AutoCAD LT for Windows 95.* He was also author of the ATC certified courseware *AutoCAD Release 13 for the Professional: Level I,* all from New Riders Publishing. Mr. Beall has been presenting CAD training seminars to architects and engineers since 1982 and is currently an Autodesk Certified Instructor (ACI) at the University of Louisville ATC. He is also a presenter for the *Mastering Today's AutoCAD* seminar series from At.A.Glance, Inc., an organization founded by Hugh Bathurst (www.awarenesslearning.com). Mr. Beall received a Bachelor of Architecture degree from the University of Cincinnati. Michael Beall assisted with several topics in *AutoCAD 14 Companion*. Contact Michael at 502.633.3994 or MBEALL_CAMP@compuserve.com.

ACKNOWLEDGMENTS

I want to thank the contributing authors for their assistance in writing *AutoCAD 14 Companion*. Without their help, this text could not have been as application-specific nor could it have been completed in the short time frame. I especially want to thank Steven H. Baldock for his valuable input and hard work on earlier editions of the AutoCAD textbooks.

I am very grateful to Gary Bertoline for his foresight in conceiving the Irwin Graphics Series and for including my efforts in it.

I would like to give thanks to the excellent editorial and production group at WCB/McGraw-Hill who gave their talents and support during this project, especially Betsy Jones, Kelley Butcher, and Beth Cigler.

A special thanks goes to Karen Collins and Bryan Chamberlain for the layout and design of *AutoCAD 14 Companion*. They were instrumental in fulfilling my objective of providing the most direct and readable format for conveying concepts.

Doug Clark and Kristine Clayton deserve credit for the creation of the "command tables" used throughout the text. Thanks for meeting the deadlines.

I appreciate the time and attention given by Mike Anderson in quality checking the Chapter Exercises with a fine-tooth comb.

I also acknowledge: my colleague and friend, Robert A. Matthews, for his support of this project and for doing his job well; Charles Grantham of Contemporary Publishing Company of Raleigh, Inc., for generosity and consultation; and Speed Scientific School Dean's Office for support and encouragement.

Special thanks, once again, to my wife, Donna, for the many hours of copy editing required to produce this and the other texts.

TRADEMARK AND COPYRIGHT ACKNOWLEDGMENTS

The following drawings used for Chapter Exercises appear courtesy of James A. Leach, *Problems in Engineering Graphics Fundamentals, Series A* and *Series B*, ©1984 and 1985 by Contemporary Publishing Company of Raleigh, Inc.: Gasket A, Gasket B, Pulley, Holder, Angle Brace, Saddle, V-Block, Bar Guide, Cam Shaft, Bearing, Cylinder, Support Bracket, Corner Brace, and Adjustable Mount. Reprinted or redrawn by permission of the publisher.

The following are registered U.S. trademarks (®) of Autodesk, Inc.: 3D Studio, 3D Studio MAX, Advanced Modeling Extension, AME, ATC, AutoCAD, Autodesk LT, Autodesk, Autodesk Animator, AutoLISP, AutoShade, AutoSurf, AutoVision, Heidi, and Mechanical Desktop. The following are currently pending U.S. trademarks (™) of Autodesk, Inc.: Autodesk Device Interface, AutoFlix, AutoSnap, DXF, FLI, FLIC, Kinetix, and ObjectARX. All AutoCAD Release 13 and Release 14 sample drawings appearing throughout the text are reprinted courtesy of Autodesk, Inc. Windows 95, Windows NT,

Notepad, WordPad, Excel, and MS-DOS are registered trademarks of Microsoft Corporation. Corel WordPerfect is a registered trademark of Corel, Inc. Norton Editor is a registered trademark of S. Reifel & Company. City Blueprint, Country Blueprint, EuroRoman, EuroRoman-oblique, PanRoman, Super-French, Romantic, Romantic-bold, Sans Serif, Sans Serif-bold, Sans Serif-oblique, Sans Serif-BoldOblique, Technic Technic-light, and Technic-bold are Type 1 fonts, copyright 1992 P. B. Payne.

LEGEND

The following special treatment of characters and fonts in the textual content is intended to assist you in translating the meaning of words or sentences in ***AutoCAD 14 Companion.***

<u>Underline</u>	Emphasis of a word or an idea.
Helvetica font	An AutoCAD prompt appearing on the <u>screen</u> at the command line or in a text window.
Italic (Upper and Lower)	An AutoCAD command, option, menu, toolbar, or dialog box name.
UPPER CASE	A file name.
UPPER CASE ITALIC	An AutoCAD system variable or a drawing aid (*OSNAP, SNAP, GRID, ORTHO*).

Anything in **Bold** represents user input:

Bold	What you should <u>type</u> or press on the keyboard.
Bold Italic	An AutoCAD <u>command</u> that you should type or <u>menu item</u> that you should select.
BOLD UPPER CASE	A <u>file name</u> that you should type.
BOLD UPPER CASE ITALIC	A <u>system variable</u> that you should type.
PICK	Move the cursor to the indicated position on the screen and press the <u>select</u> button (button #1 or left mouse button).

TABLE OF
CONTENTS

TABLE OF CONTENTS

INTRODUCTION

WHAT IS CAD?

CAD is an acronym for Computer-Aided Design or Computer-Aided Drafting. CAD allows you to accomplish design and drafting activities using a computer. A CAD software package, such as AutoCAD, enables you to create designs and generate drawings to document those designs.

Design is a broad field involving the process of making an idea into a real product or system. The design process requires repeated refinement of an idea or ideas until a solution results—a manufactured product or constructed system. Traditionally, design involves the use of sketches, drawings, renderings, 2-dimensional and 3-dimensional models, prototypes, testing, analysis, and documentation. Drafting is generally known as the production of drawings that are used to document a design for manufacturing or construction or to archive the design.

CAD is a <u>tool</u> that can be used for design and drafting activities. CAD can be used to make "rough" idea drawings, although it is more suited to creating accurate finished drawings and renderings. CAD can be used to create a 2-dimensional or 3-dimensional computer model of the product or system for further analysis and testing by other computer programs. In addition, CAD can be used to supply manufacturing equipment such as lathes, mills, laser cutters, or rapid prototyping equipment with numerical data to manufacture a product. CAD is also used to create the 2-dimensional documentation drawings for communicating and archiving the design.

The tangible result of CAD activity is usually a drawing generated by a plotter or printer but can be a rendering of a model or numerical data for use with another software package or manufacturing device. Regardless of the purpose for using CAD, the resulting drawing or model is stored in a CAD file. The file consists of numeric data in binary form usually saved to a magnetic or optical device such as a diskette, hard disk, tape, or CD.

WHY SHOULD YOU USE CAD?

Although there are other methods used for design and drafting activities, CAD offers the following advantages over other methods in many cases:

1.   Accuracy
2.   Productivity for repetitive operations
3.   Sharing the CAD file with other software programs

Accuracy

Since CAD technology is based on computers, it offers great accuracy compared to older "manual" methods of drafting and design. When you draw with a CAD system, the graphical elements, such as lines, arcs, and circles, are stored in the CAD file as numeric data. CAD systems store that numeric data with great precision. For example, AutoCAD stores values with fourteen significant digits. The value 1, for example, is stored in scientific notation as the equivalent of 1.0000000000000. This precision provides you with the ability to create designs and drawings that are 100% accurate for almost every case.

Productivity for Repetitive Operations

It may be faster to create a simple "rough" drawing, such as a sketch by hand (pencil and paper), than it would by using a CAD system. However, for larger and more complex drawings, particularly those involving similar shapes or repetitive operations, CAD methods are very efficient. Any kind of shape or operation accomplished with the CAD system can be easily duplicated since it is stored in a CAD file. In short, it may take some time to set up the first drawing and create some of the initial geometry, but any of the existing geometry or drawing setups can be easily duplicated in the current drawing or for new drawings.

Likewise, making changes to a CAD file (known as editing) is generally much faster than making changes to a traditional manual drawing. Since all the graphical elements in a CAD drawing are stored, only the affected components of the design or drawing need to be altered, and the drawing can be plotted or printed again or converted to other formats.

As CAD and the associated technology advance and software becomes more interconnected, more productive developments are available. For example, it is possible to make a change to a 3-dimensional model that automatically causes a related change in the linked 2-dimensional engineering drawing. One of the main advantages of these technological advances is productivity.

Sharing the CAD File with Other Software Programs

Of course, CAD is not the only form of industrial activity that is making technological advances. Most industries use computer software to increase capability and productivity. Since software is written using digital information and may be written for the same or similar computer operating systems, it is possible and desirable to make software programs with the ability to share data or even interconnect, possibly appearing simultaneously on one screen.

For example, word processing programs can generate text that can be imported into a drawing file, or a drawing can be created and imported into a text file as an illustration. (This book is a result of that capability.) A drawing created with a CAD system such as AutoCAD can be exported to a finite element analysis program that can read the computer model and compute and analyze stresses. CAD files can be dynamically "linked" to spreadsheets or databases in such a way that changing a value in a spreadsheet or text in a database can automatically make the related change in the drawing, or vice versa.

Another advance in CAD technology is the automatic creation and interconnectivity of a 2-dimensional drawing and a 3-dimensional model in one CAD file. With this tool, you can design a 3-dimensional model and have the 2-dimensional drawings automatically generated. The resulting set has bi-directional associativity; that is, a change in either the 2-dimensional drawings or the 3-dimensional model is automatically updated in the other.

CAD, however, may not be the best tool for every design related activity. For example, CAD may help develop ideas but probably won't replace the idea sketch, at least not with present technology. A 3-dimensional CAD model can save much time and expense for some analysis and testing but cannot replace the "feel" of an actual model, at least not until virtual reality technology is developed and refined.

With everything considered, CAD offers many opportunities for increased accuracy, productivity, and interconnectivity. Considering the speed at which this technology is advancing, many more opportunities are rapidly obtainable. However, we need to start with the basics. Beginning by learning to create an AutoCAD drawing is a good start.

WHY USE AutoCAD?

CAD systems are available for a number of computer platforms: laptops, personal computers (PCs), workstations, and mainframes. AutoCAD, offered to the public in late 1982, was one of the first PC-based CAD software products. Since that time, it has grown to be the world leader in market share for

<u>all</u> CAD products. Autodesk, the manufacturer of AutoCAD, is the world's leading supplier of PC design software and multimedia tools. At the time of this writing, Autodesk is the fifth largest software producer in the world and has three million customers in more than 150 countries.

Learning AutoCAD offers a number of advantages to you. Since AutoCAD is the most widely used CAD software, using it gives you the highest probability of being able to share CAD files and related data and information with others.

As a student, learning AutoCAD, as opposed to learning another CAD software product, gives you a higher probability of using your skills in industry. Likewise, there are more employers who use AutoCAD than any other single CAD system. In addition, learning AutoCAD as a first CAD system gives you a good foundation for learning other CAD packages because many concepts and commands introduced by AutoCAD are utilized by other systems. In some cases, AutoCAD features become industry standards. The .DXF file format, for example, was introduced by Autodesk and has become an industry standard for CAD file conversion between systems.

As a professional, using AutoCAD gives you the highest possibility that you can share CAD files and related data with your colleagues, vendors, and clients. Compatibility of hardware and software is an important issue in industry. Maintaining compatible hardware and software allows you the highest probability for sharing data and information with others as well as offering you flexibility in experimenting with and utilizing the latest technological advancements. AutoCAD provides you with the greatest compatibility in the CAD domain.

This introduction is not intended as a selling point but to remind you of the importance and potential of the task you are about to undertake. If you are a professional or a student, you have most likely already made up your mind that you want to learn to use AutoCAD as a design or drafting tool. If you have made up your mind, then you can accomplish anything. Let's begin.

GETTING STARTED

Chapter Objectives

After completing this chapter you should:

1. understand how the X, Y, Z coordinate system is used to define the location of drawing elements in digital format in a CAD drawing file;

2. understand why you should create drawings full size in the actual units with CAD;

3. be able to start AutoCAD to begin drawing;

4. recognize the areas of the AutoCAD Drawing Editor and know the function of each;

5. be able to use the five methods of entering commands;

6. be able to turn on and off the *SNAP*, *GRID*, and *ORTHO* drawing aids;

7. know how to customize the AutoCAD for Windows screen to your preferences.

CONCEPTS

Coordinate Systems

Any location in a drawing, such as the endpoint of a line, can be described in X, Y, and Z coordinate values (Cartesian coordinates). If a line is drawn on a sheet of paper, for example, its endpoints can be charted by giving the distance over and up from the lower-left corner of the sheet (Fig. 1-1).

These distances, or values, can be expressed as X and Y coordinates; X is the horizontal distance from the lower left corner (origin) and Y is the vertical distance from that origin. In a three-dimensional coordinate system, the third dimension, Z, is measured from the origin in a direction perpendicular to the plane defined by X and Y.

Two-dimensional (2D) and three-dimensional (3D) CAD systems use coordinate values to define the location of drawing elements such as lines and circles (called <u>objects</u> in AutoCAD).

In a 2D drawing, a line is defined by the X and Y coordinate values for its two endpoints (Fig 1-2).

In a 3D drawing, a line can be created and defined by specifying X, Y, and Z coordinate values (Fig. 1-3). Coordinate values are always expressed by the X value first separated by a comma, then Y, then Z.

Figure 1-1

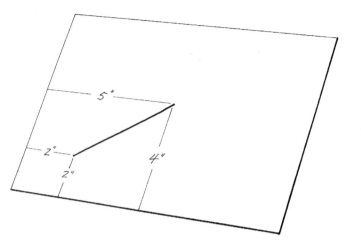

Figure 1-2

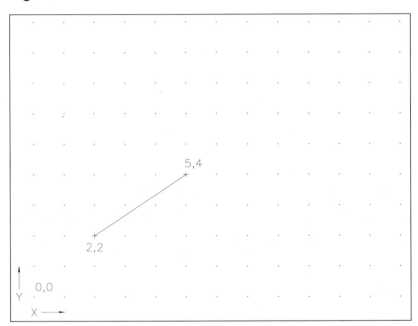

Figure 1-3

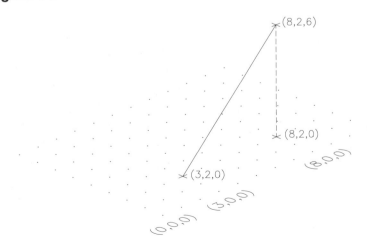

The CAD Database

A CAD (Computer-Aided Design) file, which is the electronically stored version of the drawing, keeps data in binary digital form. These digits describe coordinate values for all of the endpoints, center points, radii, vertices, etc. for all the objects composing the drawing, along with another code that describes the kinds of objects (line, circle, arc, ellipse, etc.). Figure 1-4 shows part of an AutoCAD DXF (Drawing Interchange Format) file giving numeric data defining lines and other objects. Knowing that a CAD system stores drawings by keeping coordinate data helps you understand the input that is required to create objects and how to translate the meaning of prompts on the screen.

Figure 1-4

```
LINE
  8
0
   62
        8
  10
15.0
  20
8.105789
  30
0.0
  11
15.644291
  21
8.75
  31
0.0
   0
LINE
  8
0
   62
        8
  18
```

Angles in AutoCAD

Figure 1-5

Angles in AutoCAD are measured in a <u>counter-clockwise direction</u>. Angle 0 is positioned in a positive X direction, that is, horizontally from left to right. Therefore, 90 degrees is in a positive Y direction, or straight up; 180 degrees is in a negative X direction, or to the left; and 270 degrees is in a negative Y direction, or straight down (Fig. 1-5).

The position and direction of measuring angles in AutoCAD can be changed; however, the defaults listed here are used in most cases.

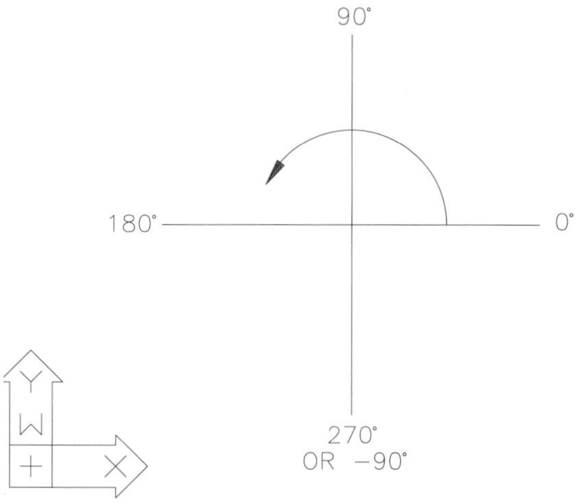

Draw True Size

Figure 1-6

When creating a drawing with pencil and paper tools, you must first determine a scale to use so the drawing will be proportional to the actual object and will fit on the sheet (Fig. 1-6). However, when creating a drawing on a CAD system, there is no fixed size drawing area. The number of drawing units that appear on the screen is variable and is assigned to fit the application.

The CAD drawing is not scaled until it is physically transferred to a fixed size sheet of paper by plotter or printer.

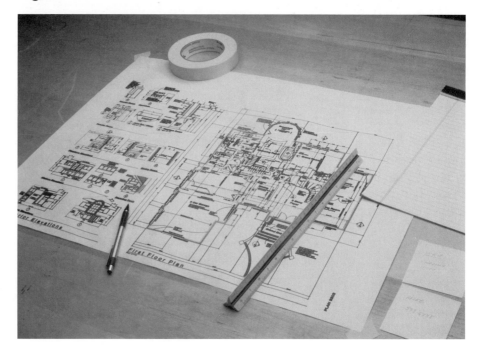

The rule for creating CAD drawings is that the drawing should be created <u>true size</u> using real-world units. The user specifies what units are to be used (architectural, engineering, etc.) and then specifies what size drawing area is needed (in X and Y values) to draw the necessary geometry.

Figure 1-7

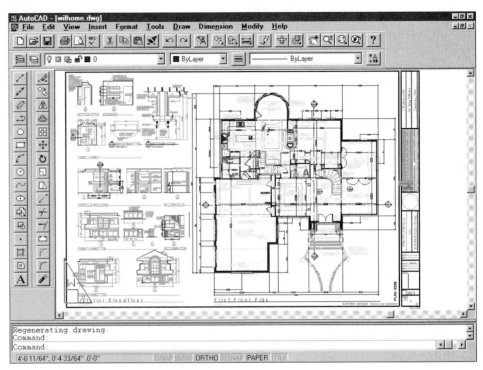

Whatever the specified size of the drawing area, it can be displayed on the screen in its entirety (Fig. 1-7) or as only a portion of the drawing area (Fig. 1-8).

Figure 1-8

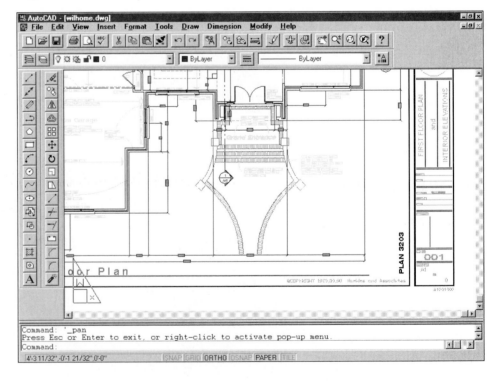

Plot to Scale

As long as a drawing exists as a CAD file or is visible on the screen, it is considered a virtual, full-sized object. Only when the CAD drawing is transferred to paper by a plotter or printer is it converted (usually reduced) to a size that will fit on a sheet. A CAD drawing can be automatically scaled to fit on the sheet regardless of sheet size; however, this action results in a plotted drawing that is not to an accepted scale (not to a regular proportion of the real object). Usually it is desirable to plot a drawing so that the resulting drawing is a proportion of the actual object size. The scale to enter as the plot scale (Fig. 1-9) is simply the proportion of the <u>plotted drawing</u> size to the <u>actual object</u>.

Figure 1-9

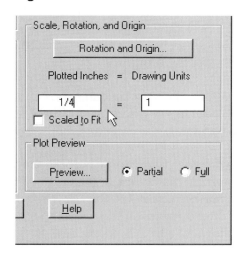

STARTING AutoCAD

Figure 1-10

Assuming that AutoCAD has been installed and configured properly for your system, you are ready to begin using AutoCAD.

To start AutoCAD for Windows 95 or Windows NT, locate the "AutoCAD R14" shortcut icon on the desktop (Fig. 1-10). Double-clicking on the icon (point the arrow and press the left mouse button quickly two times) opens AutoCAD Release 14.

If you cannot locate a shortcut icon on the desktop, press the "Start" button, highlight "Programs," and search for "AutoCAD R14" in the menu. From the list that appears, select "AutoCAD R14."

To exit AutoCAD, either select *Exit* from the *Files* pull-down menu or click on the "X" in the very upper right corner of the AutoCAD screen.

Figure 1-11

THE AutoCAD DRAWING EDITOR

The *Start Up* Dialog Box

When you start AutoCAD Release 14 for the first time, the *Start Up* dialog box appears (Fig. 1-11). This dialog box includes several tools to help you open an existing drawing or set up a new drawing, including providing access to drawing templates and a setup "Wizard." If you want to <u>start AutoCAD with the default settings</u>, you can select either of the following options:

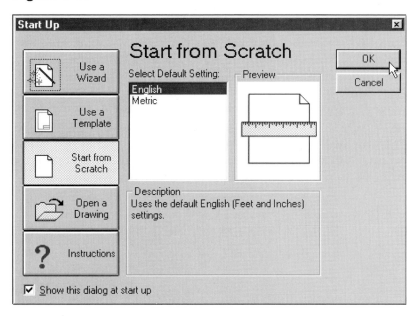

1. Select the *Start from Scratch* button on the left. Text appears in the center portion of the box displaying a choice of *English* or *Metric*. Select *English,* then press *OK.*
2. Select the *Cancel* button. Because AutoCAD's default settings are already set for use with English units, choosing *Cancel* (when starting AutoCAD) accomplishes the same action as selecting an *English* setup.

NOTE: The *Start Up* dialog box and *Create New Drawing* dialog box tools are explained fully in Chapter 6 (Basic Drawing Setup) and Chapter 13 (Advanced Drawing Setup). For the examples and exercises in Chapters 1-5, use the first method above to begin a new drawing.

After specifying your choice in the *Start Up* dialog box or the *Create New Drawing* dialog box, the Drawing Editor appears on the screen and allows you to immediately begin drawing. Figure 1-12 displays the drawing editor in AutoCAD R14 for Windows 95 and Window NT 4.0.

Figure 1-12

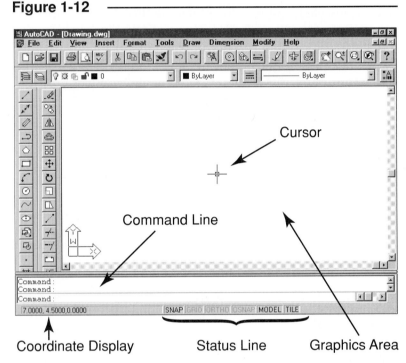

Graphics Area

The large central area of the screen is the Graphics area. It displays the lines, circles, and other objects you draw that will make up the drawing. The cursor is the intersection of the crosshairs (vertical and horizontal lines that follow the mouse or puck movements). The default size of the graphics area is 12 units (X or horizontal) by 9 units (Y or vertical). This usable drawing area (12 x 9) is called the drawing *Limits* and can be changed to any size to fit the application. As you move the cursor, you will notice the numbers in the Coordinate Display change (Fig. 1-12, bottom left).

Command Line

The Command line consists of the three text lines at the bottom of the screen (by default) and is the most important area other than the drawing itself (see Fig. 1-12). Any command that is entered or any prompt that AutoCAD issues appears here. The Command line is always visible and gives the current state of drawing activity. You should develop the habit of glancing at the Command line while you work in AutoCAD. In AutoCAD R14, the command line can be set to display any number of lines and/or moved to another location (see Customizing the AutoCAD Screen).

Pressing the F2 key opens a text window displaying the command history. This window is like an expanded Command line because it displays many more than just the 3 text lines that normally appear at the bottom of the screen. See AutoCAD Text Window, later this chapter.

Toolbars

AutoCAD Release 14 provides a variety of <u>toolbars</u> (Fig. 1-13). Each toolbar contains a number of icon buttons (tools) that can be PICKed to invoke commands for drawing or editing objects (lines, arcs, circles, etc.) or for managing files and other functions. The <u>Standard toolbar</u> is the row of icons nearest the top of the screen. The Standard toolbar contains many standard icons used in other Windows applications (like *New, Open, Save, Print, Cut, Paste,* etc.) and other icons for AutoCAD-specific functions (like *Zoom, Pan,* and *Redraw*). The <u>Object Properties toolbar</u>, located beneath the standard toolbar, is used for managing properties of objects, such as *Layers* and *Linetypes.* The <u>Draw and Modify toolbars</u> also appear (by default)

Figure 1-13

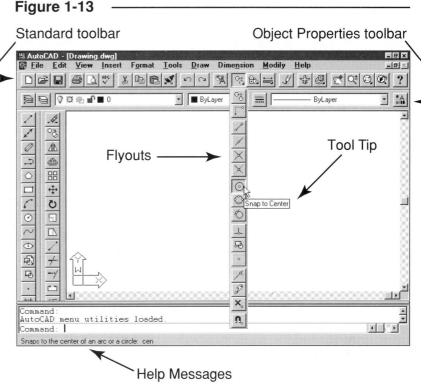

when you first use AutoCAD. As shown in Figure 1-13, the Draw and Modify toolbars (side) and the Standard and Object Properties toolbars (top) are <u>docked</u>. Many other toolbars are available and can be made to float, resize, or dock (see Customizing the AutoCAD Screen).

If you place the pointer on an any icon and wait momentarily, a <u>Tool Tip</u> and a <u>Help message</u> appear. Tool Tips pop out by the pointer and give the command name (Fig. 1-13). The Help message appears at the bottom of the screen, giving a short description of the function. <u>Flyouts</u> are groups of related icons that pop out in a row or column when one of the group is selected. PICKing any icon that has a small black triangle in its lower-right corner causes the related icons to fly out.

Pull-down Menus

The pull-down menu bar is at the top of the screen just under the title bar (Fig. 1-14). Selecting any of the words in the menu bar activates, or pulls down, the respective menu. Selecting a word appearing with an arrow activates a cascading menu with other options. Selecting a word with ellipsis (. . .) activates a dialog box (see Dialog Boxes). Words in the pull-down menus are not necessarily the same as the formal command names used when typing commands. Menus can be canceled by pressing Escape or PICKing in the graphics area. The pull-down menus <u>do not contain</u> all of the AutoCAD commands and variables but contain the most commonly used ones.

Figure 1-14

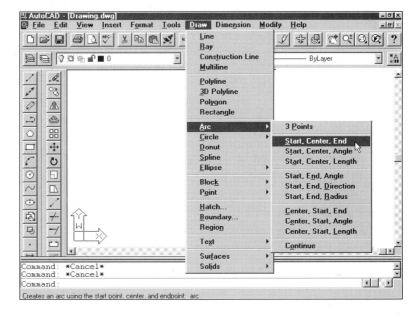

NOTE: Because the words that appear in the pull-down menus are <u>not always the same</u> as the formal commands you would type to invoke a command, AutoCAD can be confusing to learn. The Help message that appears just below the Command line (when a menu is pulled down or the pointer rests on an icon button) can be instrumental in avoiding this confusion. The Help message gives a description of the command followed by colon (:), then the <u>formal command name</u> (see Figures 1-13 and 1-14).

Screen (Side) Menu

The screen menu does not appear by default in Release 14, but can be made to appear by selecting *Preferences…* from the *Tools* pull-down menu. This selection causes the *Preferences* dialog box to appear (Fig. 1-15). Select the *Display* tab and check the *Display AutoCAD screen menu in drawing window* option. Keep in mind that this menu <u>is not needed</u> if you prefer to use the icon buttons, pull-down menus, or keyboard to enter commands.

Figure 1-15

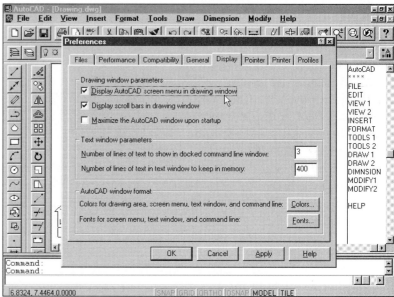

Almost all of the AutoCAD commands can be accessed through the menu system located at the right side of the screen (see Fig. 1-15). The screen menu has a tree structure; that is, other menus and commands are accessed by branching out from the root or top-level menu (Fig. 1-16). Commands (in upper and lower case) or menus (capital letters only) are selected by moving the cursor to the desired position until the word is highlighted and then pressing the PICK button. The root menu (top level) is accessed from any level of the structure by selecting the word "AutoCAD" at the top of any menu.

The commands in these menus are generally the formal command names that can also be typed at the keyboard. (Some commands names are truncated in the screen menu because only eight characters can be displayed.)

If commands are invoked by typing, pull-downs, or toolbars, the screen menu automatically changes to the current command.

Figure 1-16

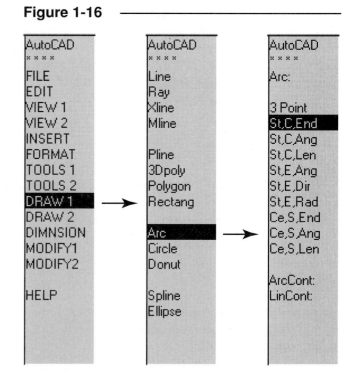

Dialog Boxes

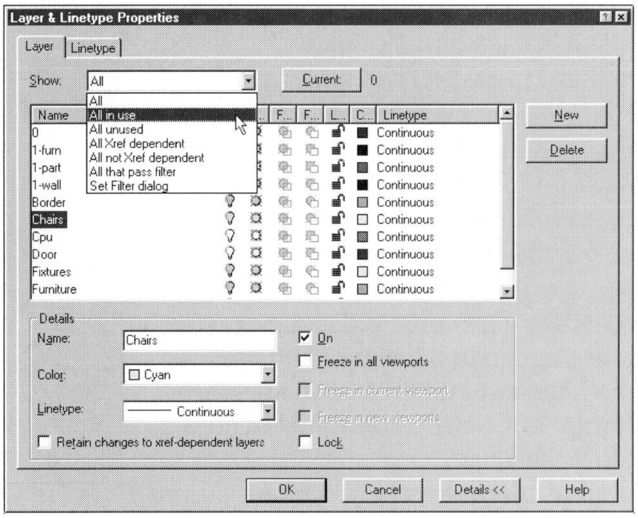

Dialog boxes provide an interface for controlling complex commands or a group of related commands. Depending on the command, the dialog boxes allow you to select among multiple options and sometimes give a preview of the effect of selections. The *Layer & Linetype Properties* dialog box (Fig. 1-17) gives complete control of layer colors, linetypes, and visibility.

Dialog boxes can be invoked by typing a command, selecting an icon button, or PICKing from the menus. For example, typing *Layer*, PICKing the *Layers* button, or selecting *Layer* from the *Format* pull-down menu causes the *Layer & Linetype Properties* dialog box to appear. In the pull-down menu, all commands that invoke a dialog box end with ellipsis points (…). If you type, sometimes you may have to type a command that begins with "DD" to invoke a dialog box; for example, typing *Ddchprop* invokes the *Change Properties* dialog box.

Figure 1-17

The basic element or smallest component of a dialog box is called a <u>tile</u>. Several types of tiles and the resulting actions of tile selection are listed below.

Button   Resembles a push button and triggers some type of action
Edit box   Allows typing or editing of a single line of text
Image tile   A button that displays a graphical image
List box   A list of text strings from which one or more can be selected
Drop-down list A text string that drops down to display a list of selections
Radio button   A group of buttons, only one of which can be turned on at a time
Checkbox   A checkbox for turning a feature on or off (displays a check mark when on)

You can change dialog box border colors (for all applications) with the Windows Control Panel. The AutoCAD screen colors can also be customized (see Customizing the AutoCAD Screen, this chapter). See File Dialog Box Functions in Chapter 2 for more information on dialog boxes that are used to access files.

New dialog boxes can be created and customized. Programming the configuration of tiles and resulting action of tile selection requires use of Dialog Control Language (DCL) to configure the tiles and AutoLISP programs to control the action of tile selection.

Status Line

The Status line is a set of informative words or symbols that gives the status of the drawing aids. The Status line appears at the very bottom of the screen (Fig. 1-12). The following drawing aids can be toggled on or off by double-clicking (quickly pressing the left mouse button twice) on the desired word or by using Function keys or Ctrl key sequences. The following drawing aids are explained here or in following chapters:

SNAP, GRID, ORTHO, OSNAP, TILE

Coordinate Display (*Coords*)

The Coordinate Display is located in the lower-left corner of the AutoCAD screen (Fig. 1-18). The Coordinate Display (*Coords*) displays the current position of the cursor in one of two possible formats explained below. This display can be very helpful when you draw because it can give the X, Y, and Z coordinate position of the cursor or give the cursor's distance and angle from the last point established. The format of *Coords* is controlled by toggling the F6 key, pressing Ctrl+D, or double-clicking on the *Coords* display (numbers) at the bottom left of the screen. *Coords* can also be toggled off.

Cursor tracking

When *Coords* is in this position, the values display the current location of the cursor in absolute (X, Y, and Z) coordinates (Fig. 1-18).

Relative polar display

This display is only possible if a draw or edit command is in use. The values give the distance and angle of the "rubberband" line from the last point established (Fig. 1-19).

Figure 1-18

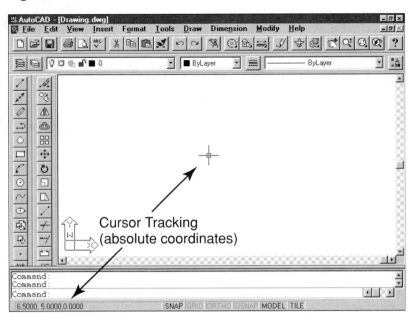

Cursor Tracking
(absolute coordinates)

Figure 1-19

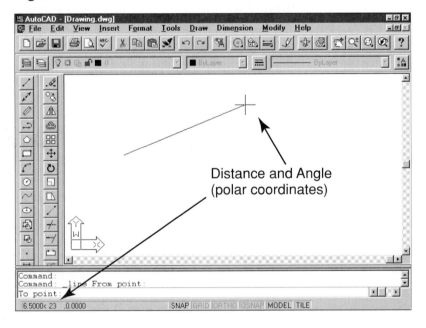

Distance and Angle
(polar coordinates)

Digitizing Tablet Menu

If you have a digitizing tablet, the AutoCAD commands are available by making the desired selection from the AutoCAD digitizing tablet menu (Fig. 1-20). The open area located slightly to the right of center is called the Screen Pointing area. Locating the digitizing puck there makes the cursor appear on the screen. Locating the puck at any other location allows you to select a command. The icons on the tablet menu are identical to the toolbar icons. Similar to the format of the toolbars, commands are located in groups such as Draw, Edit, View, Dimension, etc. The tablet menu has the columns numbered along the top and the rows lettered along the left side. The location of each command by column and row is given in the "command tables" in his book. For example, the *Line* command can be found at *10,J*.

Figure 1-20

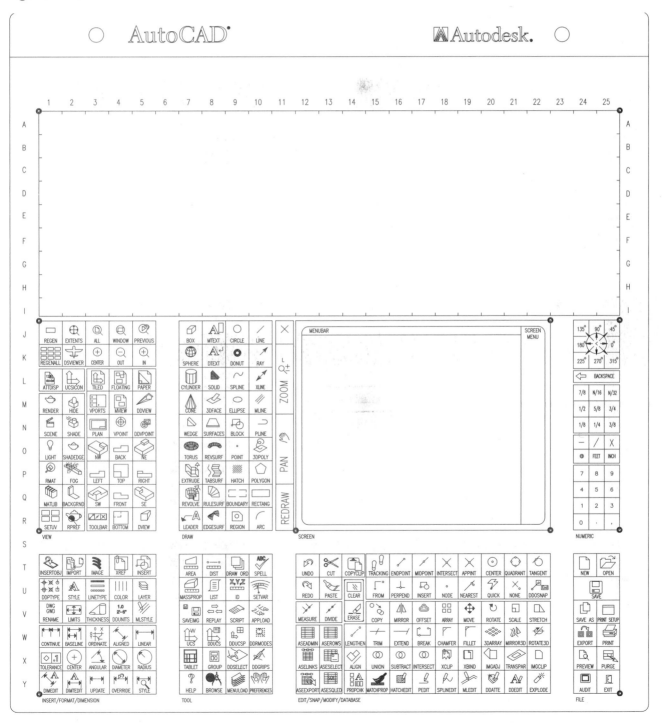

COMMAND ENTRY

Methods for Entering Commands

There are five possible methods for entering commands in AutoCAD depending on your *Preferences* setting (for the screen menu) and availability of a digitizing tablet. Generally, <u>any one</u> of the five methods can be used to invoke a particular command.

1. **Toolbars**   Select the command or dialog box by PICKing an icon (tool) from a toolbar.
2. **Pull-down menu**   Select the command or dialog box from a pull-down menu.
3. **Screen (side) menu** Select the command or dialog box from the screen (side) menu.
4. **Keyboard**   Type the command name, command alias, or accelerator (Ctrl) keys at the keyboard.*
5. **Tablet menu**   Select the command from the digitizing tablet menu (if available).

*A command alias is a one- or two-letter shortcut. Accelerator keys use the Ctrl key plus another key and are used for utility operations. The command aliases and accelerator keys are given in the command tables (see below).

All five methods of entering commands accomplish the same goals; however, one method may offer a slightly different option or advantage to another depending on the command used. A few considerations are listed below.

- Commands invoked by any method automatically change the screen menu to display the same command.
- Typing commands requires that no other commands are currently in use; therefore, the Esc (Escape) key should be used to cancel active commands before typing.
- Pull-downs are easily visible but do <u>not</u> contain all of the commands.
- The screen menus contain almost <u>all</u> AutoCAD commands; however, locating a particular command in the menu structure may take a few seconds.

All menus, including the digitizing tablet, can be customized by editing the ACAD.MNU file, and command aliases can be changed or added in the ACAD.PGP file so that command entry can be designed to your preference.

Using the "Command Tables" in This Book to Locate a Particular Command

Command tables, like the one below, are used throughout this book to show the possible methods for entering a particular command. The table shows the icon used in the toolbars and digitizing tablet, gives the selections to make for the pull-down and screen (side) menus, gives the correct spelling for entering commands and command aliases at the keyboard, and gives the command location (column, row) on the digitizing tablet menu. This example uses the *Copy* command.

COPY

Pull-down Menu	COMMAND (TYPE)	ALIAS (TYPE)	Short-cut	Screen (side) Menu	Tablet Menu
Modify *Copy*	COPY	CO	...	*MODIFY1* *Copy*	V,15

Mouse and Digitizing Puck Buttons

Depending on the type of mouse or digitizing puck used for cursor control, a different number of buttons is available. In any case, the buttons perform the following tasks:

#1 (left mouse)	**PICK**	Used to select commands or point to locations on screen.
#2 (right mouse)	**Enter**	Generally, performs the same action as the Enter or Return key on the keyboard. When some dialog boxes are visible, right-clicking produces a menu of choices affecting the active dialog box.
#3	*OSNAP*	Activates the cursor *OSNAP* menu.
#4	**Cancel**	Cancels a command.

Function Keys

Several function keys are usable with AutoCAD. They offer a quick method of turning on or off (toggling) drawing aids.

F1	*Help*	Opens a help window providing written explanations on commands and variables.
F2	*Flipscreen*	Activates a text window showing the previous command line activity (command history).
F3	*Osnap Toggle*	If Running Osnaps are set, toggling this tile temporarily turns the Running Osnaps off so that a point can be picked without using Osnaps. If no Running Osnaps are set, F3 produces the *Osnap Settings* dialog box (discussed in Chapter 7).
F4	*Tablet*	Turns the *TABMODE* variable on or off. If *TABMODE* is on, the digitizing tablet can be used to digitize an existing paper drawing into AutoCAD.
F5	*Isoplane*	When using an *Isometric* style *SNAP* and *GRID* setting, toggles the cursor (with *ORTHO* on) to draw on one of three isometric planes.
F6	*Coords*	Toggles the Coordinate Display between cursor tracking mode and off. If used transparently (during a command in operation), displays a polar coordinate format.
F7	*GRID*	Turns the *GRID* on or off (see Drawing Aids).
F8	*ORTHO*	Turns *ORTHO* on or off (see Drawing Aids).
F9	*SNAP*	Turns *SNAP* on or off (see Drawing Aids).
F10	*Status*	Turns the Status line on or off.

Control Key Sequences (Accelerator Keys)

Several Control key sequences (holding down the Ctrl key and pressing another key simultaneously) invoke regular AutoCAD commands or produce special functions. The first six key sequences (Drawing Aids) have the same duties as Function keys F3 through F9.

Drawing Aids

Ctrl+F (F3)	*Osnap Toggle*	If Running Osnaps are set, pressing Ctrl+F temporarily turns the Running Osnaps off so that a point can be PICKed without using Osnaps. If there are no Running Object Snaps set, Ctrl+F produces the *Osnap Settings* dialog box. This dialog box is used to turn on and off Running Object Snaps (discussed in Chapter 7).
Ctrl+T (F4)	*Tablet*	Turns the *TABMODE* variable on or off. If *TABMODE* is on, the digitizing tablet can be used to digitize an existing paper drawing into AutoCAD.
Ctrl+E (F5)	*Isoplane*	When using an *Isometric* style *SNAP* and *GRID* setting, toggles the cursor (with *ORTHO* on) to draw on one of three isometric planes.
Ctrl+D (F6)	*Coords*	Toggles the Coordinate Display between cursor tracking mode and off. If used during a command operation, can be toggled to a polar coordinate format.

Ctrl+G	(F7)	*GRID*	Turns the *GRID* on or off (see Drawing Aids).
Ctrl+L	(F8)	*ORTHO*	Turns *ORTHO* on or off (see Drawing Aids).
Crtl+B	(F9)	*SNAP*	Turns *SNAP* on or off (see Drawing Aids).

Windows Copy, Cut, Paste

Ctrl+C	*Copyclip*	Copies the highlighted objects to the Windows clipboard.
Ctrl+X	*Cutclip*	Cuts the highlighted objects from the drawing and copies them to the Windows clipboard.
Ctrl+V	*Pasteclip*	Pastes the clipboard contents into the AutoCAD drawing as a Block Reference.

File Operations (see Chapter 2)

Ctrl+O	*Open*	Invokes the *Open* command to open an existing drawing.
Ctrl+N	*New*	Invokes the *New* command to start a new drawing.
Ctrl+S	*Qsave*	Performs a quick save or produces the *SaveAs* dialog box if the file is not yet named.

Other Control Key Sequences

Ctrl+Z	*Undo*	Undoes the last command (see Chapter 5).
Ctrl+Y	*Redo*	Invokes the *Redo* command (see Chapter 5).
Ctrl+P	*Print/Plot*	Produces the *Print/Plot Configuration* dialog box for creating and controlling prints and plots (see Chapter 14).
Ctrl+A	*Group*	Toggles selectable *Groups* on or off.
Ctrl+J	*Enter*	Has the same function as Enter.
Ctrl+K	*PICKADD*	Changes the setting (0 or 1) of the *PICKADD* variable.

Special Key Functions

Esc	The Escape key cancels a command, menu, or dialog box or interrupts processing of plotting or hatching.
Space bar	In AutoCAD, the space bar performs the same action as the Enter key or #2 button. Only when you are entering text into a drawing does the space bar create a space.
Enter	If Enter, Spacebar, or #2 button is pressed when no command is in use (the open Command: prompt is visible), the last command used is invoked again.
Up Arrow	The up arrow key can be used to recall previous command line entries, similar to the DOSKEY feature. Pressing the up arrow places the previous command at the Command: prompt, but does not execute it. You can use the up arrow repeatedly to find a previously used command, then press Enter to execute it.
Down Arrow	After the up arrow is used to recall previously used commands in reverse order, the down arrow cycles back through those commands in forward order. The up arrow and down arrow keys are used together to locate particular command line entries.

Drawing Aids

This section gives a brief introduction to AutoCAD's Drawing Aids. For a full explanation of the related commands and options, see Chapter 6.

SNAP (**F9** or **Ctrl+B**)

SNAP is a function that forces the cursor to "snap" to a regular interval (.5 units by default), which aids in drawing geometry accurate to equal interval lengths. The *Snap* command or *Drawing Aids* dialog box allows you to specify any value for the *SNAP* interval. In Figure 1-21, the *SNAP* is set to .125 (note the values in the coordinate display).

GRID (**F7** or **Ctrl+G**)

A drawing aid called *GRID* can be used to give a visual reference of units of length. The *GRID* default value is .5 units. The *Grid* command or *Drawing Aids* dialog box allows you to change the interval to any value. The *GRID* is not part of the geometry and is not

Figure 1-21

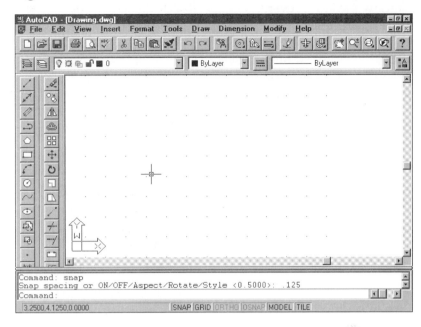

plotted. Figure 1-21 displays a *GRID* of 1.0. *SNAP* and *GRID* are independent functions—they can be turned on or off independently. However, you can force the *GRID* to have the same interval as *SNAP* by entering a *GRID* value of 0 or you can use a proportion of *SNAP* by entering a *GRID* value followed by an "X".

ORTHO (**F8** or **Ctrl+L**)

If *ORTHO* is on, lines are forced to an orthogonal alignment (horizontal or vertical) when drawing (Fig. 1-22). *ORTHO* is often helpful since so many drawings are composed mainly of horizontal and vertical lines. *ORTHO* can only be turned on or off.

Figure 1-22

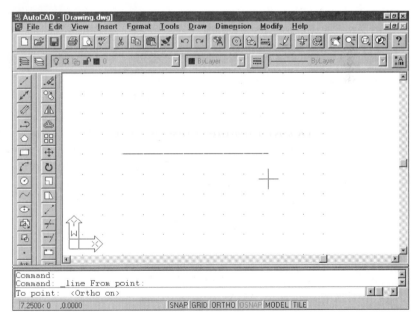

AutoCAD Text Window (F2)

Pressing the F2 key activates the *AutoCAD Text Window*, sometimes called the Command History. Here you can see the text activity that occurred at the command line—kind of an "expanded" command line. Press F2 again to close the text window.

R14

The *Edit* pull-down menu in this text window provides several options. If you highlight text in the window (Fig. 1-23), you can then *Paste to Cmdline* (command line), *Copy* it to another program such as a word processor, *Copy History* (entire command history) to another program, or *Paste* text into the window. The *Preferences* option invokes the *Preferences* dialog box (discussed later).

Figure 1-23

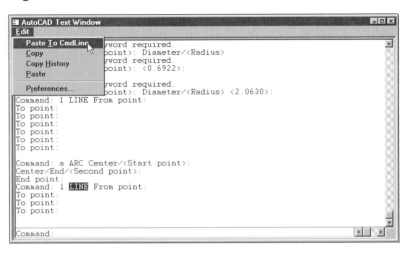

CUSTOMIZING THE AutoCAD SCREEN

Toolbars

Many toolbars are available, each with a group of related commands for specialized functions. For example, when you are ready to dimension a drawing, you can activate the Dimension toolbar for efficiency. Selecting the *View* pull-down menu, then *Toolbars...* displays the list of possible selections in the *Toolbars* dialog box (Fig. 1-24). Making a selection in the dialog box (PICKing in the checkbox) activates that toolbar. You may instead type in the *Toolbar* command or right-click on any tool (icon button) to invoke the *Toolbars* dialog box. Toolbars can be removed from the screen by clicking once on the "X" symbol in the upper-right of a floating toolbar.

Figure 1-24

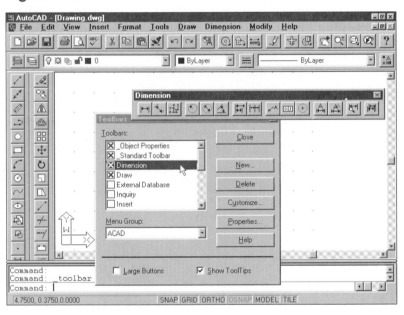

R14

By default, the Object Properties and Standard toolbars (top) and the Draw and Modify toolbars (side) are docked, whereas toolbars that are newly activated are floating (see Fig. 1-24). A floating toolbar can be easily moved to any location on the screen if it obstructs an important area of a drawing. Placing the pointer in the title background allows you to move the toolbar by holding down the left button and dragging it to a new location (Fig. 1-25). Floating toolbars can also be resized by placing the pointer on the narrow border until a two-way arrow appears, then dragging left, right, up, or down (Fig 1-26).

Figure 1-25

Figure 1-26

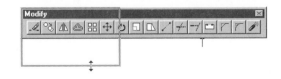

R13

A floating toolbar can be docked against any border (right, left, top, bottom) by dragging it to the desired location (Fig. 1-27). Several toolbars can be stacked in a docked position. By the same method, docked toolbars can be dragged back onto the graphics area. The Object Properties and Standard tool-bars can be moved onto the graphics area or docked on another border, although it is wise to keep these toolbars in their standard position. The Object Properties toolbar only displays its full options when located in a horizontal position.

Holding down the **Ctrl** key while dragging a toolbar near a border of the drawing window prevents the toolbar from docking. This feature enables you to float a toolbar anywhere on the screen.

Figure 1-27

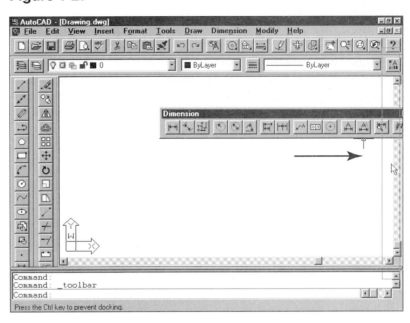

Preferences

Fonts, colors, and other features of the AutoCAD drawing editor can be customized to your liking by using the *Preferences* command and selecting the *Display* tab. Typing *Preferences* or selecting *Preferences...* from the *Tools* pull-down menu activates the dialog box shown in Figure 1-28. The screen menu can also be activated through this dialog box (first checkbox in the dialog box).

Figure 1-28

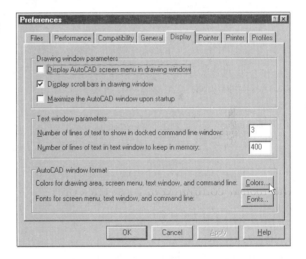

Selecting the *Color...* tile provides a dialog box for customizing the screen colors (Fig. 1-29).

Figure 1-29

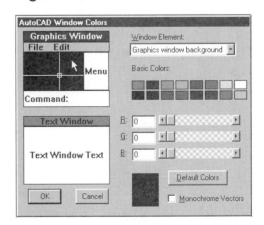

CHAPTER EXERCISES

1.   Starting and Exiting AutoCAD

Start AutoCAD by clicking the "14" shortcut icon or "AutoCAD 14" from the Programs menu. If the *Start Up* dialog box or *Create New Drawing* dialog box appears, select *Start from Scratch,* choose *English* as the default setting, and click the *OK* button. Draw a *Line.* Exit AutoCAD by selecting the *Exit* option from the *Files* pull-down menu. Answer *No* to the "Save Changes to Drawing.dwg?" prompt. Repeat these steps until you are confident with the procedure.

2.   Using Drawing Aids

Start AutoCAD. Turn on and off each of the following modes:

   SNAP, GRID, ORTHO

3.   Understanding Coordinates

Begin drawing a *Line* by PICKing a "From point:." Toggle *Coords* to display each of the <u>three</u> formats. PICK several other points at the "To point:" prompt. Pay particular attention to the coordinate values displayed for each point and visualize the relationship between that point and coordinate 0,0 (absolute value) or the last point established (relative polar value). Finish the command by pressing Enter.

4.   Using *Flipscreen*

Use *Flipscreen* (**F2**) to toggle between the text screen or window and the graphics screen.

5.   Drawing with Drawing Aids

Draw four *Lines* using <u>each</u> Drawing Aid: **GRID, SNAP, ORTHO.** Toggle on and off each of the drawing aids one at a time for each set of four *Lines.* Next, draw *Lines* using combinations of the Drawing Aids, particularly **GRID + SNAP** and **GRID + SNAP + ORTHO.**

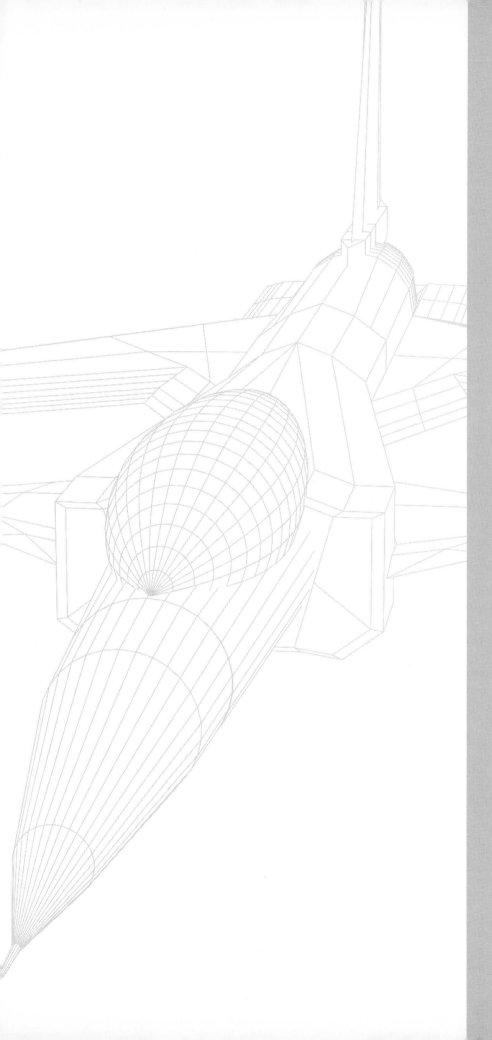

2

WORKING WITH FILES

Chapter Objectives

After completing this chapter you should be able to:

1. name drawing files;

2. use file-related dialog boxes;

3. create *New* drawings;

4. *Open* existing drawings;

5. *Save* drawings to disk;

6. use *SaveAs* to save a drawing under a different name, path, and/or format;

7. *Exit AutoCAD;*

8. practice good file management techniques;

9. use the Windows right-click shortcut menus in file dialog boxes.

AutoCAD DRAWING FILES

Naming Drawing Files

What is a drawing file? A CAD drawing file is the electronically stored data form of a drawing. The computer's hard disk is the principal magnetic storage device used for saving and restoring CAD drawing files. Diskettes, compact disks (CDs), networks, and the Internet are used to transport files from one computer to another, as in the case of transferring CAD files among clients, consultants, or vendors in industry. The AutoCAD commands used for saving drawings to and restoring drawings from files are explained in this chapter.

An AutoCAD drawing file has a name that you assign and a file extension of ".DWG." An example of an AutoCAD drawing file is:

PART-024.DWG

filename   extension

The file name you assign must be compliant with the Windows 95 or Windows NT (FAT) filename conventions; that is, it can have a <u>maximum</u> of 256 alphanumeric characters. File names and directory names (folders) can be in UPPER CASE, Title Case, or lower case letters. Characters such as _ - $ # ( ) ^ and spaces can be used in names, but other characters such as \ / : * ? < > | are not allowed. AutoCAD automatically appends the extension of .DWG to all AutoCAD-created drawing files.

NOTE: The chapter exercises and other examples in this book list file names in UPPER CASE letters for easy recognition. The file names and directory (folder) names on your system may appear as UPPER CASE, Title Case, or lower case.

Beginning and Saving an AutoCAD Drawing

When you start AutoCAD, the drawing editor appears and allows you to begin drawing even before using any file commands. If you choose, you can assign a name for the drawing when you begin with the *New* command. As you draw, you should develop the habit of saving the drawing periodically (about every 15 or 20 minutes) using *Save*. *Save* stores the drawing in its most current state to disk.

The typical drawing session would involve using *New* to begin a new drawing or using *Open* to open an existing drawing. Alternately, the *Start Up* dialog box that appears when starting AutoCAD would be used to begin a new drawing or open an existing one. *Save* would be used periodically, and *Exit* would be used for the final save and to end the session.

Figure 2-1

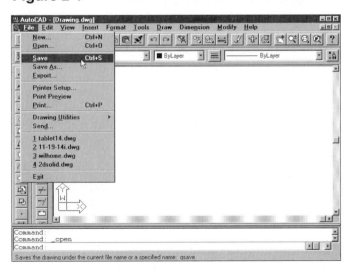

Accessing File Commands

Proper use of the file-related commands covered in this chapter allows you to manage your AutoCAD drawing files in a safe and efficient manner. Although the file-related commands can be invoked by several methods, they are easily accessible via the first pull-down menu option, *File* (Fig. 2-1). Most of the selections from this pull-down menu invoke dialog boxes for selection or specification of file names.

The Standard toolbar at the top of the AutoCAD screen has tools (icon buttons) for *New, Open,* and *Save*. File commands can also be entered at the keyboard by typing the formal command name, the command alias, or using the Ctrl keys sequences. File commands and related dialog boxes are also available from the *File* screen menu or by selection from the digitizing tablet menu.

File Dialog Box Functions

There are many dialog boxes appearing in AutoCAD that help you manage files. All of these dialog boxes operate in a similar manner. A few guidelines will help you use them. The *Save Drawing As* dialog box for Windows (Fig. 2-2) is used as an example.

Figure 2-2

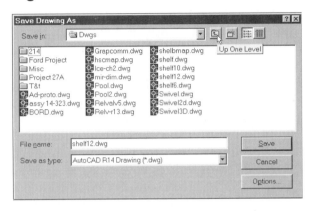

- The top of the box gives the title describing the action to be performed. It is <u>very important</u> to glance at the title before acting, especially when saving or deleting files.
- The desired file can be selected by PICKing it, then PICKing *OK*. Double-clicking on the selection accomplishes the same action. File names can also be typed in the *File name:* edit box.
- Every file name has an extension (three letters following the period) called the <u>type.</u> File types can be selected from the *Save as Type:* or *Files of Type:* section of the dialog boxes, or the desired file extension can be entered in the *File name:* edit box.
- The current folder (directory) is listed in the box near the top of the dialog box. You can select another folder (directory) or drive by using the drop-down list to display the current path or by selecting the *Up One Level* icon to the right of the list (Fig. 2-3). (Rest your pointer on an icon momentarily to make the tool tip appear.)
- A new folder (subdirectory) can be created within the current folder (directory) by selecting the *New Folder* icon.
- The last two icons to the right of the row near the top of the dialog box toggle the listing of files to a *List* or to show *Details*. The *List* option only displays file names (Fig. 2-2), whereas the *Details*

Figure 2-3

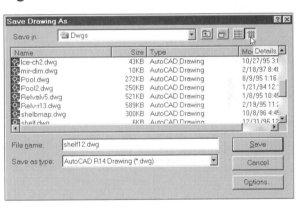

option gives file-related information such as file size, file type, and time and date last modified (Fig. 2-3). The detailed list can also be sorted alphabetically by name, by file size, alphabetically by file type, or chronologically by time and date. Do this by clicking the *Name, Type, Size,* or *Modified* tiles immediately above the list. Double-clicking on one of the tiles reverses the order of the list.
- You can resize the width of a column (*Name, Size, Type,* etc.) by moving the pointer to the "crack" between two columns, then sliding the double arrows that appear in either direction.
- Scroll bars appear at the right and/or bottom if the list is longer than one page. PICKing the arrows moves the list one character or line at a time, or sliding the button causes motion at any rate.
- As an alternative, the file name (and path) can be typed in the edit box labeled *File name:*. Make sure the name is highlighted or the blinking cursor appears in the box (move the pointer and PICK) before typing.

R14

Windows Right-Click Shortcut Menus

AutoCAD for Windows 95 or Windows NT utilizes the right-click shortcut menus that operate with software products running on the Windows 95 and NT operating systems. Activate the shortcut menus by pressing the right mouse button (right-clicking) <u>inside the file list area</u> of a dialog box. There are two menus that give you additional file management capabilities.

Right-Click, No Files Highlighted

When no files are highlighted, right-clicking produces a menu for file list display and file management options (Fig. 2-4). The menu choices are as follows:

Figure 2-4

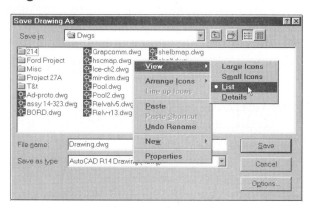

View	Shows *Large Icons* or *Small Icons* next to the file names, displays a *List* of file names only (see Fig. 2-2), or displays all file *Details* (see Fig. 2-3).
Arrange Icons	Sorts the files *by Name, by Type* (file extension), *by Size,* or *by Date.*
Paste	If the *Copy* or *Cut* function was previously used (see Right-Click, File Highlighted, Fig. 2-5), *Paste* can be used to place the copied file into the displayed folder.
Paste Shortcut	Use this option to place a *Shortcut* (to open a file) into the current folder.
Undo Rename	If the *Rename* function was previously used (see Right-Click, File Highlighted, Fig. 2-5), this action restores the original file name.
New	Creates a new *Folder, Shortcut,* or document.
Properties	Displays a dialog box listing properties of the current folder.

Right-Click, File Highlighted

When a file is highlighted, right-clicking produces a menu with options for the selected file (Fig. 2-5). The menu choices are as follows:

Figure 2-5

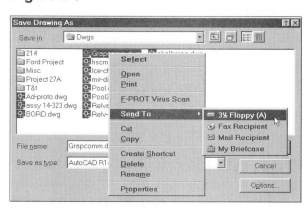

Select	Processes the file (like selecting the *OK* button) according to the dialog box function. For example, if the *SaveAs* dialog box is open, *Select* saves the highlighted file; or if the *Select File* dialog box is active, *Select* opens the drawing.
Open	Opens the application (AutoCAD or other program) and loads the selected file. <u>Do not use this function in AutoCAD unless you want to open a second AutoCAD session.</u>
Print	Sends the selected file to the configured system printer. In AutoCAD, opens the *Print/Plot Configuration* dialog box.
Send To	Copies the selected file to the selected device. <u>This is an easy way to copy a file from your hard drive to a diskette in A: drive.</u>
Cut	This option, in conjunction with *Paste,* allows you to move a file from one location to another.
Copy	In conjunction with *Paste,* allows you to copy the selected file to another location. You can copy the file to the same folder, but Windows renames the file to "Copy of . . .".

Create Shortcut	Use this option to create a *Shortcut* (to open the highlighted file) in the current folder. The Shortcut can be moved to another location using drag and drop.
Delete	Sends the selected file to the Recycle Bin.
Rename	Allows you to rename the selected file. Move your cursor to the highlighted file name, click near the letters you want to change, then type or use the backspace, delete, space, or arrow keys.
Properties	Displays a dialog box listing properties of the selected file.

Specific features of other dialog boxes are explained in the following sections.

THE *START UP* DIALOG BOX

When you start AutoCAD for the first time, the *Start Up* dialog box appears (Fig 2-6). The *Start Up* dialog box may not appear on your system if it has been suppressed (deactivated). It can be suppressed by removing the check in the box at the bottom of the dialog box labeled *Show this dialog at start up.* You can make the dialog box reappear on start up by checking *Show the Start Up dialog box* in the *Compatibility* tab of the *Preferences* dialog box.

There is no command that invokes the *Start Up* dialog box. However, the *Start Up* dialog box is identical to and <u>combines</u> the functions of the *New* and *Open* commands. Using the *New* command produces the *Create New Drawing* dialog box, which contains the *Use a Wizard, Use a Template,* and *Start from Scratch* options as seen in the *Start Up* dialog box (Fig. 2-6). Using the *Open a Drawing* option in the *Start Up* dialog performs the same action as using the *Open* command.

Figure 2-6

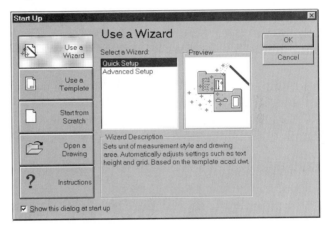

Because of the similarities of the *Start Up* dialog box and the *New* and *Open* commands, an understanding of the *Start Up* dialog box can be acquired from the following explanation of the *New* command and the *Open* command.

AutoCAD FILE COMMANDS

NEW

Pull-down Menu	COMMAND (TYPE)	ALIAS (TYPE)	Short-cut	Screen (side) Menu	Tablet Menu
File *New...*	*NEW*	...	Ctrl+N	*FILE* *New*	T,24

The *New* command produces the *Create a New Drawing* dialog box (Fig. 2-7). (Similar functions are available in the *Start Up* dialog box that may appear when you first start AutoCAD.) In this dialog box you can begin a new drawing by choosing *Use a Wizard, Use a Template,* or *Start from Scratch.* Selecting any one of these tiles changes the appearance of the center and right side of the box.

Use a Wizard

You can choose between a *Quick Setup Wizard* and an *Advanced Setup Wizard* (Fig 2-6). The *Quick Setup Wizard* prompts you to select the type of drawing *Units* you want to use and to specify the drawing *Area* (*Limits*). The *Advanced Setup* offers options for units, angular direction and measurement, area, title blocks, and layouts for paper space and model space. Both Wizards also automatically make changes to many size-related system variables. The *Quick Setup Wizard* and the *Advanced Setup Wizard* are discussed in Chapter 13, Advanced Drawing Setup.

Figure 2-7

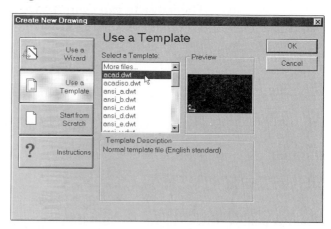

Use a Template

Use this option if you want to use another existing drawing as a template (sometimes known as a "prototype") (Fig. 2-7). A template drawing is one that already has many of the setup steps performed but contains no geometry (graphical objects).

Selecting the ACAD.DWT template begins a new drawing using the traditional AutoCAD default settings for drawing area (called *Limits*) of 12 x 9 units and other default settings. Selecting the ACAD.DWT template accomplishes the same action as using *Start from Scratch, English* option. Other template drawings are discussed in Chapter 13, Advanced Drawing Setup.

Start from Scratch

Choose from either the *English* or *Metric* default settings. Selecting *English* causes AutoCAD to use the default drawing area (called *Limits*) of 12 x 9 units (Fig. 2-8). Choosing the *Metric* option causes AutoCAD to use the metric drawing template, which has *Limits* settings for a A4 metric size sheet (297 x 210 mm).

Figure 2-8

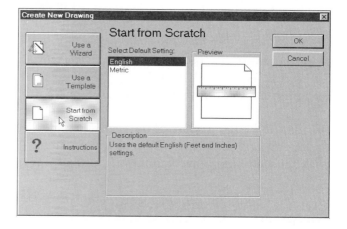

For the purposes of learning AutoCAD starting with the basic principles and commands, it is helpful to use the default AutoCAD settings when beginning a new drawing. Until you read Chapters 6 and 13, you can begin new drawings for completing the exercises and practicing the examples in Chapters 1 through 5 by any of the following methods:

Selecting the ACAD.DWT template drawing
Selecting the *English* defaults in the *Start from Scratch* option
Pressing *Cancel* in the *Setup* dialog box that appears when AutoCAD starts
Starting AutoCAD after suppressing the *Setup* dialog box

OPEN

Pull-down Menu	COMMAND (TYPE)	ALIAS (TYPE)	Short-cut	Screen (side) Menu	Tablet Menu
File Open...	OPEN	...	Ctrl+O	FILE Open	T,25

Use *Open* to select an existing drawing to be loaded into AutoCAD. Normally, you would open an existing drawing (one that is completed or partially completed) so you could continue drawing or make a print or plot. You can *Open* only one drawing at a time in AutoCAD Release 14 for Windows 95 and NT.

The *Open* command produces the *Select File* dialog box (Fig. 2-9). In this dialog box you can select any drawing from the current directory list. PICKing a drawing name from the list displays a small bitmap image of the drawing in the *Preview* tile. Select the *Open* button or double-click on the file name to open the high-lighted drawing. You could instead type the file name (and path) of the desired drawing in

Figure 2-9

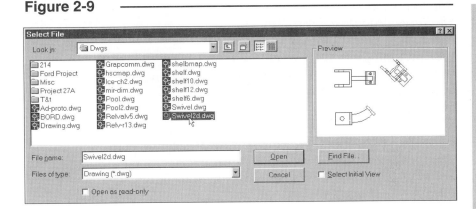

the *File name:* edit box and press Enter, but a preview of the typed entry will not appear. AutoCAD Release 12 or previous DWG files usually do not have bitmap images.

To locate a drawing in another folder (directory) or on another drive on your computer or network, select the drop-down list on top of the dialog box next to *Look in:* or use the *Up one Level* button. Remember that you can also display the file details (size, type, modified) by toggling the *Details* button. You can open a DWG (drawing), DXF (drawing interchange format), or DWT (template) file by selecting from the *Files of Type:* drop-down list.

Selecting the *Find File...* button in the *Select File* dialog box invokes the *Browse/Search* dialog box (Fig. 2-10). This box displays a small bitmap image of all drawings in the search directory. You can open a drawing by double-clicking on its image. The dialog box offers two methods to find files: *Browse* and *Search*.

Figure 2-10

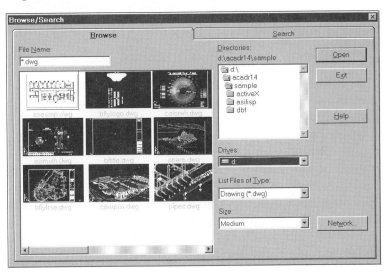

Using the *Browse* tab (Fig. 2-10), you can browse for drawing (DWG, DWT, or DXF) files in any directory. A preview image of all drawings in the selected directory (in the *Directories:* area) is displayed. The preview images can be displayed in a large, medium, or small *Size:*. Use the scroll bar along the bottom to see all drawings in the directory.

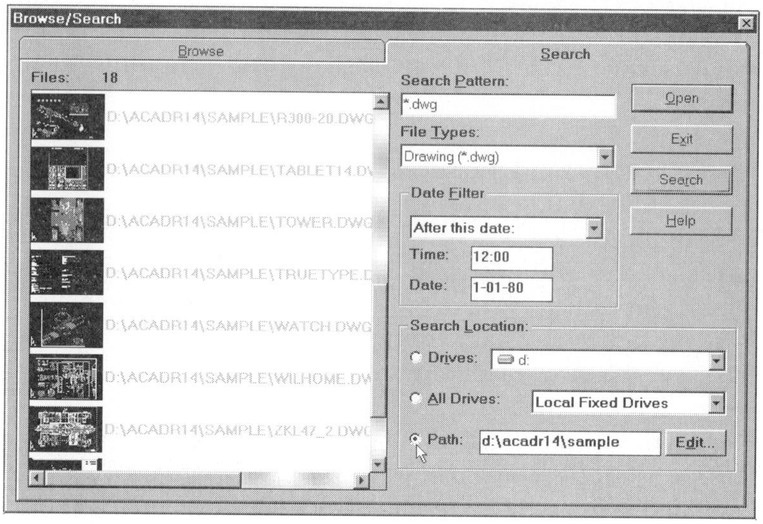

The *Search* tab of the *Browse/Search* dialog box enables you to search for files meeting specific criteria that you specify (Fig. 2-11). You can enter file names (including wildcards) in the *Search Pattern* box, and the desired drawing type can be selected from the *File Types* drop-down list. The *Date Filter* helps you search for files by *Date* and *Time* based on when the files were last saved. Any drive or path can be selected using the *Search Location* cluster.

Figure 2-11 ——————————

After specifying the search criteria, select the *Search* button. All file names found matching the criteria are displayed with the drawing preview images on the left side of the dialog box. Files can be opened by double-clicking on the name or image tile or by the *Open* button.

NOTE: It is considered poor practice to *Open* a drawing from a diskette in A: drive. Normally, drawings should be copied to a directory on a fixed (hard) drive, then opened from that directory. Opening a drawing from a diskette and performing numerous *Saves* is much slower and less reliable than operating from a hard drive. See other NOTEs at the end of the discussion on *Save* and *SaveAs*.

SAVE

Pull-down Menu	COMMAND (TYPE)	ALIAS (TYPE)	Short-cut	Screen (side) Menu	Tablet Menu
File *Save As*	SAVE	...	Ctrl+S	...	U,24-U,25

The *Save* command is intended to be used periodically during a drawing session (every 15 to 20 minutes is recommended). When *Save* is selected from a menu, the current version of the drawing is saved to disk without interruption of the drawing session. The first time an unnamed drawing is saved, the *Save Drawing As* dialog box (Figs. 2-2 and 2-3) appears, which prompts you for a drawing name. Typically, however, the drawing already has an assigned name, in which case *Save* actually performs a *Qsave* (quick save). A *Qsave* gives no prompts or options, nor does it display a dialog box.

When the file has an assigned name, using *Save* by selecting the command from a menu or icon button automatically performs a quick save (*Qsave*). Therefore, *Qsave* automatically saves the drawing in the same drive and directory from which it was opened or where it was first saved. In contrast, typing *Save* always produces the *Save Drawing As* dialog box, where you can enter a new name and/or path to save the drawing or press Enter to keep the same name and path (see *SaveAs*).

NOTE: If you want to save a drawing directly to a diskette in drive A:, <u>type</u> *Save* and specify drive A:. Even though the *SaveAs* dialog is used, <u>typing *Save*</u> does not reset the path of the current drawing for subsequent saves. (See the NOTE at the end of the explanation of *SaveAs*.) Another useful method for copying a file to a diskette in A: drive is to use the *Send To* option in the shortcut menu that appears if you right-click in the *Save, SaveAs,* or *Select File* dialog box (see Windows Right-Click Shortcut Menus and Fig. 2-5).

QSAVE

Pull-down Menu	COMMAND (TYPE)	ALIAS (TYPE)	Short-cut	Screen (side) Menu	Tablet Menu
...	QSAVE	...	Ctrl+S	FILE Qsave	...

Qsave (quick save) is normally invoked automatically when *Save* is used (see *Save*) but can also be typed. *Qsave* saves the drawing under the previously assigned file name. No dialog boxes appear nor are any other inputs required. This is the same as using *Save* (from a menu), assuming the drawing name has been assigned. However, if the drawing has not been named when *Qsave* is invoked, the *Save Drawing As* dialog box appears.

SAVEAS

Pull-down Menu	COMMAND (TYPE)	ALIAS (TYPE)	Short-cut	Screen (side) Menu	Tablet Menu
File Save As...	SAVEAS	...	...	FILE Saveas	V,24

The *SaveAs* command can fulfill four functions:

1. save the drawing file under a new name if desired;
2. save the drawing file to a new path (drive and directory location) if desired;
3. in the case of either 1 or 2, assign the new file name and/or path to the current drawing (change the name and path of the current drawing);
4. save the drawing in a format other than the default Release 14 format (but does not reassign the current drawing format for future saves).

Therefore, assuming a name has previously been assigned, *SaveAs* allows you to save the current drawing under a different name and/or path; but, beware, <u>*SaveAs* sets the current drawing name and/or path to the last one entered.</u> This dialog box is shown in Figures 2-2 and 2-3.

SaveAs can be a benefit when creating two similar drawings. A typical scenario follows. A design engineer wants to make two similar but slightly different design drawings. During construction of the first drawing, the engineer periodically saves under the name DESIGN1 using *Save*. The first drawing is then completed and *Saved*. Instead of starting a *New* drawing, *SaveAs* is used to save the current drawing under the name DESIGN2. *SaveAs* also resets the current drawing name to DESIGN2. The designer then has two separate but identical drawing files on disk which can be further edited to complete the specialized differences. The engineer continues to work on the current drawing DESIGN2.

NOTE: If you want to save the drawing to a diskette in A: drive, do not use *SaveAs*. Invoking *SaveAs* by any method resets the drawing name and path to whatever is entered in the *Save Drawing As* dialog box, so entering A:NAME would set A: as the current drive. This could cause problems because of the speed and reliability of a diskette as opposed to a hard drive. Instead, you should (1) <u>type</u> *Save* (see previous NOTE), or (2) save the drawing to the hard drive (usually C:), then close the drawing (by using *Open, New,* or *Exit*). Next, use Windows Explorer or the Windows right-click shortcut menu to copy the drawing file to A:. (See Windows Right-Click Shortcut Menus and Fig. 2-5.)

You can save an AutoCAD Release 14 drawing in several formats other than the default format (*AutoCAD R14 Drawing *.dwg*). Use the drop-down list at the bottom of the *Save* (or *SaveAs*) dialog box (Fig. 2-12) to save the current drawing as an *AutoCAD R13/LT 95 Drawing, AutoCAD R12/LT2 Drawing,* or a *Drawing Template file *.dwt.* The current drawing format is not changed for future saves. (If saving to previous versions, especially Release 12 or LT2, some information may be lost when the "downward" conversion is made. Newer objects like *Leader, Xline, Ray, Ellipse, Tolerance, Mline,* OLE objects, *Xref overlay, Mtext, Spline,* 3D solids, solid fills, and raster objects may be converted to the closest matching object in the selected format when possible.)

Figure 2-12 ——————

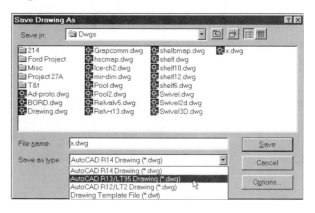

EXIT

	Pull-down Menu	COMMAND (TYPE)	ALIAS (TYPE)	Short-cut	Screen (side) Menu	Tablet Menu
	File Exit	*EXIT*	...	...	...	*Y,25*

This is the simplest method to use when you want to exit AutoCAD. If any changes have been made to the drawing since the last *Save*, *Exit* invokes a dialog box asking if you want to *Save Changes to . . .?* before ending AutoCAD (Fig. 2-13). *Yes* causes AutoCAD to use *Save*, then exit; *No* exits without saving; and *Cancel* does not exit. The *Quit* command could optionally be typed to accomplish the same action as using *Exit.*

Figure 2-13 ——————

An alternative to using the *Exit* option is to use the standard Windows methods for exiting an application. The two options are (1) select the **"X"** in the extreme upper-right corner of the AutoCAD window, or (2) select the AutoCAD logo (red) in the extreme upper-left corner of the AutoCAD window. Selecting the logo in the upper-left corner produces a pull-down menu allowing you to *Minimize, Maximize,* etc., or to *Close* the window (Fig. 12-14). Using *Close* is the same as using *Exit.*

Figure 2-14 —

QUIT

	Pull-down Menu	COMMAND (TYPE)	ALIAS (TYPE)	Short-cut	Screen (side) Menu	Tablet Menu
	...	*QUIT*	*EXIT*	...	*FILE Quit*	...

The *Quit* command accomplishes the same action as *Exit*. *Quit* discontinues (exits) the AutoCAD session and usually produces the dialog box shown in Figure 2-13 if changes have been made since the last *Save.*

RECOVER

Pull-down Menu	COMMAND (TYPE)	ALIAS (TYPE)	Short-cut	Screen (side) Menu	Tablet Menu
File *Drawing Utilities >* *Recover...*	RECOVER	...	...	*FILE* *Recover*	...

This option is used only in the case that *Open* does not operate on a particular drawing because of damage to the file. Damage to a drawing file can occur from improper exiting of AutoCAD (such as power failure) or from damage to a diskette. *Recover* can usually reassemble the file to a usable state and reload the drawing.

The *Recover* command produces the *Select File* dialog box. Select the drawing to recover. A recovery log is displayed in a text window reporting results of the recover operation (Fig. 2-15).

Figure 2-15

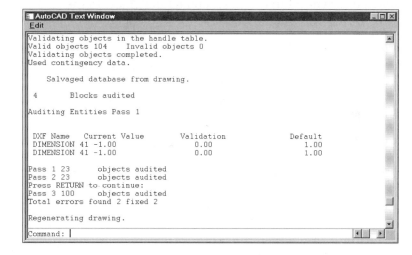

SAVETIME

Pull-down Menu	COMMAND (TYPE)	ALIAS (TYPE)	Short-cut	Screen (side) Menu	Tablet Menu
Tools *Preferences...* *General*	SAVETIME	...	...	*TOOLS 2* *Prefernc* *General*	...

SAVETIME is actually a system variable that controls AutoCAD's automatic save feature. AutoCAD automatically saves the current drawing for you at time intervals that you specify. The default time interval is 120 (minutes)! If you want to use this feature effectively, change the value stored in the *SAVETIME* variable to about 15 or 20 (minutes). A value of 0 disables this feature. The value for *SAVETIME* can also be set in the *General* tab of the *Preferences* dialog box (see Figure 2-16).

When automatic saving occurs, the current drawing is saved under the default name of AUTO.SV$ (sometimes in the Windows/Temp directory, depending on your operating system and setup). The current drawing name is not overwritten. If you want to save the drawing under the name you assigned, *Save* or *Qsave* must be used, or *SaveAs* should be used to save under a new name. The default name and path of the automatic save file can be changed in the *Files* tab of the *Preferences* dialog box under *Menu, Help, Log, and Miscellaneous File Names.*

AutoCAD Backup Files

When a drawing is saved, AutoCAD creates a file with a .DWG extension. For example, if you name the drawing PART1, using *Save* creates a file named PART1.DWG. The next time you save, AutoCAD makes a new PART1.DWG and renames the old version to PART1.BAK. One .BAK (backup) file is kept automatically by AutoCAD by default. You can disable the automatic backup function by changing the *ISAVEBAK* system variable to 0 or by accessing the *General* tab in the *Preferences* dialog box (Fig. 2-16).

Figure 2-16

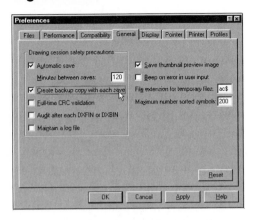

You cannot *Open* a .BAK file. It must be renamed to a .DWG file. Remember that you already have a .DWG file by the same name, so rename the extension <u>and</u> the filename. For example, PART1.BAK could be renamed to PART1OLD.DWG. Use the Windows Explorer, the *Select File* dialog box (use *Open*) with the right-click options, or use *Shell* and the DOS **ren** command to rename the file.

The .BAK files can also be deleted without affecting the .DWG files. The .BAK files accumulate after time, so you may want to periodically delete the unneeded ones to conserve disk space.

AutoCAD Drawing File Management

AutoCAD drawing files should be stored in folders (directories) used exclusively for that purpose. For example, you may use Windows Explorer to create a new folder called "Drawings" or "DWG." This could be a folder in the main C: or D: drive, or could be a subdirectory such as "C:\My Documents\ AutoCAD Drawings." You can create folders for different types of drawings or different projects. For example, you may have a folder named "DWG" with several subdirectories for each project, drawing type, or client, such as "COMPONENTS" and "ASSEMBLIES," or "In Progress" and "Completed," or "TWA," "Ford," and "Dupont." If you are working in an office or school laboratory, a directory structure most likely already has been created for you to use.

It is considered poor practice to create drawings in folders where the AutoCAD system files are kept. Beware, if you install AutoCAD and accept all defaults, when you first use *Save*, the files are saved in the "C:\Program Files\AutoCAD R14" directory.

Safety and organization are important considerations for storage of AutoCAD drawing files. AutoCAD drawings should be saved to designated folders to prevent accidental deletion of system or other important files. If you work with AutoCAD long enough or if you work in an office or laboratory, most likely many drawings are saved on the computer hard drive or network drives. It is imperative that a logical directory structure be maintained so important drawings can be located easily and saved safely.

CHAPTER EXERCISES

Start AutoCAD. If the *Start Up* dialog box appears, select *Start from Scratch*, then use the *English* default settings. NOTE: The chapter exercises in this book list file names in UPPER CASE letters for easy recognition. The file names and directory (folder) names on your system may appear as UPPER CASE, Title Case, or lower case.

1. **Create or determine the name of the folder ("working" directory) for opening and saving AutoCAD files on your computer system**

   If you are working in an office or laboratory, a folder (directory) has most likely been created for saving your files. If you have installed AutoCAD yourself and are learning on your home or office system, you should create a folder for saving AutoCAD drawings. It should have a name like "C:\ACADR14\DWG," "C:\My Documents\Dwgs," or "C:\Acadr14\Drawing Files." (HINT: Use the *SaveAs* command. The name of the folder last used for saving files appears at the top of the *SaveAs* dialog box.)

2. *Save* **and name a drawing file**

   Draw 2 vertical *Lines*. Select *Save* from the *Files* pull-down, from the Standard toolbar, or by another method. (The *SaveAs* dialog box appears since a name has not yet been assigned.) Name the drawing **"CH2 VERTICAL."**

3. **Using** *Qsave*

   Draw 2 more vertical *Lines*. Select *Save* again from the menu or by any other method except typing. (Notice that the *Qsave* command appears at the command line since the drawing has already been named.) Draw 2 more vertical *Lines* (a total of six lines now). Select *Save* again.

4. **Start a** *New* **drawing**

   Invoke *New* from the *Files* pull-down, the Standard toolbar, or by any other method. If the *Start Up* dialog box appears, select **Start from Scratch**, then use the *English* default settings. Draw 2 horizontal *Lines*. Use *Save*. Enter **"CH2 HORIZONTAL"** as the name for the drawing. Draw 2 more horizontal *Lines*, but <u>do not</u> *Save*. Continue to exercise 5.

5. *Open* **an existing drawing**

   Use *Open* to open **CH2 VERTICAL**. Notice that AutoCAD first forces you to answer *Yes* or *No* to *Save changes to CH2 HORIZONTAL.DWG?* PICK *Yes* to save the changes. Proceed to *Open* **CH2 VERTICAL.**

6. **Using** *SaveAs*

   Draw 2 inclined (angled) *Lines* in the CH2 VERTICAL drawing. Invoke *SaveAs* to save the drawing under a new name. Enter **"CH2 INCLINED"** as the new name. Notice the current drawing name displayed in the AutoCAD title bar (at the top of the screen) is reset to the new name. Draw 2 more inclined *Lines* and *Save.*

7. *Open* **an AutoCAD sample drawing**

   Open a drawing named **"COLORWH.DWG"** usually located in the "C:\Programs\ AutoCAD R14\Sample" directory. Use the *Directories* section of the dialog box to change drive and directory if necessary. Do <u>not</u> *Save* the sample drawing after viewing it. Practice the *Open* command by looking at other sample drawings in the Sample directory, such as **WILHOME.DWG** or **ZKL47_2.DWG.** Be careful not to *Save* the drawings.

8. **Check the *SAVETIME* setting**

   At the command prompt, enter *SAVETIME*. The reported value is the interval (in minutes) between automatic saves. Your system may be set to 120 (default setting). If you are using your own system, change the setting to **15** or **20.**

9. **Find and rename a backup file**

   Use the *Open* command. When the *Select File* dialog box appears, locate the folder where your drawing files are saved, then enter **"*.BAK"** in the *File name:* edit box (* is a wildcard that means "all files"). All of the .BAK files should appear. Search for a backup file that was created the last time you saved **CH2 VERTICAL.DWG,** named **CH2 VERTICAL.BAK. Right-click** on the file name. Select *Rename* from the pop-up menu and rename the file **"CH2 VERT 2.DWG."** Next enter *.DWG in the *File name:* edit box to make all the .dwg files reappear. Highlight **CH2 VERT 2.DWG** and the bitmap image should appear in the preview display. You can now *Open* the file if you wish.

10. **Find and rename the automatic save file**

    If you are familiar with Windows 95 or NT functions, use the "Find File" feature to locate the AutoCAD automatic save file. You <u>may or may not</u> find a file named AUTO?.SV$ (the last automatic save drawing file), depending on the *SAVETIME* interval setting and how long AutoCAD has been used on your computer. (Use the Windows Task Bar *Startup* button, then select *Find Files,* or use Explorer and select *Find Files or Folders...* from the *Tools* pull-down menu.) Then enter **"*.SV$"** in the *Named:* edit box. (*.SV$ lists any AutoCAD automatic save files.)

    Use the *Open* command again. Locate the folder that was indicated in the *Find File* results. Enter **"*.SV$"** in the *File name:* edit box. Using the same technique as in exercise 9, rename the file to **"AUTOSAVE.DWG."** <u>Be extremely careful not to alter any other files in the folder.</u> Highlight the **AUTOSAVE.DWG** file name and the bitmap image should appear in the preview display. You can now *Open* the file if you wish. *Exit* AutoCAD.

3

DRAW
COMMAND
CONCEPTS

Chapter Objectives

After completing this chapter you should be able to:

1. recognize drawing objects;

2. draw *Lines* <u>interactively</u>;

3. use *SNAP, GRID,* and *ORTHO* while drawing *Lines* and *Circles* interactively;

4. create *Lines* by specifying <u>absolute</u> coordinates;

5. create *Lines* by specifying <u>relative rectangular</u> coordinates;

6. create *Lines* by specifying <u>relative polar</u> coordinates;

7. create *Lines* by specifying <u>direct distance entry</u> coordinates;

8. create *Circles* by each of the five coordinate entry methods.

AutoCAD OBJECTS

Figure 3-1 —————————————

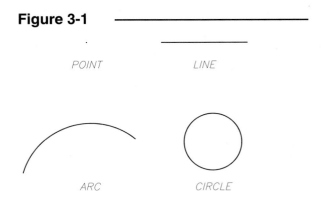

The smallest component of a drawing in AutoCAD is called an <u>object</u> (sometimes referred to as an entity). An example of an object is a *Line*, an *Arc*, or a *Circle* (Fig. 3-1). A rectangle created with the *Line* command would contain four objects.

Draw commands <u>create</u> objects. The draw command names are the same as the object names.

Simple objects are *Point*, *Line*, *Arc*, and *Circle*.

Complex objects are shapes such as *Polygon, Rectangle, Ellipse, Polyline,* and *Spline* which are created with one command (Fig. 3-2). Even though they appear to have several segments, they are <u>treated</u> by AutoCAD as one object.

Figure 3-2 —————————————

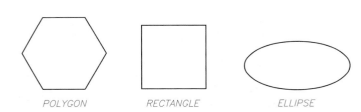

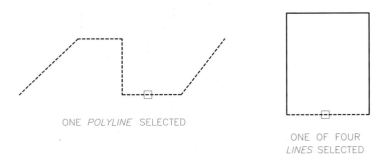

It is not always apparent whether a shape is composed of one or more objects. However, if you pick an object with the "pickbox," an object is "highlighted," or shown in a broken line pattern (Fig. 3-3). This highlighting reveals whether the shape is composed of one or several objects.

Figure 3-3 —————————————

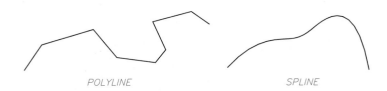

LOCATING THE DRAW COMMANDS

To invoke *Draw* commands, any of the five command entry methods can be used depending on your computer setup.

1. **Toolbars**   Select the icon from the *Draw* toolbar.
2. **Pull-down menu**   Select the command from the *Draw* pull-down menu.
3. **Screen menu**   Select the command from the *Draw* screen menu.
4. **Keyboard**   Type the command name, command alias, or accelerator keys at the keyboard (a command alias is a one- or two-letter shortcut).
5. **Tablet menu**   Select the icon from the digitizing tablet menu (if available).

For example, a draw command can be activated by PICKing its icon button from the *Draw* toolbar (Fig. 3-4).

Figure 3-4

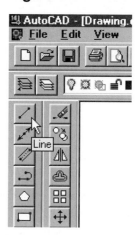

The *Draw* pull-down menu can also be used to select draw commands (Fig. 3-5). Options for a command are found on cascading menus.

Figure 3-5

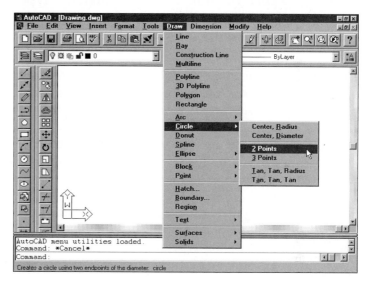

Additionally, the Screen menu can be used to select draw commands. Two menus contain all the draw commands: *DRAW1* and *DRAW2* (Fig. 3-6).

Figure 3-6

THE FIVE COORDINATE ENTRY METHODS

All drawing commands prompt you to specify points, or locations, in the drawing. For example, the *Line* command prompts you to give the "From point:" and "To point:," expecting you to specify locations for the first and second endpoints of the line. After you specify those points, AutoCAD stores the coordinate values to define the line. A 2-dimensional line in AutoCAD is defined and stored in the database as two sets of X, Y, and Z values (with Z values of 0), one for each endpoint.

There are five ways to specify coordinates; that is, there are five ways to tell AutoCAD the location of points when you draw objects.

1. **Interactive method**   PICK   Use the cursor to select points on the screen.
2. **Absolute coordinates**   X,Y   Type explicit X and Y values relative to the origin at 0,0.
3. **Relative rectangular coordinates**   @X,Y   Type explicit X and Y values relative to the last point (@ means "last point").
4. **Relative polar coordinates**   @dist<angle   Type a distance value and angle value relative to the last point (< means "angle of").
5. **Direct distance entry**   dist,direction   Type a distance value relative to the last point, indicate direction with the cursor, then press Enter.

The interactive method of coordinate entry is simple because you PICK the desired locations with your cursor when AutoCAD prompts for a point. It is generally the easiest to use for most applications. It should be used with *SNAP* or with *Object Snaps* (Chapter 7) for accuracy.

Absolute coordinates are used when AutoCAD prompts for a point and you know the exact coordinates of the desired location. You simply key in the coordinate values at the keyboard (X and Y values are separated by a comma).

Relative rectangular coordinates are similar to absolute values except the X and Y distances are given in relation to the last point, instead of being relative to the origin. AutoCAD interprets the @ symbol as the "last point." Use this type of coordinate entry when you do not know the exact absolute values, but you know the X and Y distances from the last specified point.

Relative polar coordinate entry is used whenever you want to draw a line or specify a point that is at an exact angle with respect to the last point. Interactive coordinate entry is not accurate enough to draw diagonal lines to a specific angle (unless *Object Snap* can be used). AutoCAD interprets the @ symbol as last point and the < symbol as an angular designator for the following value. So, "@ 2<90" means "from the last point, 2 units at an angle of 90 degrees."

Direct distance coordinate entry is a combination of relative polar coordinates and interactive specification because a distance value is entered at the keyboard and the angle is specified by the direction of the cursor movement (from the last point). Direct distance entry is useful primarily for orthogonal (horizontal or vertical) operations by toggling *ORTHO* on. For example, assume you wanted to draw a *Line* 7.5 units in a horizontal (positive X) direction from the last point. Using direct distance entry to establish the "To point:" (second end point), you would turn on *ORTHO* and move the cursor to the right any distance, then type "7.5" and press Enter. Alternately, you could first type the distance value, then move the cursor before pressing Enter. This method is very fast and accurate for horizontal and vertical drawing and editing operations.

DRAWING *LINES* USING THE FIVE COORDINATE ENTRY METHODS

Begin a *New* drawing to complete the *Line* exercises. If the *Start Up* dialog box or *Create New Drawing* dialog box appears, select *Start from Scratch*, choose *English* as the default setting, and click the *OK* button. The *Line* command can be activated by any one of the methods shown in the command table below.

LINE

Pull-down Menu	COMMAND (TYPE)	ALIAS (TYPE)	Short-cut	Screen (side) Menu	Tablet Menu
Draw *Line*	LINE	L	...	*DRAW 1* *Line*	J,10

Drawing Horizontal Lines

1. Draw a horizontal *Line* of 2 units length starting at point 2,2. Use the <u>interactive</u> method. See Figure 3-7.

Steps	Command Prompt	Perform Action	Comments
1.		turn on *GRID* (**F7**)	grid appears
2.		turn on *SNAP* (**F9**)	"SNAP" appears on Status line
3.	Command:	select or type **Line**	use any method
4.	LINE From point:	**PICK** location **2,2**	watch *Coords*
5.	To point:	**PICK** location **4,2**	watch *Coords*
6.	To point:	press **Enter**	completes command

The preceding steps produce a *Line* as shown in Figure 3-7.

Figure 3-7

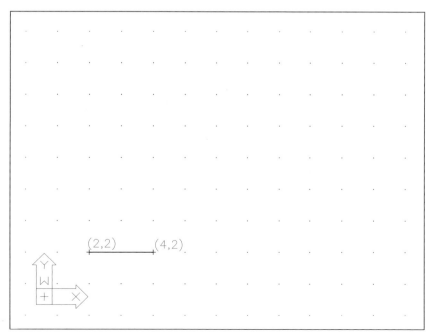

2.   Draw a horizontal *Line* of 2 units length starting at point 2,3 using <u>absolute coordinates</u>.

Steps	Command Prompt	Perform Action	Comments
1.	Command:	select *Line*	use any method
2.	LINE From point:	type **2,3** and press **Enter**	establishes the first endpoint
3.	To point:	type **4,3** and press **Enter**	a *Line* should appear
4.	To point:	press **Enter**	completes command

The above procedure produces the new *Line* above the first *Line* as shown (Fig. 3-8).

Figure 3-8

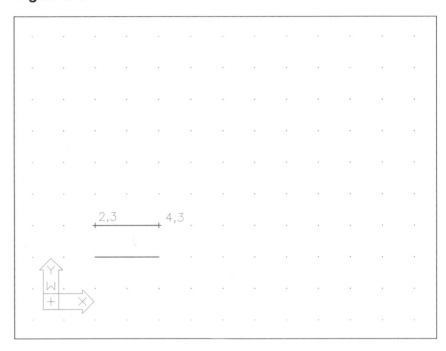

3.   Draw a horizontal *Line* of 2 units length starting at point 2,4 using <u>relative rectangular coordinates</u>.

Steps	Command Prompt	Perform Action	Comments
1.	Command:	select *Line*	use any method
2.	LINE From point:	type **2,4** and press **Enter**	establishes the first endpoint
3.	To point:	type **@2,0** and press **Enter**	@ means "last point"
4.	To point:	press **Enter**	completes command

The new *Line* appears above the previous two as shown in Figure 3-9.

Figure 3-9

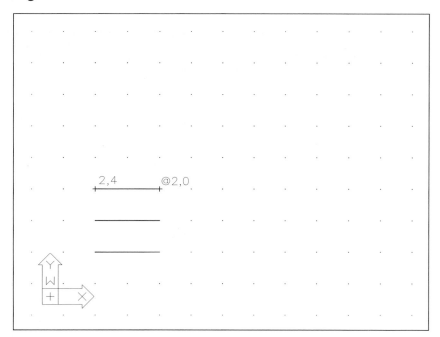

4. Draw a horizontal *Line* of 2 units length starting at point 2,5 using <u>relative polar coordinates</u>.

Steps	Command Prompt	Perform Action	Comments
1.	Command:	select *Line*	use any method
2.	LINE From point:	type **2,5** and press **Enter**	establishes the first endpoint
3.	To point:	type **@2<0** and press **Enter**	@ means "last point" < means "angle of"
4.	To point:	press **Enter**	completes command

The new horizontal *Line* appears above the other three. See Figure 3-10.

Figure 3-10

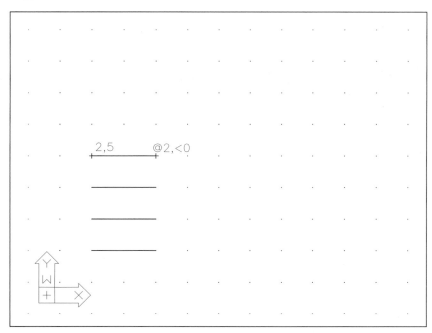

5. Draw a horizontal *Line* of 2 units length starting at point 2,6 using <u>direct distance</u> coordinate entry.

Steps	Command Prompt	Perform Action	Comments
1.	Command:	select *Line*	use any method
2.	LINE From point:	type **2,6** and press **Enter**	establishes the first endpoint
3.	To point:	turn on **ORTHO,** move the cursor to the right, type **2,** then press **Enter**	the cursor movement indicates direction, 2 is the distance
4.	To point:	press **Enter**	completes command

The new line appears above the other four as shown in Figure 3-11.

Figure 3-11 ————————————————————————————

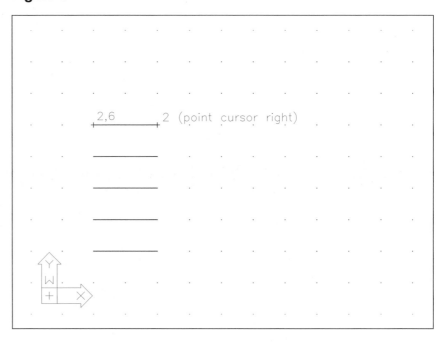

One of these methods may be more favorable than another in a particular situation. The interactive method is fast and easy, assuming that *OSNAP* is used (see Chapter 7) or that *SNAP* and *GRID* are used and set to appropriate values. *SNAP* and *GRID* are used successfully for small drawings where objects have regular interval lengths. Direct distance coordinate entry is fast, easy, and accurate when drawing horizontal or vertical lines. *ORTHO* should be on for most applications of direct distance entry. Direct distance coordinate entry is preferred for horizontal or vertical drawing and editing in cases when *OSNAP* cannot be used (no geometry to *OSNAP* to) and when *SNAP* and *GRID* are not appropriate, such as an architectural or civil engineering application. In these applications, objects are relatively large (lines are long) and are not necessarily drawn to regular intervals; therefore, it is convenient and accurate to enter exact lengths as opposed to interactively PICKing points.

Drawing Vertical Lines

Below are listed the steps for drawing vertical lines using each of the five methods of coordinate entry. The following completed problems should look like those in Figure 3-12.

Figure 3-12

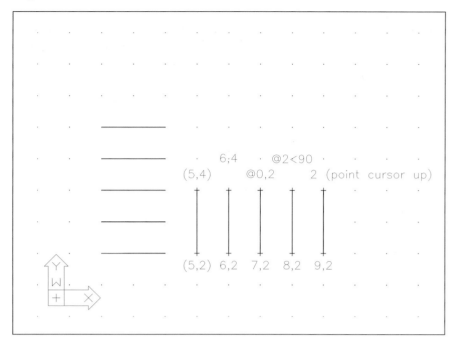

1. Draw a vertical *Line* of 2 units length starting at point 5,2 using the <u>interactive</u> method.

Steps	Command Prompt	Perform Action	Comments
1.		turn on *GRID* (**F7**)	grid appears
2.		turn on *SNAP* (**F9**)	"SNAP" appears on Status line
3.	Command:	select or type *Line*	use any method
4.	LINE From point:	**PICK** location **5,2**	watch *Coords*
5.	To point:	**PICK** location **5,4**	watch *Coords*
6.	To point:	press **Enter**	completes command

2. Draw a vertical *Line* of 2 units length starting at point 6,2 using <u>absolute coordinates</u>.

Steps	Command Prompt	Perform Action	Comments
1.	Command:	select *Line*	use any method
2.	LINE From point:	type **6,2** and press **Enter**	establishes the first endpoint
3.	To point:	type **6,4** and press **Enter**	a *Line* should appear
4.	To point:	press **Enter**	completes command

3.   Draw a vertical *Line* of 2 units length starting at point 7,2 using <u>relative rectangular coordinates</u>.

Steps	Command Prompt	Perform Action	Comments
1.	Command:	select *Line*	use any method
2.	LINE From point:	type **7,2** and press **Enter**	establishes the first endpoint
3.	To point:	type **@0,2** and press **Enter**	@ means "last point"
4.	To point:	press **Enter**	completes command

4.   Draw a vertical *Line* of 2 units length starting at point 8,2 using <u>relative polar coordinates</u>.

Steps	Command Prompt	Perform Action	Comments
1.	Command:	select *Line*	use any method
2.	LINE From point:	type **8,2** and press **Enter**	establishes the first endpoint
3.	To point:	type **@2<90** and press **Enter**	@ means "last point," < means "angle of"
4.	To point:	press **Enter**	completes command

5.   Draw a vertical *Line* of 2 units length starting at point 9,2 using <u>direct distance</u> coordinate entry.

Steps	Command Prompt	Perform Action	Comments
1.	Command:	select *Line*	use any method
2.	LINE From point:	type **9,2** and press **Enter**	establishes the first endpoint
3.	To point:	turn on **ORTHO,** move the cursor upward, type **2,** then press **Enter**	the cursor movement indicates direction, 2 is the distance
4.	To point:	press **Enter**	completes command

The method used for drawing vertical lines depends on the application and the individual, much the same as it would for drawing horizontal lines. Refer to the discussion after the horizontal lines drawing exercises.

R14

Drawing Inclined Lines

Following are listed the steps in drawing <u>inclined lines</u> using each of the five methods of coordinate entry. The following completed problems should look like those in Figure 3-13.

Figure 3-13 ────────────────────────

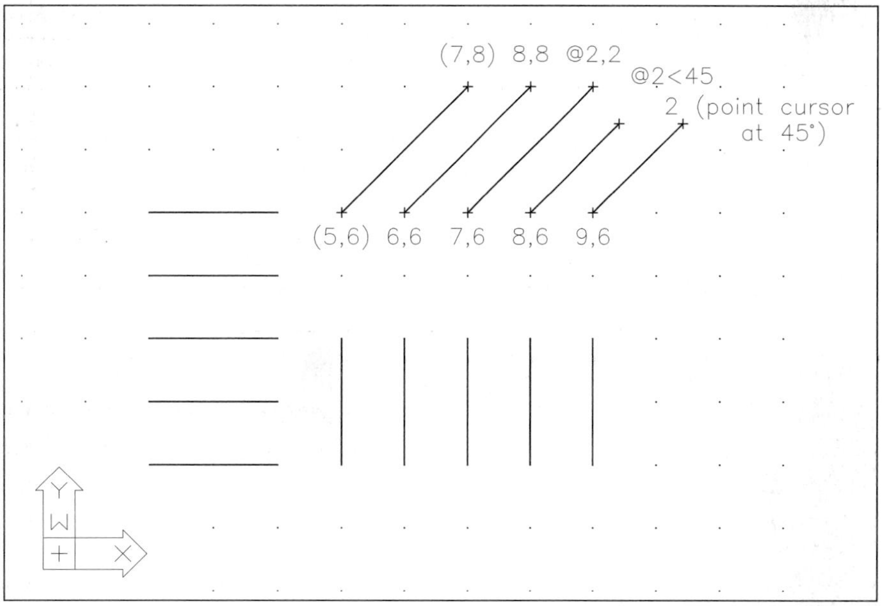

1. Draw an inclined *Line* of from 5,6 to 7,8. Use the <u>interactive</u> method.

Steps	Command Prompt	Perform Action	Comments
1.		turn on *GRID* (**F7**)	grid appears
2.		turn on *SNAP* (**F9**)	"SNAP" appears on Status line
3.		turn off *ORTHO* (**F8**)	in order to draw inclined *Lines*
4.	Command:	select or type *Line*	use any method
5.	LINE From point:	**PICK** location **5,6**	watch *Coords*
6.	To point:	**PICK** location **7,8**	watch *Coords*
7.	To point:	press **Enter**	completes command

2. Draw an inclined *Line* starting at 6,6 and ending at 8,8. Use <u>absolute coordinates</u>.

Steps	Command Prompt	Perform Action	Comments
1.	Command:	select *Line*	use any method
2.	LINE From point:	type **6,6** and press **Enter**	establishes the first endpoint
3.	To point:	type **8,8** and press **Enter**	a *Line* should appear
4.	To point:	press **Enter**	completes command

3. Draw an inclined *Line* starting at 7,6 and ending 2 units over (in a positive X direction) and 2 units up (in a positive Y direction). Use <u>relative rectangular coordinates</u>.

Steps	Command Prompt	Perform Action	Comments
1.	Command:	select *Line*	use any method
2.	LINE From point:	type **7,6** and press **Enter**	establishes the first endpoint
3.	To point:	type **@2,2** and press **Enter**	@ means "last point"
4.	To point:	press **Enter**	completes command

4. Draw an inclined *Line* of 2 units length at a 45 degree angle and starting at 8,6. Use <u>relative polar coordinates</u>.

Steps	Command Prompt	Perform Action	Comments
1.	Command:	select *Line*	use any method
2.	LINE From point:	type **8,6** and press **Enter**	establishes the first endpoint
3.	To point:	type **@2<45** and press **Enter**	@ means "last point" < means "angle of"
4.	To point	press **Enter**	completes command

5. Draw an inclined *Line* of 2 units length at a 45 degree angle starting at point 9,6. Use <u>direct distance</u> coordinate entry.

Steps	Command Prompt	Perform Action	Comments
1.		Turn on *GRID* (**F7**)	grid appears
2.		Turn on *SNAP* (**F9**)	"SNAP" appears on status line
3.	Command:	select *Line*	use any method
4.	LINE From point:	type **9,6** and press **Enter**	establishes the first endpoint
5.	To point:	turn off ***ORTHO,*** move the cursor up and to the right (the coordinate display must indicate an angle of 45), type **2,** then press **Enter**	the cursor movement indicates direction, 2 is the distance
6.	To point:	press **Enter**	completes command

Method 4, using polar values, is preferred if you are required to draw an inclined line with an exact length and angle. The interactive method and the direct distance coordinate entry method can only be used for drawing inclined lines at 45 degree angle increments (45, 135, etc.) and only when *SNAP* is on and set to equal X and Y intervals. These two methods are <u>not accurate</u> for drawing inclined lines at other than 45 degree increments. Notice that each of the first three methods draws an inclined line of 2.828 (2 * square root of 2) units length since the line is the hypotenuse of a right triangle.

DRAWING *CIRCLES* USING THE FIVE COMMAND ENTRY METHODS

Begin a *New* drawing to complete the *Circle* exercises. If the *Start Up* dialog box or *Create New Drawing* dialog box appears, select *Start from Scratch*, choose *English* as the default setting, and click the *OK* button. The *Circle* command can be invoked by any of the methods shown in the command table.

CIRCLE

Pull-down Menu	COMMAND (TYPE)	ALIAS (TYPE)	Short-cut	Screen (side) Menu	Tablet Menu
Draw *Circle >*	*CIRCLE*	*C*	...	*DRAW 1* *Circle*	*J,9*

Below are listed the steps for drawing *Circles* using the *Center, Radius* method. The circles are to be drawn using each of the four coordinate entry methods and should look like those in Figure 3-14.

Figure 3-14

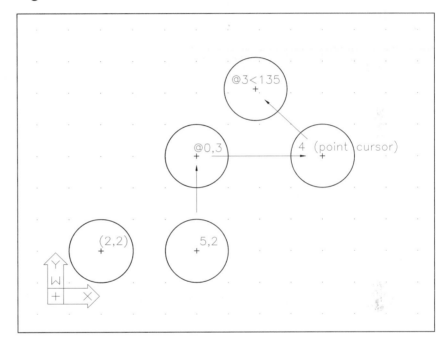

1.   Draw a *Circle* of 1 unit radius with the center at point 2,2. Use the <u>interactive</u> method.

Steps	Command Prompt	Perform Action	Comments
1.		turn on *GRID* (**F7**)	grid appears
2.		turn on *SNAP* (**F9**)	"SNAP" appears on Status line
3.		turn on *ORTHO* (**F8**)	"ORTHO" appears on Status line
4.	Command:	select or type **Circle**	use *Center, Radius* method
5.	CIRCLE 3P/2P/TTR/<Center point>:	**PICK** location **2,2**	watch *Coords*
6.	Diameter/<Radius>:	move 1 unit and **PICK**	watch *Coords*

2. Draw a circle of 1 unit radius with the center at point 5,2. Use <u>absolute coordinates</u>.

Steps	Command Prompt	Perform Action	Comments
1.	Command:	select *Circle*	use *Center, Radius* method
2.	CIRCLE 3P/2P/TTR/<Center point>:	type **5,2** and press **Enter**	a *Circle* should appear
3.	Diameter/<Radius>:	type **1** and press **Enter**	the correct *Circle* appears

3. Draw a circle of 1 unit radius with the center 3 units above the last point (previous *Circle* center). Use <u>relative rectangular coordinates</u>.

Steps	Command Prompt	Perform Action	Comments
1.	Command:	select *Circle*	use *Center, Radius* method
2.	CIRCLE 3P/2P/TTR/<Center point>:	type **@0,3** and press **Enter**	a *Circle* should appear
3.	Diameter/<Radius>:	type **1** and press **Enter**	the correct *Circle* appears

(If the new *Circle* is not above the last, type "ID" and enter "5,2." The entered point becomes the last point. Then try again.)

4. Draw a circle of 1 unit radius with the center 4 units to the right of the previous *Circle* center. Use <u>direct distance</u> coordinate entry.

Steps	Command Prompt	Perform Action	Comments
1.		turn on *ORTHO* (**F8**)	"ORTHO" appears on Status line
2.	Command:	select *Circle*	use *Center, Radius* method
3.	CIRCLE 3P/2P/TTR/<Center point>:	type **4**, move the cursor to the right, and press **Enter**	a *Circle* should appear
4.	Diameter/<Radius>:	type **1**, then press **Enter**	the correct *Circle* appears

(If the new *Circle* is not directly to the right of the last, type "ID" and enter "5,5." The entered point becomes the last point. Then try again.)

5. Draw a circle of 1 unit radius with the center 3 units at a 35 degree angle above and to the left of the previous *Circle*. Use <u>relative polar coordinates</u>.

Steps	Command Prompt	Perform Action	Comments
1.	Command:	select *Circle*	use *Center, Radius* method
2.	CIRCLE 3P/2P/TTR/<Center point>:	type @3<135 and press **Enter**	a *Circle* should appear
3.	Diameter/<Radius>:	type **1** and press **Enter**	the *Circle* appears

(If the new *Circle* is not correctly positioned with respect to the last, type "ID" and enter "9,5." The entered point becomes the last point. Then try again.)

CHAPTER EXERCISES

1. **Start a *New* Drawing**

   Start a *New* drawing. Select the ***Start from Scratch*** option and select ***English*** settings. Next, use the *Save* command and save the drawing as "**CH3EX**." Remember to *Save* often as you complete the following exercises. The completed exercise should look like Figure 3-15.

2. **Using interactive coordinate entry**

   Draw a square with sides of 2 units length. Locate the lower-left corner of the square at **2,2**. Use the *Line* command with interactive coordinate entry. (HINT: Turn on *SNAP, GRID,* and *ORTHO*.)

3. **Using absolute coordinates**

   Draw another square with 2 unit sides using the *Line* command. Enter absolute coordinates. Begin with the lower-left corner of the square at **5,2**.

4. **Using relative rectangular coordinates**

   Draw a third square (with 2 unit sides) using the *Line* command. Enter relative rectangular coordinates. Locate the lower-left corner at **8,2**.

5. **Using direct distance coordinate entry**

   Draw a fourth square (with 2 unit sides) beginning at the lower-left corner of **2,5**. Complete the square drawing *Lines* using direct distance coordinate entry. Don't forget to turn on ***ORTHO***.

6. **Using relative polar coordinates**

   Draw an equilateral triangle with sides of 2 units. Locate the lower-left corner at **5,5**. Use relative polar coordinates (<u>after</u> establishing the "From point:"). HINT: An equilateral triangle has interior angles of 60 degrees.

7. **Using interactive coordinate entry**

   Draw a *Circle* with a 1 unit <u>radius</u>. Locate the center at **9,6**. Use the interactive method. (Turn on *SNAP* and *GRID*.)

8. **Using relative rectangular, relative polar, or direct distance entry coordinates**

   Draw another *Circle* with a 2 unit <u>diameter.</u> Using any method listed above, locate the center 3 units below the previous *Circle*.

9. ***Save* your drawing**

   Use *Save*. Compare your results with Figure 3-15. When you are finished, *Exit* AutoCAD.

Figure 3-15

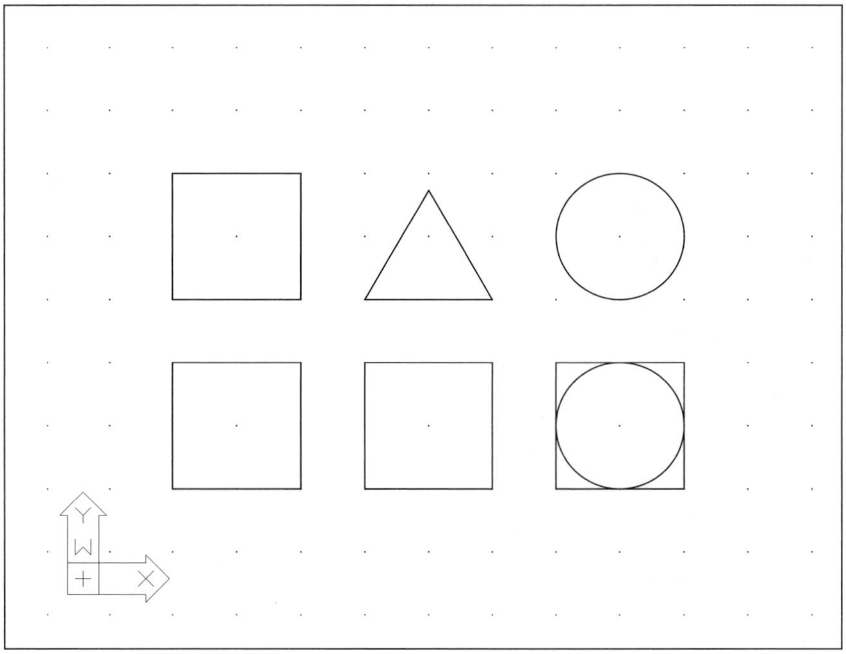

4

SELECTION SETS

Chapter Objectives

After completing this chapter you should:

1. know that Modify commands and many other commands require you to select objects;

2. be able to create a selection set using each of the specification methods;

3. be able to *Erase* objects from the drawing;

4. be able to *Move* objects from one location to another;

5. understand Noun/Verb and Verb/Noun order of command syntax.

MODIFY COMMAND CONCEPTS

Draw commands create objects. Modify commands <u>change existing objects</u> or <u>use existing objects to create new ones</u>. Examples are *Copy* an existing *Circle, Move* a *Line*, or *Erase* a *Circle*.

Since all of the Modify commands use or modify <u>existing</u> objects, you must first select the objects that you want to act on. The process of selecting the objects you want to use is called building a <u>selection set</u>. For example, if you want to *Copy, Erase,* or *Move* several objects in the drawing, you must first select the set of objects that you want to act on.

Remember that any of the five command entry methods (depending on your setup) can be used to invoke Modify commands.

1. **Toolbars**   *Modify* or *ModifyII* toolbar
2. **Pull-down menu** *Modify* pull-down menu
3. **Screen menu**   *MODIFY1* or *MODIFY2* screen menus
4. **Keyboard**   Type the command name <u>or</u> command alias
5. **Tablet menu**   Select the command icon

All of the Modify commands will be discussed in detail later, but for now we will focus on how to build selection sets.

SELECTION SETS

No matter which of the five methods you use to invoke a Modify command, you must specify a selection set during the command operation. There are two ways you can select objects: you can select the set of objects either (1) immediately before you invoke the command or (2) when the command prompts you to select objects. For example, examine the command syntax that would appear at the command line when the *Erase* command is used (method 2).

> Command: **erase**
> Select Objects:

Figure 4-1

The "Select Objects:" prompt is your cue to use any of several methods to PICK the objects to erase. As a matter of fact, every Modify command begins with the same "Select Objects:" prompt (unless you selected immediately before invoking the command).

When the "Select Objects:" prompt appears, the "crosshairs" cursor disappears and only a small, square pickbox appears at the cursor (Fig. 4-1).

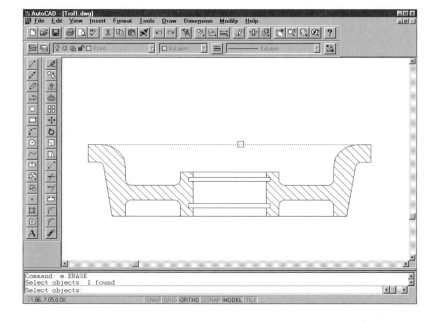

You can PICK objects using only the pickbox or any of several other methods illustrated in this chapter. (Only when you PICK objects at the "Select Objects:" prompt can you use all of the selection methods shown here. If you PICK immediately before the command, called Noun/Verb order, only the *AUto* method can be used. See Noun/Verb Syntax in this chapter.)

When the objects have been selected, they become highlighted (displayed as a broken line), which serves as a visual indication of the current selection set. Press Enter to indicate that you are finished selecting and are ready to proceed with the command.

Accessing Selection Set Options

When the "Select Objects:" prompt appears, you should select objects using the pickbox or one of a variety of other methods. Any method can be used <u>independently</u> or <u>in combination</u> to achieve the desired set of objects because object selection is a <u>cumulative</u> process.

The pickbox is the default option which can automatically be changed to a window or crossing window by PICKing in an open area (PICKing no objects). The other methods can be selected from the screen (side) menu or by typing the capitalized letters shown in the option names following.

If the screen (side) menu is displayed, you can access the object selection options from the *ASSIST* menu. *ASSIST* is displayed at the bottom of all other menus (Fig 4-2).

Figure 4-2 ——————

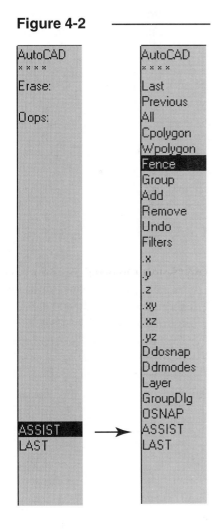

Optionally, you can type the first letter (or two in some cases) of any option explained in the next section. For example, to use the *Fence* option, type *F* at the "Select objects:" prompt. *Window Polygon* (WP) and *Crossing Polygon* (CP) are the only options that require typing two letters.

In Release 14, there are no icon buttons or pull-down menu selections that are easily available for the object selection options. However, a toolbar can be "customized" to display the object selection options.

Selection Set Options

The options for creating selection sets (PICKing objects) are shown on this and the following pages. Two *Circles* and five *Lines* (as shown in Figure 4-3) are used for every example. In each case, the circles only are selected for editing. If you want to follow along and practice as you read, draw *Circles* and *Lines* in a configuration similar to this. Then use a *Modify* command. (Press Escape to cancel the command after selecting objects.)

Figure 4-3

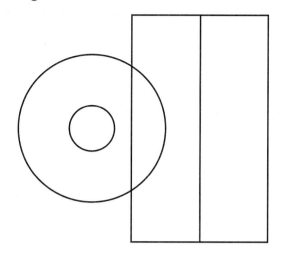

pickbox

This default option is used for selecting one object at a time. Locate the pickbox so that an object crosses through it and **PICK** (Fig. 4-4). You do not have to type or select anything to use this option.

Figure 4-4

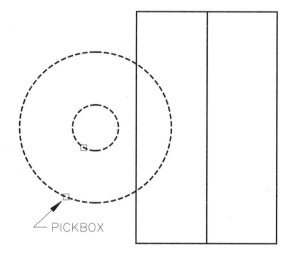

PICKBOX

AUto

To use this option, you do not have to type or select anything from the Select Objects menu or toolbars. The pickbox must be positioned in an open area so that no objects cross through it; then **PICK** to start a window. If you drag to the right, a *Window* is created (Fig. 4-5). **PICK** the other corner.

Figure 4-5

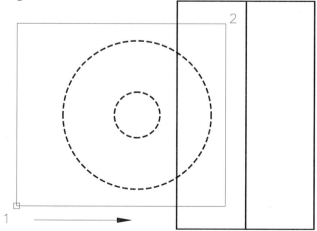

If you drag to the <u>left</u> instead, a *Crossing Window* forms (Fig. 4-6). (See Window and Crossing Window below.)

Figure 4-6

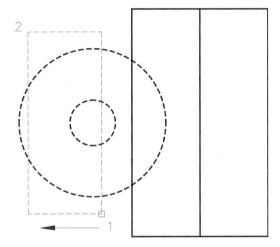

Window
Only objects <u>completely within</u> the *Window* are selected. The *Window* is a solid linetype rectangular box. Select the first and second points (diagonal corners) in any direction as shown in Figure 4-7.

Figure 4-7

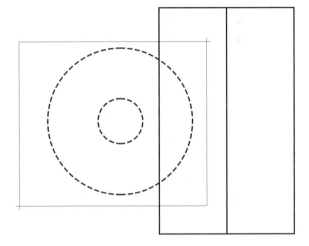

Crossing Window
All objects <u>within and crossing through</u> the window are selected. The *Crossing Window* is displayed as a broken linetype rectangular box (Fig. 4-8). Select two diagonal <u>corners</u>.

Figure 4-8

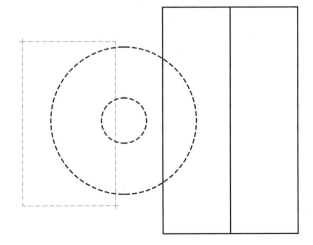

Window Polygon

The *Window Polygon* operates like a *Window,* but the box can be <u>any</u> irregular polygonal shape (Fig. 4-9). You can pick any number of corners to form any shape rather than picking just two corners to form a rectangle as with the *Window* option.

Figure 4-9

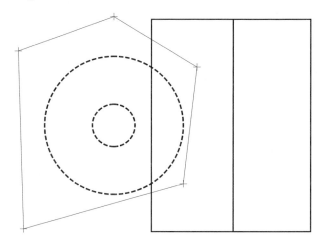

Crossing Polygon

The *Crossing Polygon* operates like a *Crossing Window,* but can have any number of corners like *Window Polygon* (Fig. 4-10).

Figure 4-10

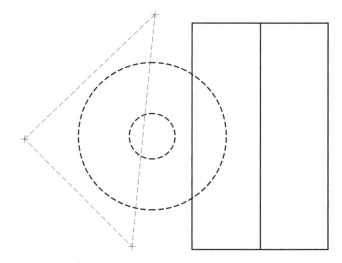

Fence

This option operates like a <u>crossing line</u>. Any objects crossing the *Fence* are selected. The *Fence* can have any number of segments (Fig. 4-11).

Figure 4-11

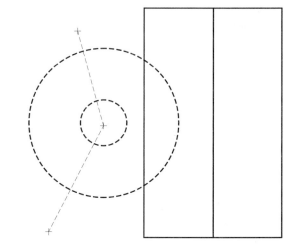

Last

This option automatically finds and selects only the last object created. *Last* does not find the last object modified (with *Move, Stretch,* etc.).

Previous

Previous finds and selects the previous selection set, that is, whatever was selected during the previous command (except after *Erase*). This option allows you to use several editing commands on the same set of objects without having to respecify the set.

ALL

This option selects all objects in the drawing except those on *Frozen* or *Locked* layers (*Layers* are covered in Chapter 12).

Remove

Selecting this option causes AutoCAD to switch to the "Remove objects:" mode. Any selection options used from this time on remove objects from the highlighted set (see Figure 4-12).

Add

The *Add* option switches back to the default "Select objects:" mode so additional objects can be added to the selection set.

Shift + #1

Holding down the **Shift** key and pressing the **#1** button simultaneously removes objects selected from the highlighted set as shown in Figure 4-12. This method is generally quicker than, but performs the same action as, *Remove*. The advantage here is that the *Add* mode is in effect unless Shift is held down.

Figure 4-12

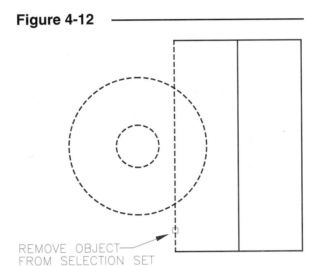

REMOVE OBJECT
FROM SELECTION SET

Group

The *Group* option selects groups of objects that were previously specified using the *Group* command. Groups are selection sets to which you can assign a name.

Ctrl + #1 (Object Cycling)

Holding down the **Ctrl** key and simultaneously pressing the **PICK** (#1) button causes AutoCAD to cycle through (highlight one at a time) two or more objects that may cross through the pickbox.

For example, if you attempted to PICK an object but other objects were in the pickbox, AutoCAD may not highlight the one you want (Fig. 4-13). In this case, hold down the **Ctrl** key and press the **PICK** button several times (the cursor can be located anywhere on the screen during cycling). All of the objects that passed through the pickbox will be highlighted one at a time. The "<Cycle on>" prompt appears at the command line. When the object you want is highlighted, press Enter and AutoCAD adds the object to the selection set and returns to the "Select objects:" prompt.

Figure 4-13

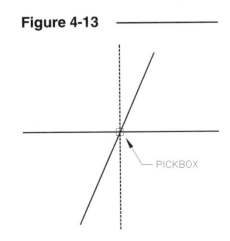

PICKBOX

SELECT

Pull-down Menu	COMMAND (TYPE)	ALIAS (TYPE)	Short-cut	Screen (side) Menu	Tablet Menu
...	*SELECT*	...	...	...	...

The *Select* command can be used to PICK objects to be saved in the selection set buffer for subsequent use with the *Previous* option. Any of the selection methods can be used to PICK the objects.

    Command: **select**
    Select Objects: **PICK** (use any selection option)
    Select Objects: **Enter** (completes the selection process)
    Command:

The selected objects become unhighlighted when you complete the command by pressing Enter. The objects become highlighted again and are used as the selection set if you use the *Previous* selection option in the next editing command.

NOUN/VERB SYNTAX

An object is the noun and a command is the verb. Noun/Verb syntax order means to pick objects (nouns) first, then use an editing command (verb) second. If you select objects first (at the open Command: prompt) and then immediately choose a Modify command, AutoCAD recognizes the selection set and passes through the "Select Objects:" prompt to the next step in the command.

Verb/Noun means to invoke a command and then select objects within the command. For example, if the *Erase* command (verb) is invoked first, AutoCAD then issues the prompt to "Select Objects:"; therefore, objects (nouns) are PICKed second. In the previous examples, and with older versions of AutoCAD, only Verb/Noun syntax order was used.

You can use either order you want (Noun/Verb or Verb/Noun) and AutoCAD automatically understands. If objects are selected first, the selection set is passed to the next editing command used, but if no objects are selected first, the editing command automatically prompts you to "Select Objects:".

If you use Noun/Verb order, you are limited to using only the *AUto* options for object selection (pickbox, window, and crossing window). You can only use the other options (e.g., *Crossing Polygon, Fence, Previous*) if you invoke the desired Modify command first, then select objects when the "Select Objects:" prompt appears.

The *PICKFIRST* variable (a very descriptive name) enables Noun/Verb syntax. The default setting is 1 (*On*). If *PICKFIRST* is set to 0 (*Off*), Noun/Verb syntax is disabled and the selection set can be specified only within the editing commands (Verb/Noun).

Setting *PICKFIRST* to 1 provides two options: Noun/Verb and Verb/Noun. You can use either order you want. If objects are selected first, the selection set is passed to the next editing command, but if no objects are selected first, the editing command prompts you to select objects.

SELECTION SETS PRACTICE

NOTE: While learning and practicing with the editing commands, it is suggested that *GRIPS* be turned off. This can be accomplished by typing in **GRIPS** and setting the *GRIPS* variable to a value of **0**. The AutoCAD Release 14 default has *GRIPS* on (set to 1). *GRIPS* is covered in Chapter 21.

Begin a *New* drawing to complete the selection set practice exercises. If the *Start Up* dialog box or *Create New Drawing* dialog box appears, select *Start from Scratch,* choose *English* as the default setting, and click the *OK* button. The *Erase, Move,* and *Copy* commands can be activated by any one of the methods shown in the command tables that follow.

Using *Erase*

Erase is the simplest editing command. *Erase* removes objects from the drawing. The only action required is the selection of objects to be erased.

ERASE

Pull-down Menu	COMMAND (TYPE)	ALIAS (TYPE)	Short-cut	Screen (side) Menu	Tablet Menu
Modify *Erase*	ERASE	E	...	*MODIFY1* *Erase*	V,14

1. Draw several *Lines* and *Circles*. Practice using the object selection options with the *Erase* command. The following sequence uses the pickbox, Window, and Crossing Window.

STEPS	COMMAND PROMPT	PERFORM ACTION	COMMENTS
1.	Command:	type *E* and press **Spacebar**	*E* is the alias for *Erase,* Spacebar can be used like Enter
2.	ERASE Select Objects:	use pickbox to select one or two objects	objects are highlighted
3.	Select Objects:	type *W* or use a *Window,* then select more objects	objects are highlighted
4.	Select Objects:	type *C* or use a *Crossing Window,* then select objects	objects are highlighted
5.	Select Objects:	press **Enter**	objects are erased

2. Draw several more *Lines* and *Circles*. Practice using the *Erase* command with the *AUto Window* and *AUto Crossing Window* options as indicated below.

STEPS	COMMAND PROMPT	PERFORM ACTION	COMMENTS
1.	Command:	select the *Modify* pull-down, then **Erase**	
2.	ERASE Select Objects:	use pickbox to **PICK** an open area, drag *Window* to the <u>right</u> to select objects	objects inside Window are highlighted
3.	Select Objects:	**PICK** an open area, drag *Crossing Window* to the <u>left</u> to Select Objects	objects inside and crossing through Window are highlighted
4.	Select Objects:	press **Enter**	objects are erased

Using *Move*

The *Move* command specifically prompts you to (1) select objects, (2) specify a "basepoint," or point to move <u>from</u>, and (3) specify a "second point of displacement," or point to move <u>to</u>.

MOVE

Pull-down Menu	COMMAND (TYPE)	ALIAS (TYPE)	Short-cut	Screen (side) Menu	Tablet Menu
Modify *Move*	*MOVE*	*M*	...	*MODIFY2* *Move*	*V,19*

1. Draw a *Circle* and two *Lines*. Use the *Move* command to practice selecting objects and to move one *Line* and the *Circle* as indicated in the following table.

STEPS	COMMAND PROMPT	PERFORM ACTION	COMMENTS
1.	Command:	type *M* and press **Spacebar**	*M* is the command alias for *Move*
2.	MOVE Select Objects:	use pickbox to select one *Line* and the *Circle*	objects are highlighted
3.	Select Objects:	press **Spacebar** or **Enter**	
4.	Base point or displacement:	**PICK** near the *Circle* center	basepoint is the handle, or where to move <u>from</u>
5.	Second point of displacement:	**PICK** near the other *Line*	second point is where to move <u>to</u>

2. Use *Move* again to move the *Circle* back to its original position. Select the *Circle* with the *Window* option.

STEPS	COMMAND PROMPT	PERFORM ACTION	COMMENTS
1.	Command:	select the *Modify* pull-down, then *Move*	
2.	MOVE Select Objects:	type *W* and press **Spacebar**	select only the circle; object is highlighted
3.	Select Objects:	press **Spacebar** or **Enter**	
4.	Base point or displacement:	**PICK** near the *Circle* center	basepoint is the handle, or where to move <u>from</u>
5.	Second point of displacement:	**PICK** near the original location	second point is where to move <u>to</u>

Using *Copy*

The *Copy* command is similar to *Move* because you are prompted to (1) select objects, (2) specify a "base-point," or point to copy <u>from,</u> and (3) specify a "second point of displacement," or point to copy <u>to.</u>

COPY

Pull-down Menu	COMMAND (TYPE)	ALIAS (TYPE)	Short-cut	Screen (side) Menu	Tablet Menu
Modify *Copy*	*COPY*	*CO*	...	*MODIFY1* *Copy*	*V,15*

1. Using the *Circle* and 2 *Lines* from the previous exercise, use the *Copy* command to practice selecting objects and to make copies of the objects as indicated in the following table.

STEPS	COMMAND PROMPT	PERFORM ACTION	COMMENTS
1.	Command:	type *CO* and press **Enter** or **Spacebar**	*CO* is the command alias for *Copy*
2.	COPY Select objects:	type *F* and press **Enter**	*F* invokes the *Fence* selection option
3.	First fence point:	**PICK** a point near two *Lines*	starts the "fence"
4.	Undo/<Endpoint of line>:	**PICK** a second point across the two *Lines*	lines are highlighted
5.	Undo/<Endpoint of line>:	Press **Enter**	completes *Fence* option
6.	Select objects:	Press **Enter**	completes object selection
7.	<Base point or displace-ment>/Multiple:	**PICK** between the *Lines*	basepoint is the handle, or where to copy <u>from</u>
8.	Second point of displacement:	enter *@2<45* and press **Enter**	copies of the lines are created 2 units in distance at 45 degrees from the original location

2. Practice removing objects from the selection set by following the steps given in the table below.

STEPS	COMMAND PROMPT	PERFORM ACTION	COMMENTS
1.	Command:	type *CO* and press **Enter** or **Spacebar**	*CO* is the command alias for *Copy*
2.	COPY Select objects:	**PICK** in an open area near the right of your drawing, drag Window to the left to select all objects	all objects within and crossing the Window are highlighted
3.	Select objects:	hold down **Shift** and **PICK** all highlighted objects except one *Circle*	holding down Shift while PICKing removes objects from the selection set
4.	Select objects:	press **Enter**	only one circle is highlighted
5.	<Base point or displacement>Multiple:	**PICK** near the circle's center	basepoint is the handle, or where to copy from
6.	Second point of displacement:	turn on *ORTHO*, move the cursor to the right, type **3** and press **Enter**	a copy of the circle is created 3 units to the right of the original circle

NOUN/VERB Command Syntax Practice

1. Practice using the *Move* command using Noun/Verb syntax order by following the steps in the table below.

STEPS	COMMAND PROMPT	PERFORM ACTION	COMMENTS
1.	Command:	**PICK** one *Circle*	circle becomes high-lighted (grips may appear if not disabled)
2.	Command:	type *M* and press **Enter** or **Spacebar**	*M* is the command alias for *Move*
3.	MOVE 1 found Base point or displacement:	**PICK** near the *Circle's* center	command skips the "Select objects" prompt and proceeds
4.	Second point of displacement:	enter **@-3,-3** and press **Enter**	circle is moved -3 X units and -3 Y units from the original location

2. Practice using Noun/Verb syntax order by selecting objects for *Erase*.

STEPS	COMMAND PROMPT	PERFORM ACTION	COMMENTS
1.	Command:	**PICK** one *Line*	line becomes high-lighted (grips may appear if not disabled)
2.	Command:	type *E* and press *Enter* or **Spacebar**	E is the command alias for *Erase*
3.	ERASE 1 found		command skips the "Select objects" prompt and erases highlighted line

CHAPTER EXERCISES

Open drawing **CH3EX** that you created in Chapter 3 Exercises. Turn off *SNAP* (**F9**) to make object selection easier.

1. **Use the pickbox to select objects**

   Invoke the *Erase* command by any method. Select the lower left square with the pickbox (Fig. 4-14, highlighted). Each *Line* must be selected individually. Press **Enter** to complete *Erase*. Then use the *Oops* command to unerase the square. (Type *Oops*.)

Figure 4-14 ─────────

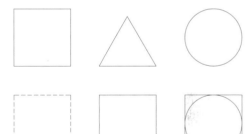

2. **Use the *AUto* Window and *Auto* Crossing Window**

   Invoke *Erase*. Select the center square on the bottom row with the *AUto* Window and select the equilateral triangle with the *AUto* Crossing Window. Press **Enter** to complete the *Erase* as shown in Figure 4-15. Use *Oops* to bring back the objects.

Figure 4-15 ─────────

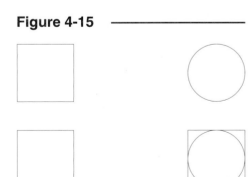

3. **Use the *Fence* selection option**

   Invoke *Erase* again. Use the *Fence* option to select all the vertical *Lines* and the *Circle* from the squares on the bottom row. Complete the *Erase* (see Fig. 4-16). Use *Oops* to unerase.

Figure 4-16

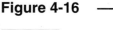

4. **Use the *ALL* option and deselect**

   Use *Erase*. Select all the objects with *ALL*. Remove the four *Lines* (shown highlighted in Fig. 4-17) from the selection set by pressing **Shift** while PICKing. Complete the *Erase* to leave only the four *Lines*. Finally, use *Oops*.

Figure 4-17

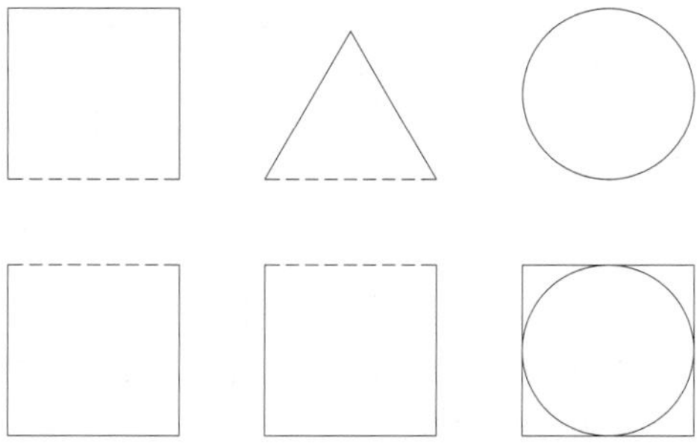

5. **Using Noun/Verb selection**

   <u>Before</u> invoking *Erase,* use the pickbox or *AUto* window to select the triangle. (Make sure no other commands are in use.) <u>Then</u> invoke *Erase*. The triangle should disappear. Retrieve the triangle with *Oops*.

6. **Using *Move* with *Wpolygon***

   Invoke the *Move* command by any method. Use the *WP* option (*Window Polygon*) to select only the *Lines* comprising the triangle. Turn on *SNAP* and PICK the lower-left corner as the "Base point:". *Move* the triangle up 1 unit. (See Fig. 4-18.)

Figure 4-18

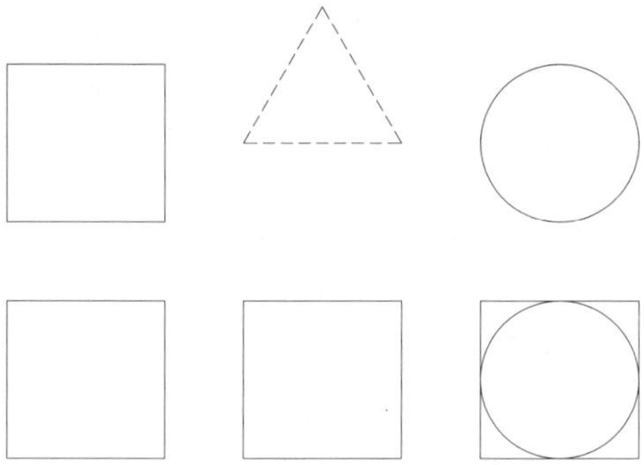

7. **Using *Previous* with *Move***

   Invoke *Move* again. At the "Select Objects:" prompt, type *P* for *Previous*. The triangle should highlight. Using the same basepoint, move the triangle back to its original position.

8. ***Exit*** AutoCAD and do <u>not</u> save changes.

5

HELPFUL COMMANDS

Chapter Objectives

After completing this chapter you should be able to:

1. find *Help* for any command or system variable;

2. use *Oops* to unerase objects;

3. use *U* to undo one command or use *Undo* to undo multiple commands;

4. set a *Mark* and use *Undo Back* to undo all commands until the marker is encountered;

5. *Redo* commands that were undone;

6. regenerate the drawing with *Regen*.

CONCEPTS

There are several commands that do not draw or edit objects in AutoCAD, but are intended to assist you in using AutoCAD. These commands are used by experienced AutoCAD users and are particularly helpful to the beginner. The commands, as a group, are not located in any one menu, but are scattered in several menus.

HELP

Pull-down Menu	COMMAND (TYPE)	ALIAS (TYPE)	Short-cut	Screen (side) Menu	Tablet Menu
Help	*HELP*	*?*	*F1*	*HELP* *Help*	*Y,7*

Help gives you an explanation for any AutoCAD command or system variable as well as help for using the menus and toolbars. *Help* displays a window which gives a variety of methods for finding the information that you need. There is even help for using help!

Help can be used two ways: (1) entered as a command at the open Command: prompt or (2) used transparently while a command is currently in use.

1. If the *Help* command is entered at an open Command: prompt (when no other commands are in use), the *Help* window appears (Fig. 5-1).

Figure 5-1 ————————————

2. When *Help* is used transparently (when a command is in use), it is context sensitive; that is, help on the current command is given automatically. For example, Figure 5-2 displays the window that appears if *Help* is invoked during the *Line* command. (If typing a transparent command, an ' [apostrophe] symbol is typed as a prefix to the command, e.g., *'HELP* or *'?*. If you PICK *Help* from the menus or press **F1**, it is automatically transparent.)

Much of the text that appears in the window can be PICKed to reveal another level of help on that item. This feature, called Hypertext, is activated by moving the pointer to a word (usually underlined and in a green color) or an icon. When the pointer changes to a small hand (Fig. 5-2), click on the item to activate the new information.

Figure 5-2 ————————————

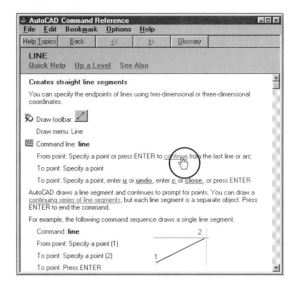

If you use *Help* when no commands are in use (at the open Command: prompt), the *Help Topics: AutoCAD Help* dialog box is displayed. There are three tabs in this dialog box described below, *Contents*, *Index*, and *Find*.

Contents

Several levels of the *Contents* tab are available, offering an overwhelming amount of information (Fig. 5-3). Each main level can be opened to reveal a second (chapter level) or third level of information. The main levels are:

Figure 5-3

Using Help	Get instruction on how to use the *Help* system and the on-line documentation.
Tutorial	Several tutorials are given to assist you to learn AutoCAD.
How To. . .	Select from a list of several main topics such as *Set up a Drawing*, *Using Grips*, or *Modify a Dimension*. Another level is available for each of these topics.
Command Reference	The following informative sections are available: *Commands* *System Variables* *Menus* *Toolbars* *Utilities* *Standard Libraries* *AutoCAD Graphical Objects*
User's Guide	Seventeen chapters and three appendices are given, each with several sections explaining concepts and commands for using AutoCAD.
Installation Guide	Eight chapters can be accessed to give information on installing, configuring, and optimizing AutoCAD.
Customization Guide	This section includes information such as how to customize linetypes, shapes, fonts, menus, dialog boxes, and how to write AutoLISP programs.
ActiveX Automation	This section gives information concerning a new programming interface that allows you to manipulate AutoCAD objects from an application controller such as Visual Basic or Excel.

Figure 5-4

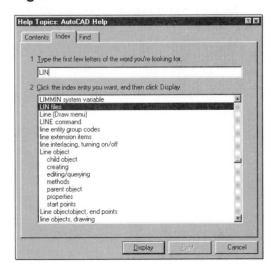

Index

In this section you can type two or three letters of a word that you want information about (Fig. 5-4). As you type, the list below displays the available contents beginning with the letters that you have entered. The word you type can be a command, system variable, term, or concept used in AutoCAD. Once the word you want information on is found, press Enter or click the *Display* button to invoke a new *AutoCAD Help* box.

Find

Find operates like the *Index* function, except you can enter several words describing the topic you want help with (Fig. 5-5). There are three steps to using the *Find* function: (1) type in word or words in the top edit box, (2) select a topic from the list in the center to narrow your search, (3) select a topic from the bottom list and click *Display*. An *AutoCAD Help* box appears.

How To... or To...

Because so much information and so many topics are listed in the *Help* utility, it may take several attempts to locate the information you are looking for. Often your search results in information about programming features or advanced topics that you did not intend to find. Generally, if you are looking for information on how to use AutoCAD to create and edit objects, use any listing that begins with **To...** (do something), or use the **How To...** section of the *Contents* tab. In either case, the resulting *AutoCAD User's Guide* window gives step-by-step information explaining the procedure for accomplishing a task (Fig. 5-6). This window appears near the side of the screen so you can draw or edit in the graphics area while reading the instructions in the Help box. This is generally the most useful help for beginning AutoCAD users.

The simplest way to invoke the *AutoCAD User's Guide* window is by following these steps:

1. Invoke *Help* and select the *Index* tab;
2. Type the command you want help for in the top edit box, <u>then press Enter</u>.
3. The *Topics Found* box appears displaying *AutoCAD User's Guide (AUG)* selections (Fig. 5-7). Make a selection.
4. The *AutoCAD User's Guide* appears on the side of the screen (Fig 5-6).

Figure 5-5

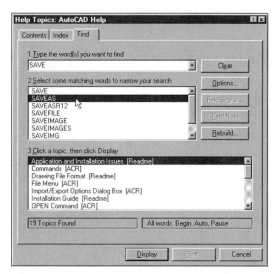

Figure 5-6

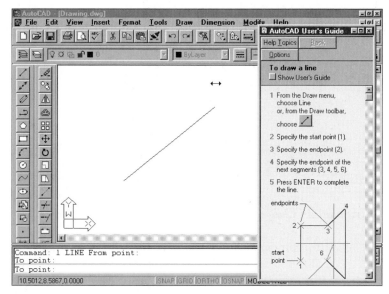

Figure 5-7

OOPS

Pull-down Menu	COMMAND (TYPE)	ALIAS (TYPE)	Short-cut	Screen (side) Menu	Tablet Menu
...	OOPS	...	...	MODIFY1 Erase Oops:	...

The *Oops* command unerases whatever was erased with the last *Erase* command. *Oops* does not have to be used immediately after the *Erase*, but can be used at any time after the *Erase*. *Oops* is typically used after an accidental erase. However, *Erase* could be used intentionally to remove something from the screen temporarily to simplify some other action. For example, you can *Erase* a *Line* to simplify PICKing a group of other objects to *Move* or *Copy*, and then use *Oops* to restore the erased *Line*.

Oops can be used to restore the original set of objects after the *Block* or *Wblock* command is used to combine many objects into one object (explained in Chapter 20).

The *Oops* command is available only from the side menu; there is no icon button or option available in the pull-down menu. *Oops* appears on the side menu only if the *Erase* command is typed or selected. Otherwise, *Oops* must be typed.

U

Pull-down Menu	COMMAND (TYPE)	ALIAS (TYPE)	Short-cut	Screen (side) Menu	Tablet Menu
Edit Undo	U	...	Crtl+Z	EDIT Undo	T,12

The *U* command undoes only the last command. *U* means "undo one command." If used after *Erase*, it unerases whatever was just erased. If used after *Line*, it undoes the group of lines drawn with the last *Line* command. Both *U* and *Undo* do not undo inquiry commands (like *Help*), the *Plot* command, or commands that cause a write-to-disk, such as *Save*.

If you type the letter **U**, select the icon button, or select **Undo** from the pull-down menu or screen (side) menu, only the last command is undone. Typing **Undo** invokes the full *Undo* command. The *Undo* command is explained next.

UNDO

Pull-down Menu	COMMAND (TYPE)	ALIAS (TYPE)	Short-cut	Screen (side) Menu	Tablet Menu
...	UNDO	...	...	ASSIST Undo	...

The full *Undo* command (as opposed to the *U* command) must be typed. This command has the same effect as the *U* command in that it undoes the previous command(s). However, the *Undo* command allows you to undo multiple commands in reverse chronological order. For example, you can use *Undo* and enter a value of **5** to undo the last 5 commands you used. Entering a value (specifying the number of commands to undo) is the default option of *Undo*. The command syntax is as follows:

Command: **undo**
Auto/Control/BEgin/End/Mark/Back/<Number>:

All of the *Undo* options are listed next.

<number>

Enter a value for the number of commands to *Undo*. This is the default option.

Mark

This option sets a marker at that stage of the drawing. The marker is intended to be used by the *Back* option for future *Undo* commands.

Back

This option causes *Undo* to go back to the last marker encountered. Markers are created by the *Mark* option. If a marker is encountered, it is removed. If no marker is encountered, beware, because *Undo* goes back to the <u>beginning of the session</u>. A warning message appears in this case.

BEgin

This option sets the first designator for a group of commands to be treated as one *Undo*.

End

End sets the second designator for the end of a group.

Auto

If *On*, *Auto* treats each command as one group; for example, several lines drawn with one *Line* command would all be undone with *U*.

Control

This option allows you to disable the *Undo* command or limit it to one undo each time it is used.

REDO

Pull-down Menu	COMMAND (TYPE)	ALIAS (TYPE)	Short-cut	Screen (side) Menu	Tablet Menu
Edit Redo	REDO	...	Crtl+Y	EDIT Redo	U,12

The *Redo* command undoes an *Undo*. *Redo* must be used as the <u>next</u> command after the *Undo*. The result of *Redo* is as if *Undo* was never used. Remember, *U* or *Undo* can be used anytime, but *Redo* has no effect unless used immediately after *U* or *Undo*.

REDRAW

Pull-down Menu	COMMAND (TYPE)	ALIAS (TYPE)	Short-cut	Screen (side) Menu	Tablet Menu
View Redraw	REDRAW	R	...	VIEW 1 Redraw	...

Redraw erases the "blips" from the screen if *BLIPMODE* is on. If *BLIPMODE* is on, small crosses appear on the screen any time you PICK a point during a command such as when selecting objects, drawing *Lines*, etc., or creating windows. The "blips" remain on the screen until a viewing command is used (such as *Zoom*), or a regeneration occurs, or until *Redraw* is used. In AutoCAD Release 14, *BLIPMODE* is off by default, so there is little need to use *Redraw* when *BLIPMODE* is off.

In previous releases of AutoCAD, *BLIPMODE* is on by default. Therefore, if you work with drawings created in previous versions of AutoCAD, the "blips" may appear. The *BLIPMODE* variable setting is stored in the drawing file. *BLIPMODE* can be controlled by typing *BLIPMODE* at the command prompt or using the *Drawing Aids* dialog box from the *Tools* pull-down menu.

Redrawall is needed only when Viewports are being used (see Chapters 19 and 29).

REGEN

Pull-down Menu	COMMAND (TYPE)	ALIAS (TYPE)	Short-cut	Screen (side) Menu	Tablet Menu
View *Regen*	*REGEN*	*RE*	...	VIEW 1 *Regen*	*J,1*

The *Regen* command reads the database and redisplays the drawing accordingly. A *Regen* is caused by some commands automatically. Occasionally, the *Regen* command is required to update the drawing to display the latest changes made to some system variables. For example, if you change the style in which *Point* objects appear (displayed as a dot, an "X," or other options), you should then *Regen* in order to display the *Points* according to the new setting. (*Points* are explained in Chapter 8.) *Regenall* is used to regenerate all viewports when several viewports are being used.

CHAPTER EXERCISES

Begin a *New* drawing. If the *Start Up* dialog box or *Create New Drawing* dialog box appears, select **Start from Scratch**, choose **English** as the default setting, and click the **OK** button. Complete the following exercises.

1. *Help*

   Use *Help* by any method to find information on the following commands. Use the *Contents* tab and select the *Command Reference* to locate information on each command. (Use *Help* at the open Command: prompt, not during a command in use.) Read the text screen for each command.

   New, Open, Save, SaveAs
   Oops, Undo, U

2. **Context-sensitive *Help***

   Invoke each of the commands listed below. When you see the first prompt in each command, enter '*Help* or '*?* (transparently) or select *Help* from the menus or Standard toolbar. Read the explanation for each prompt. Select a Hypertext item in each screen (underlined and green in color).

   Line, Arc, Circle, Point

3. *Oops*

   Draw 3 vertical **Lines**. **Erase** one line; then use **Oops** to restore it. Next **Erase** two *Lines*, each with a separate use of the *Erase* command. Use **Oops**. Only the last *Line* is restored. **Erase** the remaining two *Lines*, but select both with a Window. Now use **Oops** to restore both *Lines* (since they were *Erased* at the same time).

4. **Using delayed *Oops***

   Oops can be used at any time, not only immediately after the *Erase*. Draw several horizontal **Lines** near the bottom of the screen. Draw a **Circle** on the **Lines**. Then **Erase** the *Circle*. Use **Move**, select the *Lines* with a Window, and displace the *Lines* to another location above. Now use **Oops** to make the *Circle* reappear.

5. ***U***

   Press the letter ***U*** (make sure no other commands are in use). The *Circle* should disappear (*U* undoes the last command—*Oops*). Do this repeatedly to *Undo* one command at a time until the *Circle* and *Lines* are in their original position (when you first created them).

6. ***Undo***

   Use the ***Undo*** command and select the ***Back*** option. Answer ***Yes*** to the warning message. This action should *Undo* everything.

   Draw a vertical ***Line***. Next, draw a square with four ***Line*** segments (all drawn in the same *Line* command). Finally, draw a second vertical ***Line***. ***Erase*** the first *Line*.

   Now <u>type</u> ***Undo*** and enter a value of **3**. You should have only one *Line* remaining. *Undo* reversed the following three commands:

Erase	The first vertical *Line* was unerased.
Line	The second vertical *Line* was removed.
Line	The four *Lines* comprising the square were removed.

7. ***Redo***

   Invoke ***Redo*** immediately after the *Undo* (from the previous exercise). The three commands are redone. *Redo* must be used immediately after *Undo*.

8. ***Exit*** AutoCAD and answer ***No*** to "Save Changes to...?"

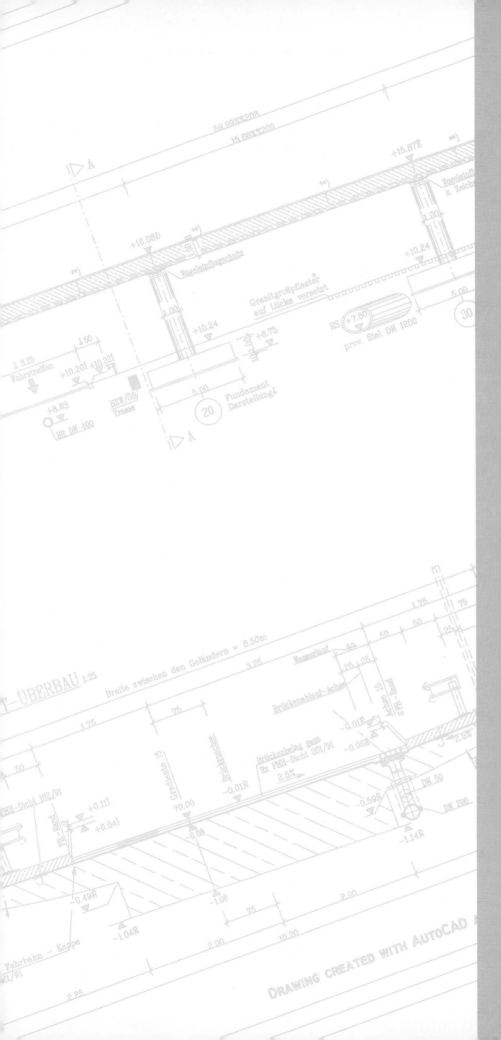

6

BASIC DRAWING SETUP

Chapter Objectives

After completing this chapter you should:

1. know the basic steps for setting up a drawing;

2. be able to specify the desired *Units, Angles* format, and *Precision* for the drawing;

3. be able to specify the drawing *Limits;*

4. know how to specify the *Snap* increment;

5. know how to specify the *Grid* increment;

6. know how to use the *Setup* and *Create New Drawing* dialog boxes to accomplish basic drawing setup.

STEPS FOR BASIC DRAWING SETUP

Assuming the general configuration (dimensions and proportions) of the geometry to be created is known, the following steps are suggested for setting up a drawing:

1. Determine and set the *Units* that are to be used.
2. Determine and set the drawing *Limits;* then *Zoom All.*
3. Set an appropriate *Snap* value.
4. Set an appropriate *Grid* value to be used.

These additional steps for drawing setup are discussed in Chapter 13, Advanced Drawing Setup.

5. Change the *LTSCALE* value based on the new *Limits.*
6. Create the desired *Layers* and assign appropriate *linetype* and *color* settings.
7. Create desired *Text Styles* (optional).
8. Create desired *Dimension Styles* (optional).
9. Create a title block and border (optional).

When you start AutoCAD Release 14 for the first time or use the *New* command to begin a new drawing, the *Start Up* or *Create New Drawing* dialog box appears. Each of these dialog boxes makes available three options for setting up a new drawing. The three options represent three levels of automation/preparation for drawing setup. Only the *Start from Scratch* option is described in this chapter. The *Start from Scratch* option allows you to step through each of the individual commands (listed above) to set up a drawing to your specifications.

It might occur that using the automated *Quick Setup Wizard* or *Advanced Setup Wizard* available in the *Start Up* and *Create New Drawing* dialog boxes would be the fastest and easiest methods to set up a new drawing; however, the automated settings may or may not be exactly what you want or need. Surprisingly, the Wizards do not perform the operations needed to correctly set up a drawing for plotting or printing to a standard scale. As a third alternative, template drawings can be an efficient method for beginning a new drawing since many of the setup steps can be prepared in the template drawing. You must first understand the individual setup commands to be able to create a template drawing or to modify Wizard settings for your needs. The Wizards and template drawings are discussed in detail in Chapter 13.

Setting up a new drawing usually involves both the use of the *Start Up* or *Create New Drawing* dialog box and the individual setup commands listed above and described in the Setup Commands section.

THE *START UP* AND *CREATE NEW DRAWING* DIALOG BOXES

To "manually" achieve the first four steps for drawing setup, use the *Start from Scratch* option of the *Create New Drawing* dialog box.

When you start AutoCAD for the first time, the *Start Up* dialog appears (Fig 6-1). (The *Start Up* dialog box can be suppressed by removing the check in the box at the bottom of the dialog labeled *Show this dialog at start up.* You can make the dialog box reappear on start up by checking *Show the Start Up dialog box* in the *Compatibility* tab of the *Preferences* dialog box.)

Figure 6-1

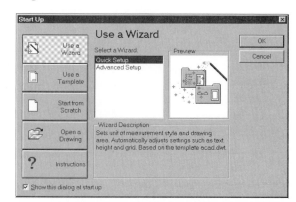

When you use the *New* command, the *Create New Drawing* dialog box appears. This is essentially the same as the *Start Up* dialog box that may appear when you begin AutoCAD, except the *Open a Drawing* option is not included in the *Create New Drawing* dialog box.

The *Start Up* and *Create New Drawing* dialog boxes offer the following three options that are intended to assist you in setting up a drawing.

Use a Wizard

The *Quick Setup Wizard* prompts you to select the type of drawing *Units* you want to use and to specify the drawing *Area* (*Limits*). The *Advanced Setup* offers options for units, angular direction and measurement, area, title blocks, and layouts for paper space and model space. Both Wizards also automatically make changes to many size-related system variables. See Chapter 13, Advanced Drawing Setup, for more information on this option.

Use a Template

Use this option if you want to use an existing template drawing (.DWT file) as a starting point. A template drawing is one that already has many of the setup steps performed but contains no geometry (graphical objects). See Chapter 13, Advanced Drawing Setup, for more information on this option.

Start from Scratch

Use the *Start from Scratch* button and the *English* option (Fig. 6-2) if you want to begin with the traditional AutoCAD default drawing settings, then determine your own system variable values using the individual setup commands such as *Units, Limits, Snap,* and *Grid*. Use the *Metric* option for setting up a metric drawing for plotting on a metric A3 sheet (drawing area of 420mm x 297mm).

Figure 6-2

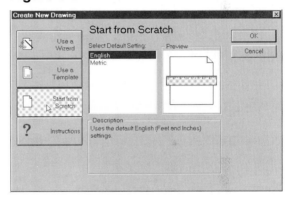

English
Selecting the *English* option causes AutoCAD to use the template drawing ACAD.DWT, which has the traditional AutoCAD default settings (see Table of *Start from Scratch English* Settings). This template drawing is essentially the same drawing used as the default in previous releases of AutoCAD, with settings such as drawing area of 12 x 9 units, etc. As an alternative to selecting the *English* defaults in the *Start from Scratch* option, a drawing can be started using the AutoCAD defaults by any of the following methods:

Selecting the ACAD.DWT template drawing with the *Use a Template* option
Pressing *Cancel* in the *Setup* dialog box that appears when AutoCAD starts
Starting AutoCAD after suppressing the *Setup* dialog box

The following table lists the AutoCAD default settings related to drawing setup. Many of these settings such as *Units, Limits, Snap,* and *Grid* are explained in detail in the following section.

R14

Table of *Start from Scratch English* Settings (ACAD.DWT Defaults)

Related Command	Description	System Variable	Default Setting
Units	drawing units	*LUNITS*	2 (decimal)
Limits	drawing area	*LIMMAX*	12.0000,9.0000
Snap	snap increment	*SNAPUNIT*	.5000
Grid	grid increment	*GRIDUNIT*	.5000
LTSCALE	linetype scale	*LTSCALE*	1.0000
DIMSCALE	dimension scale	*DIMSCALE*	1.0000
Text, Dtext, Mtext	text height	*TEXTSIZE*	.2000

The *English* default setting is recommended if you want to print or plot a drawing to scale on other than a standard A size U.S. architectural or engineering sheet. (See Chapter 13 for information on plotting to scale.)

Metric
Selecting the *Metric* option causes AutoCAD to use the metric template drawing, ACADISO.DWT, in which many of the drawing setups and system variable settings are different from the ACAD.DWT drawing, as shown below.

Table of *Start from Scratch Metric* Settings

Related Command	Description	System Variable	Default Setting
Units	drawing units	*LUNITS*	2 (decimal)
Limits	drawing area	*LIMMAX*	420.0000, 297.0000
Snap	snap increment	*SNAPUNIT*	10.0000
Grid	grid increment	*GRIDUNIT*	10.0000
LTSCALE	linetype scale	*LTSCALE*	1.0000
DIMSCALE	dimension scale	*DIMSCALE*	1.0000
Text, Dtext, Mtext	text height	*TEXTSIZE*	2.5000
Hatch	hatch pattern scale	*HPSIZE*	1.0000

The metric drawing setup is intended to be used with ISO linetypes and ISO hatch patterns, which are pre-scaled for these *Limits*, hence the *LTSCALE* and hatch pattern scale of 1. The individual dimensioning variables for arrow size, dimension text size, gaps and extensions, etc. are changed so the dimensions are drawn correctly with a *DIMSCALE* of 1.

The *Metric* option is recommended for plotting to scale on an A3 metric sheet size. If you want to use the *Metric* option for drawing setup and to plot or print on standard U.S. architectural and engineering sheets, the *Limits* should be changed to be proportional to A, B, C, D, or E size sheets. (See Chapter 13 for information on plotting to scale.)

SETUP COMMANDS

UNITS and
DDUNITS

Pull-down Menu	COMMAND (TYPE)	ALIAS (TYPE)	Short-cut	Screen (side) Menu	Tablet Menu
Format *Units...*	*UNITS or* *DDUNITS*	*UN or* *—UN*	...	*FORMAT* *Ddunits*	V,4

The *Units* command allows you to specify the type and precision of linear and angular units as well as the direction and orientation of angles to be used in the drawing. The current setting of *Units* determines the display of values by the coordinates display (*Coords*) and in some dialog boxes. (In previous releases of AutoCAD, the setting for *Units Precision* also affected the display of values in dialog boxes, which caused problems in some dialog boxes. *Precision* does not affect display of values in dialog boxes in Release 14.)

You can select *Units* ... from the *Format* pull-down or type *Ddunits* (or command alias *UN*) to invoke the *Units Control* dialog box (Fig. 6-3). Type *Units* (or *—UN*) to produce a text screen (Fig. 6-4).

Figure 6-3

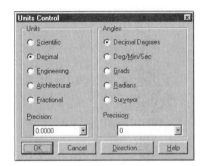

The linear and angular units options are displayed in the dialog box format (Fig. 6-3) and in command line format (Fig. 6-4). The choices for both linear and angular *Units* are shown in the figures.

Figure 6-4

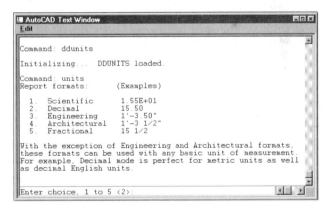

Units	Format	
1. Scientific	1.55E + 01	Generic decimal units with an exponent
2. Decimal	15.50	Generic decimal usually used for applications in metric or decimal inches
3. Engineering	1'-3.50"	Explicit feet and decimal inches with notation, one unit equals one inch
4. Architectural	1'-3 1/2"	Explicit feet and fractional inches with notation, one unit equals one inch
5. Fractional	15 1/2	Generic fractional units

Precision

When setting *Units*, you should also set the precision. *Precision* is the number of places to the right of the decimal or the denominator of the smallest fraction to display. The precision is set by making the desired selection from the *Precision* pop-up list in the *Units Control* dialog box (Fig. 6-3) or by keying in the desired selection in command line format.

Precision controls only the <u>display</u> of *Coords*. The <u>actual</u> <u>precision</u> of the drawing database is always the same in AutoCAD, that is, 14 significant digits.

Angles

You can specify a format other than the default (decimal degrees) for expression of angles. Format options for angular display and examples of each are shown in Figure 6-3 (dialog box format).

Figure 6-5 ──────

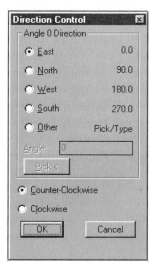

The orientation of angle 0 can be changed from the default position (east or X positive) to other options by selecting the *Direction* tile in the *Units Control* dialog box. This produces the *Direction Control* dialog box (Fig. 6-5). Alternately, the *Units* command can be typed to select these options in command line format.

The direction of angular measurement can be changed from its default of counter-clockwise to clockwise. The direction of angular measurement affects the direction of positive and negative angles in commands such as *Array Polar*, *Rotate* and dimension commands that <u>measure</u> angular values but does not change the direction *Arcs* are <u>drawn</u>, which is always counter-clockwise.

Keyboard Input of *Units* Values

When AutoCAD prompts for a point or a distance, you can respond by entering values at the keyboard. The values can be in <u>any format</u>—integer, decimal, fractional, or scientific—regardless of the format of *Units* selected.

You can <u>type in explicit feet and inch values only if *Architectural* or *Engineering*</u> units have been specified as the drawing units. For this reason, specifying *Units* is the first step in setting up a drawing.

Type in explicit feet or inch values by using the ' (apostrophe) symbol after values representing feet and the " (quote) symbol after values representing inches. If no symbol is used, the values are understood by AutoCAD to be <u>inches</u>.

Feet and inches input <u>cannot</u> contain a blank, so a hyphen (-) must be typed between inches and fractions. For example, with *Architectural* units, key in *6'2-1/2"*, which reads "six feet two and one-half inches." The standard engineering and architectural format for dimensioning, however, places the hyphen between feet and inches (as displayed by the default setting for the *Coords* display).

The *UNITMODE* variable set to **1** changes the display of *Coords* to remind you of the correct format for <u>input</u> of feet and inches (with the hyphen between inches and fractions) rather than displaying the standard format for feet and inch notation (standard format, *UNITMODE* of **0**, is the default setting). If options other than *Architectural* or *Engineering* are used, values are read as generic units.

LIMITS

Pull-down Menu	COMMAND (TYPE)	ALIAS (TYPE)	Short-cut	Screen (side) Menu	Tablet Menu
Format *Drawing Limits*	*LIMITS*	...	...	*FORMAT* *Limits*	*V,2*

The *Limits* command allows you to set the size of the drawing area by specifying the lower left and upper-right corners in X,Y coordinate values.

Command: *limits*
Reset Model space limits
ON/OFF/<Lower left corner> <0,0 or current values>: **x,y** or **Enter** (Enter an X,Y value or accept the 0,0 default—normally use 0,0 as lower-left corner.)
Upper right corner <12,9>: **x,y** (Enter new values to change upper-right corner to allow adequate drawing area.)

The default *Limits* values in AutoCAD are 12 and 9; that is, 12 units in the X direction and 9 units in the Y direction (Fig. 6-6). Starting a drawing by any of the following methods (of the *Setup* or *Create New Drawing* dialog boxes) results in *Limits* of 12 x 9:

Selecting the ACAD.DWT template drawing
Pressing *Cancel* in the *Setup* dialog box that appears when AutoCAD starts
Starting AutoCAD after suppressing the *Setup* dialog box
Selecting the *English* defaults in the *Start from Scratch* option
Accepting the default values for the *Area* in the *Quick Setup Wizard*
Accepting the default values for the *Area* in the *Advanced Setup Wizard*

Figure 6-6

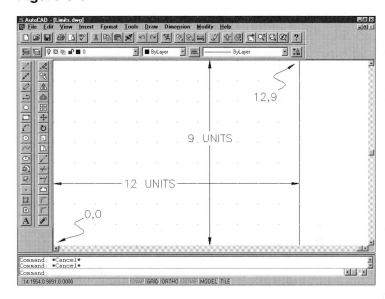

If the *GRID* is turned on, the dots are displayed only over the *Limits*. The AutoCAD screen (default configuration) displays additional area on the right past the *Limits*. The units are generic decimal units that can be used to represent inches, feet, millimeters, miles, or whatever is appropriate for the intended drawing. Typically, however, decimal units are used to represent inches or millimeters. If the default units are used to represent inches, the default drawing size would be 12 by 9 inches.

Remember that when a CAD system is used to create a drawing, the geometry should be drawn <u>full size</u> by specifying dimensions of objects in <u>real-world units</u>. A completed CAD drawing or model is virtually an exact dimensional replica of the actual object. Scaling of the drawing occurs only when plotting or printing the file to an actual fixed-size sheet of paper.

Before beginning to create an AutoCAD drawing, determine the size of the drawing area needed for the intended geometry. After setting *Units*, appropriate *Limits* should be set in order to draw the object or geometry to the <u>real-world size in the actual units</u>. There are no practical maximum or minimum settings for *Limits*.

The X,Y values you enter as *Limits* are understood by AutoCAD as values in the units specified by the *Units* command. For example, if you previously specified *Architectural units*, then the values entered are understood as inches unless the notation for feet (') is given (**240,180** or **20',15'** would define the same coordinate). Remember, you can type in explicit feet and inch values only if *Architectural* or *Engineering* units have been specified as the drawing units.

If you are planning to plot the drawing to scale, *Limits* should be set to a proportion of the <u>sheet size</u> you plan to plot on. For example, setting limits to 22 by 17 (2 times 11 by 8.5) would allow enough room for drawing an object about 20" by 15" and allow plotting at 1/2 size on the 11" x 8.5" sheet. Simply stated, <u>set *Limits* to a proportion of the paper</u>.

ON/OFF

If the *ON* option of *Limits* is used, limits checking is activated. Limits checking prevents you from drawing objects outside of the limits by issuing an outside-limits error. This is similar to drawing "off the paper." Limits checking is *OFF* by default.

Limits also defines the display area for *GRID* as well as the minimum area displayed when a *Zoom All* is used. *Zoom All* forces the full display of the *Limits*. *Zoom All* can be invoked by typing **Z** (command alias) then **A** for the *All* option.

Changing *Limits* does <u>not</u> automatically change the display. As a general rule, you should make a habit of invoking a *Zoom All* <u>immediately following</u> a change in *Limits* to display the area defined by the new limits (Fig. 6-7).

When you <u>reduce</u> *Limits* while *Grid* is *ON*, it is apparent that a change in *Limits* does not automatically change the display. In this case, the area covered by the grid is reduced in size as *Limits* are reduced, yet the display remains unchanged.

Figure 6-7

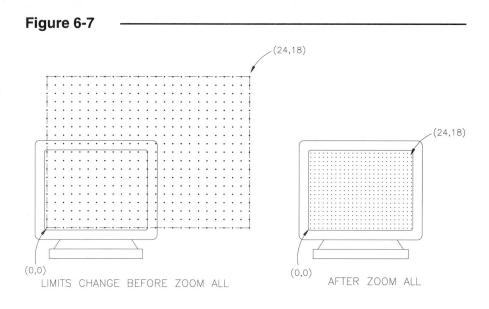

If you are already experimenting with drawing in different *Linetypes*, a change in *Limits* and *Zoom All* affects the display of the hidden and dashed lines. The *LTSCALE* variable controls the spacing of non-continuous lines. The *LTSCALE* is often changed proportionally with changes in *Limits*.

SNAP

Pull-down Menu	COMMAND (TYPE)	ALIAS (TYPE)	Short-cut	Screen (side) Menu	Tablet Menu
Tools *Drawing Aids...*	*SNAP*	*SN*	*F9 or* *Ctrl+B*	*TOOLS 2* *Ddrmodes*	*W,10*

SNAP, when activated by pressing **F9** or double-clicking on *SNAP* in the Status line, forces the cursor position to regular increments. This function can be of assistance to you by making it faster and more accurate for creating and editing objects. The default *Snap* setting is **0.5**. The *Snap* command is used to set the value for these invisible snap increments. *Snap* spacing can be set to any value.

Since *Snap* controls the cursor position, the value of *Snap* should be set to the <u>interactive</u> accuracy desired. As a general rule, you specify the *Snap* value to be that of the <u>common</u> dimensional length expected in the drawing. For example, if the common dimensional length in the drawing is 1/2", or you intend for dimensional accuracy of the drawing to be to the nearest 1/2", the *Snap* command is used to change the snap spacing to **1/2** or **.5**. In this way the cursor always "snaps" to .5 increments.

In some cases such as architectural or civil engineering applications, where dimensional increments are small with relation to the size of the geometry, there is no practical <u>interactive</u> accuracy or there may be no <u>common</u> dimensional length; therefore, some users prefer not to use *Snap*.

The *Snap* command is easily typed, displaying the options in command line format. The *Drawing Aids* dialog box (Fig. 6-8) can be invoked by menu selection or by typing *Ddrmodes* or the command alias, *RM*. Enter the desired value for *Snap* in the *X Spacing* box and press *Enter*. The *Y Spacing* value automatically changes to match.

Figure 6-8

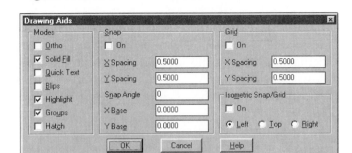

The command line format is as follows:

    Command: **snap**
    Snap spacing or ON/OFF/Aspect/Rotate/Style <current value>: **(value or letter)**
    (enter a value or option)

ON/OFF
Selecting *ON* or *OFF* accomplishes the same action as toggling the **F9** key, pressing **Ctrl+B,** or double-clicking **SNAP** on the Status Line. Typically, *SNAP* should be *ON* for drawing and editing but turned *OFF* to make object selection easier (the cursor moves smoothly to any location with *SNAP OFF*).

Aspect
The *Aspect* option allows specification of unequal X and Y spacing for *SNAP*. This action can also be accomplished in the *Drawing Aids* dialog box by entering different values for *X Spacing* and *Y Spacing*.

Rotate
SNAP can also be *Rotat*ed about any point and set to any angle. When *SNAP* has been rotated, the *GRID, ORTHO,* and "crosshairs" automatically follow this alignment. This action facilitates creating objects oriented at the specified angle, for example, creating an auxiliary view or drawing part of a floor plan at an angle. To accomplish this, use the *R* option in command line format or set *Snap Angle* and *X Base* and *Y Base* (point to rotate about) in the dialog box.

Style
The *Style* option allows switching between a *Standard* snap pattern (the default square or rectangular) and an *Isometric* snap pattern. If using the dialog box, toggle *Isometric Snap/Grid On*. When the *SNAP Style* or *Rotate* angle is changed, the *GRID* automatically aligns with it.

Keep in mind that using the *Quick Setup Wizard* or *Advanced Setup Wizard* causes AutoCAD to automatically calculate *Snap* and *Grid* values for you. In some cases this is helpful, but in other cases this is a hindrance and <u>can cause unintentional errors</u> in geometry creation and editing. Because the Wizard uses a formula to calculate *Snap* and *Grid* values based on the *Area* (*Limits*) settings you specify, the resulting *Snap* and *Grid* can be <u>unusable</u> in some cases and should be changed by using the *Snap* and *Grid* commands or the *Drawing Aids* dialog box. For example, using the Wizard to set up a drawing to plot at a scale of 1=1 on a standard 17 x 11, 18 x 12, or 22 x 17 sheet size results in *Snap* and *Grid* settings of .6, .7, and .9, respectively. (See Chapter 13, Advanced Drawing Setup.)

GRID

Pull-down Menu	COMMAND (TYPE)	ALIAS (TYPE)	Short-cut	Screen (side) Menu	Tablet Menu
Tools *Drawing Aids...*	*GRID*	...	*F7 or* *Ctrl+G*	*TOOLS 2* *Ddrmodes*	*W,10*

GRID is visible on the screen, whereas *SNAP* is invisible. *GRID* is only a <u>visible</u> display of some regular interval. *GRID* and *SNAP* can be <u>independent</u> of each other. In other words, each can have separate spacing settings and the active state of each (*ON, OFF*) can be controlled independently. The *GRID* <u>follows</u> the *SNAP* if *SNAP* is rotated or changed to *Isometric Style*. Although the *GRID* spacing can be different than that of *SNAP,* it can also be forced to follow *SNAP* by using the *Snap* option. The default *GRID* setting is **0.5**.

The *GRID* <u>cannot</u> be plotted. It is <u>not</u> comprised of *Point* objects and therefore is not part of the current drawing. *GRID* is only a visual aid.

Grid can be accessed by command line format (shown below) or set via the *Drawing Aids* dialog box (Fig. 6-8). The dialog box is invoked by menu selection or by typing *Ddrmodes* or *RM.* The dialog box allows only *X Spacing* and *Y Spacing* input for *Grid.*

Command: **grid**
Grid spacing(X) or ON/OFF/Snap/Aspect <current value>: **(value or letter)**
(enter a value or option)

Grid Spacing (X)
If you supply a value for the *Grid spacing, GRID* is displayed at that spacing regardless of *SNAP* spacing. If you key in an *X* as a suffix to the value (for example, **2X**), the *GRID* is displayed as that value <u>times</u> the *SNAP* spacing (for example, "2 times" *SNAP*).

ON/OFF
The *ON* and *OFF* options simply make the *GRID* visible or not (like toggling the **F7** key, pressing **Ctrl+G**, or double-clicking **GRID** on the Status line).

Snap
The *Snap* option of the *Grid* command forces the *GRID* spacing to equal that of *SNAP,* even if *SNAP* is subsequently changed.

Aspect
The *Aspect* option of *GRID* allows different X and Y spacing (causing a rectangular rather than a square *GRID*).

Remember that using the *Quick Setup Wizard* or *Advanced Setup Wizard* causes AutoCAD to automatically calculate *Snap* and *Grid* values for you, which can be helpful, but can also be nonproductive in some cases (see Chapter 13, Advanced Drawing Setup). If you want to use values other than the Wizard calculated values, use the *Snap* and *Grid* commands or the *Drawing Aids* dialog box to change them to a more useful increment.

CHAPTER EXERCISES

1. A drawing is to be made to detail a mechanical part. The part is to be manufactured from sheet metal stock; therefore, only one view is needed. The overall dimensions are 18 by 10 inches, accurate to the nearest .125 inch. Complete the steps for drawing setup:

    A. The drawing will be automatically "scaled to fit" the paper (no standard scale).

        1. Begin a *New* drawing. When the *Start Up* or *Create New Drawing* dialog box appears, select *Start from Scratch*. Select the *English* default settings.
        2. *Units* should be *Decimal*. Set the *Precision* to **0.000**.
        3. Set *Limits* in order to draw full size. Make the lower-left corner **0,0** and the upper right at **24,18**. This is a 4 by 3 proportion and should allow space for the part, a title block, border, and dimensions or notes.
        4. *Zoom All*. (Type **Z** for *Zoom;* then type **A** for *All*.)
        5. Set the *GRID* to **1**.
        6. Set *SNAP* to **.125**.
        7. Save this drawing as **CH6EX1A** (to be used again later). (When plotting at a later time, "Scale to Fit" can be specified.)

    B. The drawing will be plotted to scale on engineering A or B size paper (11" by 8.5" or 22" by 17").

        1. Begin a *New* drawing. When the *Start Up* or *Create New Drawing* dialog box appears, select *Start from Scratch*. Select the *English* default settings.
        2. *Units* should be *Decimal*. Set the *Precision* to **0.000**.
        3. Set *Limits* to the paper size (or a proportion thereof), making the lower-left corner **0,0** and the upper-right at **22,17**. This allows space for drawing full size and for a title block, border, and dimensions or notes.
        4. *Zoom All*. (Type **Z** for *Zoom;* then type **A** for *All*.)
        5. Set the *GRID* to **1**.
        6. Set *SNAP* to **.125**.
        7. *Save* this drawing as **CH6EX1B** (to be used again later). (When plotting, a scale of 1=1 can be specified to plot on 22" by 17" paper, or a scale of 1/2=1 can be specified to plot on 11" by 8.5" paper.)

2. A drawing is to be prepared for a house plan. Set up the drawing for a floor plan that is approximately 50' by 30'. Assume the drawing is to be automatically "Scaled to Fit" the sheet (no standard scale).

    A. Begin a *New* drawing. When the *Start Up* or *Create New Drawing* dialog box appears, select *Start from Scratch*. Select the *English* default settings.
    B. Set *Units* to *Architectural*. Set the *Precision* to **0'-0 1/4"**. Each unit equals 1 inch.
    C. Set *Limits* to **0,0** and **80',60'**. Use the ' (apostrophe) symbol to designate feet. Otherwise, enter **0,0** and **960,720** (size in inch units is: 80x12=960 and 60x12=720).
    D. *Zoom All*. (Type **Z** for *Zoom;* then type **A** for *All*.)
    E. Set *GRID* to **24** (2 feet).
    F. Set *SNAP* to **6** (anything smaller would be hard to PICK).
    G. *Save* this drawing as **CH6EX2**.

3.   A multiview drawing of a mechanical part is to be made. The part is 125mm in width, 30mm in height, and 60mm in depth. The plot is to be made on an A3 metric sheet size (420mm x 297mm). The drawing will use ISO linetypes and ISO hatch patterns, so AutoCAD's *Metric* default settings can be used.

    A.   Begin a *New* drawing. When the *Start Up* or *Create New Drawing* dialog box appears, select *Start from Scratch*. Select the *Metric* default settings.

    B.   *Units* should be *Decimal*. Set the *Precision* to **0.00**.

    C.   Calculate the space needed for three views. If *Limits* are set to the sheet size, there should be adequate space for the views. Make sure the lower-left corner is at **0,0** and the upper right is at **420,297**. (Since the *Limits* are set to the sheet size, a plot can be made later at 1:1 scale.)

    D.   Change *SNAP* to **5**.

    E.   *Save* this drawing as **CH6EX3** (to be used again later).

4.   Assume you are working in an office that designs many mechanical parts in metric units. However, the office uses a standard laser jet printer for 11" x 8.5" sheets. Since AutoCAD does not have a setup for metric drawings on non-metric sheets, it would help to carry out the steps for drawing setup and save the drawing as a template to be used later.

    A.   Begin a *New* drawing. When the *Start Up* or *Create New Drawing* dialog box appears, select *Start from Scratch*. Select *English* default settings.

    B.   Set the *Units Precision* to **0.00**.

    C.   Change the *Limits* to match an 11" x 8.5" sheet. Make the lower-left corner **0,0** and the upper right **279,216** (11 x 8.5 times 25.4, approximately). (Since the *Limits* are set to the sheet size, plots can easily be made at 1:1 scale.)

    D.   *Zoom All*. (Type *Z* for *Zoom*, then *A* for *All*).

    E.   Change the *Snap* to **2**.

    F.   Change *Grid* to **10**.

    G.   At the command prompt, type *LTSCALE*. Change the value to **12**.

    H.   *Save* the drawing as **A-METRIC**.

5.   Assume you are commissioned by the local parks and recreation department to provide a layout drawing for a major league sized baseball field. Follow these steps to set up the drawing:

    A.   Begin a *New* drawing. When the *Start Up* or *Create New Drawing* dialog box appears, **select** *Start from Scratch,* then *English* default units.

    B.   Set the *Units* to *Architectural* and the *Precision* to **1/2"**.

    C.   Set *Limits* to an area of **512' x 384'** (make sure you key in the apostrophe to designate feet).

    D.   Use the *Snap* command. Change the *Snap* value to **10'** (don't forget the apostrophe).

    E.   Use the *Grid* command and change the value to **20'**. Ensure *SNAP* and *GRID* are on.

    F.   *Save* the drawing and assign the name **BALL FIELD CH6**.

7

OBJECT SNAP

Chapter Objectives

After completing this chapter you should:

1. understand the importance of accuracy in CAD drawings;

2. know the function of each of the Object Snap *(OSNAP)* modes;

3. be able to recognize the AutoSnap Marker symbols;

4. be able to invoke *OSNAP*s for single point selection;

5. be able to operate Running Object Snap modes;

6. know that you can toggle Running Object Snap off to specify points not at object features;

7. know that you can use *OSNAP* any time AutoCAD prompts for a point.

CAD ACCURACY

Because CAD databases store drawings as digital information with great precision (fourteen numeric places in AutoCAD), it is possible, practical, and desirable to create drawings that are 100% accurate; that is, a CAD drawing should be created as an exact dimensional replica of the actual object. For example, lines that appear to connect should actually connect by having the exact coordinate values for the matching line endpoints. Only by employing this precision can dimensions placed in a drawing automatically display the exact intended length, or can a CAD database be used to drive CNC (Computer Numerical Control) machine devices such as milling machines or lathes, or can the CAD database be used for rapid prototyping devices such as Stereo Lithography Apparatus. With CAD/CAM technology (Computer-Aided Design/Computer-Aided Manufacturing), the CAD database defines the configuration and accuracy of the finished part. Accuracy is critical. Therefore, in no case should you create CAD drawings with only visual accuracy such as one might do when sketching using the "eyeball method."

OBJECT SNAP

AutoCAD provides a capability called "Object Snap", or *OSNAP* for short, that enables you to "snap" to existing object endpoints, midpoints, centers, intersections, etc. When an *OSNAP* mode (*Endpoint, Midpoint, Center, Intersection,* etc.) is invoked, you can move the cursor <u>near</u> the desired object feature (endpoint, midpoint, etc.) and AutoCAD locates and calculates the coordinate location of the desired object feature.

Available Object Snap modes are:

> *Center*
> *Endpoint*
> *Insert*
> *Intersection*
> *Midpoint*
> *Nearest*
> *Node* (Point)
> *Perpendicular*
> *Quadrant*
> *Tangent*
> *From*
> *Apparent Intersection* (for 3D use)

For example, when you want to draw a *Line* and connect its endpoint to an existing *Line,* you can invoke the *Endpoint OSNAP* mode at the "To point:" prompt, then snap <u>exactly</u> to the desired line end by moving the cursor near it and PICKing (Fig. 7-1). *OSNAP*s can be used for any draw or modify operation—whenever AutoCAD prompts for a point (location).

Figure 7-1

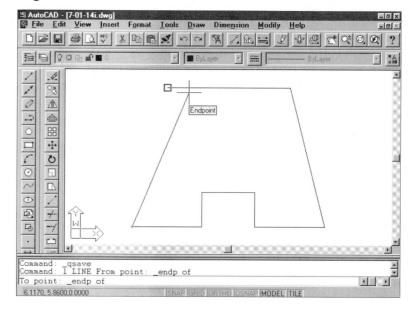

A new Release 14 feature called "AutoSnap" displays a "Snap Marker" indicating the particular object feature (endpoint, midpoint, etc.) when you to move the cursor <u>near</u> an object feature. Each *OSNAP* mode (*Endpoint, Midpoint, Center, Intersection*, etc.) has a distinct symbol (AutoSnap Marker) representing the object feature. This innovation allows you to preview and confirm the snap points before you PICK them (Fig. 7-2). The table in Figure 7-19 gives each Object Snap mode, the related AutoSnap Marker, and the icon button used to activate a single Object Snap selection.

Figure 7-2

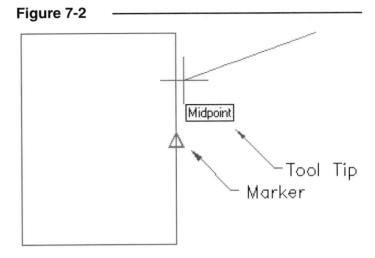

AutoSnap provides two other aids for previewing and confirming *OSNAP* points before you PICK. A "Snap Tip" appears shortly after the Snap Marker appears (hold the cursor still and wait one second). The Snap Tip gives the name of the found *OSNAP* point, such as an *Endpoint, Midpoint, Center,* or *Intersection,* (Fig. 7-2). In addition, a "Magnet" draws the cursor to the snap point if the cursor is within the confines of the Snap Marker. This Magnet feature helps confirm that you have the desired snap point before making the PICK.

A visible target box, or "Aperture," <u>can be displayed</u> at the cursor (invisible by default) whenever an *OSNAP* mode is in effect (Fig. 7-3). The Aperture is a square box larger than the pickbox (default size of 10 pixels square). Technically, this target box (visible or invisible) must be located <u>on an object</u> before a Snap Marker and related Snap Tip appear. In pre-

Figure 7-3

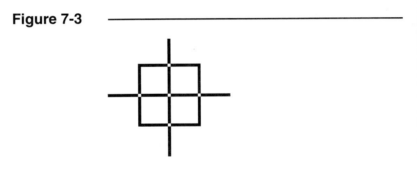

vious versions of AutoCAD, the Aperture is always displayed. Snap Markers, Snap Tips, and the Magnet feature are not available in previous releases; therefore, *OSNAP* selections sometimes result in unexpected snap points.

The settings for the Aperture, Snap Markers, Snap Tips, and Magnet are controlled in the *AutoSnap* tab of the *Osnap Settings* dialog box (discussed later).

The Object Snap modes are explained in the next section. Each mode, and its relation to the AutoCAD objects, is illustrated.

Object Snaps must be activated in order for you to "snap" to the desired object features. Two methods for activating Object Snaps, Single Point Selection and Running Object Snaps, are discussed in the sections following Object Snap Modes. The *OSNAP* modes (*Endpoint, Midpoint, Center, Intersection,* etc.) operate identically for either method.

OBJECT SNAP MODES

AutoCAD provides the following Object Snap Modes.

Center

This *OSNAP* option finds the center of a *Circle, Arc,* or *Donut*. You must PICK the *Circle* <u>object</u>, not where you think the center is.

Figure 7-4

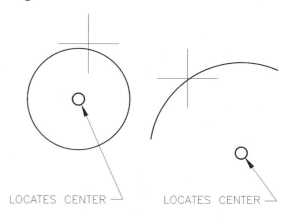

LOCATES CENTER   LOCATES CENTER

Endpoint

The *Endpoint* option snaps to the endpoint of a *Line, Pline, Spline,* or *Arc*. PICK the object <u>near</u> the desired end.

Figure 7-5

LOCATES ENDPOINT

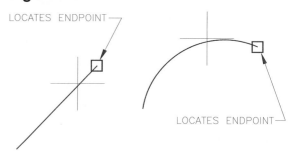

LOCATES ENDPOINT

Insert

This option locates the insertion point of *Text* or a *Block*. PICK anywhere on the *Block* or line of *Text*.

Figure 7-6

LOCATES INSERTION POINT

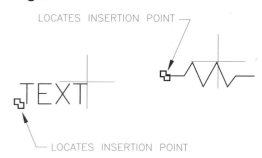

TEXT

LOCATES INSERTION POINT

Intersection

Using this option causes AutoCAD to calculate and snap to the intersection of any two objects. You can locate the cursor (Aperture) so that <u>both</u> objects pass near (through) it, or you can PICK each object <u>individually</u>.

Figure 7-7

LOCATES INTERSECTION

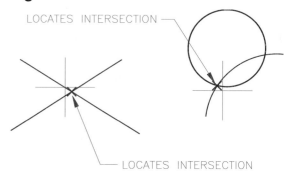

LOCATES INTERSECTION

Even if the two objects that you PICK do not physically intersect, you can PICK each one individually with the *Intersection* mode and AutoCAD will find the <u>extended</u> intersection.

Figure 7-8

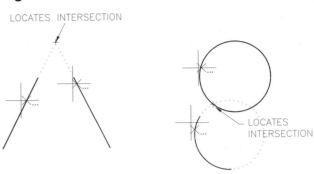

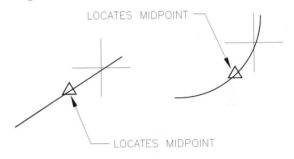

Midpoint

The *Midpoint* option snaps to the point of a *Line* or *Arc* that is <u>halfway</u> between the endpoints. PICK anywhere on the object.

Figure 7-9

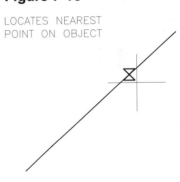

Nearest

The *Nearest* option locates the point on an object nearest to the <u>cursor position</u>. Place the cursor center nearest to the desired location, then PICK.

Figure 7-10

Nearest <u>cannot</u> be used effectively with *ORTHO* because *OSNAPs* override *ORTHO*. In other words, using *Nearest* to locate a "To point:" will not produce an orthogonal line if *ORTHO* is on because *OSNAPs* take priority.

Node

This option snaps to a <u>Point</u> object. The *Point* must be within the Aperture (visible or invisible Aperture).

Figure 7-11

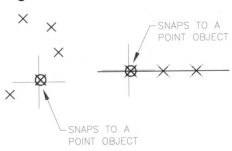

Perpendicular

Use this option to snap perpendicular to the selected object. PICK anywhere on a *Line* or straight *Pline* segment.

Figure 7-12

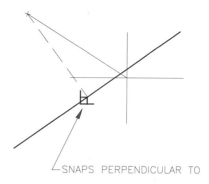

Quadrant

The *Quadrant* option snaps to the 0, 90, 180, or 270 degree quadrant of a *Circle*. PICK <u>nearest</u> to the desired *Quadrant*.

Figure 7-13

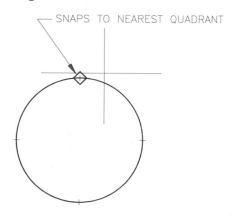

Tangent

This option calculates and snaps to a tangent point of an *Arc* or *Circle*. PICK the *Arc* or *Circle* as near as possible to the expected *Tangent* point.

Figure 7-14

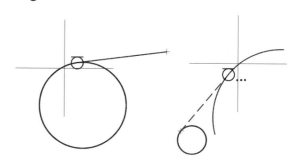

From

The *From* option is designed to let you snap to a point <u>relative</u> to another point using relative rectangular, relative polar, or direct distance entry coordinates. There are two steps: first select a "Basepoint:" (coordinates or another *OSNAP* may be used); then select an "Offset:" (enter relative rectangular, relative polar, or direct distance entry coordinates).

Figure 7-15

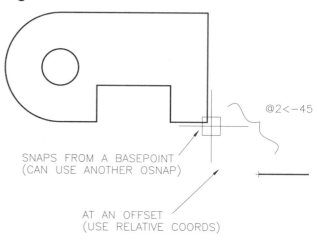

Apparent intersection

Use this option when you are working with a 3D drawing and want to snap to a point in space where two objects appear to intersect (from your viewpoint) but do not actually physically intersect.

Object Snaps must be activated in order for you to "snap" to the desired object features. Two methods for activating Object Snaps, Single Point Selection and Running Object Snaps, are discussed next. The *OSNAP* modes (*Endpoint, Midpoint, Center, Intersection,* etc.) operate identically for either method.

Figure 7-16

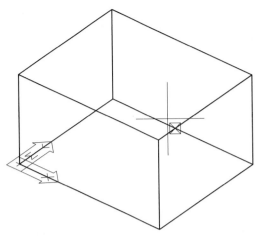

SNAPS TO APPARENT INTERSECTION IN 3D SPACE

OSNAP SINGLE POINT SELECTION

OBJECT SNAPS (SINGLE POINT)	Pull-down Menu	COMMAND (TYPE)	ALIAS (TYPE)	Short-cut	Screen (side) Menu	Tablet Menu	Cursor Menu (button 3 or shift+button 2)
	...	END, MID, etc. (first three letters)	...	...	****(asterisks)	T,16 - U,21	Endpoint Midpoint, etc.

There are many methods for invoking *OSNAP* modes for single point selection, as shown in the command table. If you prefer to type, enter only the <u>first three letters</u> of the *OSNAP* mode at the "From point: " prompt, "To point:" prompt, or <u>any time AutoCAD prompts for a point</u>.

Object Snaps are available from the Standard toolbar (Fig. 7-17). If desired, a separate *Object Snap* toolbar can be activated to float or dock on the screen by using the *View* pull-down menu and selecting *Toolbars...* , then *Object Snap* (see the docked toolbar on the right, Fig. 7-17). The advantage of invoking the *Object Snap* toolbar is that only one PICK is required for an *OSNAP* mode, whereas the Standard toolbar requires two PICKs because the *OSNAP* icons are on a "flyout."

Figure 7-17

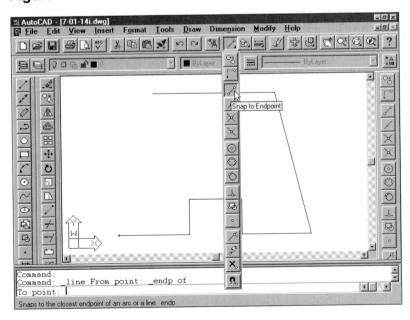

In addition to these options, a special menu called the <u>cursor menu</u> can be used. The cursor menu pops up at the <u>current location</u> of the cursor and replaces the cursor when invoked (Fig. 7-18). This menu is activated as follows:

2-button mouse   press **Shift + #2** (hold down the Shift key while clicking the right mouse button)
3-button mouse   press **#3**
digitizing puck   press **#3**

Figure 7-18 ─────────────

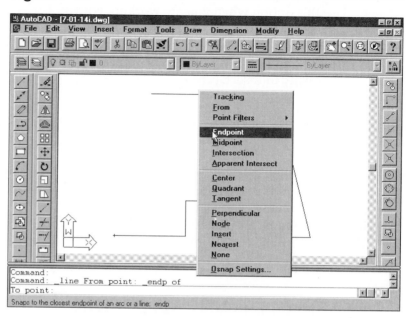

<u>With any of these methods, *OSNAP* modes are active only for selection of a single point.</u> If you want to *OSNAP* to another point, you must select an *OSNAP* mode again for the second point. With this method, the desired *OSNAP* mode is selected <u>transparently</u> (invoked during another command operation) immediately before selecting a point when prompted. In other words, whenever you are prompted for a point (for example, the "From point:" prompt of the *Line* command), select or type an *OSNAP* option. Then PICK near the desired object feature (endpoint, center, etc.) with the cursor. AutoCAD "snaps" to the feature of the object and uses it for the point specification. Using *OSNAP* in this way allows the *OSNAP* mode to operate only for that <u>single point selection</u>.

For example, when using *OSNAP* during the *Line* command, the command line reads as shown:

> Command: **Line**
> From Point: **Endpoint** of (**PICK**)
> To point: **Endpoint** of (**PICK**)
> To point: **Enter** (completes the command)

If you prefer using *OSNAP* for single point specification, it may be helpful to associate the tools (icons) with the Marker that appears on the drawing when you use the *OSNAP* mode. The table in Figure 7-19 displays the *OSNAP* icons and the respective Markers.

Figure 7-19 ─────────────

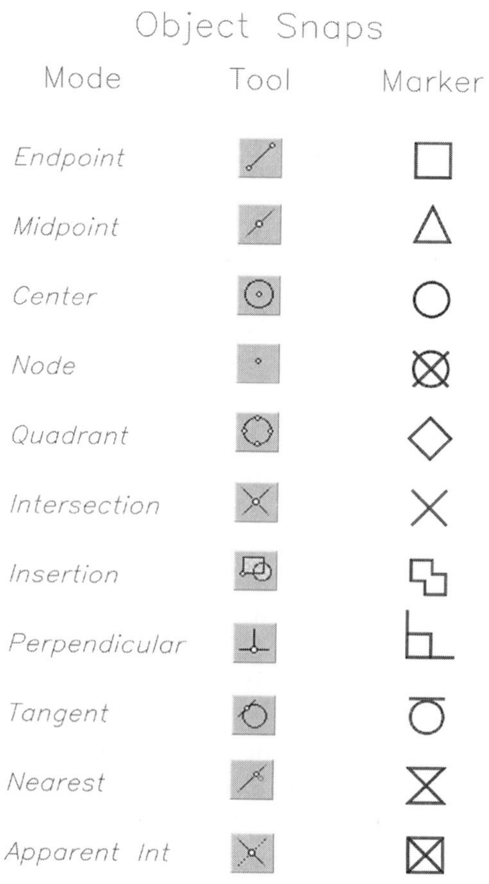

Object Snaps

Mode	Tool	Marker
Endpoint		□
Midpoint		△
Center		○
Node		⊗
Quadrant		◇
Intersection		✕
Insertion		⬓
Perpendicular		⊥
Tangent		◯
Nearest		⋉
Apparent Int		⊠

OSNAP RUNNING MODE

OBJECT SNAPS (RUNNING)	Pull-down Menu	COMMAND (TYPE)	ALIAS (TYPE)	Short-cut	Screen (side) Menu	Tablet Menu	Cursor Menu (button 3 or Shift+button 2)
	Tools Object Snap Settings...	*OSNAP or* -OSNAP	*OS or* -OS	Ctrl+F	*TOOLS 2 Ddosnap*	U,22	*Osnap Settings...*

Running Object Snap

Another method for using Object Snap is called "Running Object Snap" because one or more *OSNAP* modes (*Endpoint, Center, Midpoint,* etc.) can be turned on and kept running indefinitely. This method can obviously be more productive because you do not have to continually invoke an *OSNAP* mode each time you need to use one. For example, suppose you have several *Endpoint*s to connect. It would be most efficient to turn on the running *Endpoint OSNAP* mode and leave it running during the multiple selections. This is faster than continually selecting the *OSNAP* mode each time before you PICK.

You can even have several *OSNAP* modes running at the same time. A common practice is to turn on the *Endpoint, Center, Midpoint* modes simultaneously. In that way, if you move your cursor near a *Circle*, the *Center* Marker appears; if you move the cursor near the end or the middle of a *Line,* the *Endpoint* or the *Midpoint* mode Markers appear.

With versions of AutoCAD previous to Release 14, Running Object Snap is not as useful and is often dangerous because there were no Markers to indicate which *OSNAP* mode would be used for a particular PICK. Therefore, when multiple modes were running simultaneously, users did not know exactly if the selection would find an *Endpoint, Center,* or *Midpoint* until after the PICK. With Release 14, Running Object Snap is much more effective because the AutoSnap Markers make using multiple modes completely predictable.

Accessing Running Object Snap

All features of Running Object Snaps are controlled by the *Osnap Settings* dialog box (Fig. 7-20). This dialog box can be invoked by the following methods (see Command Table above):

1. type the *OSNAP* command;
2. type *OS,* the command alias;
3. select *Object Snap Settings...* from the *Tools* pull-down menu;
4. select *Osnap Settings...* from the bottom of the cursor menu (Shift + #2 button);
5. select the *Object Snap Settings* icon button from the *Object Snap* toolbar;
6. if Running Object Snaps have not yet been set, use an *OSNAP* toggle (F3, Ctrl+F, or double-clicking *OSNAP* at the Status line).

The *Osnap Settings* dialog box has two tabs: *Running Osnap* and *AutoSnap.*

Figure 7-20

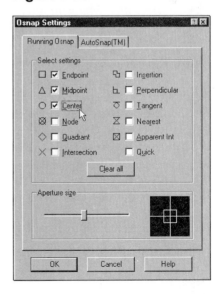

Running Osnap **Tab**

Use the *Running Osnap* tab to select the desired Object Snap settings. Try using three of four commonly used modes together, such as *Endpoint, Midpoint, Center,* and *Intersection.* The AutoSnap Markers indicate which one of the modes would be used as you move the cursor near different object features. Using similar modes simultaneously, such as *Center, Quadrant,* and *Tangent,* can sometimes lead to difficulties since it requires the cursor to be placed almost in an exact snap spot, or in some cases it may not find one of the modes (*Quadrant* overrides *Center*). In these cases, the tab key can be used to cycle through the options (see Object Snap Cycling).

Quick

This option only operates <u>in addition to</u> other selected *Running Osnap* modes. Using the *Quick* option causes AutoCAD to snap to the first snap point found. In other words, when on, *Quick* finds the quickest of the active options.

Aperture size

Notice that the *Aperture Size* can be adjusted through this dialog box (or through the use of the *Aperture* command). Whether the Aperture is visible or not (see Fig. 7-21, *Display aperture box*), its size determines how close the cursor must be to an object before the Marker appears and the displayed *OSNAP* mode takes effect. In other words, the smaller the Aperture, the closer the cursor must be to the object before a Marker appears and the *OSNAP* mode has an effect, and the larger the Aperture, the further the cursor can be from the object while still having an effect. The Aperture default size is 10 pixels square. The Aperture size influences *Single Point Osnap* selection as well as *Running Osnaps.*

AutoSnap **Tab**

The *AutoSnap* tab in the *Osnap Settings* dialog box (Fig. 7-21) gives you control over the visual aspects of the following object snap features.

Figure 7-21

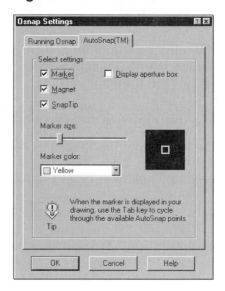

Marker

When this box is checked, an AutoSnap Marker appears when the cursor is moved near an object feature. Each *OSNAP* mode (*Endpoint, Center, Midpoint,* etc.) has a unique Marker (see Fig. 7-19). Normally this box should be checked, especially when using Running Object Snap, because the Markers help confirm when and which *OSNAP* mode is in effect.

Magnet

The Magnet feature causes the cursor to "lock" onto the object feature (*Endpoint, Center, Midpoint,* etc.) when the cursor is within the confines of the Marker. The Magnet helps confirm the exact location that will be snapped to for the subsequent PICK.

Snap Tip

The Snap Tip (similar to a Tool Tip) is helpful for beginners because it gives a verbal indication of the *OSNAP* mode in effect (*Endpoint, Center, Midpoint,* etc.) when the cursor is near the object feature. Experienced users may want to turn off the Snap Tips once the Marker symbols are learned.

Display aperture box

This checkbox controls the <u>visibility</u> of the Aperture. Whether the Aperture is visible or not, its size determines how close the cursor must be to an object before a Marker appears and the displayed *OSNAP* mode takes effect. (See *Aperture size* in the *Running Osnap* dialog box.)

Marker size

This control determines the size of the Markers. Remember that if the Magnet is on, the cursor "locks" to the object feature when the cursor is within the confines of the Magnet. Therefore, the larger the Marker, the greater the Magnet effective area, and the smaller the Marker, the smaller the Magnet effective area.

Marker color

Use this drop-down box to select a color for the Markers. You may want to change Marker colors if you change the color of the Drawing Editor background. For example, if the background is changed to white, you may want to change the Marker color to dark blue instead of yellow.

Running Object Snap Toggle

Another feature of Release 14 that makes Running Object Snap effective is Osnap Toggle. If you need to PICK a point without using *OSNAP*, use the Osnap Toggle to temporarily override (turn off) the modes. With Running Osnaps temporarily off, you can PICK any point without AutoCAD forcing your selection to an *Endpoint, Center,* or *Midpoint,* etc. When you toggle Running Object Snap on again, AutoCAD remembers which modes were previously set.

The following methods can be used to toggle Running Osnaps on and off:

1. double-click the word *OSNAP* that appears on the Status line (at the bottom of the screen)
2. press **F3**
3. press **Ctrl+F**

None

The *None OSNAP* option is a Running Osnap <u>override effective for only one PICK</u>. *None* is similar to the Running Object Snap Toggle, except it is effective for one PICK, then Running Osnaps automatically come back on without having to use a toggle. If you have *OSNAP* modes running but want to deactivate them for a <u>single</u> PICK, use *None* in response to "From point:" or other point selection prompt. In other words, using *None* <u>during a draw or edit command</u> overrides any Running Osnaps for that single point selection. *None* can be typed at the command prompt (when prompted for a point) or can be selected from the bottom of the Object Snap toolbar.

Object Snap Cycling

In cases when you have multiple Running Osnaps set, and it is difficult to get the desired AutoSnap Marker to appear, you can use the <u>Tab</u> key to cycle through the possible *OSNAP* modes for the high-lighted object. In other words, pressing the Tab key makes AutoCAD highlight the object nearest the cursor, then cycles through the possible snap Markers (for running modes that are set) that affect the object.

For example, when the *Center* and *Quadrant* modes are both set as Running Osnaps, moving the cursor near a *Circle* always makes the *Quadrant* Marker appear but never the *Center* Marker. In this case, pressing the Tab key highlights the *Circle,* then cycles through the four *Quadrant* and one *Center* snap candidates.

TRACKING

Tracking is a new feature in Release 14 that assists you to locate points that are orthogonally aligned with other object features. *Tracking* is used in conjunction with Object Snap. When you are prompted for a point, such as the "From point:" of a *Line* or the center of a *Circle*, you can turn on *Tracking* by one of several methods. With *Tracking* on, you can restrict the point selection to be horizontally or vertically aligned with an *Endpoint, Center, Midpoint*, etc. You can use *Tracking* again to further restrict the point selection to be vertically or horizontally aligned with a second *Endpoint, Center, Midpoint*, etc. In this way, it is easy to find the center of a rectangle by *Tracking* the *Midpoint* of two sides. *Tracking* is explained fully in Chapter 15, Draw Commands II.

OSNAP APPLICATIONS

OSNAP can be used any time AutoCAD prompts you for a point. This means that you can invoke an *OSNAP* mode during any draw or modify command as well as during many other commands. *OSNAP* provides you with the potential to create 100% accurate drawings with AutoCAD. Take advantage of this feature whenever it will improve your drawing precision. Remember, any time you are prompted for a point, use *OSNAP* if it can improve your accuracy.

CHAPTER EXERCISES

1. *OSNAP* **Single Point Selection**

    Open the **CH6EX1A** drawing and begin con-
    structing the sheet metal part. Each unit in
    the drawing represents one inch.

    A. Create four *Circles*. All *Circles* have a
        radius of **1.685**. The *Circles'* centers are
        located at **5,5**, **5,13**, **19,5**, and **19,13**.

    B. Draw four *Lines*. The *Lines* (highlighted
        in Figure 7-22) should be drawn on the
        outside of the *Circles* by using the
        Quadrant *OSNAP* mode as shown for
        each *Line* endpoint.

    C. Draw two *Lines* from the *Center* of the
        existing *Circles* to form two diagonals as
        shown in Figure 7-23.

    D. At the *Intersection* of the diagonals create
        a *Circle* with a **3** unit radius.

Figure 7-22 ─────────────

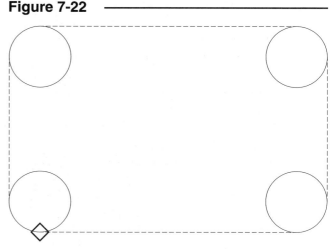

Figure 7-23 ─────────────

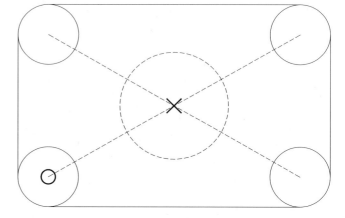

E. Draw two *Lines*, each from the *Intersection* of the diagonals to the *Midpoint* of the vertical *Lines* on each side. Finally, construct four new *Circles* with a radius of **.25,** each at the *Center* of the existing ones (Fig. 7-24).

F. *SaveAs* **CH7EX1**.

Figure 7-24 ─────────

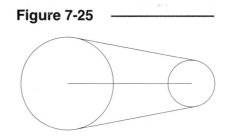

2. *OSNAP* **Single Point Selection**

A multiview drawing of a mechanical part is to be constructed using the A-METRIC drawing. All dimensions are in millimeters, so each unit equals one millimeter.

A. *Open* **A-METRIC** from Chapter 6 Exercises. Draw a *Line* from **60,140** to **140,140**. Create two *Circles* with the centers at the *Endpoints* of the *Line,* one *Circle* having a <u>diameter</u> of **60** and the second *Circle* having a diameter of **30**. Draw two *Lines Tangent* to the *Circles* as shown in Figure 7-25. *SaveAs* **PIVOTARM CH7**.

Figure 7-25 ─────────

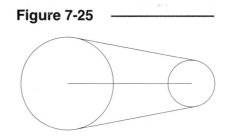

B. Draw a vertical *Line* down from the far left *Quadrant* of the *Circle* on the left. Specify relative polar (**@100<270**) or direct distance entry coordinates to make the *Line* 100 units. Draw a horizontal *Line* **125** units from the last *Endpoint* using relative polar or direct distance entry coordinates. Draw another *Line* between that *Endpoint* and the *Quadrant* of the *Circle* on the right. Finally, draw a horizontal *Line* from point **30,70** and *Perpendicular* to the vertical *Line* on the right.

Figure 7-26 ─────────

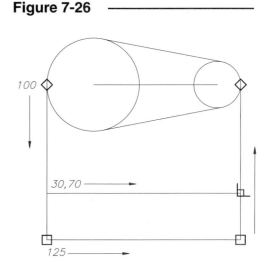

C. Draw two vertical *Lines* from the *Intersections* of the horizontal *Line* and *Circles* and *Perpendicular* to the *Line* at the bottom. Next, draw two *Circles* concentric to the previous two and with diameters of **20** and **10** as shown in Figure 7-27.

Figure 7-27 ————

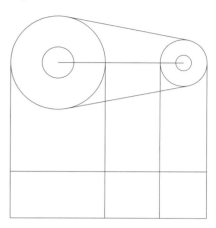

D. Draw four more vertical *Lines* as shown in Figure 7-28. Each *Line* is drawn from the new *Circles' Quadrant* and *Perpendicular* to the bottom line. Next, draw a miter *Line* from the *Intersection* of the corner shown to **@150<45**. *Save* the drawing for completion at a later time as another chapter exercise.

Figure 7-28 ————

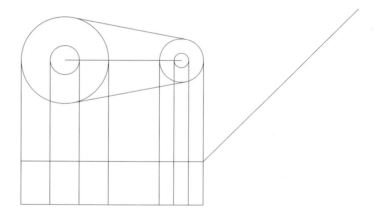

3. **Running Osnap**

   Create a cross-sectional view of a door header composed of two 2 x 6 wooden boards and a piece of 1/2″ plywood. (The dimensions of a 2 x 6 are actually 1-1/2″ x 5-3/8″.)

   A. Begin a *New* drawing and assign the name **HEADER**. Draw four vertical lines as shown in Figure 7-29.

   B. Use the *OSNAP* command or select *Object Snap Settings...* from the *Tools* pull-down menu and turn on the *Endpoint* and *Intersection* modes.

Figure 7-29 ————

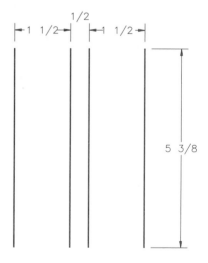

C. Draw the remaining lines as shown in Figure 7-30 to complete the header cross-section. Don't forget to turn off the running *OSNAP* modes when you are finished by selecting *Clear All* in the *Osnap Settings* dialog box. *Save* the drawing.

Figure 7-30 ——

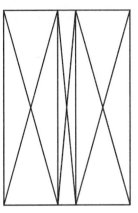

4. **Running Object Snap and** *OSNAP* **toggle**

Figure 7-31 ——————

Assume you are commissioned by the local parks and recreation department to provide a layout drawing for a major league sized baseball field. Lay out the location of bases, infield, and outfield as follows.

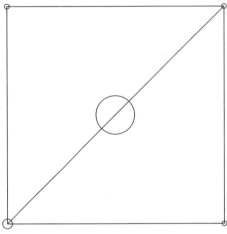

A. *Open* the **BALL FIELD CH6.DWG** that you set up in Chapter 6, Exercise 7. Make sure *Limits* are set to 512',384', *Snap* is set to 10', and *Grid* is set to 20'. Ensure *SNAP* and *GRID* are on. Use *SaveAs* to save and rename the drawing to **BALL FIELD CH7**.

B. Begin drawing the baseball diamond by using the *Line* command and using direct distance coordinate entry. **PICK** the "From point:" at **20',20'** (watch *Coords*). Draw the foul line to first base by turning on *ORTHO*, move the cursor to the right (along the X direction) and enter a value of **90'** (don't forget the apostrophe to indicate feet). At the "To point:" prompt, continue by drawing a vertical *Line* of **90'**. Continue drawing a square with **90'** between bases (see Fig. 7-31).

C. Invoke the *Osnap Settings* dialog box by any method. Turn on the *Endpoint*, *Midpoint*, and *Center* object snaps.

D. Use the *Circle* command to create a "base" at the lower-right corner of the square. At the "3P/2P/TTR/<Center Point>:" prompt, **PICK** the *Line Endpoint* at the lower-right corner of the square. Enter a *Radius* of **1'** (don't forget the apostrophe). Since Running Object Snaps are on, AutoCAD should display the Marker (square box) at each of the *Line Endpoints* as you move the cursor near; therefore, you can easily "snap" the center of the bases (*Circles*) to the corners of the square. Draw *Circles* of the same *Radius* at second base (upper-right corner), and third base (upper-left corner of the square). Create home plate with a *Circle* of a **2'** *Radius* by the same method (see Fig. 7-31).

E. Draw the pitcher's mound by first drawing a *Line* between home plate and second base. **PICK** the "From point:" at home plate (*Center* or *Endpoint*), then at the "To point:" prompt, PICK the *Center* or the *Endpoint* at second base. Construct a *Circle* of **8'** *Radius* at the *Midpoint* of the newly constructed diagonal line to represent the pitcher's mound (see Fig. 7-31).

F. *Erase* the diagonal *Line* between home plate and second base. Draw the foul lines from first and third base to the outfield. For the first base foul line, construct a **Line** with the "From point:" at the **Endpoint** of the existing first base line or **Center** of the base. Move the cursor (with *ORTHO* on) to the right (X positive) and enter a value of **240'** (don't forget the apostrophe). Press **Enter** to complete the *Line* command. Draw the third base foul line at the same length (in the positive Y direction) by the same method. (see Fig. 7-32).

Figure 7-32 ───────────────

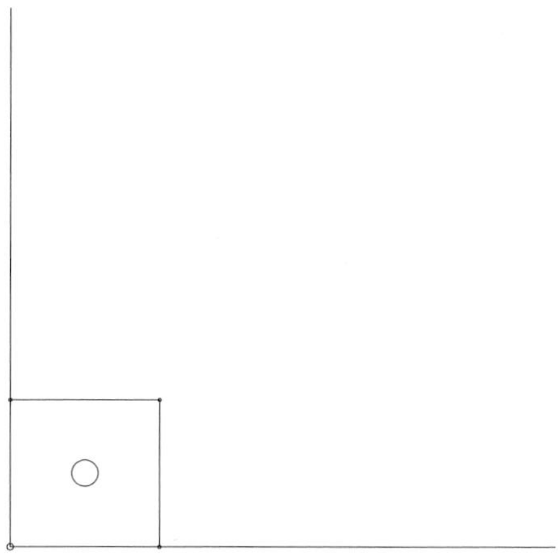

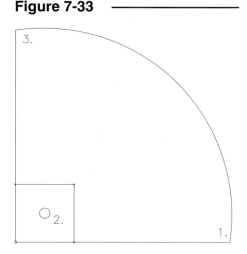

G. Draw the home run fence by using the *Arc* command from the *Draw* pull-down menu. Select the **Start, Center, End** method. **PICK** the end of the first base line (**Endpoint** object snap) for the **Start** of the *Arc* (Fig. 7-33, point 1.), **PICK** the pitcher's mound (**Center** object snap) for the **Center** of the *Arc* (point 2.), and the end of the third base line (**Endpoint** object snap) for the **End** of the *Arc* (point 3.). Don't worry about *ORTHO* in this case because *OSNAP* overrides *ORTHO*.

Figure 7-33 ───────────────

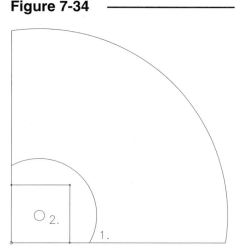

H. Now turn off **ORTHO**. Construct an *Arc* to represent the end of the infield. Select the **Arc Start, Center, End** method from the *Draw* pull-down menu. For the **Start** point of the *Arc,* toggle Running Osnaps <u>off</u> by pressing **F3, Ctrl+F,** or double-clicking the word *OSNAP* on the Status line and **PICK** location **140', 20', 0'** on the first base line (watch *Coords* and ensure *SNAP* and *GRID* are on). (See Figure 7-34, point 1.) Next, toggle Running Osnaps back on, and **PICK** the **Center** of the pitcher's mound as the **Center** of the *Arc* (point 2.). Third, toggle Running Osnaps off again and **PICK** the **End** point of the *Arc* on the third base line (point 3.).

Figure 7-34 ───────────────

I. Lastly, the pitcher's mound should be moved to the correct distance from home plate. Type **M** (the command alias for *Move*) or select **Move** from the *Modify* pull-down menu. When prompted to "Select objects:" **PICK** the *Circle* representing pitcher's mound. At the "Basepoint or displacement:" prompt, toggle Running Osnaps <u>on</u> and **PICK** the **Center** of the mound. At the "Second point of displacement:" prompt, enter **@3'2"<235**. This should reposition the pitcher's mound to the regulation distance from home plate (60'-6"). Compare your drawing to Figure 7-34. *Save* the drawing (as BALL FIELD CH7).

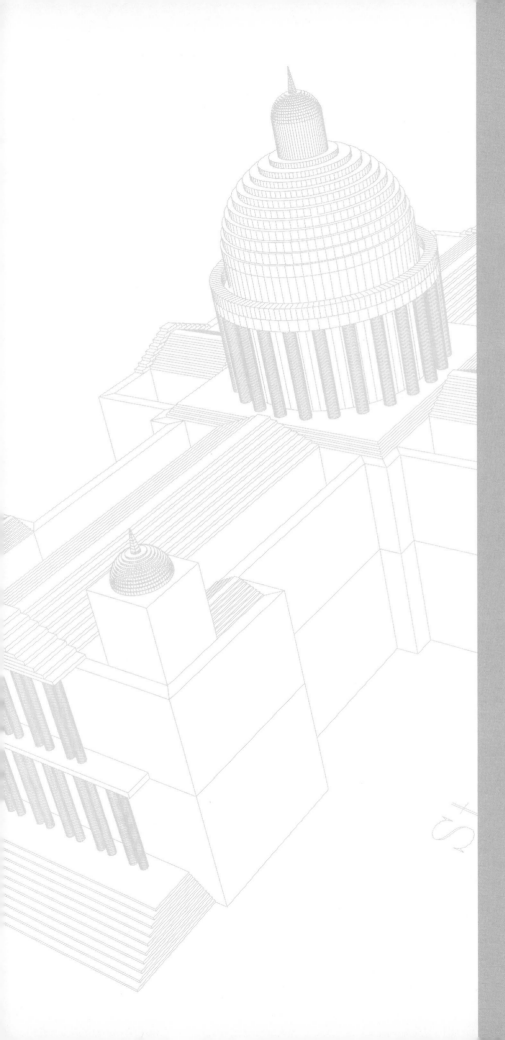

DRAW COMMANDS I

Chapter Objectives

After completing this chapter you should:

1. know where to locate and how to invoke the draw commands;

2. be able to draw *Lines;*

3. be able to draw *Circles* by each of the five options;

4. be able to draw *Arcs* by each of the eleven options;

5. be able to create *Point* objects and specify the *Point Style;*

6. be able to create *Plines* with width and combined of line and arc segments.

CONCEPTS

Draw Commands—Simple and Complex

Draw commands create objects. An object is the smallest component of a drawing. The draw commands listed immediately below create simple objects and are discussed in this chapter. Simple objects <u>appear</u> as one entity.

   Line, Circle, Arc, Point

Other draw commands create more complex shapes. Complex shapes appear to be composed of several components, but each shape is usually <u>one object</u>. An example of an object that is one entity but usually appears as several segments is listed below and is also covered in this chapter:

   Pline

Other draw commands discussed in Chapter 15 (listed below) are a combination of simple and complex shapes:

   Xline, Ray, Polygon, Rectangle, Donut, Spline, Ellipse, Divide, and *Measure*

Figure 8-1 —

Draw Command Access

As a review from Chapter 3, Draw Command Basics, remember that any of the five methods can be used to access the draw commands: *Draw* toolbar (Fig. 8-1), *Draw* pull-down menu (Fig. 8-2), *DRAW1* and *DRAW2* screen menus, keyboard entry of the command or alias, and digitizing tablet icons.

Figure 8-2 ⎯⎯⎯⎯⎯⎯⎯⎯⎯⎯⎯⎯

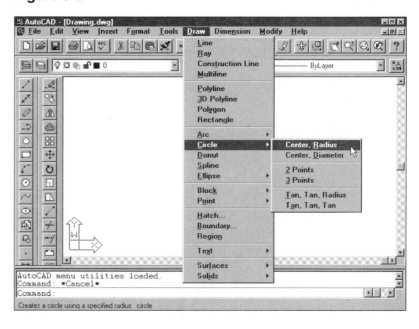

Coordinate Entry

When creating objects with draw commands, AutoCAD always prompts you to indicate points (such as endpoints, centers, radii) to describe the size and location of the objects to be drawn. An example you are familiar with is the *Line* command, where AutoCAD prompts for the "From point:." Indication of these points, called <u>coordinate entry</u> can be accomplished by five formats (for 2D drawings):

1. **Interactive**   **PICK** points on screen with input device
2. **Absolute coordinates**   **X,Y**
3. **Relative rectangular coordinates**   **@X,Y**
4. **Relative polar coordinates**   **@distance<angle**
5. **Direct distance entry**   **dist,direction** (Type a distance value relative to the last point, indicate direction with the cursor, then press Enter.)

Any of these methods can be used <u>whenever</u> AutoCAD prompts you to specify points. (For practice with these methods, see Chapter 3, Draw Command Basics.)

Also keep in mind that you can specify points interactively using *OSNAP* modes as discussed in Chapter 7. *OSNAP* modes can be used <u>whenever</u> AutoCAD prompts you to select points.

COMMANDS

LINE

Pull-down Menu	COMMAND (TYPE)	ALIAS (TYPE)	Short-cut	Screen (side) Menu	Tablet Menu
Draw *Line*	LINE	L	...	DRAW 1 Line	J,10

This is the fundamental drawing command. The *Line* command creates straight line segments; each segment is an object. One or several line segments can be drawn with the *Line* command.

Command: **line**
From point: **PICK** or (**coordinates**) (A point can be designated by interactively selecting with the input device or by entering coordinates. If using the input device, the *Coords* display can be viewed to locate the current cursor position. If entering coordinates, any format is valid.)
To point: **PICK** or (**coordinates**) (Again, device input or keyboard input can be used. If using the input device to select, *ORTHO* (**F8**) can be toggled *ON* to force vertical or horizontal lines.)
To point: **PICK** or (**coordinates**) or **Enter** (Line segments can continually be drawn. Press Enter to complete the command.)
Command:

Figure 8-3

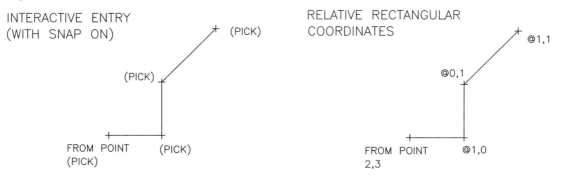

INTERACTIVE ENTRY
(WITH SNAP ON)

(PICK)

(PICK)

FROM POINT   (PICK)
(PICK)

RELATIVE RECTANGULAR
COORDINATES

@1,1

@0,1

FROM POINT   @1,0
2,3

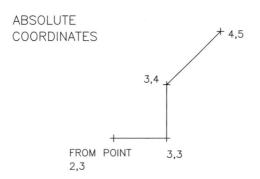

ABSOLUTE
COORDINATES

4,5

3,4

FROM POINT   3,3
2,3

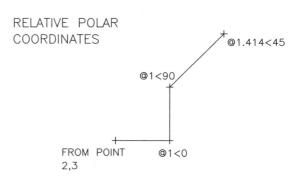

RELATIVE POLAR
COORDINATES

@1.414<45

@1<90

FROM POINT   @1<0
2,3

Figure 8-3 shows four examples of creating the same *Line* segments using different methods of coordinate entry.

Refer to Chapter 3, Draw Command Basics, for examples of drawing vertical, horizontal, and inclined lines using the four formats for coordinate entry.

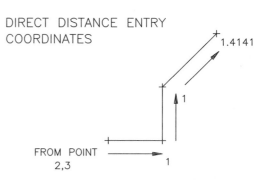

DIRECT DISTANCE ENTRY
COORDINATES

1.4141

1

FROM POINT
2,3   1

CIRCLE

Pull-down Menu	COMMAND (TYPE)	ALIAS (TYPE)	Short-cut	Screen (side) Menu	Tablet Menu
Draw *Circle >*	CIRCLE	C	...	*DRAW 1* *Circle*	J,9

The *Circle* command creates one object. Depending on the option selected, you can provide two or three points to define a *Circle*. As with all commands, the command line prompt displays the possible options:

Command: *circle*
3P/2P/TTR/<Center point>: **PICK** or (**coordinates**), (**option**). (PICKing or entering coordinates designates the center point for the circle. You can enter *"3P," "2P"* or *"TTR"* for another option.)

As with many other commands, when typing the command, the default and other options are displayed on the command line. The default option always appears in brackets "<option>:." The other options can be invoked by typing the indicated uppercase letter(s).

All of the options for creating *Circles* are available from the *Draw* pull-down menu and from the *DRAW 1* screen (side) menu. The tool (icon button) for only the *Center, Radius* method is included in the *Draw* toolbar. Tools for the other explicit *Circle* and *Arc* options are available, but only if you customize your own toolbar.

The options, or methods, for drawing *Circles* are listed below. Each figure gives several possibilities for each option, with and without *OSNAPs*.

Center, Radius
Specify a center point, then a radius. (Fig. 8-4).

The *Radius* (or *Diameter*) can be specified by entering values or by indicating a length interactively (PICK two points to specify a length when prompted). As always, points can be specified by PICKing or entering coordinates. Watch *Coords* for coordinate or distance (polar format) display. *OSNAPs* can be used for interactive point specification.

Figure 8-4

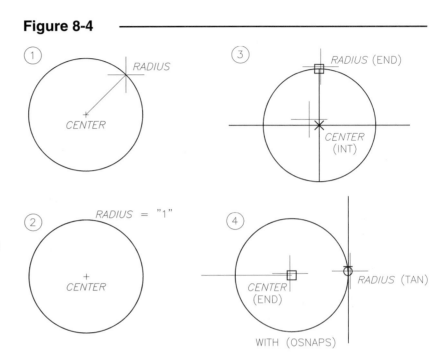

Center, Diameter
Specify the center point, then the diameter (Fig. 8-5).

Figure 8-5

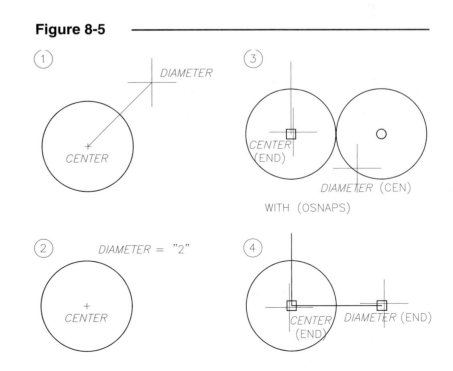

2 Points

The two points specify the location and diameter.

The *Tangent OSNAPs* can be used when selecting points with the *2 Point* and *3 Point* options as shown in Figures 8-6 and 8-7.

Figure 8-6

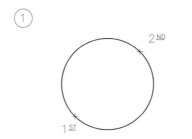

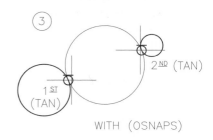

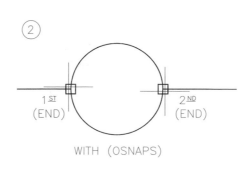

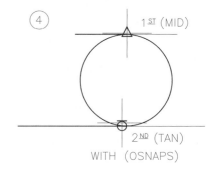

3 Points

The *Circle* passes through all three points specified.

Figure 8-7

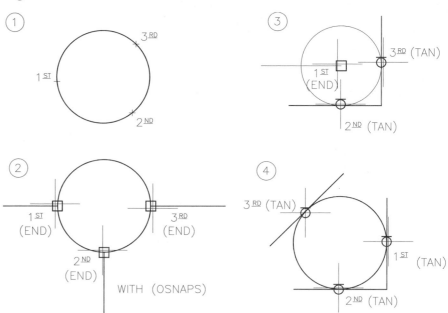

Tangent, Tangent, Radius
Specify two objects for the
Circle to be tangent to; then
specify the radius.

The *TTR* (Tangent, Tangent,
Radius) method is extremely
efficient and productive. The
OSNAP Tangent modes are
automatically invoked (the
aperture is displayed on the
cursor). This is the only draw
command option that auto-
matically calls *OSNAPs*.

Figure 8-8

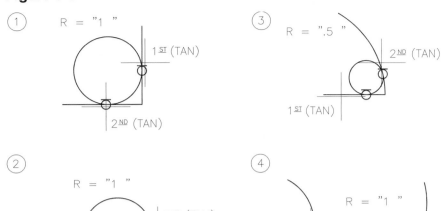
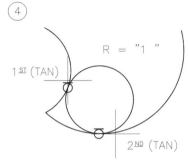
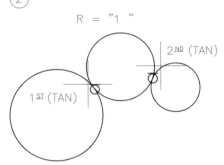

ARC

Pull-down Menu	COMMAND (TYPE)	ALIAS (TYPE)	Short-cut	Screen (side) Menu	Tablet Menu
Draw Arc >	ARC	A	...	DRAW 1 Arc	R,10

An arc is part of a circle; it is a regular curve of less than 360 degrees. The *Arc* command in AutoCAD
provides eleven options for creating arcs. An *Arc* is one object. *Arcs* are always drawn by default in a
counter-clockwise direction. This occurrence forces you to decide in advance which points should be
designated as *Start* and *End* points (for options requesting those points). For this reason, it is often easier
to create arcs by another method, such as drawing a *Circle* and then using *Trim* or using the *Fillet*
command. (See Use *Arcs* or *Circles*? at the end of this section on *Arcs*.) The *Arc* command prompt is:

    Command: **arc**
    Center/<Start point>: **PICK** or (**coordinates**), or **C** (Interactively select or enter coordinates in any
    format for the start point. Type "C" to use the *Center* option instead.)

The prompts displayed by AutoCAD are different depending on which option is selected. At any time
while using the command, you can select from the options listed on the command line by typing in the
capitalized letter(s) for the desired option.

Alternately, to use a particular option of the *Arc* command, you can select from the *Draw* pull-down
menu or from the *DRAW 1* screen (side) menu. The tool (icon button) for only the *3 Points* method is
included in the *Draw* toolbar. The other tools are available, but only if you customize your own toolbar.
These options require coordinate entry of points in a specific order.

3Points
Specify three points through which the *Arc* passes (Fig. 8-9).

Figure 8-9

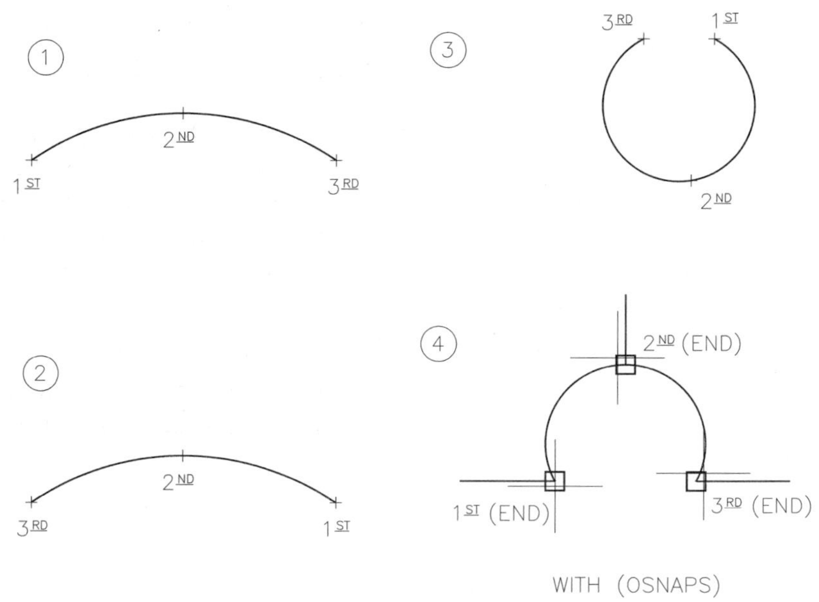

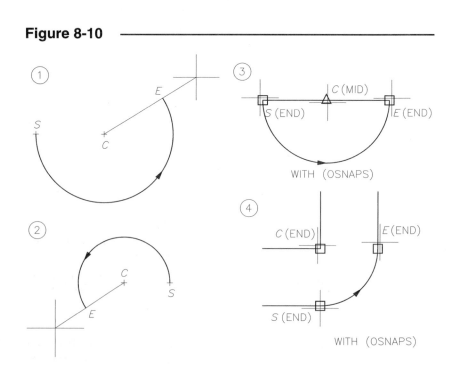

Start, Center, End
The radius is defined by the first two points that you specify (Fig. 8-10).

Figure 8-10

Start, Center, Angle

The angle is the <u>included</u> angle between the sides from the center to the endpoints. A <u>negative</u> angle can be entered to generate an *Arc* in a <u>clockwise</u> direction.

Figure 8-11

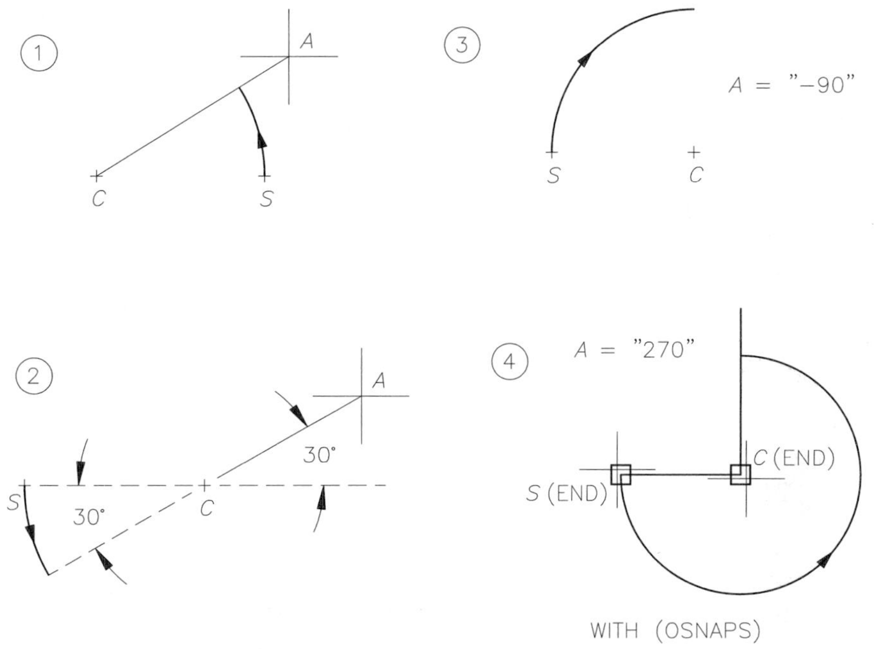

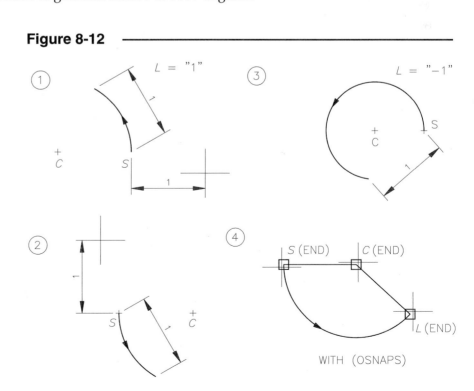

Start, Center, Length

Length means length of chord. The length of chord is between the start and the other point specified. A negative chord length can be entered to generate an *Arc* of 180+ degrees.

Figure 8-12

Start, End, Angle

The included angle is between the sides from the center to the endpoints. Negative angles generate clockwise *Arcs*.

Figure 8-13

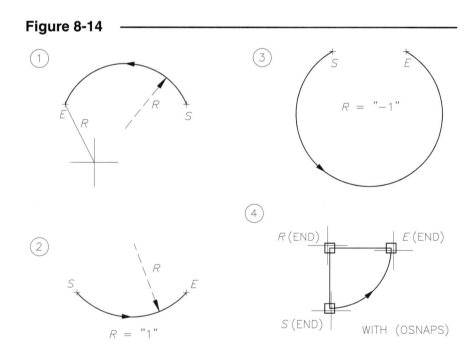

Start, End, Radius

The radius can be PICKed or entered as a value. A negative radius value generates an *Arc* of 180+ degrees.

Figure 8-14

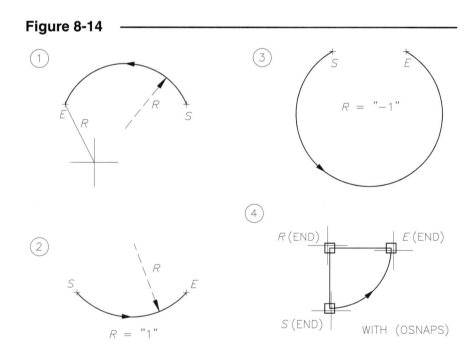

Start, End, Direction
The direction is tangent to the start point.

Figure 8-15

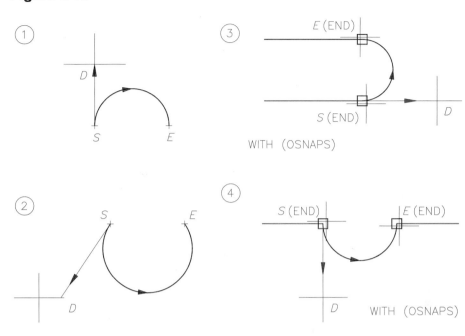

Center, Start, End
This option is like *Start, Center, End* but in a different order.

Figure 8-16

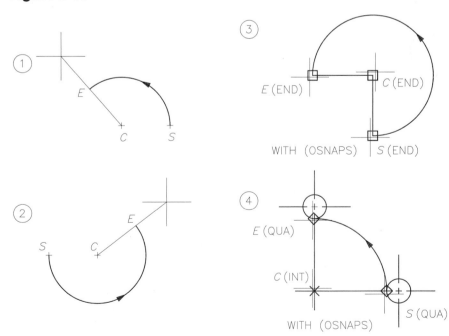

Center, Start, Angle
This option is like *Start, Center, Angle* but in a different order.

Figure 8-17 ————————————————————————

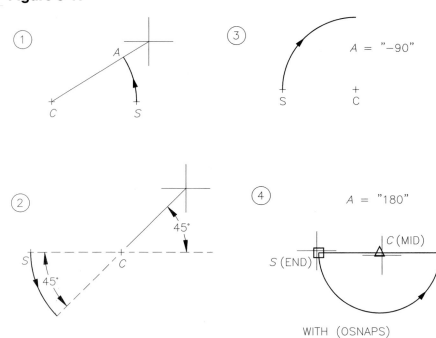

Center, Start, Length
This is similar to the *Start, Center, Length* option but in a different order. *Length* means length of chord.

Figure 8-18 ————————————————————————

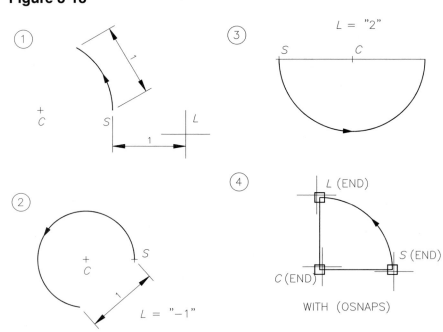

Continue

The new *Arc* continues from and is tangent to the last point. The only other point required is the endpoint of the *Arc*. This method allows drawing *Arcs* tangent to the preceding *Line* or *Arc*.

Figure 8-19

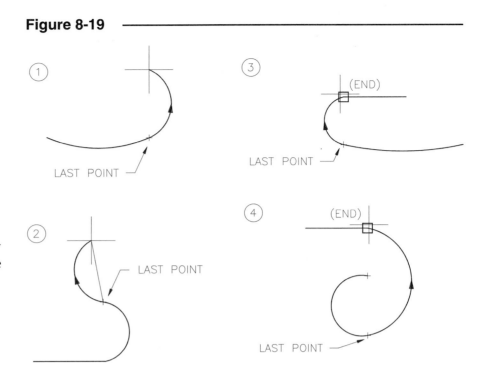

Arcs are always created in a counter-clockwise direction. This fact must be taken into consideration when using any method except the *3-Point*, the *Start, End, Direction,* and the *Continue* options. The direction is explicitly specified with *Start, End, Direction* and *Continue* methods, and direction is irrelevant for *3-Point* method.

As usual, points can be specified by PICKing or entering coordinates. Watch *Coords* to display coordinate values or distances. *OSNAPs* can be used when PICKing. The *Endpoint, Intersection, Center, Midpoint,* and *Quadrant OSNAP* options can be used with great effectiveness. The *Tangent OSNAP* option cannot be used effectively with most of the *Arc* options. The *Radius, Direction, Length,* and *Angle* specifications can be given by entering values or by PICKing with or without *OSNAPs*.

Use *Arcs* or *Circles*?

Although there are sufficient options for drawing *Arcs*, usually it is easier to use the *Circle* command followed by *Trim* to achieve the desired arc. Creating a *Circle* is generally an easier operation than using *Arc* because the counter-clockwise direction does not have to be considered. The unwanted portion of the circle can be *Trimmed* at the *Intersection* of or *Tangent* to the connecting objects using *OSNAP*. The *Fillet* command can also be used instead of the *Arc* command to add a fillet (arc) between two existing objects (see Chapter 10, Modify Commands II).

POINT

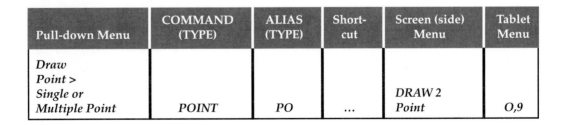

	Pull-down Menu	COMMAND (TYPE)	ALIAS (TYPE)	Short-cut	Screen (side) Menu	Tablet Menu
	Draw Point > Single or Multiple Point	*POINT*	*PO*	...	DRAW 2 Point	O,9

A *Point* is an object that has no dimension; it only has location. A *Point* is specified by giving only one coordinate value or by PICKing a location on the screen.

Figure 8-20 compares *Points* to *Line* and *Circle* objects.

Figure 8-20

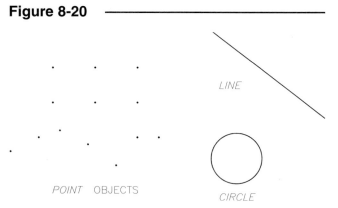

> Command: **point**
> Point: **PICK** or (**coordinates**) (Select a location for the *Point* object.)
> Command:

Points are useful in construction of drawings to locate points of reference for subsequent construction or locational verification. The *Node OSNAP* option is used to snap to *Point* objects.

POINT OBJECTS

LINE

CIRCLE

Points are drawing objects and therefore appear in prints and plots. The default "style" for points is a tiny "dot." The *Point Style* dialog box can be used to define the format you choose for *Point* objects (see *Ddptype* next).

The *Draw* pull-down menu offers the *Single Point* and the *Multiple Point* options. The *Single Point* option creates one *Point*, then returns to the command prompt. This option is the same as using the *Point* command by any other method. Selecting *Multiple Point* continues the *Point* command until you press the Escape key.

If you draw with *BLIPMODE* on, small cross marks appear when you create *Point* objects. Use *Redraw* to make the "blips" disappear to view the *Points* more clearly. This problem may occur if you are editing pre-Release 14 drawings because *BLIPMODE* is on by default (saved in the individual drawing files) and the default *Point* style is a "dot." Therefore, the "blips" conceal the *Points*.

DDPTYPE

Pull-down Menu	COMMAND (TYPE)	ALIAS (TYPE)	Short-cut	Screen (side) Menu	Tablet Menu
Format *Point Style...*	DDPTYPE	...	...	DRAW 2 Point Ddptype:	U,1

The *Point Style* dialog box (Fig. 8-21) is available only through the methods listed in the command table above. This dialog box allows you to define the format for the display of *Point* objects. The selected style is applied immediately to all newly created *Point* objects. The *Point Style* controls the format of *Points* for printing and plotting as well as for the computer display.

Figure 8-21

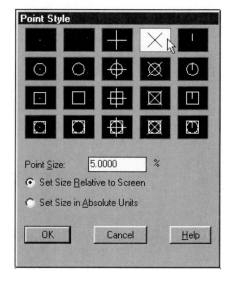

Changing the format of *Point* objects <u>does not automatically update existing *Point* objects</u> in the drawing. A regeneration must occur before existing *Points* are displayed in the new style. Use the *Regen* command immediately after changing the *Point Style* to cause a regeneration of the drawing, which updates the display of *Points* to the new style.

You can set the *Point Size* in *Absolute* units or *Relative to Screen* (default option). The *Relative to Screen* option keeps the *Points* the same size on the display when you *Zoom* in and out, whereas setting *Point Size* in *Absolute Units* gives you control over the size of *Points* for prints and plots. The selected *Point Style* is stored in the *PDMODE* variable.

PLINE

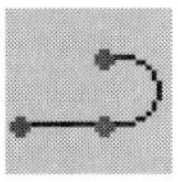

Pull-down Menu	COMMAND (TYPE)	ALIAS (TYPE)	Short-cut	Screen (side) Menu	Tablet Menu
Draw Polyline	*PLINE*	*PL*	...	DRAW 1 *Pline*	N,10

A *Pline* (or *Polyline*) has special features that make this object more versatile than a *Line*. Three features are most noticeable when first using *Plines*:

1. A *Pline* can have a specified *width*, whereas a *Line* has no width.
2. Several *Pline* segments created with one *Pline* command are treated by AutoCAD as <u>one</u> object, whereas individual line segments created with one use of the *Line* command are individual objects.
3. A *Pline* can contain arc segments.

Figure 8-22 illustrates *Pline* versus *Line* and *Arc* comparisons.

The *Pline* command begins with the same prompt as *Line*; however, <u>after</u> the "From point:" is established, the *Pline* options are accessible.

Command: *pline*
From point: **PICK** or (**coordinates**)
Arc/Close/Halfwidth/Length/Undo/Width/ <Endpoint of Line>: **PICK** or (**coordinates**) or (**letter**)

The options and descriptions follow.

Figure 8-22

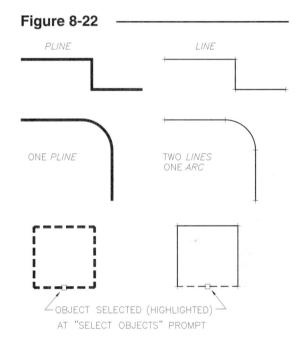

Width
You can use this option to specify starting and ending widths. Width is measured perpendicular to the centerline of the *Pline* segment (Fig. 8-23). *Plines* can be tapered by specifying different starting and ending widths. See NOTE at the end of this section.

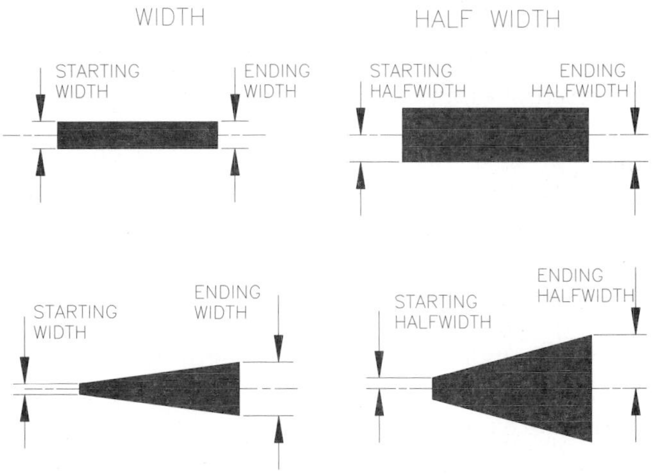

Halfwidth
This option allows specifying half of the *Pline* width. *Plines* can be tapered by specifying different starting and ending widths (Fig. 8-23).

Figure 8-23

Arc

This option (by default) creates an arc segment in a manner similar to the *Arc Continue* method (Fig. 8-24). Any of several other methods are possible (see *Pline Arc* Segments).

Figure 8-24

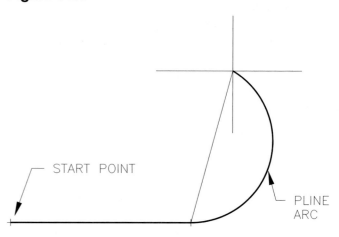

START POINT

PLINE ARC

Close

The *Close* option creates the closing segment connecting the first and last points specified with the current *Pline* command as shown in Figure 8-25.

This option can also be used to close a group of connected *Pline* segments into one continuous *Pline*. (A *Pline* closed by PICKing points has a specific start and endpoint.) A *Pline Closed* by this method has special properties if you use *Pedit* for *Pline* editing or if you use the *Fillet* command with the *Pline* option (see *Fillet* in Chapter 10 and *Pedit* in Chapter 16).

Figure 8-25

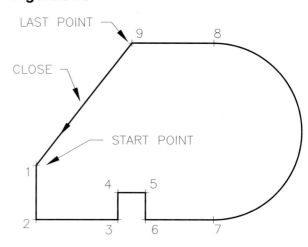

LAST POINT

CLOSE

START POINT

Length

Length draws a *Pline* segment at the same angle as and connected to the previous segment and uses a length that you specify. If the previous segment was an arc, *Length* makes the current segment tangent to the ending direction (Fig. 8-26).

Undo

Use this option to *Undo* the last *Pline* segment. It can be used repeatedly to undo multiple segments.

NOTE: If you change the *Width* of a *Pline,* be sure to respond to <u>both prompts</u> for width ("Starting width:" and "Ending width:") before you draw the first *Pline* segment. It is easy to hastily PICK the endpoint of the *Pline* segment after specifying the "Starting width:" instead of responding with a value (or Enter) for the "Ending width:". In this case (if you PICK at the "Ending Width:" prompt), AutoCAD understands the line length that you interactively specified to be the ending width you want for the next line segment (you can PICK two points in response to the "Starting width:" or "Ending width:" prompts).

Figure 8-26

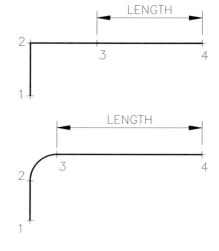

LENGTH

LENGTH

```
Command: PLINE
From point:
Current line-width is 0.0000
Arc/Close/Halfwidth/Length/Undo/Width/<Endpoint of line>: w
Starting width <0.0000>: .2
Ending width <0.2000>:  Enter a value or press the Enter key; do not PICK the endpoint of the next
```
Pline segment.

Pline Arc Segments

When the *Arc* option of *Pline* is selected, the prompt changes to provide the various methods for construction of arcs:

Angle/CEnter/CLose/Direction/Halfwidth/Line/Radius/Second pt/Undo/Width/<Endpoint of Arc>:

Angle
You can draw an arc segment by specifying the included angle (a negative value indicates a clockwise direction for arc generation).

CEnter
This option allows you to specify a specific center point for the arc segment.

CLose
This option closes the *Pline* group with an arc segment.

Direction
Direction allows you to specify an explicit starting direction rather than using the ending direction of the previous segment as a default.

Line
This switches back to the line options of the *Pline* command.

Radius
You can specify an arc radius using this option.

Second pt
Using this option allows specification of a 3-point arc.

Because a shape created with one *Pline* command is <u>one object</u>, manipulation of the shape is generally easier than with several objects. For some applications, one *Pline* shape can have advantages over shapes composed of several objects (see *Offset*, Chapter 10). Editing *Plines* is accomplished by using the *Pedit* command. As an alternative, *Plines* can be "broken" back down into individual objects with *Explode*.

Drawing and editing *Plines* can be somewhat involved. As an alternative, you can draw a shape as you would normally with *Line, Circle, Arc, Trim*, etc., and then <u>convert</u> the shape to one *Pline* object using *Pedit*. (See Chapter 16 for details on converting *Lines* and *Arcs* to *Plines*.)

CHAPTER EXERCISES

Create a *New* drawing. When the *Create New Drawing* dialog box appears, select **Start from Scratch** and select *English* settings. Set *SNAP* to **.25** and set *GRID* to **1**. *Save* the drawing as **CH8EX**. For each of the following problems, *Open* **CH8EX**, complete one problem, then use *SaveAs* to give the drawing a new name.

Figure 8-27

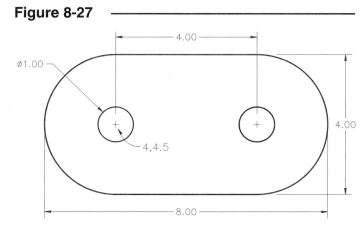

1. *Open* **CH8EX**. Create the geometry shown in Figure 8-27. Start the first *Circle* center at point **4,4.5** as shown. Do not copy the dimensions. *SaveAs* **LINK**. (HINT: Locate and draw the two small *Circles* first. Use *Arc, Start, Center, End* or *Center, Start, End* for the rounded ends.)

2. *Open* **CH8EX**. Create the geometry as shown in Figure 8-28. Do not copy the dimensions. Assume symmetry about the vertical axis. *SaveAs* **SLOTPLATE CH8**.

Figure 8-28

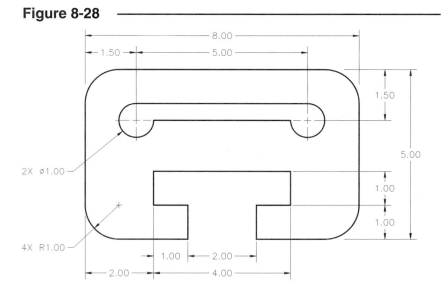

3. *Open* **CH8EX**. Create the shapes shown in Figure 8-29. Do not copy the dimensions. *SaveAs* **CH8EX3**.

   Draw the *Lines* at the bottom first, starting at coordinate **1,3**. Then create *Point* objects at **5,7**, **5.4,7**, **5.8,7**, etc. Change the *Point Style* to an X and *Regen*. Use the *NODe OSNAP* mode to draw the inclined *Lines*. Create the *Arc* on top with the *Start, End, Direction* option.

Figure 8-29

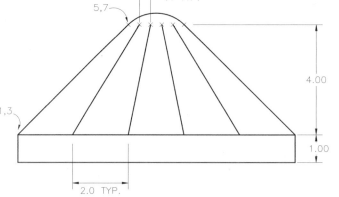

4. *Open* **CH8EX**. Create the shape shown in Figure 8-30. Draw the two horizontal *Lines* and the vertical *Line* first by specifying the endpoints as given. Then create the *Circle* and *Arcs*. *SaveAs* **CH8EX4**.

   HINT: Use the *Circle 2P* method with *Endpoint* OSNAPs. The two upper *Arcs* can be drawn by the *Start, End, Radius* method.

Figure 8-30

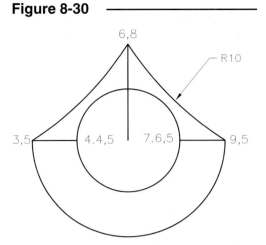

5. *Open* **CH8EX**. Draw the shape shown in Figure 8-31. Assume symmetry along a vertical axis. Start by drawing the two horizontal *Lines* at the base.

   Next, construct the side *Arcs* by the *Start, Center, Angle* method (you can specify a negative angle). The small *Arc* can be drawn by the *3P* method. Use OSNAPs when needed (especially for the horizontal *Line* on top and the *Line* along the vertical axis). *SaveAs* **CH8EX5**.

Figure 8-31

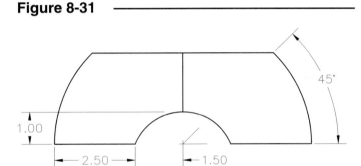

6. *Open* **CH8EX**. Complete the geometry in Figure 8-32. Use the coordinates to establish the *Lines*. Draw the *Circles* using the *Tangent, Tangent, Radius* method. *SaveAs* **CH8EX6**.

Figure 8-32

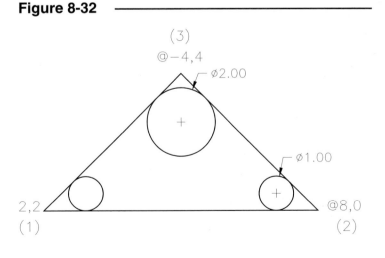

7. *Pline*

Create the shape shown in Figure 8-33. Draw the outside shape with <u>one continuous</u> *Pline* (with 0.00 width). When finished, *SaveAs* **PLINE1**.

Figure 8-33

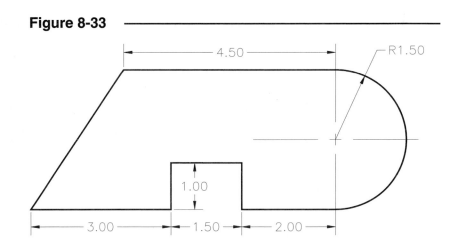

9

MODIFY
COMMANDS I

Chapter Objectives

After completing this chapter you should:

1. know where to locate and how to invoke the *Modify* commands;

2. be able to *Erase* objects from the drawing;

3. be able to *Move* objects from a base-point to a second point of displace-ment;

4. know how to *Rotate* objects about a basepoint;

5. be able to enlarge or reduce objects with *Scale*;

6. be able to *Stretch* selected objects;

7. be able to change the length of *Lines* and *Arcs* with *Lengthen*;

8. be able to *Trim* away parts of objects at cutting edges;

9. be able to *Extend* objects to selected boundary edges;

10. know how to use the four *Break* options.

CONCEPTS

Draw commands are used to create new objects. *Modify* commands are used to change existing objects or to use existing objects to create new and similar objects. The *Modify* commands covered in this chapter only <u>change existing objects</u>. The commands listed below are covered in this chapter, while other *Modify* commands are covered in Chapter 10 and Chapter 16:

    Erase, Move, Rotate, Scale, Stretch, Lengthen, Trim, Extend, and *Break*

Modify commands can be invoked by any of the five command entry methods: toolbar icons, pull-down menus, screen menus, keyboard entry, and digitizing tablet menu. The *Modify* toolbar is visible and docked to the right side of the screen by default in Release 14 (Fig. 9-1).

The digitizing tablet uses the same icons that are displayed for the toolbars.

Figure 9-1

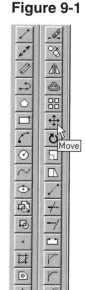

The *Modify* pull-down menu (Fig. 9-2) contains commands that <u>change</u> existing geometry and that use existing geometry to create new but similar objects. All of the *Modify* commands are contained in this single pull-down menu.

Figure 9-2

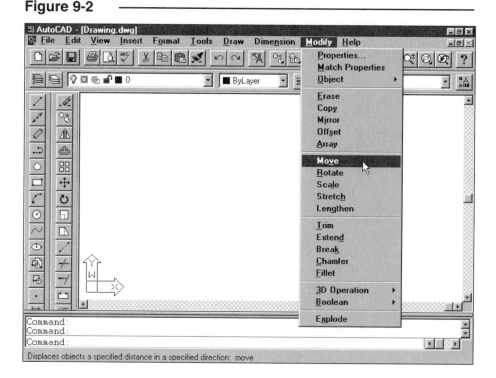

The *MODIFY1* and *MODIFY2* screen menus branch to the individual commands and options (Fig. 9-3).

Since all *Modify* commands affect or use existing geometry, the first step in using any modify command is to construct a selection set (see Chapter 4). This can be done by one of two methods:

1. Invoking the desired command and then creating the selection set in response to the "Select Objects:" prompt (Verb/Noun syntax order) using any of the select object options;

2. Selecting the desired set of objects with the pickbox or *Auto* Window or Crossing Window <u>before</u> invoking the edit command (Noun/Verb syntax order).

Figure 9-3

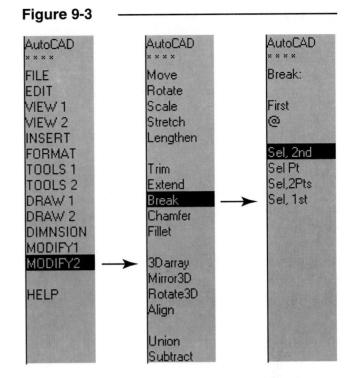

The first method allows use of any of the selection options (*Last, All, WPolygon, Fence,* etc.), while the latter method allows <u>only</u> the use of the pickbox and *Auto* Window and Crossing Window.

COMMANDS

ERASE

Pull-down Menu	COMMAND (TYPE)	ALIAS (TYPE)	Short-cut	Screen (side) Menu	Tablet Menu
Modify *Erase*	*ERASE*	*E*	...	*MODIFY1* *Erase*	*V,14*

The *Erase* command deletes the objects you select from the drawing. Any of the object selection methods can be used to highlight the objects to *Erase*. The only other required action is for you to press *Enter* to cause the erase to take effect.

    Command: **Erase**
    Select Objects: **PICK** (Use any object selection method.)
    Select Objects: **PICK** (Continue to select desired objects.)
    Select Objects: **Enter** (Confirms the object selection process and causes *Erase* to take effect.)
    Command:

If objects are erased accidentally, *U* can be used immediately following the mistake to undo one step, or *Oops* can be used to bring back into the drawing whatever was *Erased* the last time *Erase* was used (see Chapter 5). If only part of an object should be erased, use *Trim* or *Break*.

MOVE

Pull-down Menu	COMMAND (TYPE)	ALIAS (TYPE)	Short-cut	Screen (side) Menu	Tablet Menu
Modify *Move*	*MOVE*	*M*	...	*MODIFY2* *Move*	*V,19*

Move allows you to relocate one or more objects from the existing position in the drawing to any other position you specify. After selecting the objects to *Move*, you must specify the "base-point" and "second point of displacement." You can use any of the five coordinate entry methods to specify these points. Examples are shown in Figure 9-4.

Figure 9-4

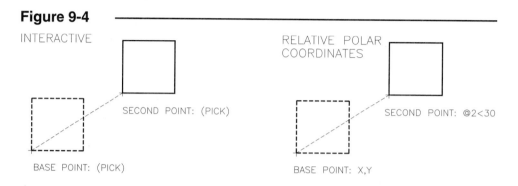

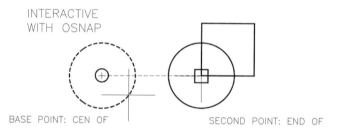

Command: **move**
Select Objects: **PICK**   (Use any of the object selection methods.)
Select Objects: **PICK** (Continue to select other desired objects.)
Select Objects: **Enter** (Press Enter to indicate selection of objects is complete.)
Base point or displacement: **PICK** or (**coordinates**) (This is the point to <u>move from</u>. Select a point to use as a "handle." An *Endpoint* or *Center*, etc., can be used.)
Second point of displacement: **PICK** or (**coordinates**) (This is the point to <u>move to</u>. *OSNAP*s can also be used here.)
Command:

Keep in mind that *OSNAP*s can be used when PICKing any point. It is often helpful to toggle *ORTHO ON* (**F8**) to force the *Move* in a horizontal or vertical direction.

If you know a specific distance and an angle that the set of objects should be moved, relative rectangular or relative polar coordinates can be used. In the following sequence, relative polar coordinates are used to move objects 2 units in a 30 degree direction (see Fig. 9-4, coordinate entry).

Command: **move**
Select Objects: **PICK**
Select Objects: **PICK**
Select Objects: **Enter**
Base point or displacement: **X,Y (coordinates)**
Second point of displacement: **@2<30**
Command:

Direct distance coordinate entry can be used effectively with *Move*. For example, assume you wanted to move the right side view of a multiview drawing 20 units to the right using direct distance entry (Fig. 9-5). First, invoke the *Move* command and select all of the objects comprising the right side view in response to the "Select Objects:" prompt. The command sequence is as follows:

Figure 9-5

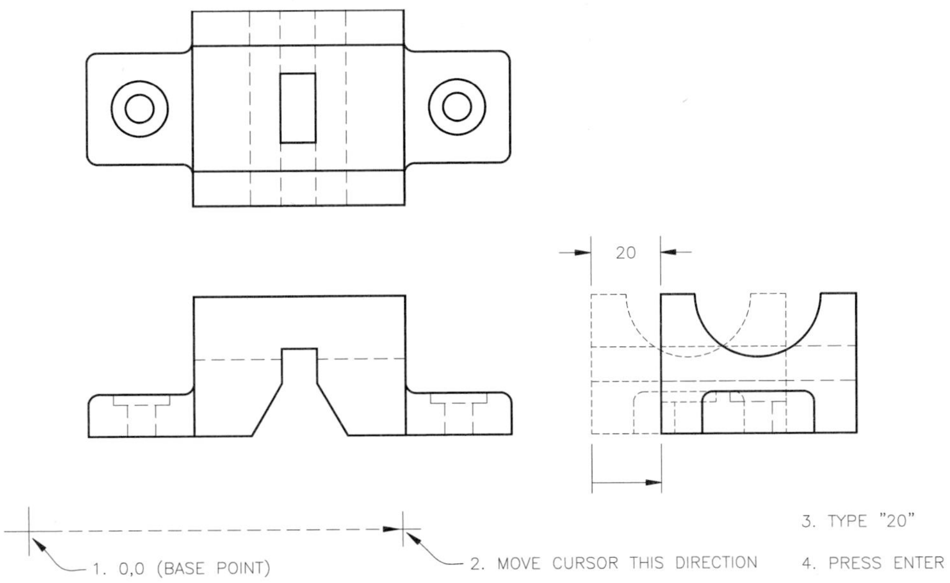

1. 0,0 (BASE POINT)
2. MOVE CURSOR THIS DIRECTION
3. TYPE "20"
4. PRESS ENTER

Command: *Move*
Select Objects:
PICK
Select Objects: **Enter**
Base point or displacement: **0,0** (or any value)
Second point of displacement: **20** (with *ORTHO* on, move cursor to right), then press **Enter**

In response to the "Base point or displacement:" prompt, PICK a point or enter any coordinate pairs or single value. At the "Second point of displacement:" prompt, move the cursor to the right any distance (with *ORTHO* on), then type "20" and press Enter. The value specifies the distance, and the cursor location from the last point indicates the direction of movement.

In the previous example, note that <u>any value</u> can be entered in response to the "Base point or displacement:" prompt. If a single value is entered ("3" for example), AutoCAD recognizes it as direct distance entry. The point designated is 3 units from the last PICK point in the direction specified by wherever the cursor is at the time of entry.

With the relative coordinate entry methods (relative rectangular, relative polar, and direct distance entry), the designated basepoint is of no consequence since the "second point of displacement" is always relative.

ROTATE

Pull-down Menu	COMMAND (TYPE)	ALIAS (TYPE)	Short-cut	Screen (side) Menu	Tablet Menu
Modify *Rotate*	*ROTATE*	*RO*	...	*MODIFY2* *Rotate*	*V,20*

Selected objects can be rotated to any position with this command. After selecting objects to *Rotate*, you select a "basepoint" (a point to rotate about) then specify an angle for rotation.

AutoCAD rotates the selected objects by the increment specified from the original position (Fig. 9-6). For example, specifying a value of **45** would *Rotate* the selected objects 45 degrees counter-clockwise from their current position; a value of **-45** would *Rotate* the objects 45 degrees in a clockwise direction.

Figure 9-6 ———————

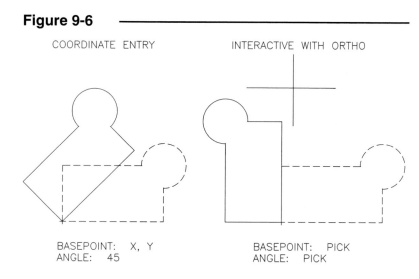

COORDINATE ENTRY   INTERACTIVE WITH ORTHO

BASEPOINT: X, Y   BASEPOINT: PICK
ANGLE: 45   ANGLE: PICK

    Command: **rotate**
    Select Objects: **PICK** or (**coordi-
    nates**) (Select the objects to rotate.)
    Select Objects: **Enter** (Indicate
    completion of the object selection.)
    Base point: **PICK** or
    (**coordinates**) (Select the point to <u>rotate about</u>.)
    <Rotation angle>Reference: **PICK** or (**value**) or (**coordinates**) (Enter a value for the number of
    degrees to rotate or interactively rotate the object set.)
    Command:

The basepoint is often selected interactively with *OSNAP*s. When specifying the angle for rotation, a value (incremental angle) can be typed. Alternately, if you want to **PICK** an angle, *Coords* (in the polar format) displays the current angle of the "rubberband" line. Turning on *ORTHO* forces the rotation to a 90 degree increment. The *List* command can be used to report the angle of an existing *Line* or other object (discussed in Chapter 17).

The *Reference* option can be used to specify a vector as the original angle before rotation (Fig. 9-7). This vector can be indicated interactively (*OSNAP*s can be used) or entered as an angle using keyboard entry. Angular values that you enter in response to the "New angle:" prompt are understood by AutoCAD as <u>absolute</u> angles for the *Reference* option only.

Figure 9-7 ———————

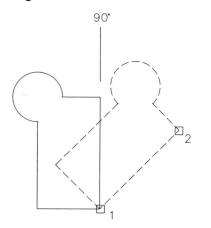

REFERENCE ANGLE: PICK 1,2
NEW ANGLE: 90°

    Command: **rotate**
    Select Objects: **PICK** or (**coordinates**) (Select the objects to
    rotate.)
    Select Objects: **Enter** (Indicates completion of object selection.)
    Base point: **PICK** or (**coordinates**) (Select the point to rotate
    about.)
    <Rotation angle>Reference: **R** (Indicates the *Reference* option.)
    Reference angle<0>: **PICK** or (**value**) (PICK the first point of the
    vector.)
    Second point: **PICK** (Indicates the second point defining the vector.)
    New angle: **PICK** or (**value**) (Indicates the new angle value.)
    Command:

SCALE

Pull-down Menu	COMMAND (TYPE)	ALIAS (TYPE)	Short-cut	Screen (side) Menu	Tablet Menu
Modify *Scale*	*SCALE*	*SC*	...	*MODIFY2* *Scale*	*V,21*

The *Scale* command is used to increase or decrease the size of objects in a drawing. The *Scale* command does not normally have any relation to plotting a drawing to scale.

After selecting objects to *Scale*, AutoCAD prompts you to select a "Base point:", which is the <u>stationary point</u>. You can then scale the size of the selected objects interactively or enter a scale factor. Using interactive input, you are presented with a "rubberband" line connected to the basepoint. Making the "rubberband" line longer or shorter than 1 unit increases or decreases the scale of the selected objects by that proportion; for example, pulling the "rubberband" line to two units length increases the scale by a factor of two (Fig. 9-8).

Figure 9-8

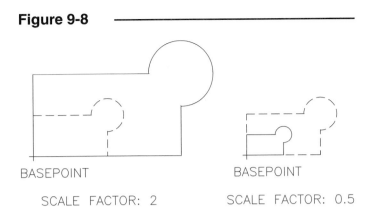

BASEPOINT

SCALE FACTOR: 2

BASEPOINT

SCALE FACTOR: 0.5

> Command: **scale**
> Select Objects: **PICK** or (**coordinates**) (Select the objects to scale.)
> Select Objects: **Enter** (Indicates completion of the object selection.)
> Base point: **PICK** or (**coordinates**) (Select the stationary point.)
> Scale factor<Reference>: **PICK** or (**value**) or (**coordinates**) (Enter a value for the scale factor or interactively scale the set of objects.)
> Command:

It may be desirable in some cases to use the ***Reference*** option to specify a value or two points to use as the reference length. This length can be indicated interactively (*OSNAPs* can be used) or entered as a value. This length is used for the subsequent reference length that the rubberband uses when interactively scaling. For example, if the reference distance is 2, then the "rubberband" line must be stretched to a length greater than 2 to increase the scale of the selected objects.

Scale normally should <u>not</u> be used to change the scale of an entire drawing in order to plot on a specific size sheet. CAD drawings should be created <u>full size</u> in <u>actual units</u>.

STRETCH

Pull-down Menu	COMMAND (TYPE)	ALIAS (TYPE)	Short-cut	Screen (side) Menu	Tablet Menu
Modify *Stretch*	*STRETCH*	S	...	*MODIFY2* *Stretch*	*V,22*

Objects can be made longer or shorter with *Stretch*. The power of this command lies in the ability to *Stretch* groups of objects while retaining the connectivity of the group (Fig. 9-9). When *Stretched*, *Lines* and *Plines* become longer or shorter and *Arcs* change radius to become longer or shorter. *Circles* do not stretch; rather, they move if the center is selected within the Crossing Window.

Objects to *Stretch* are not selected by the typical object selection methods but are indicated by a <u>Crossing Window or Crossing Polygon only</u>. The *Crossing Window* or *Polygon* should be created so the objects to *stretch* <u>cross</u> through the window. *Stretch* actually moves the object endpoints that are located within the Crossing Window.

Figure 9-9

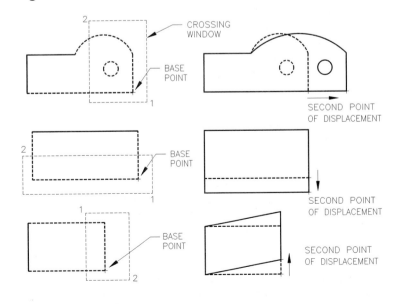

Following is the command sequence for *Stretch*.

> Command: **stretch**
> Select object(s) to stretch by crossing-window or -polygon. . .
> Select Objects:
> First Corner: **PICK**
> Other Corner: **PICK**
> Select Objects: **Enter**
> Base point or displacement: **PICK** or (**coordinates**) (Select a point to use as the point to stretch <u>from</u>.)
> Second point of displacement: **PICK** or (**coordinates**) (Select a point to use as the point to stretch <u>to</u>.)
> Command:

Stretch can be used to lengthen one object while shortening another. Application of this ability would be repositioning a door or window on a wall (Fig. 9-10).

In summary, the *Stretch* command <u>stretches objects that cross</u> the selection Window and <u>moves objects that are completely within</u> the selection Window, as shown in Figure 9-10.

Figure 9-10

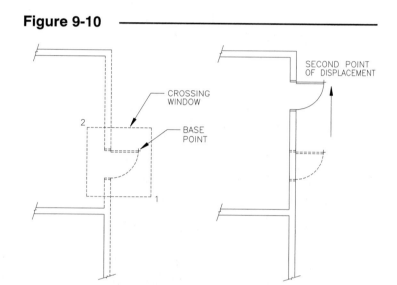

LENGTHEN

Pull-down Menu	COMMAND (TYPE)	ALIAS (TYPE)	Short-cut	Screen (side) Menu	Tablet Menu
Modify *Lengthen*	LENGTHEN	LEN	...	*MODIFY2* *Lengthen*	W,14

Lengthen changes the length (longer or shorter) of linear objects and arcs. No additional objects are required (as with *Trim* and *Extend*) to make the change in length. Many methods are provided as displayed in the command prompt.

Figure 9-11

> Command: **lengthen**
> Delta/Percent/Total/DYnamic/<Select object>:

LENGTHEN DELTA

BEFORE   AFTER

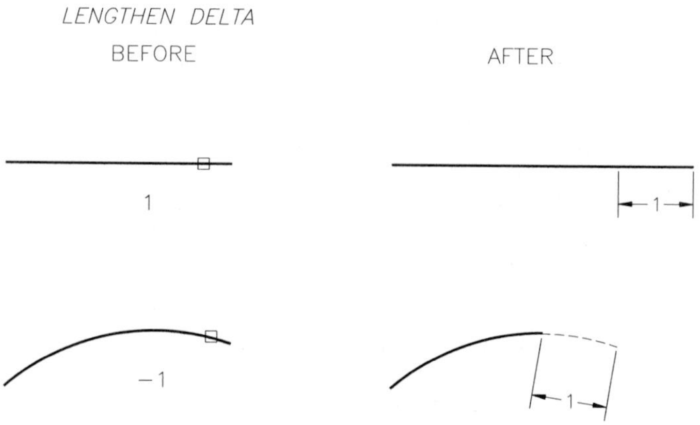

Select object
Selecting an object causes AutoCAD to report the current length of that object. If an *Arc* is selected, the included angle is also given.

DElta
Using this option returns the prompt shown next. You can change the current length of an object (including an arc) by any increment that you specify. Entering a positive value increases the length by that amount, while a negative value decreases the current length. The end of the object that you select changes while the other end retains its current endpoint (Fig. 9-11).

Angle/<Enter delta length>: (`value`) or `a`

The *Angle* option allows you to change the <u>included</u> <u>angle</u> of an arc (the length along the curvature of the arc can be changed with the *Delta* option). Enter a positive or negative value (degrees) to add or subtract to the current included angle, then select the end of the object to change (Fig. 9-12).

Figure 9-12

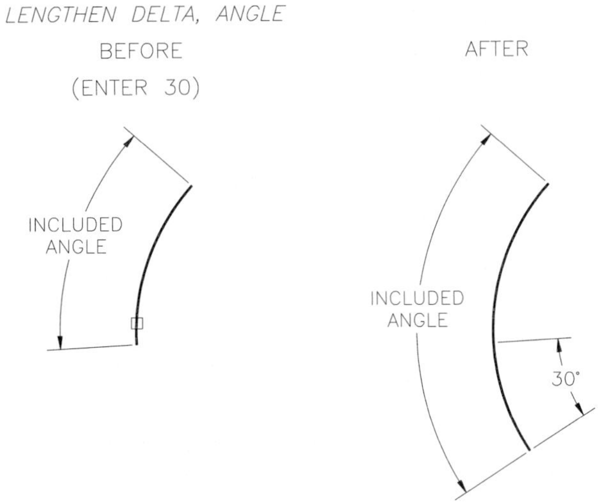

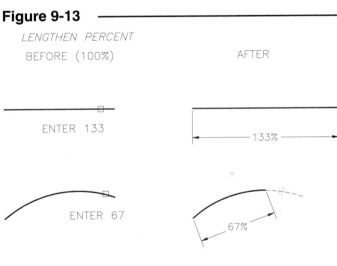

Percent
Use this option if you want to change the length by a percentage of the current total length. For arcs, the percentage applied affects the length and the included angle equally, so there is no *Angle* option. A value of greater than 100 increases the current length, and a value of less than 100 decreases the current length. Negative values are not allowed. The end of the object that you select changes.

Figure 9-13

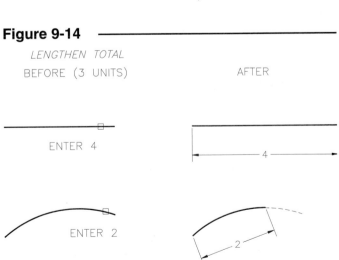

Total
This option lets you specify a value for the new total length. Simply enter the value and select the end of the object to change. The angle option is used to change the total included angle of a selected arc.

Angle/<Enter total length>: (`value`) or `a`

Figure 9-14

LENGTHEN TOTAL
BEFORE (3 UNITS)   AFTER

ENTER 4 |——————— 4 ———————|

ENTER 2   2

DYnamic

This option allows you to change the length of an object by dynamic dragging. Select the end of the object that you want to change. Object snaps can be used.

Figure 9-15

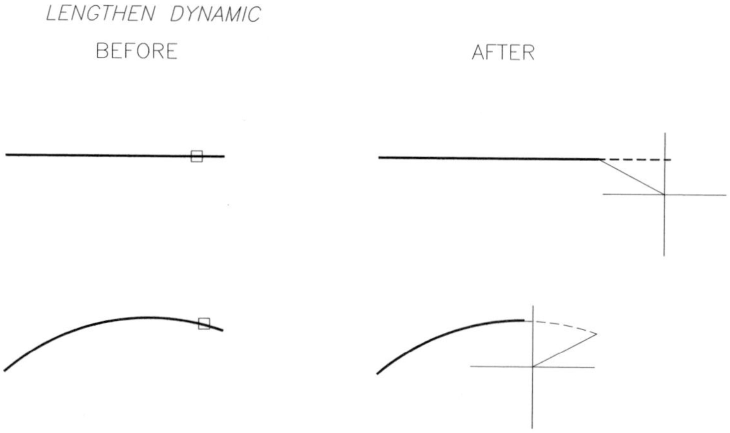

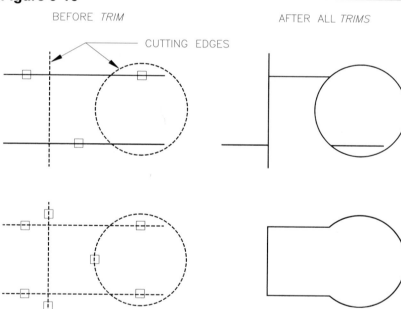

LENGTHEN DYNAMIC

BEFORE   AFTER

TRIM

Pull-down Menu	COMMAND (TYPE)	ALIAS (TYPE)	Short-cut	Screen (side) Menu	Tablet Menu
Modify *Trim*	*TRIM*	*TR*	...	*MODIFY2* *Trim*	*W,15*

The *Trim* command allows you to trim (shorten) the end of an object back to the intersection of another object (Fig. 9-16). The middle section of an object can also be *Trimmed* between two intersecting objects. There are two steps to this command: first, PICK one or more "cutting edges" (existing objects); then PICK the object or objects to *Trim* (portion to remove). The cutting edges are highlighted after selection. Cutting edges themselves can be trimmed if they intersect other cutting edges, but lose their highlight when trimmed.

```
Command: trim
Select cutting edges: (Projmode
= UCS, Edgemode = No extend)
Select Objects: PICK (Select an
object to use as a cutting edge.)
Select Objects: PICK
Select Objects: Enter
```

Figure 9-16

BEFORE *TRIM*   AFTER ALL *TRIMS*

CUTTING EDGES

<Select object to trim>/Project/Edge/Undo: **PICK** (Select the end of an object to trim.)
<Select object to trim>/Project/Edge/Undo: **PICK**
<Select object to trim>/Project/Edge/Undo: **Enter**
Command:

Edgemode

The *Edgemode* option can be set to *Extend* or *No extend*. In the *Extend* mode, objects that are selected as trimming edges will be <u>imaginarily extended</u> to serve as a cutting edge. In other words, lines used for trimming edges are treated as having infinite length (Fig. 9-17). The *No extend* mode only considers the actual length of the object selected as trimming edges.

Figure 9-17

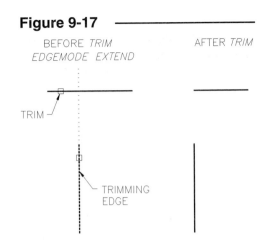

```
Command: trim
Select cutting edges: (Projmode = UCS, Edgemode = No
extend)
Select Objects: PICK
Select Objects: Enter
<Select object to trim>/Project/Edge/Undo: e
Extend/No extend <No extend>: e
<Select object to trim>/Project/Edge/Undo: PICK
<Select object to trim>/Project/Edge/Undo: Enter
Command:
```

Projectmode

The *Projectmode* switch controls how *Trim* and *Extend* operate in 3D space. *Projectmode* affects the projection of the "cutting edge" and "boundary edge." The three options are described here:

```
<Select object to trim>/Project/Edge/Undo: p
None/Ucs/View <UCS>:
```

None

The *None* option does not project the cutting edge. This mode is used for normal 2D drawing when all objects (cutting edges and objects to trim) lie in the current drawing plane. You can use this mode to *Trim* in 3D space when all objects lie in the plane of the current UCS or in planes parallel to it and the objects physically intersect. You can also *Trim* objects not in the UCS if the objects physically intersect, are in the same plane, and are not perpendicular to the UCS. In any case, the <u>objects must physically intersect</u> for trimming to occur.

The other two options (*UCS* and *View*) allow you to trim objects that do not physically intersect in 3D space.

UCS

This option projects the cutting edge perpendicular to the UCS. Any objects crossing the "projected" cutting edge in 3D space can be trimmed. For example, selecting a *Circle* as the cutting edge creates a projected cylinder used for cutting (Fig. 9-18).

Figure 9-18

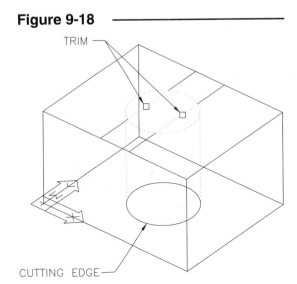

View

The *View* option allows you to trim objects that <u>appear</u> to intersect from the current viewpoint. The objects do not have to physically intersect in 3D space. The cutting edge is projected perpendicularly to the screen (parallel to the line of sight). This mode is useful for trimming "hidden edges" of a wireframe model to make the surfaces appear opaque (Fig. 9-19).

Undo

The *Undo* option allows you to undo the last *Trim* in case of an accidental trim.

Figure 9-19

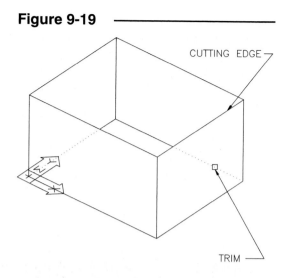

EXTEND

Pull-down Menu	COMMAND (TYPE)	ALIAS (TYPE)	Short-cut	Screen (side) Menu	Tablet Menu
Modify *Extend*	*EXTEND*	EX	...	*MODIFY2* *Extend*	W,16

Extend can be thought of as the opposite of *Trim*. Objects such as *Lines*, *Arcs*, and *Plines* can be *Extended* until intersecting another object called a "boundary edge" (Fig. 9-20). The command first requires selection of <u>existing</u> objects to serve as "boundary edge(s)" which become highlighted; then the objects to extend are selected. Objects extend until, and only if, they eventually intersect a "boundary edge." An *Extended* object acquires a new endpoint at the boundary edge intersection.

Figure 9-20

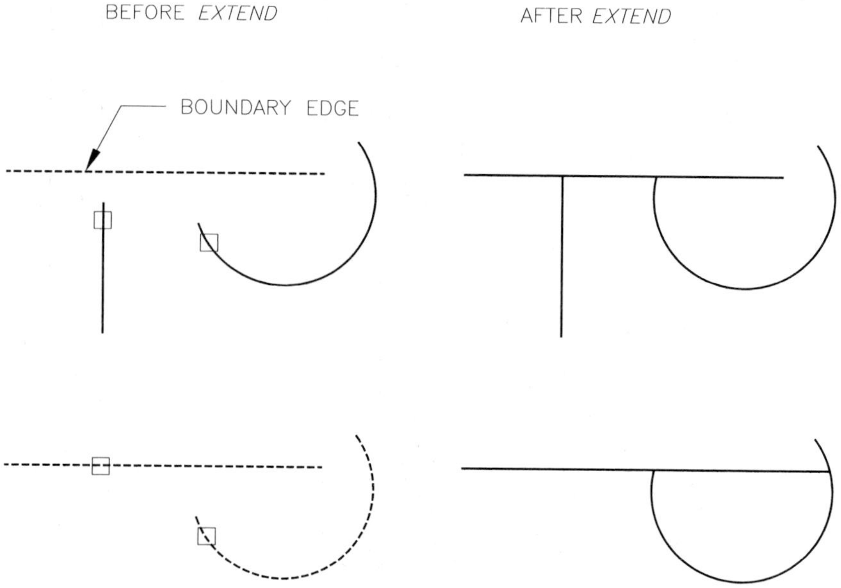

BEFORE *EXTEND*   AFTER *EXTEND*

BOUNDARY EDGE

```
Command: extend
Select boundary edges:
(Projmode = UCS, Edgemode
= No extend)
Select Objects: PICK
Select Objects: PICK
Select Objects: Enter
<Select object to extend>/Project/Edge/Undo: PICK  (Select object to extend.)
<Select object to extend>/Project/Edge/Undo: PICK
<Select object to extend>/Project/Edge/Undo: Enter
Command:
```

Edgemode/Projectmode

The *Edgemode* and *Projectmode* switches operate identically to their function with the *Trim* command. Use *Edgemode* with the *Extend* option if you want a boundary edge object to be imaginarily extended (Fig. 9-21).

Figure 9-21

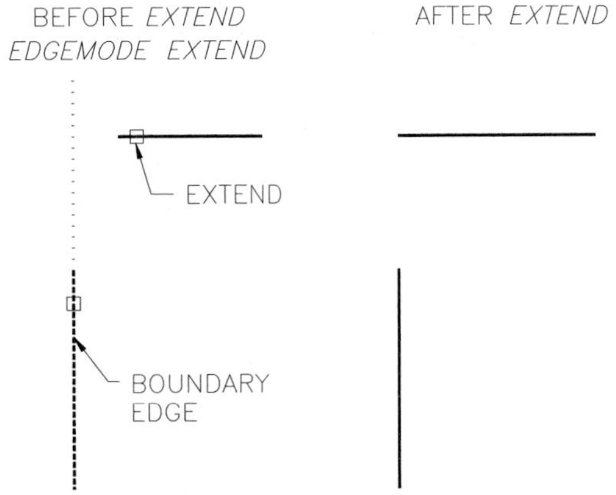

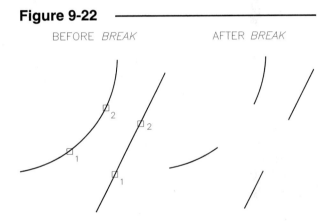

BREAK

Pull-down Menu	COMMAND (TYPE)	ALIAS (TYPE)	Short-cut	Screen (side) Menu	Tablet Menu
Modify *Break*	*BREAK*	*BR*	...	*MODIFY2* *Break*	*W,17*

Break allows you to break a space in an object or break the end off an object. You can think of *Break* as a partial erase. If you choose to break a space in an object, the space is created between two points that you specify (Fig. 9-22). In this case, the *Break* creates two objects from one.

Figure 9-22

If *Break*ing a circle (Fig. 9-23), the break is in a <u>counterclockwise</u> direction from the first to the second point specified.

Figure 9-23

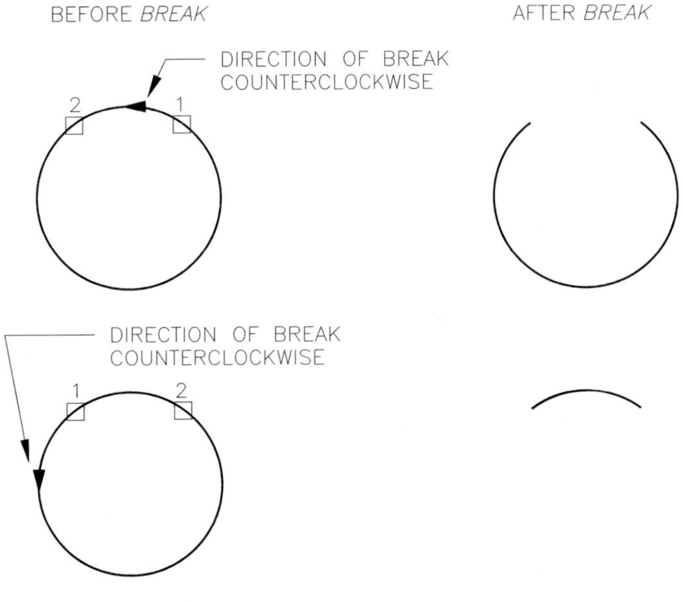

R14

If you want to *Break* the end off a *Line* or *Arc*, the first point should be specified at the point of the break and the second point should be <u>just off the end</u> of the *Line* or *Arc* (Fig. 9-24).

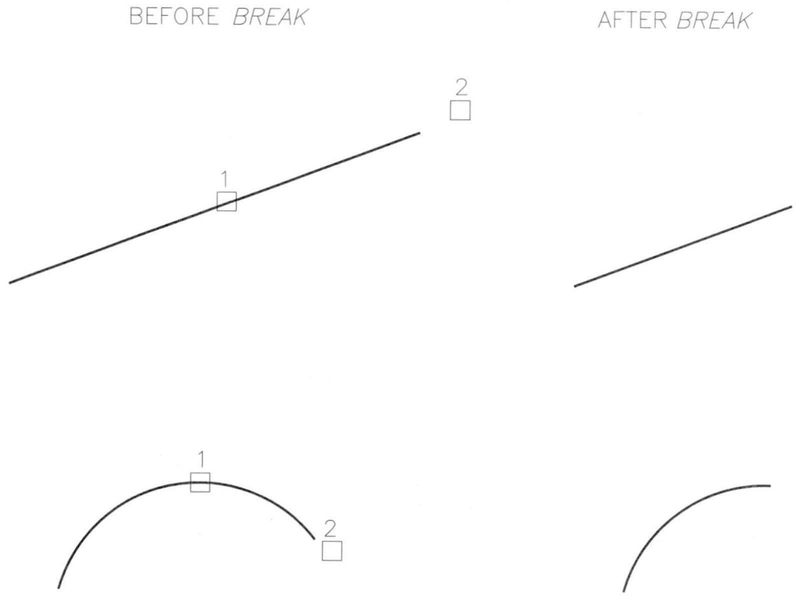

There are four options for *Break*. All four options are available <u>only from the screen menu</u>, not from the *Modify* toolbar, *Modify* pull-down menu, or digitizing tablet menu. If you are not using the screen menu, the desired option is selected by keying the letter *F* (for *First point*) or symbol @ (for "last point"). The four options are explained next in detail.

Figure 9-24

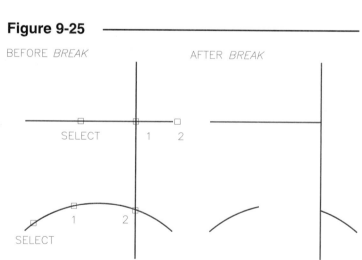

BEFORE *BREAK*   AFTER *BREAK*

Select, Second
This method (the default method) has two steps: select the object to break; then select the second point of the break. The first point used to select the object is <u>also</u> the first point of the break. (See Figs. 9-22, 9-23, 9-24.)

> Command: **break**
> Select Objects: **PICK** or (**coordinates**) (This is the first point of the break.)
> Enter second point (or F for first point): **PICK** or (**coordinates**) (This is the second point of the break.)
> Command:

Select, 2 Points
This method uses the first selection only to indicate the <u>object</u> to *Break*. You then specify the point that is the first point of the *Break,* and the next point specified is the second point of the *Break*. This option can be used with *OSNAP Intersection* to achieve the same results as *Trim*.

Selecting this option from the screen menu automatically sequences through the correct prompts (see Fig. 9-25).

Figure 9-25

BEFORE *BREAK*   AFTER *BREAK*

SELECT   1   2

SELECT   1   2

If you are typing, the command sequence is as follows:

> Command: **break**
> Select Objects: **PICK** or (**coordinates**) (Select the object to break.)
> Enter second point (or F for first point): **F** (Indicates respecification for the first point.)

Enter first point: **PICK** or (**coordinates**) (Select the first point of the break.)
Enter second point: **PICK** or (**coordinates**) (This is the second point of the break.)
Command:

Select Point
This option breaks one object into two separate objects with <u>no space</u> between. You specify only <u>one</u> point with this method. When you select the object, the point indicates the object <u>and</u> the point of the break. When prompted for the second point, the @ symbol (translated as "last point") is specified. This second step is done automatically only if selected from the screen menu. If you are typing the command, the @ symbol ("last point") must be typed:

    Command: **break**
    Select Objects: **PICK** or (**coordinates**) (This PICKs the object and specifies the first point of the break.)
    Enter second point (or F for first point): **@** (Indicates the break to be at the last point.)

The resulting object should <u>appear</u> as before; however, it has been transformed into <u>two</u> objects with matching endpoints (Fig. 9-26).

Figure 9-26

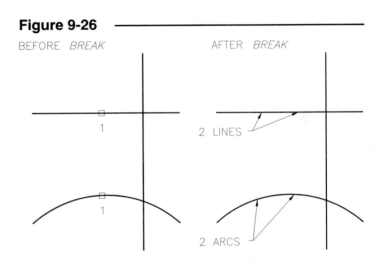

Select, First
This option creates a break with <u>no space,</u> like the *1 Point* option; however, you can select the object you want to *Break* first and the point of the *Break* next (Fig. 9-27).

Figure 9-27

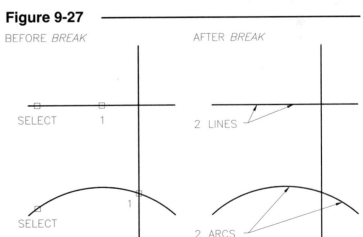

If you are typing the command, the sequence is as follows:

    Command: **break**
    Select Objects: **PICK** or (**coordinates**) (Select the object to break.)
    Enter second point (or F for first point): **F** (Indicates respecification for the first point.)
    Enter first point: **PICK** or (**coordinates**) (Select the first point of the break.)
    Enter second point: **@** (The second point of the break is the last point.)
    Command:

CHAPTER EXERCISES

1. *Move*

Begin a *New* drawing and create the geometry in Figure 9-28 A using *Lines* and *Circles*. If desired, set *SNAP* to .25 to make drawing easy and accurate.

For practice, turn *SNAP OFF* (**F9**). Use the *Move* command to move the *Circles* and *Lines* into the positions shown in illustration B. *OSNAPs* are required to *Move* the geometry accurately (since *SNAP* is off). *Save* the drawing as **MOVE1**.

Figure 9-28

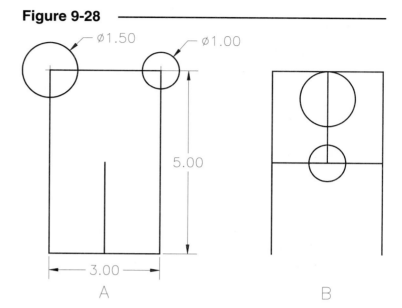

A

B

2. *Rotate*

Figure 9-29

Begin a *New* drawing and create the geometry in Figure 9-29 A.

Rotate the shape into position shown in step B. *SaveAs* **ROTATE1**.

Use the *Reference* option to *Rotate* the box to align with the diagonal *Line* as shown in C. *SaveAs* **ROTATE2**.

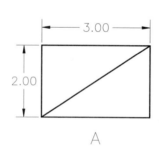

A

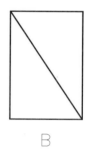

B

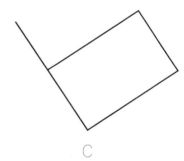

C

3. *Scale*

Open **ROTATE1** to again use the shape shown in Figure 9-30 B and *SaveAs* **SCALE1**. *Scale* the shape by a factor of **1.5**.

Open **ROTATE2** to again use the shape shown in C. Use the *Reference* option of *Scale* to increase the scale of the three other *Lines* to equal the length of the original diagonal *Line* as shown. (HINT: *OSNAPs* are required to specify the *Reference length* and *New length*.) *SaveAs* **SCALE2**.

Figure 9-30

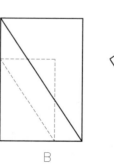

B

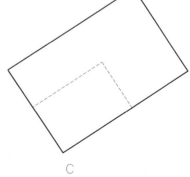

C

4. *Stretch*

A design change has been requested. *Open* the **SLOTPLATE CH8** drawing and make the following changes.

A. The top of the plate (including the slot) must be moved upward. This design change will add 1″ to the total height of the Slot Plate. Use *Stretch* to accomplish the change, as shown in Figure 9-31. Draw the Crossing Window as shown.

Figure 9-31 ─────────────

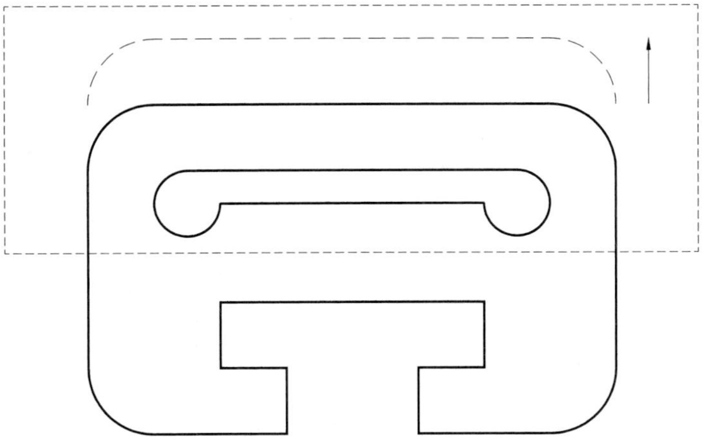

B. The notch at the bottom of the plate must be adjusted slightly by relocating it .50 units to the right, as shown in Figure 9-32. Draw the Crossing Window as shown. Use *SaveAs* to reassign the name to **SLOTPLATE 2**.

Figure 9-32 ─────────────

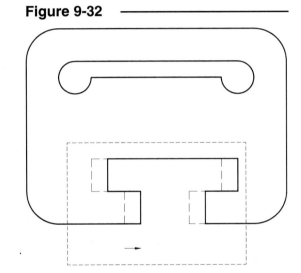

5. *Trim*

A. Create the shape shown in Figure 9-33 A. *SaveAs* **TRIM-EX**.

B. Use *Trim* to alter the shape as shown in B. *SaveAs* **TRIM1**.

C. *Open* **TRIM-EX** to create the shapes shown in C and D using *Trim*. *SaveAs* **TRIM2** and **TRIM3**.

6. *Extend*

Open each of the drawings created as solutions for Figure 9-33 (**TRIM1, TRIM2,** and **TRIM3**). Use *Extend* to return each of the drawings to the original form shown in Figure 9-33 A. *SaveAs* **EXTEND1, EXTEND2,** and **EXTEND3**.

Figure 9-33 ─────────────

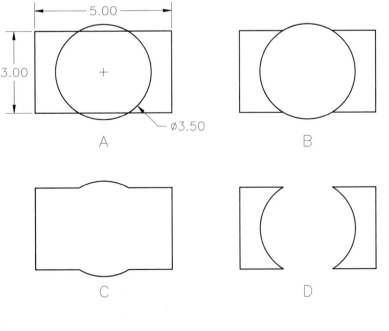

7. *Break*

   A. Create the shape shown in Figure 9-34 A. *SaveAs* **BREAK-EX**.

   B. Use *Break* to make the two breaks as shown in B. *SaveAs* **BREAK1**. (HINT: You may have to use *OSNAP*s to create the breaks at the *Intersections* or *Quadrants* as shown.)

   C. Open **BREAK-EX** each time to create the shapes shown in C and D with the *Break* command. *SaveAs* **BREAK2** and **BREAK3**.

Figure 9-34

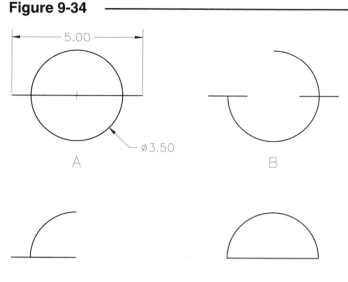

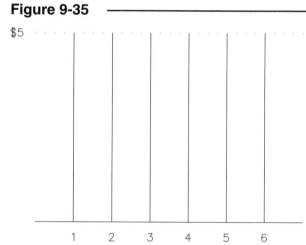

8. *Lengthen*

   Five of your friends went to the horse races, each with $5.00 to bet. Construct a simple bar graph (similar to Fig. 9-35) to illustrate how your friends' wealth compared at the beginning of the day.

   Modify the graph with *Lengthen* to report the results of their winnings and losses at the end of the day. The reports were as follows: friend 1 made $1.33 while friend 2 lost $2.40; friend 3 reported a 150% increase and friend 4 brought home 75% of the money; friend 5 ended the day with $7.80 and friend 6 came home with $5.60. Use the *Lengthen* command with the appropriate options to change the line lengths accordingly.

   Who won the most? Who lost the most? Who was closest to even? Enhance the graph by adding width to the bars and other improvements as you wish (similar to Fig. 9-36). *SaveAs* **FRIENDS GRAPH**.

Figure 9-35

$5

1 2 3 4 5 6

Figure 9-36

$5

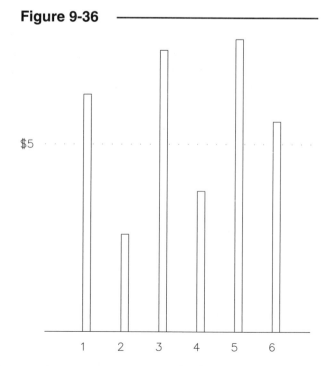

1 2 3 4 5 6

9. *Trim, Extend*

A. *Open* the **PIVOTARM CH7** drawing from the Chapter 7 Exercises. Use *Trim* to remove the upper sections of the vertical *Lines* connecting the top and front views. Compare your work to Figure 9-37.

Figure 9-37

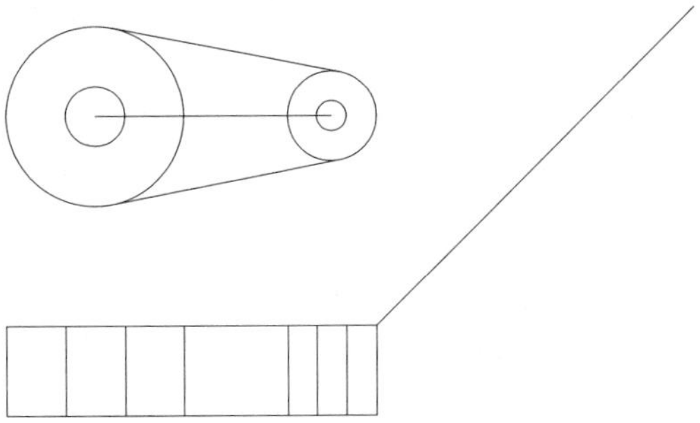

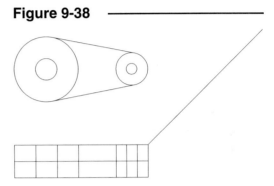

B. Next, draw a horizontal *Line* in the front view between the *Midpoints* of the vertical *Line* on each end of the view, as shown in Figure 9-38. *Erase* the horizontal *Line* in the top view between the *Circle* centers.

Figure 9-38

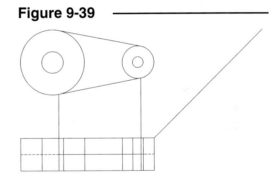

C. Draw vertical *Lines* from the *Endpoints* of the inclined *Line* (side of the object) in the top view down to the bottom *Line* in the front view, as shown in Figure 9-39. Use the two vertical lines as *Cutting edges* for *Trimming* the *Line* in the middle of the front view, as shown highlighted.

Figure 9-39

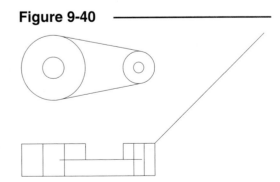

D. Finally, *Erase* the vertical lines used for *Cutting edges* and then use *Trim* to achieve the object as shown in Figure 9-40. Use *SaveAs* and name the drawing **PIVOTARM CH9**.

Figure 9-40

10. **GASKETA**

Begin a *New* drawing. Set *Limits* to **0,0** and **8,6**. Set *SNAP* and *GRID* values appropriately. Create the Gasket as shown in Figure 9-41. Construct only the gasket shape, not the dimensions or centerlines. (HINT: Locate and draw the four .5″ diameter *Circles,* then create the concentric .5″ radius arcs as full *Circles,* then *Trim.* Make use of *OSNAPs* and *Trim* whenever applicable.) *Save* the drawing and assign the name **GASKETA.**

Figure 9-41

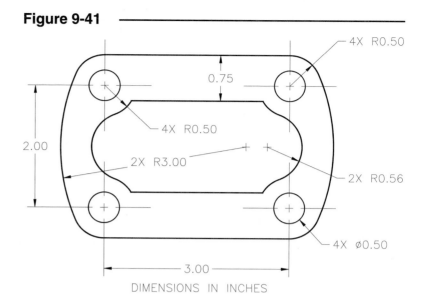

11. **Chemical Process Flow Diagram**

Begin a *New* drawing and recreate the chemical process flow diagram shown in Figure 9-42. Use *Line, Circle, Arc, Trim, Extend, Scale, Break,* and other commands you feel necessary to complete the diagram. Because this is diagrammatic and will not be used for manufacturing, dimensions are not critical, but try to construct the shapes with proportional accuracy. *Save* the drawing as **FLOWDIAG.**

Figure 9-42

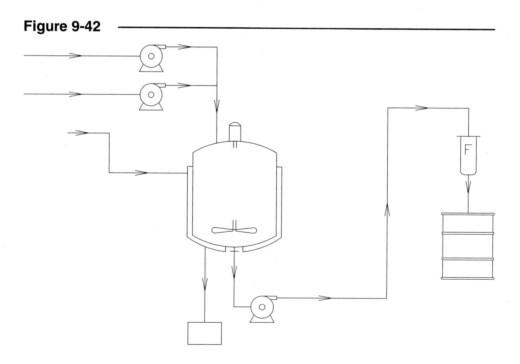

10

MODIFY COMMANDS II

Chapter Objectives

After completing this chapter you should:

1. be able to *Copy* objects;

2. be able to make *Mirror* images of selected objects;

3. know how to create parallel copies of objects with *Offset*;

4. be able to make *rectangular* and *polar Arrays* of existing objects;

5. be able to create a *Fillet* between two objects;

6. be able to create a *Chamfer* between two objects.

CONCEPTS

Modify commands use existing objects. The *Modify* commands covered in Chapter 9 (Modify Commands I) change existing objects. The *Modify* commands covered in this chapter <u>use existing objects to create new and similar objects</u>. For example, the *Copy* command prompts you to select an object (or set of objects), then creates an identical object (or set). The commands covered in this chapter are:

> *Copy, Mirror, Offset, Array, Fillet,* and *Chamfer*

These commands are invoked by the same methods and are found in the same locations in AutoCAD Release 14 as those commands discussed in Chapter 9, Modify Commands I. Please refer to Chapter 9 for information on locating commands in the *Modify* toolbar, *Modify* pull-down menu, and *MODIFY1* and *MODIFY2* screen menus.

COMMANDS

COPY

Pull-down Menu	COMMAND (TYPE)	ALIAS (TYPE)	Short-cut	Screen (side) Menu	Tablet Menu
Modify *Copy*	*COPY*	*CO or CP*	...	*MODIFY1* *Copy*	*V,15*

Copy creates a duplicate set of the selected objects and allows placement of those copies. The *Copy* operation is like the *Move* command, except with *Copy* the original set of objects remains in its original location. You specify a "Base point:" (point to copy <u>from</u>) and a "Second point of displacement:" (point to copy <u>to</u>). See Figure 10-1.

Figure 10-1 ——————————————

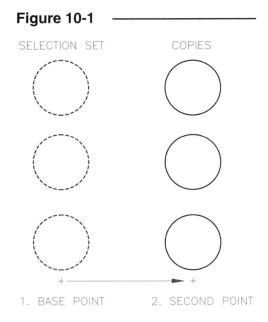

The command syntax for *Copy* is as follows.

    Command: copy
    Select Objects: PICK (Select objects to be copied.)
    Select Objects: Enter (Indicates completion of the
    selection set.)
    <Base point or displacement>Multiple: PICK or (coordi-
    nates) (This is the point to copy from. Select a point,
    usually on the object, to use as a "handle" or reference point.
    OSNAPs can be used. Coordinates in any format can also
    be entered.)
    Second point of displacement: PICK or (coordinates)
    (This is the point to copy to. Select a point. OSNAPs can be
    used. Coordinates in any format can be entered.)
    Command:

In many applications it is desirable to use *OSNAP* options to PICK the "Base point:" and the "Second point of displacement:" (Fig. 10-2).

Figure 10-2 ────────────

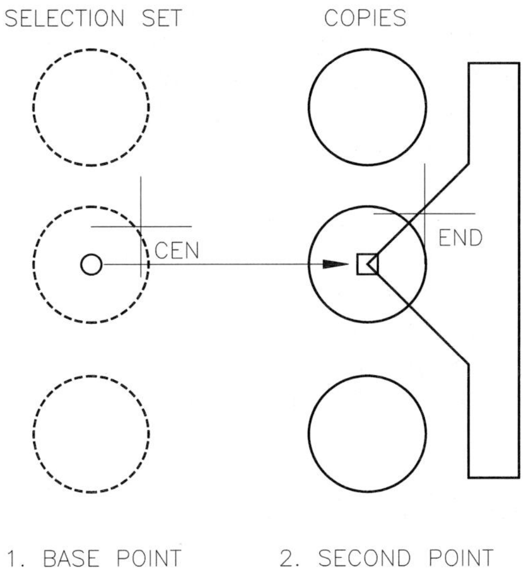

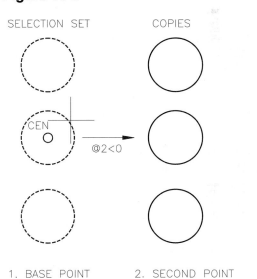

SELECTION SET   COPIES

CEN   END

1. BASE POINT   2. SECOND POINT

Alternately, you can PICK the "Base point:" and enter relative rectangular coordinates, relative polar coordinates, or direct distance entry to specify the "Second point of displacement:".

Figure 10-3 ────────────

SELECTION SET   COPIES

CEN   @2<0

1. BASE POINT   2. SECOND POINT

The *Copy* command has a *Multiple* option. The *Multiple* option allows creating and placing multiple copies of the selection set.

Command: *copy*
Select Objects: **PICK**
Select Objects: **Enter**
<Base point or displacement>/Multiple: *m*
Base point: **PICK**
Second point of displacement: **PICK**
Second point of displacement: **PICK**
Second point of displacement: **Enter**
Command:

Figure 10-4 ────────────

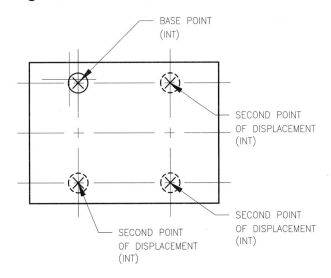

BASE POINT
(INT)

SECOND POINT
OF DISPLACEMENT
(INT)

SECOND POINT
OF DISPLACEMENT
(INT)

SECOND POINT
OF DISPLACEMENT
(INT)

MIRROR

Pull-down Menu	COMMAND (TYPE)	ALIAS (TYPE)	Short-cut	Screen (side) Menu	Tablet Menu
Modify *Mirror*	*MIRROR*	*MI*	...	*MODIFY1* *Mirror*	*V,16*

This command creates a mirror image of selected existing objects. You can retain or delete the original objects ("old objects"). After selecting objects, you create two points specifying a "rubberband line," or "mirror line," about which to *Mirror*.

The length of the mirror line is unimportant since it represents a vector or axis (Fig. 10-5).

Figure 10-5 ───────────────────────

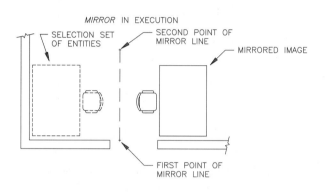

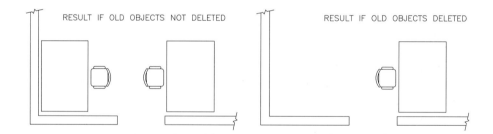

Command: **mirror**
Select Objects: **PICK** (Select object or group of objects to mirror.)
Select Objects: **Enter** (Press Enter to indicate completion of object selection.)
First point of mirror line: **PICK** or (**coordinates**) (Draw first endpoint of line to represent mirror axis by PICKing or entering coordinates.)
Second point of mirror line: **PICK** or (**coordinates**) (Select second point of mirror line by PICKing or entering coordinates.)
Delete old objects? <N> **Enter** or **Y** (Press Enter to yield both sets of objects or enter Y to keep only the mirrored set.)
Command:

If you want to *Mirror* only in a vertical or horizontal direction, toggle *ORTHO* (**F8**) *On* before selecting the "Second point of mirror line."

Mirror can be used to draw the other half of a symmetrical object, thus saving some drawing time (Fig. 10-9).

Figure 10-6 ───────────────────────

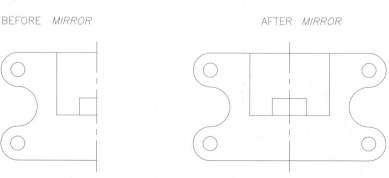

BEFORE *MIRROR*   AFTER *MIRROR*

You can control whether <u>text</u> is mirrored by using the *MIRR-TEXT* variable. Type *MIRR-TEXT* at the command prompt and change the value to 0 if you do <u>not</u> want the text to be reflected; otherwise, the default setting of 1 mirrors text along with other selected objects (Fig. 10-7). Dimensions are <u>not</u> affected by the *MIRR-TEXT* variable and therefore are not reflected (that is, if the dimensions are associative). (See Chapter 26 for details on dimensions.)

Figure 10-7

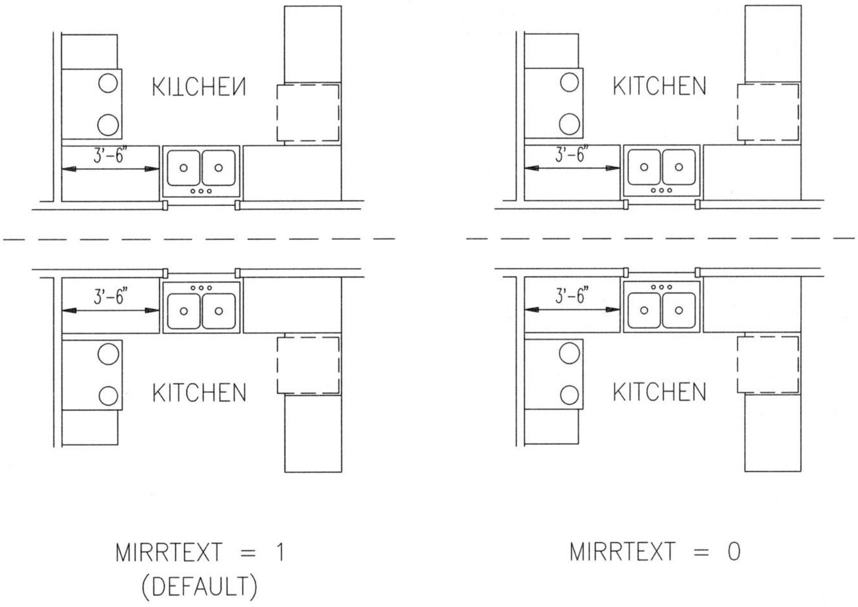

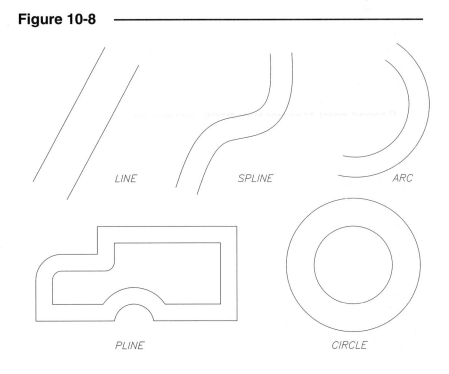

MIRRTEXT = 1
(DEFAULT)

MIRRTEXT = 0

OFFSET

Pull-down Menu	COMMAND (TYPE)	ALIAS (TYPE)	Short-cut	Screen (side) Menu	Tablet Menu
Modify *Offset*	*OFFSET*	*O*	...	*MODIFY1* *Offset*	*V,17*

Offset creates a <u>parallel copy</u> of selected objects. Selected objects can be *Lines*, *Arcs*, *Circles*, *Plines*, or other objects. *Offset* is a very useful command that can increase productivity greatly, particularly with *Plines*.

Depending on the object selected, the resulting *Offset* is drawn differently (Fig. 10-8). *Offset* creates a parallel copy of a *Line* equal in length and perpendicular to the original. *Arcs* and *Circles* have a concentric *Offset*. *Offsetting* closed *Plines* or *Splines* results in a complete parallel shape.

Figure 10-8

LINE   SPLINE   ARC

PLINE   CIRCLE

Two options are available with *Offset*: (1) *Offset* a specified *distance* and (2) *Offset through* a specified point (Fig. 10-9).

Distance

The *Distance* option command sequence is as follows:

> Command: **offset**
> Offset distance or Through <Through>: (**value**) or **PICK** (Indicate the distance to offset by entering a value or interactively picking two points.)
> Select objects to offset: **PICK** (Select an object to make a parallel copy of. Only one object can be selected.)
> Side to offset? **PICK** or (**coordinates**) (Select which side of the selected object for offset object to be drawn.)
> Select objects to offset: **PICK** or **Enter** (*Offset* can be used repeatedly to offset at the <u>same</u> distance as the previously specified value. Pressing Enter completes command.)
> Command:

Through

The command sequence for *Offset through* is as follows:

> Command: **offset**
> Offset distance or Through <Through>: **T**
> Select objects to offset: **PICK** (Select an object to make a parallel copy of. Only one object can be selected.)
> Through point: **PICK** or (**coordinates**) (Select a point for offset object to be drawn through. Coordinates can be entered or point can be selected with or without *OSNAP*s.)
> Select objects to offset: **PICK** or **Enter** (*Offset* can be used repeatedly to specify a <u>different</u> through point. Pressing Enter completes command.)
> Command:

Notice the power of using *Offset* with closed *Pline* or *Spline* shapes (Fig. 10-10). Because a *Pline* or a *Spline* is one object, *Offset* creates one complete smaller or larger "parallel" object. Any closed shape composed of *Line*s and *Arc*s can be converted to one *Pline* object (see *Pedit*, Chapter 16).

Figure 10-9 ───────────────────

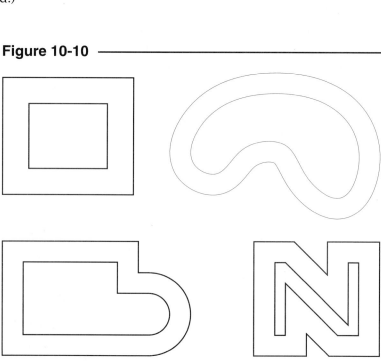

Figure 10-10 ───────────────────

ARRAY

Pull-down Menu	COMMAND (TYPE)	ALIAS (TYPE)	Short-cut	Screen (side) Menu	Tablet Menu
Modify *Array*	*ARRAY*	*AR*	...	*MODIFY1* *Array*	*V,18*

The *Array* command creates either a *Rectangular* or a *Polar* (circular) pattern of existing objects that you select. The pattern could be created from a single object or from a group of objects. *Array* copies a duplicate set of objects for each "item" in the array.

Rectangular

This option creates an *Array* of the selection set in a pattern composed of <u>rows</u> and <u>columns</u>. The command syntax for a rectangular *Array* is given next:

Command: **array**
Select Objects: **PICK** (Select objects to be arrayed.)
Select Objects: **Enter** (Indicates completion of object selection.)
Rectangular or Polar array (<R>/P): **R** (Indicates rectangular.)
Number of rows (---) <1>: **(value)** (Enter value for number of rows.)
Number of columns (|||) <1>: **(value)** (Enter value for columns.)
Unit cell or distance between rows (---): **(value)** (Enter a value for the distance from any <u>point</u> on one object to the

Figure 10-11

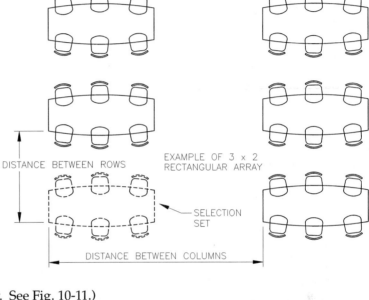

EXAMPLE OF 3 x 2 RECTANGULAR ARRAY

DISTANCE BETWEEN ROWS

SELECTION SET

DISTANCE BETWEEN COLUMNS

same point on an object in the adjacent row. See Fig. 10-11.)
Distance between columns (|||): **(value)** (Enter a value for the distance from any *point* on one object to the same point on an object in the adjacent column. See Fig. 10-11.)
Command:

The *Array* is generated in a rectangular pattern with the original set of objects as the lower-left "cornerstone." Negative values can be entered (distance between rows and columns) to generate and *Array* in a -X or -Y direction.

If you want to specify distances between cells <u>interactively</u>, use the *Unit cell* method (see Fig. 10-12).

Unit cell or distance between rows (---):
PICK (Pick point for first point of cell.)
Other corner: **PICK** (Pick point for other corner of cell.)

Figure 10-12

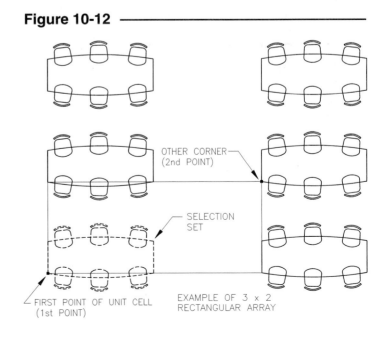

OTHER CORNER (2nd POINT)

SELECTION SET

FIRST POINT OF UNIT CELL (1st POINT)

EXAMPLE OF 3 x 2 RECTANGULAR ARRAY

A rectangular *Array* can be created at an angle by first <u>rotating the *SNAP*</u> to the desired angle and then creating the *Array* as shown in Figure 10-13. See Chapter 6 for information on the *Rotate* option of the *Snap* command.

Figure 10-13 ───────────────────

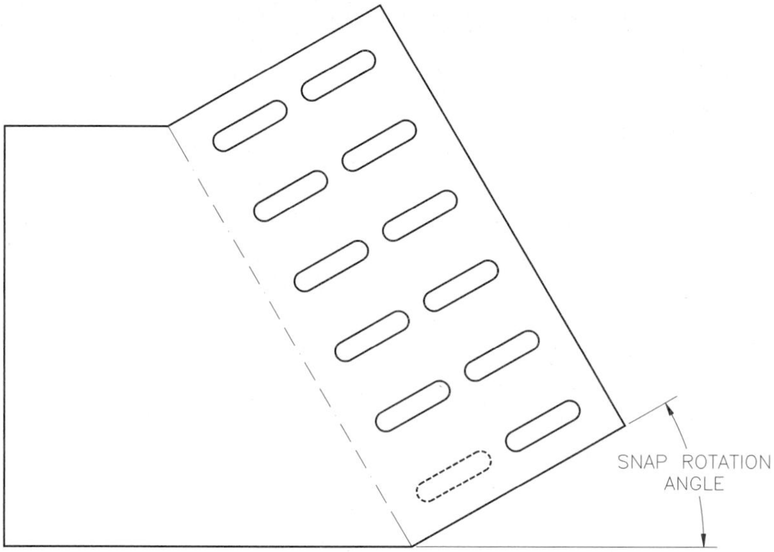

SNAP ROTATION
ANGLE

Polar

This option creates a <u>circular</u> pattern of the selection set with any number of copies or "items." The number of items specified <u>includes the original</u> selection set. You also specify the center of array, angle to generate the array through, and orientation of "items:"

    Command: **array**
    Select Objects: **PICK** (Select objects to be arrayed.)
    Select Objects: **Enter** (Indicates object selection completed.)
    Rectangular or Polar array (<R>/P): **P** (Indicates polar array.)
    Center point of array: **PICK** (Select point for array to be generated around.)
    Number of items: (**value**) (Enter value for number of copies <u>including</u> original selection set.)
    Angle to fill (+=ccw, -=cw) <360>: **Enter** or (**value**) (Press Enter for full circular array; enter value for
    less than 360 degree array; enter negative value for clockwise generation of array.)
    Rotate objects as they are copied? <Y> **Enter** or **N** (Press Enter for rotation of objects about center; N
    for keeping objects in original orientation.)

Figure 10-14 illustrates a *Polar Array* created by accepting the default values (360 degrees and "Rotate objects as they are copied").

Figure 10-14 ───────────────────

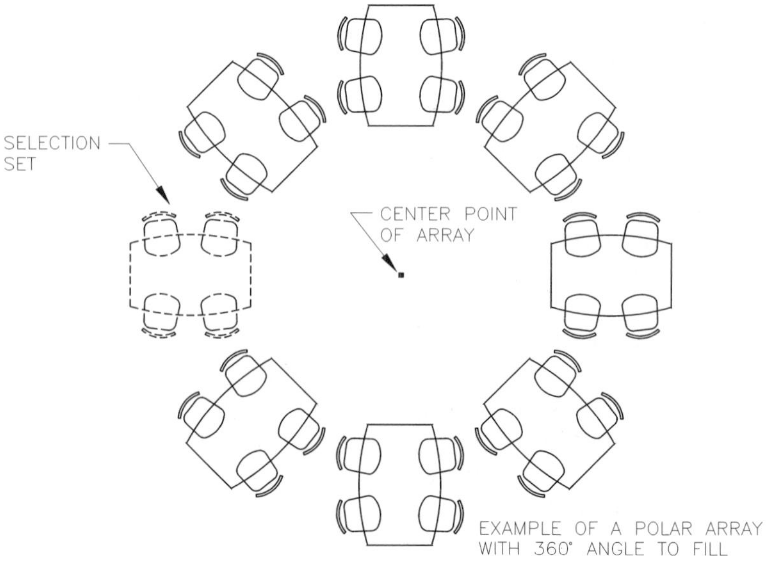

SELECTION
SET

CENTER POINT
OF ARRAY

EXAMPLE OF A POLAR ARRAY
WITH 360° ANGLE TO FILL

Figure 10-15 shows a *Polar Array* through 180 degrees with objects rotated.

In some designs, you may know the <u>angle between objects</u> in a circular pattern rather than the <u>number of objects</u> in the pattern. You can create a *Polar Array* by specifying the "Angle between items:" instead of the "Number of items:" if you choose. This is done by first entering a null response (press **Enter**) to the "Number of items:" prompt. AutoCAD proceeds to prompt you for the "Angle to fill:", then asks for the "Angle between items:"

```
Command: array
Select Objects: PICK
Select Objects: Enter
Rectangular or Polar array (<R>/P): P
Center point of array: PICK
Number of items: Enter (do not specify a value)
Angle to fill (+=ccw, -=cw) <360>: Enter or (value)
Angle between items: (value)
Rotate objects as they are copied? <Y> Enter or N
Command:
```

Figure 10-15

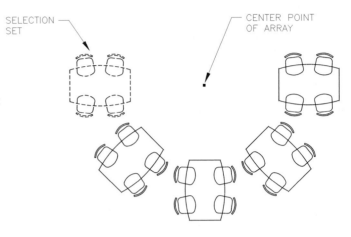

EXAMPLE OF A POLAR ARRAY
WITH 180° ANGLE TO FILL
AND 5 ITEMS TO ARRAY

SELECTION SET

CENTER POINT OF ARRAY

Figure 10-16 illustrates the results of answering *No* to the prompt "Rotate objects as they are rotated?" Note that the "items" remain in the same orientation (not rotated). Also notice the position of the *Array* with relation to the center point of the array.

When a *Polar Array* is created and objects are <u>not</u> rotated, the set of objects may not seem to rotate about the specified center, as shown in Figure 10-16. AutoCAD uses a single reference point for the *Array*—that of the <u>last</u> object in the selection set. If you PICK the objects one at a time (with the pickbox), the <u>last</u> one selected provides the reference point. If you use a Window or other method for selection, the reference point is arbitrary. This usually creates a non-symmetrical circular *Array* about the selected center point. This occurrence is no problem for *Polar Arrays* where the selection set is rotated.

Figure 10-16

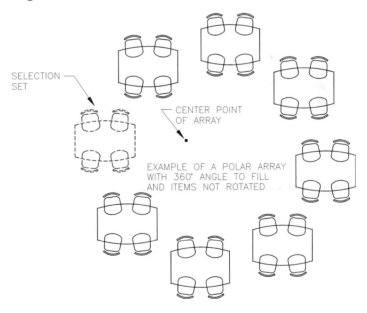

SELECTION SET

CENTER POINT OF ARRAY

EXAMPLE OF A POLAR ARRAY
WITH 360° ANGLE TO FILL
AND ITEMS NOT ROTATED

The solution is to create a *Point* object in the <u>center of the selection set</u> and select it <u>last</u>. This generates the selection set in a circle about the assigned center. (Alternately, you can make the selection set into a *Block* and use the center of the *Block* as the *Insertion* point. See Chapter 20).

FILLET

Pull-down Menu	COMMAND (TYPE)	ALIAS (TYPE)	Short-cut	Screen (side) Menu	Tablet Menu
Modify *Fillet*	*FILLET*	*F*	...	*MODIFY2* *Fillet*	*W,19*

The *Fillet* command automatically rounds a sharp corner (intersection of two *Lines, Arcs, Circles,* or *Pline* vertices) with a radius. You only specify the radius and select the objects' ends to be *Filleted.* The objects to fillet do <u>not</u> have to completely intersect but can overlap. You can specify whether or not the objects are automatically extended or trimmed as necessary (Fig. 10-17).

The *Fillet* command is used first to input the desired *radius* (if other than the default 0.5000 value) and a second time to select the objects to *Fillet.*

Figure 10-17 —————————————————

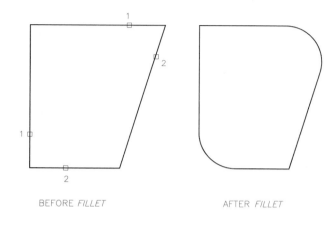

BEFORE *FILLET*   AFTER *FILLET*

      Command: **fillet**
      (TRIM mode) Current fillet radius=0.5000
      Polyline/Radius/Trim/<Select first object>: **r** (Indicates the radius option.)
      Enter fillet radius <0.5000>: (**value**) or **PICK** (Enter a value for the desired fillet radius or select two
      points to interactively specify the radius.)
      Command:

No fillet will be drawn at this time. Repeat the *Fillet* command to select objects to fillet.

      Command: **fillet**
      (TRIM mode) Current fillet radius=0.2500
      Polyline/Distances/Trim/<Select first object>: **PICK** (Select one *Line, Arc,* or *Circle* near the point where
      fillet should be created.)
      Select second object: **PICK** (Select second object to fillet near the fillet location.)
      Command:

The fillet is created at the corner selected.

Treatment of *Arcs* and *Circles* with *Fillet* is shown here. Note that the objects to *Fillet* do not have to intersect but can overlap.

Figure 10-18 ———————————————

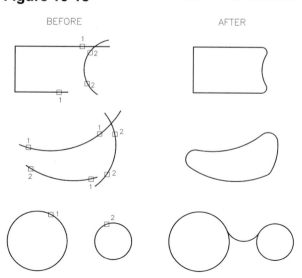

BEFORE   AFTER

Since trimming and extending are done automatically by *Fillet* when necessary, using *Fillet* with a *radius* of *0* has particular usefulness. Using *Fillet* with a *0 radius* creates clean, sharp corners even if the original objects overlap or do not intersect (Fig. 10-19).

Figure 10-19

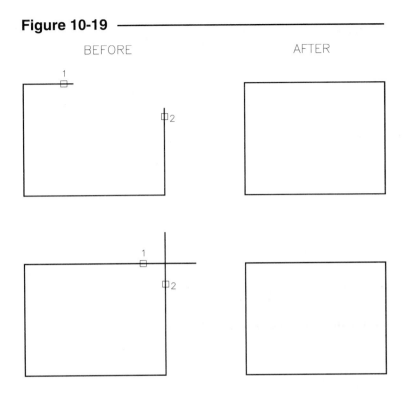

If parallel objects are selected, the *Fillet* is automatically created to the correct radius (Fig. 10-20). Therefore, parallel objects can be filleted at any time <u>without specifying a radius value</u>.

Figure 10-20

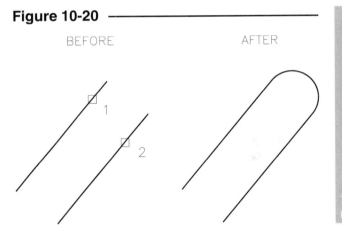

Polyline

Fillets can be created on *Polylines* in the same manner as two *Line* objects. Use *Fillet* as you normally would for *Lines*. However, if you want a fillet equal to the specified radius to be added to <u>each vertex</u> of the *Pline* (except the endpoint vertices), use the *Polyline* option; then select anywhere on the *Pline* (Fig. 10-21).

Command: `fillet`
Polyline/Distances/<Select first object>: `P` (Invokes prompts for the *Polyline* option.)
Select 2D polyline: `PICK` (Select the desired polyline.)

Figure 10-21

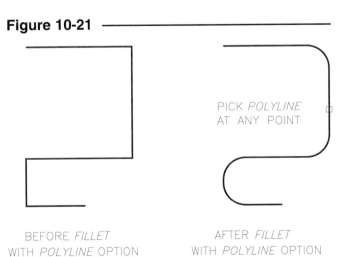

Closed Plines created by the *Close* option of *Pedit* react differently with *Fillet* than *Plines* connected by PICKing matching endpoints. Figure 10-22 illustrates the effect of *Fillet Polyline* on a *Closed Pline* and on a connected *Pline.*

Figure 10-22

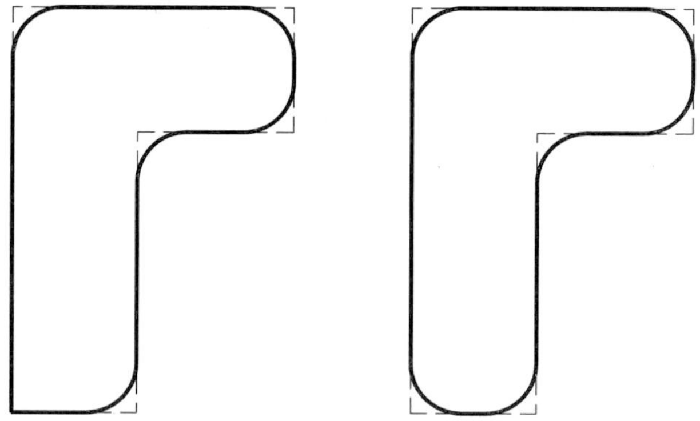

PLINE CONNECTED WITH PICK
AFTER FILLET

CLOSED PLINE
AFTER FILLET

Trim/Notrim

In the previous figures, the objects are shown having been filleted in the *Trim* mode—that is, with automatically trimmed or extended objects to meet the end of the new fillet radius. The *Notrim* mode creates the fillet without any extending or trimming of the involved objects (Fig. 10-23). Note that the command prompt indicates the current mode (as well as the current radius) when *Fillet* is invoked.

Figure 10-23

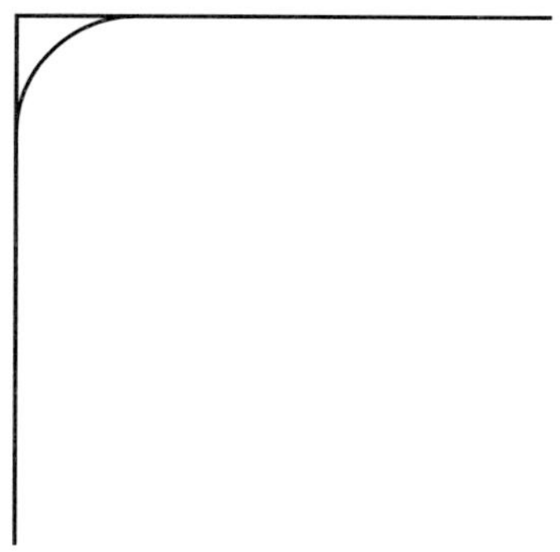

```
Command: fillet
(NOTRIM mode) Current fillet radius=2.000
Polyline/Radius/Trim/<Select first object>: T (Invokes Trim/No trim
option.)
Trim/No trim <Trim>: N (Sets Fillet to No trim.)
Polyline/Radius/Trim/<Select first object>: PICK
Select second object: PICK
Command:
```

FILLET   NOTRIM

The *Notrim* mode is helpful in situations similar to that in Figure 10-24. In this case, if the intersecting *Lines* were trimmed when the first fillet was created, the longer *Line* (highlighted) would have to be redrawn in order to create the second fillet.

Changing the *Trim/Notrim* option sets the *TRIMMODE* system variable to 0 (*Notrim*) or 1 (*Trim*). This variable controls trimming for both *Fillet* and *Chamfer,* so if you set *Fillet* to *Notrim, Chamfer* is also set to *Notrim.*

Figure 10-24

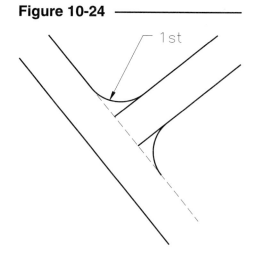

FILLET   NOTRIM

CHAMFER

	Pull-down Menu	COMMAND (TYPE)	ALIAS (TYPE)	Short-cut	Screen (side) Menu	Tablet Menu
	Modify Chamfer	*CHAMFER*	*CHA*	...	*MODIFY2 Chamfer*	*W,18*

Chamfering is a manufacturing process used to replace a sharp corner with an angled surface. In AutoCAD, *Chamfer* is commonly used to change the intersection of two *Lines* or *Plines* by adding an angled line. The *Chamfer* command is similar to *Fillet*, but rather than rounding with a radius or "fillet," an angled line is automatically drawn at the distances (from the existing corner) that you specify.

Chamfers can be created by <u>two methods</u>: *Distance* (specify <u>two distances</u>) or *Angle* (specify a <u>distance and an angle</u>). The current method and the previously specified values are displayed at the Command: prompt along with the options:

```
Command: chamfer
(TRIM mode) Current chamfer Dist1 = 0.0000, Dist2 = 0.0000
Polyline/Distance/Angle/Trim/Method/<Select first line>:
```

Method
Use this option to indicate which of the two methods you want to use: *Distance* (specify 2 distances) or *Angle* (specify a distance and an angle).

Distance
The *Distance* option is used to specify the two values applied <u>when the *Distance Method* is used</u> to create the chamfer. The values indicate the distances from the corner (intersection of two lines) to each chamfer endpoint (Fig. 10-25). Use the *Chamfer* command once to specify *Distances* and again to draw the chamfer.

Figure 10-25

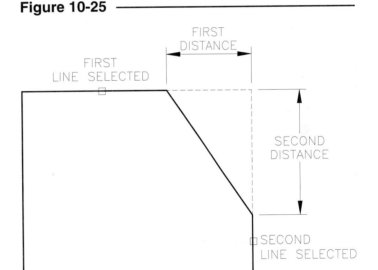

```
Command: chamfer
(TRIM mode) Current chamfer Dist1 = 0.0000, Dist2 = 0.0000
Polyline/Distance/Angle/Trim/Method/<Select first line>: d (Indicates the distance option.)
Enter first chamfer distance <0.0000>: (value) or PICK (Enter a value for the distance from the existing
corner to the endpoint of chamfer on first line or select two points to interactively specify the distance.)
Enter second chamfer distance <value of first distance>: Enter, (value) or PICK (Press Enter to use
same distance as first distance, enter a value, or PICK as before.)
Command:
```

Chamfer with distances of 0 reacts like *Fillet* with a radius of 0; that is, overlapping corners can be automatically trimmed and non-intersecting corners can be automatically extended (using the *Trim* mode).

Angle
The *Angle* option allows you to specify the values that are used for the *Angle Method*. The values specify a distance along the first line and an angle from the first line (Fig. 10-26).

Figure 10-26

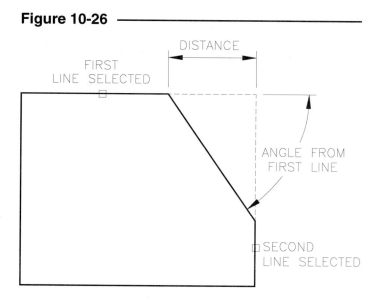

```
Polyline/Distance/Angle/Trim/Method/<Select first line>: a
Enter chamfer length on the first line <1.0000>: (value)
Enter chamfer angle from the first line <60.0000>: (value)
```

Trim/Notrim
The two lines selected for chamfering do not have to intersect but can overlap or not connect. The *Trim* setting automatically trims or extends the lines selected for the chamfer (Fig. 10-27), while the *Notrim* setting adds the chamfer without changing the length of the selected lines. These two options are the same as those in the *Fillet* command (see *Fillet*). Changing these options sets the *TRIMMODE* system variable to 0 (*Notrim*) or 1 (*Trim*). The variable controls trimming for both *Fillet* and *Chamfer*.

Figure 10-27

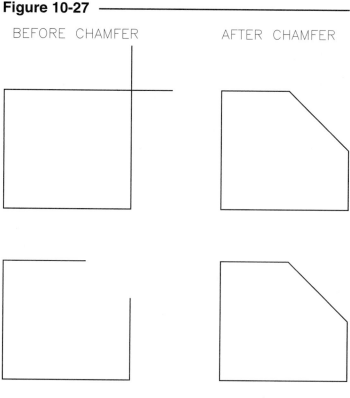

BEFORE CHAMFER   AFTER CHAMFER

CHAMFER WILL
TRIM OR EXTEND

Polyline

The *Polyline* option of *Chamfer* creates chamfers on all vertices of a *Pline*. All vertices of the *Pline* are chamfered with the supplied distances (Fig. 10-28). The first end of the *Pline* that was <u>drawn</u> takes the first distance. Use *Chamfer* without this option if you want to chamfer only one corner of a *Pline*. This is similar to the same option of *Fillet* (see *Fillet*).

Figure 10-28

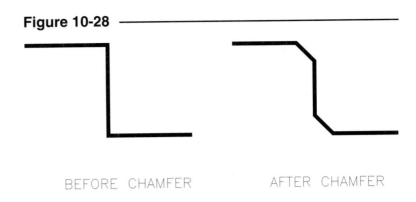

BEFORE CHAMFER   AFTER CHAMFER

A POLYLINE CHAMFER EXECUTES
A CHAMFER AT EACH VERTEX

CHAPTER EXERCISES

1. *Copy*

   Begin a *New* drawing. Set the *Limits* to **24,18**. Set *SNAP* to **.5** and *GRID* to **1**. Create the sheet metal part composed of four *Lines* and one *Circle* as shown in Figure 10-29. The lower-left corner of the part is at coordinate **2,3**.

   Use *Copy* to create two copies of the rectangle and hole in a side-by-side fashion as shown in Figure 10-30. Allow 2 units between the sheet metal layouts. *SaveAs* **PLATES**.

   A. Use *Copy* to create the additional 3 holes equally spaced near the other corners of the part as shown in Figure 10-30 A. *Save* the drawing.

   B. Use *Copy Multiple* to create the hole configuration as shown in Figure 10-30 B. *Save*.

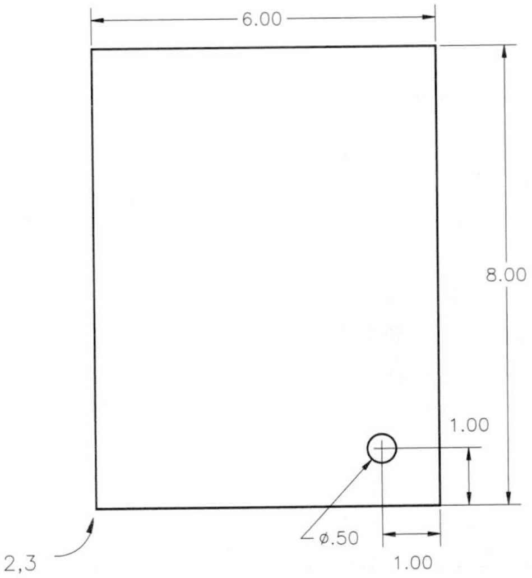

   C. Use *Copy* to create the hole placements as shown in C. Each hole <u>center</u> is **2** units at **125** degrees from the previous one (use relative polar coordinates). *Save* the drawing.

Figure 10-29

6.00

8.00

1.00

2,3

ø.50   1.00

Figure 10-30

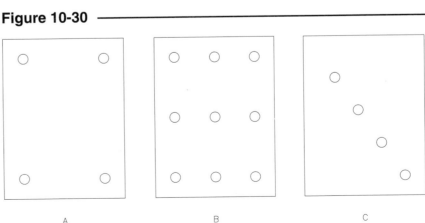

A   B   C

2. *Mirror*

A manufacturing cell is displayed in
Figure 10-31. The view is from above,
showing a robot <u>centered</u> in a work
station. The production line requires 4
cells. Begin by starting a *New* drawing,
setting *Units* to *Engineering* and *Limits* to
40' x 30'. It may be helpful to set *SNAP* to
6". Draw one cell to the dimensions indicated. Begin at the indicated coordinates
of the lower-left corner of the cell.
SaveAs **MANFCELL.**

Figure 10-31 ─────────────

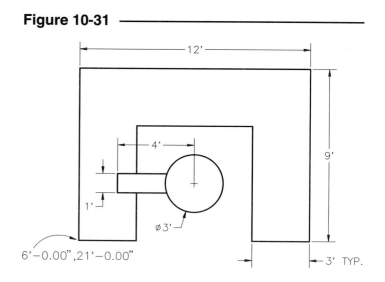

Use *Mirror* to create the other three manufacturing cells as shown in Figure 10-32. Ensure that
there is sufficient space between the cells as indicated. Draw the two horizontal *Lines* representing the walkway as shown. *Save.*

Figure 10-32 ─────────────────────────

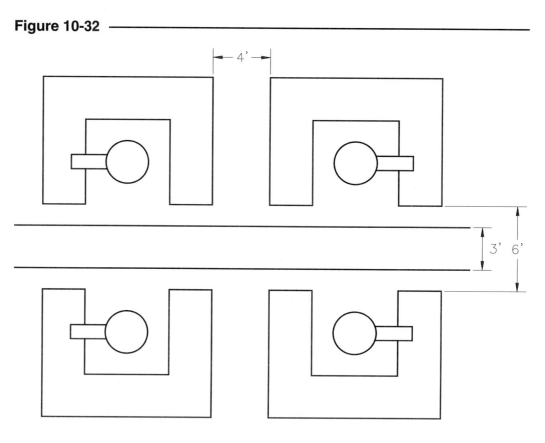

3. *Offset*

Create the electrical schematic as shown in Figure 10-33. Draw *Lines* and *Plines*, then *Offset* as needed. Use *Point* objects with an appropriate (circular) *Point Style* at each of the connections as shown. Check the *Set Size to Absolute Units* radio button (in the *Point Style* dialog box) and find an appropriate size for your drawing. (Remember to use *Regen* after changing *Point Style* or size.) Because this is a schematic, you can create the symbols by approximating the dimensions. Omit the text. *Save* the drawing as **SCHEMATIC1**. Create a *Plot* and check *Scaled to Fit*.

Figure 10-33 —

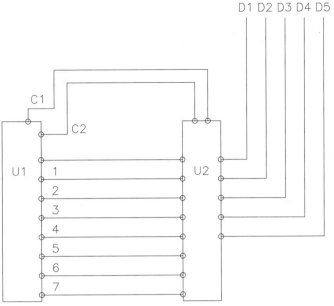

4. *Array, Polar*

Begin a *New* drawing. Select *Start from Scratch*, *English* defaults. Create the starting geometry for a Flange Plate as shown in Figure 10-34. *SaveAs* **ARRAY**.

Figure 10-34 —

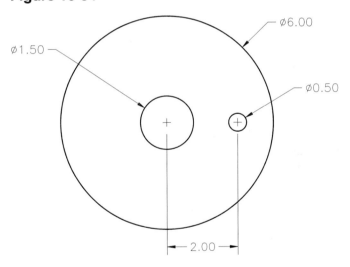

A. Create the *Polar Array* as shown in Figure 10-35 A. *SaveAs* **ARRAY1**.

B. *Open* **ARRAY**. Create the *Polar Array* as shown in Figure 10-35 B. *SaveAs* **ARRAY2**. (HINT: Use a negative angle to generate the *Array* in a clockwise direction.)

Figure 10-35 —

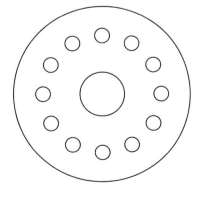
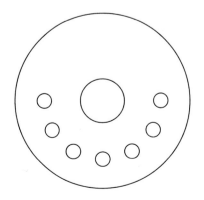

A   B

5. *Array, Rectangular*

Begin a *New* drawing. Select *Start from Scratch, English* defaults. Use *Save* and assign the name **LIBRARY DESKS**. Create the *Array* of study carrels (desks) for the library as shown in Figure 10-36. The room size is 36' x 27' (set *Units* and *Limits* accordingly). Each carrel is **30" x 42"**. Design your own chair. Draw the first carrel (highlighted) at the indicated coordinates. Create the *Rectangular Array* so that the carrels touch side to side and allow a 6' aisle for walking between carrels (not including chairs).

Figure 10-36

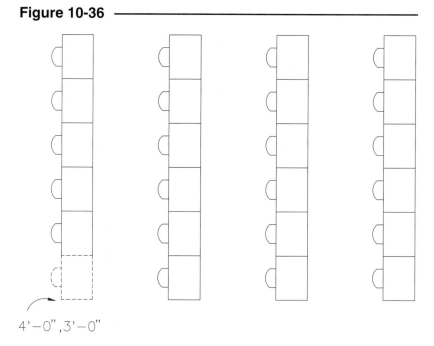

4'−0",3'−0"

6. *Array, Rectangular*

Create the bolt head with one thread as shown in Figure 10-37 A. Create the small line segment at the end (0.40 in length) as a *Pline* with a *Width* of .02. *Array* the first thread (both crest and root lines, as indicated in B) to create the schematic thread representation. There is **1** row and **10** columns with **.2** units between each. Add the *Lines* around the outside to complete the fastener. *Erase* the *Pline* at the small end of the fastener (Fig. 10-37 B) and replace it with a *Line*. *Save* as **BOLT**.

Figure 10-37

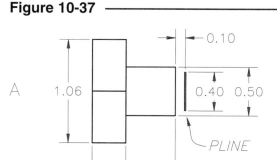

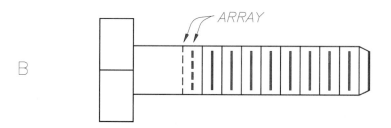

7. *Fillet*

Create the "T" Plate shown in Figure 10-38. Use *Fillet* to create all the fillets and rounds as the last step. When finished, *Save* the drawing as **T-PLATE** and make a plot.

Figure 10-38

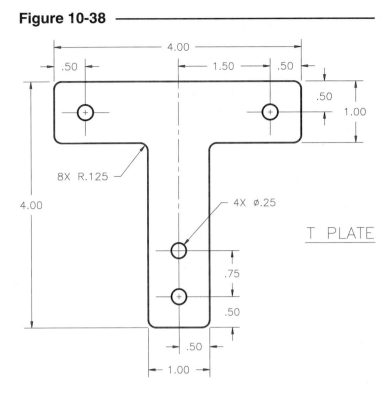

8. *Copy, Fillet*

Create the Gasket shown in Figure 10-39. Use the **Start from Scratch**, **English** settings when you begin. Include center-lines in your drawing. *Save* the drawing as **GASKETB**.

Figure 10-39

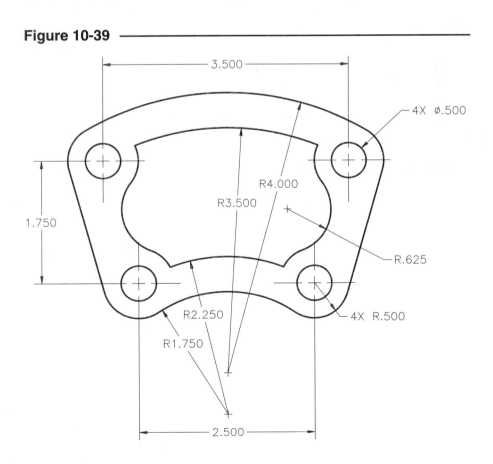

9. *Array*

   Create the perforated plate as shown Figure 10-40. *Save* the drawing as **PERFPLAT**. (HINT: For the *Rectangular Array* in the center, create 100 holes, then *Erase* four holes, one at each corner for a total of 96. The two circular hole patterns contain a total of 40 holes each.)

Figure 10-40

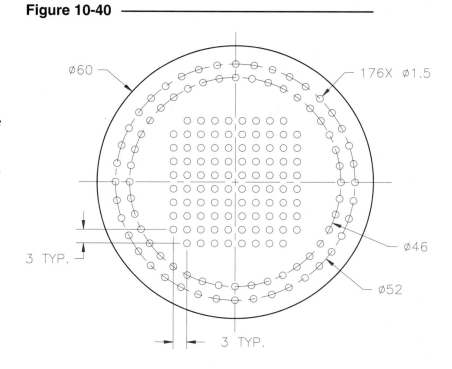

10. *Chamfer*

    Begin a *New* drawing and select **Start from Scratch, English** defaults. Set the **Limits** to **279,216**. Next, type *Zoom* and use the *All* option. Set *Snap* to **2** and *Grid* to **10**. Create the Catch Bracket shown in Figure 10-41. Draw the shape with all vertical and horizontal *Lines*, then use *Chamfer* to create the six chamfers. *Save* the drawing as **CBRACKET** and create a plot.

Figure 10-41

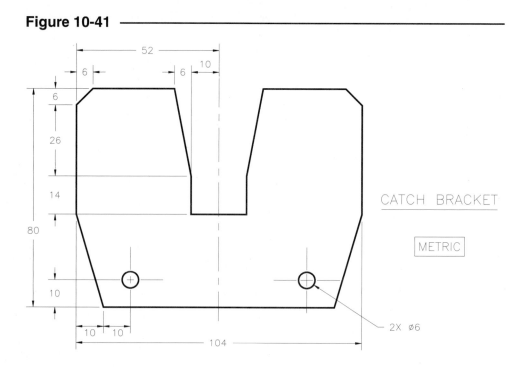

CATCH BRACKET

METRIC

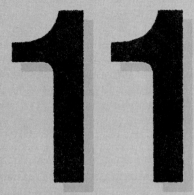

VIEWING
COMMANDS

Chapter Objectives

After completing this chapter you should:

1. understand the relationship between the drawing objects and the display of those objects;

2. be able to use all of the *Zoom* options to view areas of a drawing;

3. be able to *Pan* the display about your screen;

4. be able to *save* and *restore Views*;

5. know how to use *Viewres* to change the display resolution for curved shapes;

6. be able to use *Aerial View* to *Pan* and *Zoom*.

CONCEPTS

The accepted CAD practice is to draw full size using actual units. Since the drawing is a virtual dimensional replica of the actual object, a drawing could represent a vast area (several hundred feet or even miles) or a small area (only millimeters). The drawing is created full size with the actual units, but it can be displayed at any size on the screen. Consider also that CAD systems provide for a very high degree of dimensional precision, which permits the generation of drawings with great detail and accuracy.

Combining those two CAD capabilities (great precision and drawings representing various areas), a method is needed to view different and detailed segments of the overall drawing area. In AutoCAD the commands that facilitate viewing different areas of a drawing are *Zoom, Pan,* and *View.*

The Viewing commands are found in the *View* pull-down menu (Fig. 11-1).

Figure 11-1 ─────────────────────

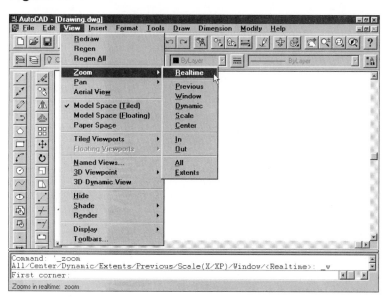

The Standard toolbar contains a group of tools (icon buttons) for the Viewing commands located near the right end of the toolbar (Fig. 11-2). The *Realtime* options of *Pan* and *Zoom, Zoom Previous,* and *Aerial View* each has an icon permanently displayed on the toolbar, whereas the other *Zoom* options are located on flyouts.

Like other commands, the Viewing commands can also be invoked by typing the command or command alias, using the screen menu (*VIEW1*), or using the digitizing tablet menu (if available).

The commands discussed in this chapter are:

    Zoom, Pan, View, and *Dsviewer* (Aerial View)

Figure 11-2 ─────────────

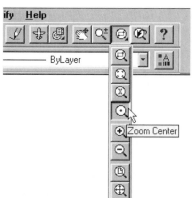

AutoCAD Release 13 Versions C3 and C4 introduced "real-time" panning and zooming (Version C3 for AutoCAD for Windows 3.1 and 3.11, and Version C4 for Windows 95 and Windows NT). Two new commands were introduced: *Rtpan* and *Rtzoom.* However, because the two commands were added after the first version of Release 13, they were not included in the menus. Now with AutoCAD Release 14, the *Realtime* options are the default options for both the *Pan* and *Zoom* commands. *Realtime* options of *Pan* and *Zoom* allow you to <u>interactively</u> change the drawing display by moving the cursor in the drawing area.

COMMANDS

ZOOM

Pull-down Menu	COMMAND (TYPE)	ALIAS (TYPE)	Short-cut	Screen (side) Menu	Tablet Menu
View *Zoom >*	ZOOM	Z	...	*VIEW 1* *Zoom*	K,11-M,11 or J,2-K,5

To <u>Zoom</u> in means to magnify a small segment of a drawing and to <u>Zoom</u> out means to display a larger area of the drawing. *Zooming* does <u>not</u> change the <u>size</u> of the drawing; *Zooming* changes only the area that is <u>displayed</u> on the screen. All objects in the drawing have the same dimensions before and after *Zooming*. Only your display of the objects changes.

The *Zoom* options described next can be typed or selected by any method. Unlike most commands, AutoCAD provides a tool (icon button) for each <u>option</u> of the *Zoom* command. If you type *Zoom*, the options can be invoked by typing the first letter of the desired option or pressing Enter for the *Realtime* option. The command prompt appears as follows.

```
Command: zoom
All/Center/Dynamic/Extents/Previous/Scale(X/XP)/Window/<Realtime>:
```

Realtime (RTZOOM)

Realtime is the default option of *Zoom*. If you <u>type</u> *Zoom*, just press Enter to activate the *Realtime* option. Alternately, you can type *Rtzoom* to invoke this option directly.

Figure 11-3

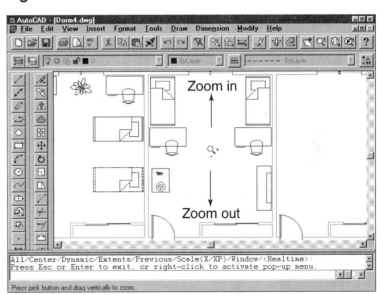

With the *Realtime* option you can interactively zoom in and out with vertical cursor motion. When you activate the *Realtime* option, the cursor changes to a magnifying glass with a plus (+) and a minus (-) symbol. Move the cursor to any location on the drawing area and hold down the PICK (left) mouse button. Move the cursor up to zoom in and down to zoom out (Fig. 11-3). Horizontal movement has no effect. You can zoom in or out repetitively. Pressing Esc or Enter exits *Realtime Zoom* and returns to the Command: prompt.

The current drawing window is used to determine the zooming factor. Moving the cursor half the window height (from the center to the top or bottom) zooms in or out to a zoom factor of 100%. Starting at the bottom or top allows zooming in or out (respectively) at a factor of 200%.

If you have zoomed in or out repeatedly, you can reach a zoom limit. In this case the plus (+) or minus (-) symbol disappears when you press the left mouse button. To zoom further with *Rtzoom*, first type *Regen*.

R14

Pressing the right mouse button (or button #2 on digitizing pucks) produces a small cursor menu with other viewing options (Fig. 11-4). This same menu and its options can also be displayed by right-clicking during *Realtime Pan*. The options of the cursor menu are described below.

Figure 11-4

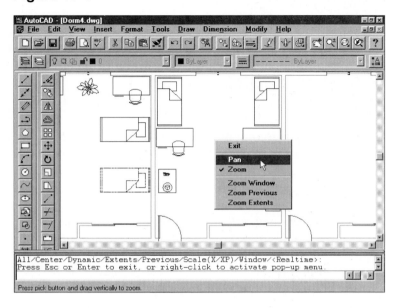

Exit
Select *Exit* to exit *Realtime Pan* or *Zoom* and return to the Command: prompt. The Escape or Enter key can be used to accomplish the same action.

Pan
This selection switches to the *Realtime* option of *Pan*. A check mark appears here if you are currently using *Realtime Pan*. See *Pan* next in this chapter.

Zoom
A check mark appears here if you are currently using *Realtime Zoom*. If you are using the *Realtime* option of *Pan*, check this option to switch to *Realtime Zoom*.

Zoom Window
Select this option if you want to display an area to zoom in to by specifying a *Window*. This feature operates differently but accomplishes the same action as the *Window* option of *Zoom*. With this option, window selection is made with one PICK. In other words, PICK for the first corner, hold down the left button and drag, then release to establish the other corner. With the *Window* option of the *Zoom* command, PICK once for each of two corners. See *Window* next page.

Zoom Previous
Selecting this option automatically displays the area of the drawing that appeared on the screen immediately before using the *Realtime* option of *Zoom* or *Pan*. Using this option successively has no effect on the display. This feature is different from the *Previous* option of the *Zoom* command, in which 10 successive previous views can be displayed.

Zoom Extents
Use this option to display the entire drawing in its largest possible form in the Drawing Editor. This option is identical to the *Extents* option of the *Zoom* command. See *Extents* next page.

Window

 To *Zoom* with a *Window* is to draw a rectangular window around the desired viewing area. You PICK a first and a second corner (diagonally) to form the rectangle. The windowed area is magnified to fill the screen (Fig. 11-5). It is suggested that you draw the window with a 4 x 3 (approximate) proportion to match the screen proportion. If you type *Zoom* or *Z*, *Window* is an <u>automatic</u> option so you can begin selecting the first corner of the window after issuing the *Zoom* command <u>without</u> indicating the *Window* option as a separate step.

Figure 11-5 ──────────────

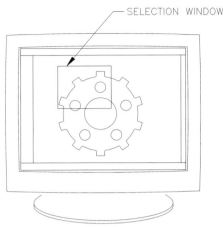
SELECTION WINDOW

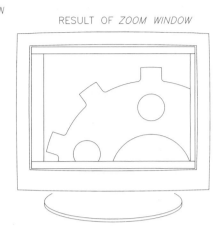
RESULT OF *ZOOM WINDOW*

All

 This option displays <u>all of the objects</u> in the drawing <u>and all of the Limits</u>. In Figure 11-6, notice the effects of *Zoom All*, based on the drawing objects and the drawing *Limits*.

Figure 11-6 ──────────────

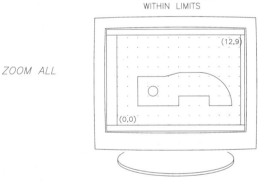
ALL OBJECTS DRAWN WITHIN LIMITS
(12,9)
(0,0)
ZOOM ALL

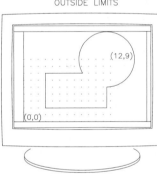
DRAWING OBJECTS OUTSIDE LIMITS
(12,9)
(0,0)

Extents

 This option results in the largest possible display of <u>all of the objects</u>, <u>disregarding</u> the *Limits*. (*Zoom All* includes the *Limits*.) See Figure 11-6.

New in Release 14, invoking *Zoom Extents* causes AutoCAD to execute two steps: (1) a true (pre-

ZOOM EXTENTS
(12,9)
(0,0)
(12,9)
(0,0)

Release 13) *Zoom Extents* is performed which causes the geometry to be maximized so it is "pressed" against the edge of the Drawing Editor window, and (2) a *Zoom Scale .95X* is performed which causes the geometry to be scaled down slightly to fit just within the window (see the *Scale* option). Previous to Release 14, using *Zoom Extents* alone made it difficult to select geometry against the edge of the window, so *Zoom Scale .95X* had to be invoked by the user to make the geometry more accessible for selection.

R14

Scale (X/XP)

This option allows you to enter a scale factor for the desired display. The value that is entered can be relative to the full view (*Limits*) or to the current display (Fig. 11-7). A value of **1, 2**, or **.5** causes a display that is 1, 2, or .5 times the size of the *Limits*, centered on the current display. A value of **1X, 2X,** or **.5X** yields a display 1, 2, or .5 times the size of the <u>current display</u>. If you are using paper space and viewing model space, a value of **1XP, 2XP,** or **.5XP** yields a model space display scaled 1, 2, or .5 times paper space units.

If you type *Zoom* or *Z*, you can enter a scale factor at the *Zoom* command prompt without having to type the letter *S*.

Figure 11-7

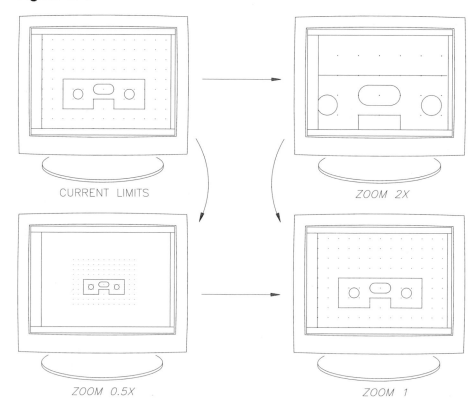

CURRENT LIMITS

ZOOM 2X

ZOOM 0.5X

ZOOM 1

In

Zoom In magnifies the current display by a factor of 2X (2 times the current display). Using this option is the same as entering a *Zoom Scale* of 2X.

Out

Zoom Out makes the current display smaller by a factor of .5X (.5 times the current display). Using this option is the same as entering a *Zoom Scale* of .5X.

Center

First, specify a location as the center of the zoomed area; then specify either a *Magnification factor* (see *Scale X/XP*), a *Height* value for the resulting display, or PICK two points forming a vertical to indicate the height for the resulting display (Fig. 11-8).

Previous

Selecting this option automatically changes to the previous display. AutoCAD saves the previous ten displays changed by *Zoom*, *Pan*, and *View*. You can successively change back through the previous ten displays with this option.

Figure 11-8

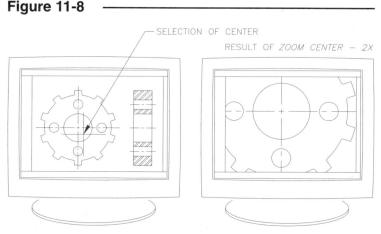

SELECTION OF CENTER

RESULT OF *ZOOM CENTER* – 2X

Dynamic

With this option you can change the display from one windowed area in a drawing to another without using *Zoom All* (to see the entire drawing) as an intermediate step. *Zoom Dynamic* causes the screen to display the drawing *Extents* bounded by a box. The current view or window is bounded by a box in a broken-line pattern. The viewbox is the box with an "X" in the center which can be moved to the desired location (Fig. 11-9). The desired location is selected by pressing **Enter** (not the PICK button as you might expect).

Pressing **PICK** allows you to resize the viewbox (displaying an arrow instead of the "X") to the desired size. Move the mouse or puck left and right to increase and decrease the window size. Press **PICK** again to set the size and make the "X" reappear.

Figure 11-9

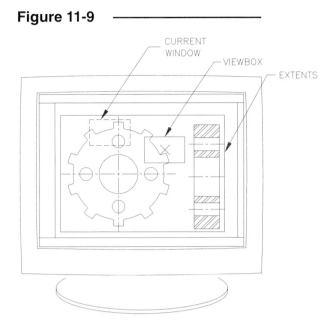

When changing the display of a drawing from one windowed area to another, *Zoom Dynamic* can be a fast method to use because (1) a *Zoom All* is not required as an intermediate step and (2) the viewbox can be moved <u>while</u> AutoCAD is "redrawing" the display.

Zoom Is Transparent

Zoom is a <u>transparent</u> command, meaning it can be invoked while another command is in operation. You can, for example, begin the *Line* command and at the "To point:" prompt *'Zoom* with a *Window* to better display the area for selecting the endpoint. This transparent feature is automatically entered if *Zoom* is invoked by the screen, pull-down, or tablet menus or toolbar icons, but if typed it must be prefixed by the **'** (apostrophe) symbol, e.g., "To point: *'Zoom.*" If *Zoom* has been invoked transparently, the >> symbols appear at the command prompt before the listed options as follows:

```
LINE from point: PICK
To point: 'zoom
>>All/Center/Dynamic/Extents/Previous/Scale(X/XP)/Window/<Realtime>:
```

PAN

Pull-down Menu	COMMAND (TYPE)	ALIAS (TYPE)	Short-cut	Screen (side) Menu	Tablet Menu
View *Pan*	*PAN or* *-PAN*	*P or -P*	...	*VIEW 1* *Pan*	*N,11-P,11*

Using the *Pan* command by typing or selecting from the tool, screen, or digitizing tablet menu produces the *Realtime* version of *Pan*. The *Point* option is available from the pull-down menu or by typing "*-Pan.*" The other "automatic" *Pan* options (*Left, Right, Up, Down*) can only be selected from the pull-down menu.

The *Pan* command is useful if you want to move (pan) the display area slightly without changing the <u>size</u> of the current view window. With *Pan*, you "drag" the drawing across the screen to display an area outside of the current view in the Drawing Editor.

R14

Realtime (RTPAN)

Realtime is the default option of *Pan*. It is invoked by selecting *Pan* by any method, including typing *Pan* or *Rtpan*. *Realtime Pan* allows you to interactively pan by "pulling" the drawing across the screen with the cursor motion. After activating the command, the cursor changes to a hand cursor. Move the hand to any location on the drawing, then hold the PICK (left) mouse button down and drag the drawing around on the screen to achieve the desired view (Fig. 11-10). When you release the mouse button, panning is discontinued. You can move the hand to another location and pan again without exiting the command.

Figure 11-10

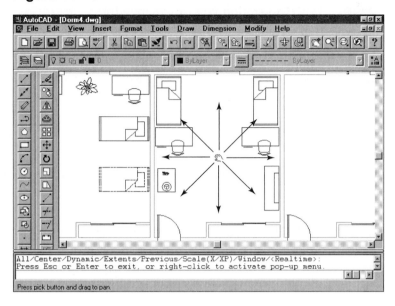

You must press Esc or Enter or use *Exit* from the pop-up cursor menu to exit *Realtime Pan* and return to the Command: prompt. The following command prompt appears during *Realtime Pan*.

> Command: **pan**
> Press Esc or Enter to exit, or right-click to activate pop-up menu.

If you press the right mouse button (#2 button), a small cursor menu pops up. This is the same menu that appears during *Realtime Zoom* (see Fig. 11-4). This menu provides options for you to *Exit, Zoom* or *Pan* (realtime), *Zoom Window, Zoom Previous,* or *Zoom Extents.* See *Zoom* earlier in this chapter.

Point

The *Point* option is available only through the pull-down menu or by typing "-*Pan*" (some commands can be typed with a - [hyphen] prefix to invoke the command line version of the command without dialog boxes, etc.). This is the same version of *Pan* that was used in the first shipping of Release 13 and previous versions of AutoCAD. Using this option produces the following command prompt.

> Command: **-pan**
> Displacement: **PICK** or (**value**)
> Second point: **PICK** or (**value**)
> Command:

Figure 11-11

The "Displacement:" can be thought of as a "handle," or point to move <u>from</u>, and the "Second point:" as the new location to move <u>to</u> (Fig. 11-11). You can PICK each of these points with the cursor.

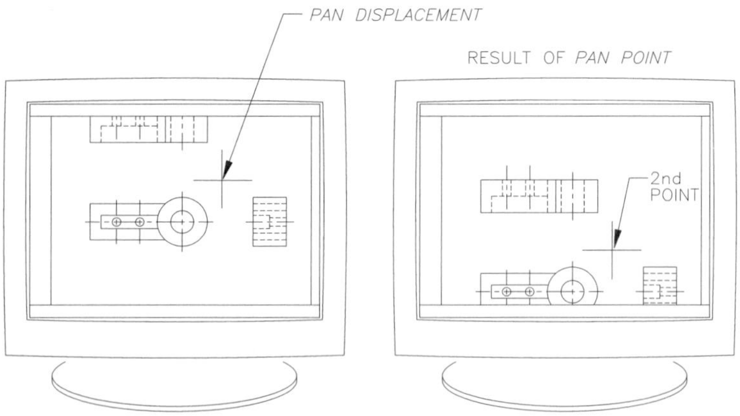

PAN DISPLACEMENT

RESULT OF *PAN POINT*

2nd POINT

You can also enter coordinate values rather than interactively PICKing points with the cursor. The command syntax is as follows:

Command: **pan**
Pan Displacement: **0,-2**
Second point: **Enter**
Command:

Entering coordinate values allows you to *Pan* to a location outside of the current display. If you use the interactive method (PICK points), you can *Pan* only within the range of whatever is visible on the screen.

The following *Pan* options are available from the *View* pull-down menu only.

Left
Automatically pans to the left, equal to about 1/8 of the current drawing area width.

Right
Pans to the right—similar to but opposite of *Left*.

Up
Automatically pans up, equal to about 1/8 of the current drawing area height.

Down
Pans down—similar to but opposite of *Up*.

Pan Is Transparent
Pan, like *Zoom*, can be used as a transparent command. The transparent feature is automatically entered if *Pan* is invoked by the screen, pull-down, or tablet menus or icons. However, if you type *Pan* during another command operation, it must be prefixed by the ' (apostrophe) symbol.

Command: **LINE**
From point: **PICK**
To point: **'pan**
>>Press Esc or Enter to exit, or right-click to activate pop-up menu. **Enter**
Resuming LINE command.
To point:

Scroll Bars

Figure 11-12 ——

You can also use the horizontal and vertical scroll bars directly below and to the right of the graphics area to pan the drawing (Fig. 11-12). These scroll bars pan the drawing in one of three ways: (1) click on the arrows at the ends of the scroll bar, (2) move the thumb wheel, or (3) click inside the scroll bar. The scroll bars can be turned off (removed from the screen) by using the *Display* tab in the *Preferences* dialog box from the *Tools* pull-down menu.

Thumb wheel

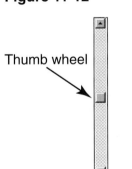

VIEW and
DDVIEW

Pull-down Menu	COMMAND (TYPE)	ALIAS (TYPE)	Short-cut	Screen (side) Menu	Tablet Menu
View *Named Views...*	*VIEW or* *DDVIEW*	*V or -V*	...	*VIEW 1* *Ddview*	*M,5*

The *View* command provides options for you to *Save* the current display window and *Restore* it at a later time. For typical applications, you would first *Zoom* in to achieve a desired display, then use *View, Save* to save that display under an assigned name. Later in the drawing session, the named *View* can be *Restored* any number of times. This method is preferred to continually *Zooming* in and out to display the same areas repeatedly. *Restoring* a named *View* requires no regeneration time. A *View* can be sent to the plotter as separate display as long as the view is named and saved. The options of the *View* command are listed below:

?   Displays the list of named views.

Delete   Deletes one or more saved views that you enter.

Restore   Displays the named view you request.

Save   Saves the current display as a *View* with a name you assign.

Window   Allows you to specify a window in the current display and save it as a named *View*.

To *Save* a *View*, first use *Zoom* or *Pan* to achieve the desired display; then invoke the *View* command and the *Save* option as follows:

    Command: **view**
    ?/Delete/Restore/Save/Window: **save**
    View name to save: **name** (Assign a descriptive name for the view.)

For assurance, check to see if the new *View* has been saved by displaying the list by using the *?* option. The *?* option causes a text screen to appear with the list of named (saved) views. Use *Flipscreen* to return to the Drawing Editor after reading the list.

To *Restore* a named *View*, assuming a different area of the drawing is currently displayed, use the *Restore* option and supply the name of the desired *View*.

View can also be used transparently, but only if typed at the keyboard. Remember to preface the command with the ' (apostrophe) symbol, i.e., '*View*.

A feature of saving *Views* is that a named *View* can be plotted by toggling the *View* option in the *Plot Configuration* dialog box. Plotting is discussed in detail in Chapter 14.

Figure 11-13 ——————

Typing the *View* command is generally quick and easy. Selecting *Ddview* from the pull-down or tablet menu invokes the *View Control* dialog box (Fig. 11-13), which is helpful but requires a number of steps to operate. A word of caution when selecting from the list of named views to restore: make sure that you select (1) the desired view <u>and</u> (2) the *Restore* tile, <u>then</u> (3) the *OK* tile.

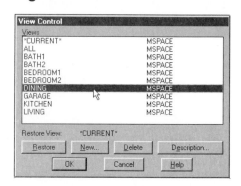

The *View Control* dialog box provides the same options that the *View* command offers. The *New* option causes the *Define New View* dialog box to pop up, giving access to the *Save* option (Fig. 11-14).

Figure 11-14

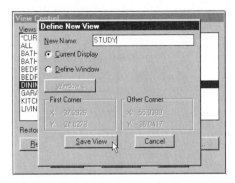

DSVIEWER

Pull-down Menu	COMMAND (TYPE)	ALIAS (TYPE)	Short-cut	Screen (side) Menu	Tablet Menu
View *Aerial View*	*DSVIEWER*	*AV*	...	VIEW 1 *Dsviewer*	*K,2*

Aerial View is a tool for navigating in a drawing. It can be used to *Zoom* and *Pan* while in another command. Invoking the *Dsviewer* command displays the window shown in Figure 11-15. The options and operations are described next.

Figure 11-15

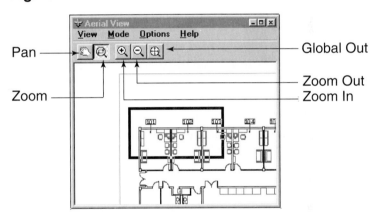

Pan

Zoom

Global Out

Zoom Out
Zoom In

Zoom

This sets the viewer in *Zoom Window* mode. PICKing two corners <u>inside the viewer</u> causes that windowed area to be displayed <u>in the main graphics area</u> (Fig. 11-16).

Figure 11-16

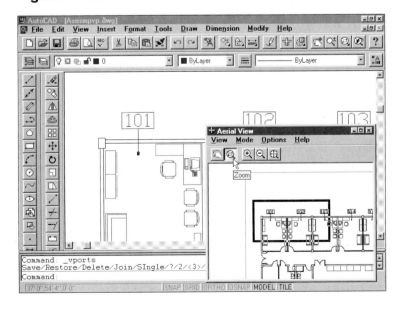

Pan

Selecting this icon or choosing
Pan from the *View* menu in
Aerial View sets the viewbox in
the *Pan* mode. If you move the
viewbox to any location <u>inside
the viewer</u>, the selected area of
the drawing is displayed in the
main graphics area (Fig. 11-17).

Figure 11-17

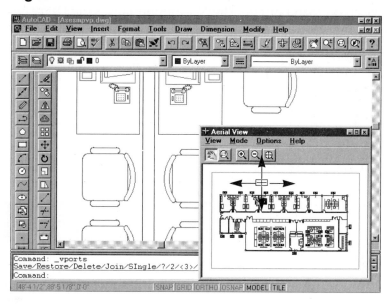

There are two ways to operate the *Pan* option: (1) move the viewbox to a new location and PICK (single-click) to achieve a <u>static display</u> of the viewbox objects in the main graphics area, or (2) hold down the left mouse button and move the viewbox around in the viewer to <u>dynamically display</u> objects in the main graphics area.

This second method is the most unique and useful feature of *Aerial View*. It allows you to dynamically move the viewbox around in viewer while the magnified area is displayed <u>instantaneously and dynamically</u> inside the viewer window. This feature is especially helpful for locating small details in the drawing, like reading text. You can change the size of the viewbox with the *Zoom* option.

Zoom In

PICKing this option causes the <u>display in the viewer</u> to be magnified 2X. This does <u>not</u> affect the display in the main graphics area.

Zoom Out

This option causes the <u>display in the viewer</u> to be reduced to .5X. This does <u>not</u> affect the display in the main graphics area.

Global

Selecting this option will resize the <u>drawing display in the viewer</u> to its maximum size.

VIEWRES

Pull-down Menu	COMMAND (TYPE)	ALIAS (TYPE)	Short-cut	Screen (side) Menu	Tablet Menu
Tools *Preferences...* *Performance*	VIEWRES	...	...	...	...

Viewres controls the resolution of curved shapes for the <u>screen display</u> only. Its purpose is to speed regeneration time by displaying curved shapes as linear approximations of curves; that is, a curved shape (such as an *Arc, Circle,* or *Ellipse*) appears as several short, straight line segments (Fig. 11-18). The drawing database and the plotted drawing, however, always define a true curve. The range of *Viewres* is from 1–20,000 with 100 being the default. The higher the value (called "circle zoom percentage"), the more accurate the display of curves and the slower the regeneration time. The lower the value, the faster is the regeneration time but the more "jagged" the curves. A value of 500 to 1,000 is suggested for most applications. The command syntax is shown here.

Figure 11-18

VIEWRES 10   VIEWRES 1000

```
Command: Viewres
Do you want fast zooms? <Y> Enter
Enter circle zoom percent (1–20000) <100>: 1000
Command:
```

If you answer *No* to *Do you want fast zooms?*, AutoCAD causes a regeneration after every *Pan, Zoom,* or *View* command. Otherwise, regenerations occur only when *Zoom*ing extremely far in or out.

UCSICON

Pull-down Menu	COMMAND (TYPE)	ALIAS (TYPE)	Short-cut	Screen (side) Menu	Tablet Menu
View *Display >* *UCS Icon >*	UCSICON	...	...	*View 2* *UCSicon*	L,2

The icon that appears in the lower-left corner of the AutoCAD Drawing Editor is the Coordinate System Icon (Fig. 11-19). The icon is sometimes called the "UCS icon" because it automatically orients itself to the new location of a coordinate system that you create, called a "USC" (User Coordinate System).

Figure 11-19

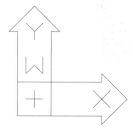

The *Ucsicon* command controls the appearance and positioning of the Coordinate System Icon. It is important to display the icon when working with 3-dimensional drawings so that you can more easily visualize the orientation of the X, Y, and Z coordinate system. However, when you are working with 2-dimensional drawings, it is not necessary to display this icon. (By now you are accustomed to normal orientation of the positive X and Y axes—X positive is to the right and Y positive is up.)

You can use the *Off* option of the *Ucsicon* command to remove the icon from the display. Use *Ucsicon* again with the *On* option to display the icon again. The command prompt is as follows:

```
Command: ucsicon
ON/OFF/All/Noorigin/ORigin <ON>: off
Command:
```

You can also control the visibility of the icon using the *View* pull-down menu. Select *Display*, then *UCS Icon*, then remove the check by the word *On*. Make sure you turn on the Coordinate System Icon when you begin working in 3D.

CHAPTER EXERCISES

1. *Zoom Extents, Realtime, Window, Previous, Center*

    Open the sample drawing supplied with AutoCAD called **ASESMP.DWG**. If your system is set up with the "Full" installation of AutoCAD, the drawing is located in the **C:\Program Files\ AutoCAD R14\Sample** directory. If you loaded only the "Typical" installation, you can copy the drawing from the AutoCAD R14 CD, Acad\Sample directory. The drawing shows an office building layout (Fig. 11-20). Use *SaveAs* and rename the drawing to **ZOOM TEST** and locate it in your working folder (directory). (When you view sample drawings, it is a good idea to copy them to another name so you do not accidentally change the original drawings.)

    A. First, use *Zoom Extents* to display the office drawing as large as possible on your screen. Next, use *Zoom Realtime* and zoom in to the center of the layout. You should be able to see rooms 123 and 124 just below the center of your screen. Right-click and use *Zoom Extents* from the cursor pop-up menu to display the entire layout again. Press **Esc**, **Enter**, or right-click and select *Exit* to exit *Realtime*.

Figure 11-20 ——————————————

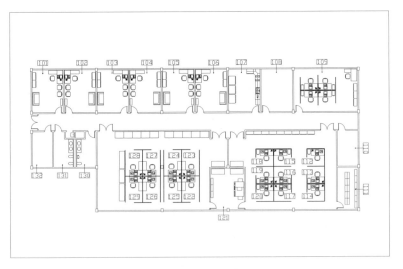

B. Use *Zoom Window* to closely examine room 123. What items in magenta color are on the desk? (HINT: The small solid shape is a digitizer puck.)

Figure 11-21 ───

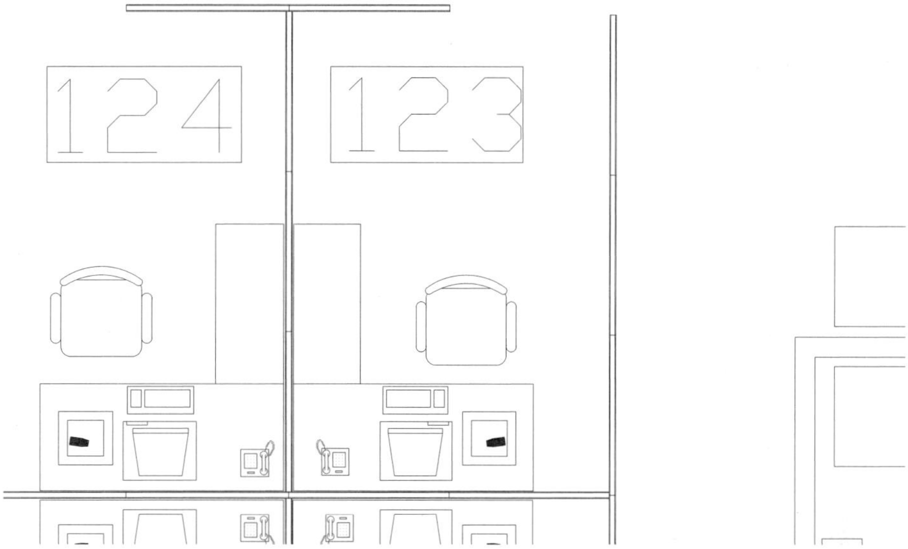

C. Use *Zoom Center*. **PICK** the telephone on the desk next to the computer as the center. Specify a height of **24 (2')**. You should see a display showing only the telephone. How many keys are on the telephone's number pad (*Point* objects)?

D. Next use *Zoom Previous*. Do you see the same display of rooms 123 and 124 as before (Fig. 11-21)? Use *Zoom Previous* repeatedly until you see the office drawing *Extents*.

2. *Pan Realtime*

A. Using the same drawing as exercise 1 (ZOOM TEST.DWG, originally ASESMP.DWG), use *Zoom Extents* to ensure you can view the entire drawing. Next *Zoom* with a *Window* to an area about the size of two or three rooms.

B. Invoke the *Realtime* option of *Pan*. *Pan* about the drawing and find the kitchen. How many vending machines are in the kitchen (along the left wall)?

3. *Realtime Pan* and *Zoom*

In the following exercise, use *Realtime Pan* and *Zoom* in concert to examine specific details of the office layout.

A. Find the copy machine for the office. What room number is it located in? (HINT: It is centrally located near the majority of the offices.)

B. Assume you were choosing an office for yourself in the office complex. What completely empty office (room number) would you choose, assuming you also wanted to be close to the kitchen?

C. *Pan* and/or *Zoom* back to the kitchen. How many burners are on the stove? What appliance is located next to the sinks?

4. *Viewres, Zoom Dynamic*

    A. *Open* the **TABLET14** drawing from the **Sample** folder (C:\Program Files\AutoCAD R14\ Sample). Use *SaveAs*, name the new drawing **TABLET TEST**, and locate it in your working folder.

    B. Use *Zoom* with the **Window** option (you can try *Realtime Zoom*, but it may be too slow depending on the speed of your system). Zoom in to the first several command icons located in the middle-left section of the tablet menu. Your display should reveal icons for *Zoom* and other viewing commands. *Pan* or *Zoom* in or out until you see the *Zoom* icons clearly.

    C. Notice how the *Zoom* command icons are shown as polygons instead of circles? Use the *Viewres* command and change the value to **1000**. Do the icons now appear as circles? Use *Zoom Realtime* again. Does the new setting change the speed of zooming? Why do you think *Viewres* was originally set to 100 instead of 1,000? Change *Viewres* back to **100**.

    D. Use *Zoom Dynamic* and try moving the viewbox immediately. Can you move the box before the drawing is completely regenerated? Change the viewbox size so it is approximately equal to the size of one icon. Zoom in to the command in the lower-right corner of the tablet menu. What is the command? What is the command in the lower-left corner? For a drawing of this complexity and file size, which option of *Zoom* is faster and easier—*Realtime* or *Dynamic*?

    E. If you wish, you can delete the **TABLET TEST** drawing from your directory.

5. *View*

    A. *Open* the **ZOOM TEST** drawing that you used in previous exercises. *Zoom* in to the office that you will be moving into (the empty room to the right of the kitchen). Make sure you can see the entire office. Use the *View* command and *Save* the view as **108**. Next, *Zoom* or *Pan* to room 101. *Save* the display as a *View* named **101**.

    B. Use the *View Control* dialog box (type *Ddview* or select *Named Views* from the *View* pull-down menu). *Restore* view **108**.

6. *Zoom All, Extents*

    A. Begin a *New* drawing. Turn on the *Snap* (**F9**) and *GRID* (**F7**). Draw two *Circles*, each with a **1.5** unit *radius*. The *Circle* centers are at **3,5** and at **5,5**. See Figure 11-22.

    B. Use *Zoom All*. Does the display change? Now use *Zoom Extents*. What happens? Now use *Zoom All* again. Which option **always** shows all of the *Limits*?

Figure 11-22 ——————————————

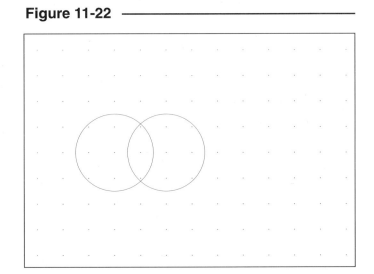

    C. Draw a *Circle* with the center at **10,10** and with a *radius* of **5**. Now use *Zoom All*. Notice the *GRID* only appears on the area defined by the *Limits*. Can you move the cursor to 0,0? Now use *Zoom Extents*. What happens? Can you move the cursor to 0,0?

    D. *Erase* the large *Circle*. Use *Zoom All*. Can you move the cursor to 0,0? Use *Zoom Extents*. Can you find point 0,0?

    E. *Exit* AutoCAD and discard changes.

12

LAYERS, LINETYPES, COLORS, AND OBJECT PROPERTIES

Chapter Objectives

After completing this chapter you should:

1. understand the strategy of grouping related geometry with *Layers*;
2. be able to create *Layers*;
3. be able to assign *Color* and *Linetype* to *Layers*;
4. be able to control a layer's properties and visibility settings (*On, Off, Freeze, Thaw, Lock, Unlock*);
5. be able to select specific layers from a long list using the *Set Layer Filters* dialog box;
6. be able to assign *Color* and *Linetype* to objects;
7. be able to set *LTSCALE* to adjust the scale of linetypes globally;
8. be able to change an object's properties (layer, color, linetype, and linetype scale) with the *Object Properties* toolbar, the *Modify* and *Change Properties* dialog boxes, and with *Match Properties*.

CONCEPTS

In CAD, layers are used to group related objects in a drawing. Objects (*Lines, Circles, Arcs*, etc.) that are created to describe one component, function, or process of a drawing are perceived as related information and, therefore, are typically drawn on one layer. A single CAD drawing is generally composed of several components and, therefore, several layers. Use of layers provides you with a method to control visible features of the components of a drawing. For each layer, you can control its color on the screen, the linetype it will be displayed with, and its visibility setting (on or off). Only visible layers will plot.

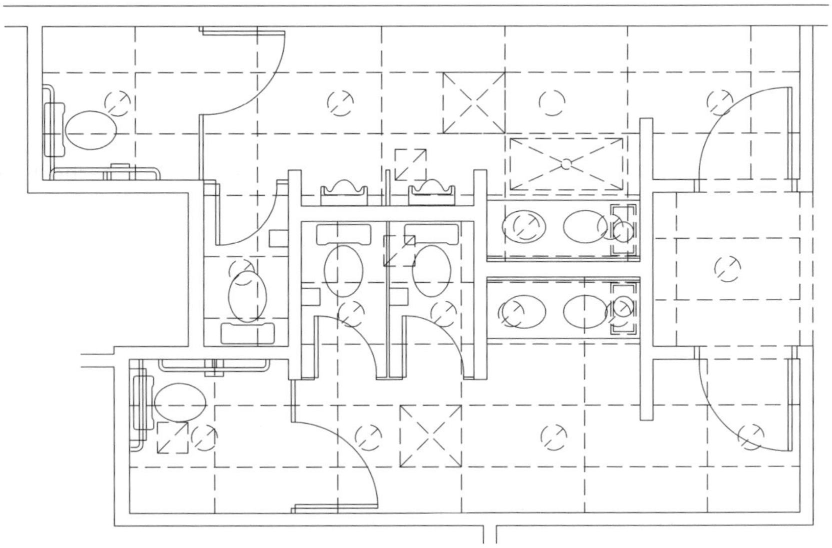

Layers in a CAD drawing can be compared to clear overlay sheets on a manual drawing. For example, in a CAD architectural drawing, the floor plan can be drawn on one layer, electrical layout on another, plumbing on a third layer, and HVAC (heating, ventilating, and air conditioning) on a fourth layer (Fig. 12-1). Each layer of a CAD drawing can be assigned a different color, linetype, and visibility setting, similar to the way clear overlay sheets on a manual drawing can be used. Layers can be temporarily turned *OFF* or *ON* to simplify drawing and editing like overlaying or removing the clear sheets. For example, in the architectural CAD

Figure 12-1

drawing, only the floor plan layer can be made visible while creating the electrical layout, but can later be cross-referenced with the HVAC layout by turning its layer on. Final plots can be made of specific layers for the subcontractors and one plot of all layers for the general contractor by controlling the layers' visibility before plotting.

AutoCAD allows you to create a practically unlimited number of layers. You must assign a name to each layer when you create it. The layer names should be descriptive of the information on the layer.

Assigning Colors and Linetypes

There are two strategies for assigning colors and linetypes in a drawing—assign to layers or assign to objects:

Assign colors and linetypes to layers
Usually, layers are assigned a color and a linetype so that all objects drawn on a single layer have the same color and linetype. Assigning colors and linetypes to layers is called *BYLAYER* color and linetype setting. Using the *BYLAYER* method makes it visually apparent which objects are related (on the same layer). All objects on the same layer have the same linetype and color.

Assign colors and linetypes to individual objects
Alternately, you can assign colors and linetypes to specific objects, overriding the layer's color and linetype setting. This method is fast and easy for small drawings and works well when layering schemes are not used or for particular applications. However, using this method makes it difficult to see which layers the objects are located on.

Object Properties

Another way to describe the assignment of color and linetype properties is to consider the concept of Object Properties. Each object has properties such as a layer, a color, and a linetype. The color and linetype for each object can be designated as *BYLAYER* or as a specific color or linetype. Object properties for the two drawing strategies (schemes) are described as follows:

Object Properties	1. *BYLAYER* Drawing Scheme	2. Object-Specific Drawing Scheme
Layer Assignment	Layer name descriptive of geometry on the layer	Layer name descriptive of geometry on the layer
Color Assignment	*BYLAYER*	*Red, Green, Blue, Yellow,* or other specific color setting
Linetype Assignment	*BYLAYER*	*Continuous, Hidden, Center,* or other specific linetype setting

It is recommended that beginners use only one method for assigning colors and linetypes. After gaining some experience, it may be desirable to combine the two methods only for specific applications. Usually, the *BYLAYER* method is learned first, and the object-specific color and linetype assignment method is used when layers are not needed or when complex applications are needed.

Color assignment is of particular importance because the colors (assigned either to layers or to objects) control the pens or line thickness for plotting and printing. In the *Print/Plot Configuration* dialog box (see Chapter 14), *Pen Assignments* are designated based on the object's color on the screen. Each screen color can be assigned to use a separate pen or line thickness for plotting or printing so that all objects that appear in one color will be plotted with the same pen or line thickness.

LAYER, LINETYPE, AND COLOR COMMANDS

LAYER

Pull-down Menu	COMMAND (TYPE)	ALIAS (TYPE)	Short-cut	Screen (side) Menu	Tablet Menu
Format *Layer...*	*LAYER or* *-LAYER*	*LA or -LA*	...	*FORMAT* *Layer*	*U,5*

The easiest way to gain complete layer control is through the *Layer* tab of the *Layer & Linetype Properties* dialog box (Fig. 12-2). The *Layer* tab of the *Layer & Linetype Properties* dialog box is invoked by using the icon button (shown above), typing the *Layer* command or *LA* command alias, or selecting *Layer* from the *Format* pull-down or screen menu.

Figure 12-2

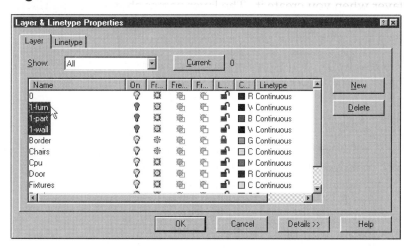

This dialog box allows full control for all layers in a drawing. Layers existing in the drawing appear in the list at the central area (only Layer 0 will appear for new drawings created from standard templates such as ACAD.DWT). New layers can be created by selecting the *New* button near the upper-right corner of the dialog box.

All properties and visibility settings of layers can be controlled by either highlighting the layer name and then selecting tiles such as *Current* or *Delete* or by PICKing one of the icons for a layer such as the light bulb icon (*On, Off*), sun/snowflake icon (*Thaw/Freeze*), *Freeze in the Current Viewport* icon, *Freeze in New Viewports* icon, padlock icon (*Lock/Unlock*), *Color* tile, or the *Linetype*.

Typical Windows 95 dialog box features apply as explained in this paragraph. Multiple layers can be highlighted (see Fig. 12-2) by holding down the Ctrl key while PICKing (to highlight one at a time) or holding down the Shift key while PICKing (to select a range of layers between and including two selected names). Right-clicking in the list area displays a cursor menu allowing you to *Select All* or *Clear All* names in the list (see Fig. 12-4). You can rename a layer by clicking twice <u>slowly</u> on the name (this is the same as PICKing an already highlighted name). The column widths can be changed by moving the pointer to the "crack" between column headings until double arrows appear (Fig. 12-3).

Figure 12-3 ——————————————

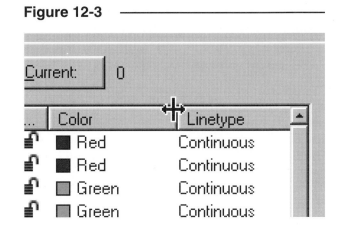

A particularly useful feature is the ability to sort the layers in the list by any one of the headings (*Name, On/Off, Freeze/Thaw, Linetype,* etc.) by clicking on the heading tile. For example, you can sort the list of names in alphabetical order (or reverse order) by clicking once (or twice) on the *Name* heading above the list of names, or you may want to display all of the *Thawed* layers together at the top of the list or all layers of similar colors together.

The *Details* tile near the bottom-right corner of the dialog box displays the details (properties and visibility settings) of the highlighted layer in the list (Fig. 12-4). This area is basically an <u>alter-native</u> method of controlling layer properties and visibility settings (instead of using the icons in the list area).

Figure 12-4 ——————————————

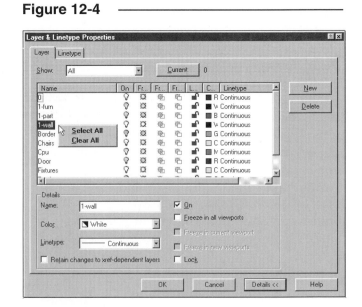

As an alternative to the dialog box, the *Layer* command can be used in command line format by typing *Layer* with a - (hyphen) prefix. The command line format of *Layer* shows all of the available options:

Command: **-layer**
?/Make/Set/New/ON/OFF/Color/LType/Freeze/Thaw/Lock/Unlock:

Detailed explanations of the options for controlling layer properties and visibility settings, whether using the *Layer* tab of the *Layer & Linetype Properties* dialog box or using *-Layer* in command line format, are listed next.

Current or *Set*

To *Set* a layer as the *Current* layer is to make it the active drawing layer. Any objects created with draw commands are created on the *Current* layer. You can, however, edit objects on any layer, but draw only on the current layer. Therefore, if you want to draw on the FLOORPLAN layer (for example), use the *Set* or *Current* option. If you want to draw with a certain *Color* or *Linetype*, set the layer with the desired *Color* and *Linetype* as the *Current* layer. Any layer can be made current, but only <u>one layer at a time</u> can be current.

To set the current layer with the *Layer* tab of the *Layer & Linetype Properties* dialog box (Fig. 12-2, 12-4), select the desired layer from the list and then select the *Current* tile. Since only one layer can be current, it may be necessary to "deselect" highlighted layer names from the list until only one is highlighted. Alternately, if you are typing, use the *Set* option of the *-Layer* command to make a layer the current layer.

ON, OFF

If a layer is *ON*, it is visible. Objects on visible layers can be edited or plotted. Layers that are *OFF* are not visible. Objects on layers that are *OFF* will not plot and cannot be edited (unless the *ALL* selection option is used, such as *Erase, All*). It is not advisable to turn the current layer *OFF*.

Freeze, Thaw

Freeze and *Thaw* override *ON* and *OFF*. *Freeze* is a more protected state than *OFF*. Like being *OFF*, a frozen layer is not visible nor can its objects be edited or plotted. Objects on a frozen layer cannot be accidentally *Erased* with the *All* option. *Freezing* also prevents the layer from being considered when *Regen*s occur. *Freezing* unused layers speeds up computing time when working with large and complex drawings. *Thawing* reverses the *Freezing* state. Layers can be *Thawed* and also turned *OFF*. Frozen layers are not visible even though the light bulb icon is on.

Lock, Unlock

Layers that are *Locked* are protected from being edited but are still visible and can be plotted. *Locking* a layer prevents its objects from being changed even though they are visible. Objects on *Locked* layers cannot be selected with the *All* selection option (such as *Erase, All*). Layers can be *Locked* and *OFF*.

Freeze in Current Viewport, Freeze in New Viewports

 These options are used when paper space viewports exist in the drawing. Using these options, you can control what geometry (layers) appears in specific viewports.

Color and *Linetype* **Properties**

Layers have properties of *Color* and *Linetype* such that (generally) an object that is drawn on, or changed to, a specific layer assumes the layer's linetype and color. Using this scheme (*BYLAYER*) enhances your ability to see what geometry is related by layer. It is also possible, however, to assign specific color and linetype to objects that will override the layer's color and linetype (see *Linetype* and *Color* commands and Changing Object Properties).

Color

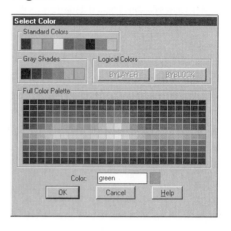

Figure 12-5

☐ Green   Selecting one of the small color boxes in the list area of the *Layer* tab of the *Layer & Linetype Properties* dialog box causes the *Select Color* dialog box to pop up (Fig. 12-5). The desired color can then be selected or the name or color number (called the ACI—AutoCAD Color Index) can be typed in the edit box. This action retroactively changes the color assigned to a layer. All objects on the layer with *BYLAYER* setting change to the new layer color. Objects with specific color assigned (not *BYLAYER*) are not affected. Alternately, the *Color* option of the *Layer* command can be typed to enter the color name or ACI number. The actual number of colors that are available depends on the type of monitor and graphics controller card that are configured. Although most setups allow up to 16 million colors and more, only 256 colors are shown in the standard ACI pallet in AutoCAD.

Linetype

To set a layer's linetype, select the *Linetype* (word such as *Continuous* or *Hidden*) in the layer list of the *Layer* tab, which in turn invokes the *Select Linetype* dialog box (Fig. 12-6). Select the desired linetype from the list. Alternately, the *Linetype* option of the *Layer* command can be typed. Similar to changing a layer's color, all objects on the layer with *BYLAYER* linetype assignment are retroactively displayed in the selected layer linetype while non-*BYLAYER* objects remain unchanged.

Figure 12-6

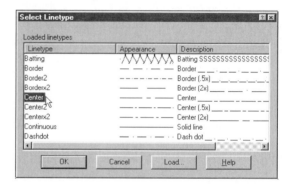

The default AutoCAD template drawing as supplied by Autodesk has only one linetype available (*Continuous*). Before you can use other linetypes in a drawing, you load the linetypes by selecting the *Load* tile (see *Linetype* command) or by using a default template drawing that has the desired linetypes already loaded.

New

The *New* option allows you to make new layers. There is only one layer in the AutoCAD default drawing (ACAD.DWT) as it is provided to you "out of the box." That layer is Layer 0. Layer 0 is a part of every AutoCAD drawing because it cannot be deleted. You can, however, change the *Linetype* and *Color* of Layer 0 from the defaults (continuous linetype and color #7 white). Layer 0 is generally used as a construction layer or for geometry not intended to be included in the final draft of the drawing. Layer 0 has special properties when creating *Blocks* (Chapter 20).

You should create layers for each group of related objects and assign appropriate layer names for that geometry. You can use up to 31 characters (no spaces) for layer names. There is practically no limit to the number of layers that can be created (although 32,767 has been found to be the actual limit). To create layers in the *Layer* tab of the dialog box, select the *New* tile. A new layer named "Layer1" (or other number) then appears in the list with the default color (*White*) and linetype (*Continuous*). The new name initially appears in the rename mode so you can immediately assign a more appropriate and descriptive name for the layer. You can create many new names quickly by typing (renaming) the first layer name, then typing a comma before other names. A comma forces a new "blank" layer name to appear. Colors and linetypes should be assigned as the next step.

If you want to create layers by typing, the *New* and *Make* options of the *Layer* command can be used. *New* allows creation of one or more new layers. *Make* allows creation of one layer (at a time) and sets it as the current layer.

Delete

The *Delete* tile allows you to delete layers. <u>Only layers with no geometry can be deleted.</u> You cannot delete a layer that has objects on it nor can you delete Layer 0, the current layer, or layers that are part of externally referenced (*Xref*) drawings. If you attempt to *Delete* such a layer, accidentally or intentionally, a warning appears (Fig. 12-7).

Figure 12-7 ———————

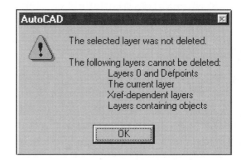

Show

The *Show* drop-down list determines what names appear in the list of layers in the *Layer* tab (Fig. 12-8). The choices are as follows:

Figure 12-8 ———————

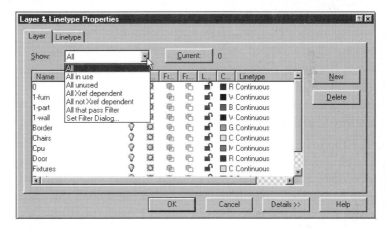

All	shows all layers in the drawing
All in use	shows all layers that have objects on them
All unused	shows only layers without geometry
All Xref dependent	shows only layers that are part of externally referenced drawings
All not Xref	shows all layers except those in externally referenced drawings
All that pass filter	shows the layers that are designated in the *Set Layer Filters* dialog box
Set Filter Dialog...	invokes the *Set Layer Filters* dialog box

Set Layer Filters **Dialog Box**

When working on drawings with a large number of layers, it is time consuming to scroll through the list of layer names in the *Layer* tab of the *Layer & Linetype Properties* dialog box or in the Layer Control drop-down list on the Object Properties toolbar. The *Show:* drop-down list in the *Layer* tab of the *Layer & Linetype Properties* dialog box controls names that appear in the list in that dialog box but not the Layer Control drop-down list. The *Set Layer Filters* dialog box enables you to specify criteria for displaying a selected set of layers in both lists. For example, you may want to list only layer names that are *Thawed*. By using layer filters, you can shorten the layer list by filtering out the layers that do not meet this criterion.

Figure 12-9 ———————

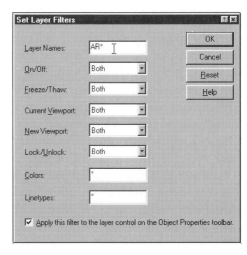

R14

Selecting the *Set Filter dialog...* option from the bottom of the *Show*: list produces the *Set Layer Filters* dialog box (Fig. 12-9). Eight criteria can be specified in the *Set Layer Filters* dialog box. The criteria define <u>what you want to see</u> in the layer listing. When the criteria are set to *Both* or *, all layer names are displayed.

The three boxes displaying a * symbol (asterisk) enable you to enter specific values. These edit boxes allow you to list only layers that match particular names, colors or linetypes. As an example, if you set the *Layer Names* filter to **AR***, only those layers whose names begin with "AR" will be displayed in the layer listing (see Fig. 12-9).

The other criteria are in pairs, enabling you to choose one or *"Both"* of the settings. When a criteria is set to *Both*, you will see layers listed that match both items in the pair, such as layers that are *On* <u>and</u> *Off*. Selecting one of the pair displays only layer names meeting that condition. Selecting *Thawed* from the *Freeze/Thaw* box, for example, causes a display of only layers that that are thawed.

When a layer filter setting is enabled, the *All that pass filter* option is automatically displayed in the *Show*: list in the *Layer* tab of the *Layer & Linetype Properties* dialog box. If you want the filter to also be applied to the Layer Control drop-down list in the Object Properties toolbar, select the checkbox at the bottom of the *Set Layer Filters* dialog box.

Only one layer filter configuration can be applied at a time, and a specific configuration cannot be saved to a file.

Layer Control Drop-down List

The Object Properties toolbar contains a drop-down list for making layer control quick and easy (Fig. 12-10). The window normally displays the current layer's name, visibility setting, and properties. When you pull down the list, all the layers (unless otherwise specified in the *Set Layer Filter* dialog box) and their settings are displayed. Selecting any layer <u>name</u> makes it current. Clicking on any of the visibility/properties icons changes the layers' setting as described previously. Several layers can be changed in one "drop." You cannot change a layer's color or linetype nor can you create new layers using this drop-down list.

Figure 12-10 ———

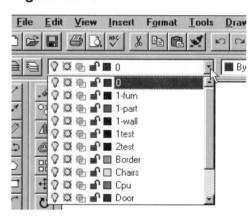

Displaying an Object's Properties and Visibility Settings

When the Layer Control drop-down list in the Object Properties toolbar is in the normal position (not "dropped down"), it can be used to display an <u>object's</u> layer properties and visibility settings. Do this by selecting an object (with the pickbox, Window, or Crossing Window) when no commands are in use. When an object is selected, the Object Properties toolbar displays the layer, color, and linetype <u>of the selected object</u> (Fig. 12-11).

Figure 12-11 ————————————

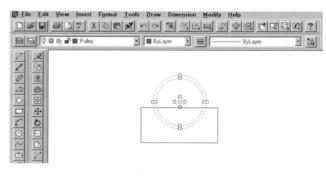

The Layer Control box displays the selected object's layer name and layer settings rather than that of the current layer. The Color Control and Linetype Control boxes in the Object Properties toolbar (just to the right) also temporarily reflect the color and linetype properties of the selected object. Pressing the Escape key causes the list boxes to display the current layer name and settings again, as would normally

R14

R13

R14

R14

be displayed. If more than one object is selected and the objects are on different layers or have different color and linetype properties, the list boxes go blank until the objects become unhighlighted or a command is used.

These sections of the Object Properties toolbar have another important new feature that allows you to change a highlighted object's (or set of objects') properties. See Changing Object Properties near the end of this chapter.

Make Object's Layer Current

A new feature in Release 14 can be used to make a desired layer current simply by selecting <u>any object on the layer</u>. This option is generally faster than using the *Layer & Linetype Properties* dialog box or *-Layer* command. The *Make Object's Layer Current* feature is particularly useful when you want to draw objects on the same layer as other objects you see but are not sure of the layer name. This feature (not a formal command) is only available by selecting the icon on the far-left of the Object Properties toolbar (Fig. 12-11). It cannot be invoked by any other method.

There are two steps in the procedure to make an object's layer current: (1) select the icon and (2) select any object on the desired layer. The selected object's layer becomes current and the new current layer name immediately appears in the Layer Control list box in the Object Properties toolbar.

LINETYPE

	COMMAND (TYPE)	ALIAS (TYPE)	Short-cut	Screen (side) Menu	Tablet Menu
Pull-down Menu					
Format *Linetype*	*LINETYPE or* *-LINETYPE*	*LT or -LT*	...	*FORMAT* *Linetype*	U,3

Invoking this command presents the *Linetype* tab of the *Layer & Linetype Properties* dialog box (Fig. 12-12). Even though this looks similar to the dialog box used for assigning linetypes to layers (shown earlier in Figure 12-6), beware!

When linetypes are selected using the *Linetype* command or *Linetype* tab of the *Layer & Linetype Properties* dialog box, they are <u>assigned to objects</u>—not to layers. That is, selecting a linetype by this manner causes all objects <u>from that time on</u> to be drawn using that linetype, regardless of the layers that they are on (unless the *BYLAYER* type is PICKed). In contrast, selecting linetypes during the *Layer* command (using the *Layer* tab of the *Layer & Linetype Properties* dialog box) results in assignment of linetypes to <u>layers</u>. Remember that using both of these methods for color and linetype assignment in one drawing can be very confus-

Figure 12-12 ————————

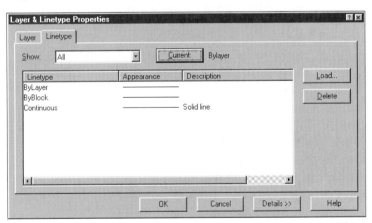

ing until you have some experience using both methods.

To draw an object in a specific linetype, simply select the linetype from the list and PICK the *OK* tile. That linetype stays in effect (all objects are drawn with that linetype) until another linetype is selected. If you want to draw objects in the layer's assigned linetype, select *BYLAYER*.

Load

The *Linetype* tab of the *Layer & Linetype Properties* dialog box also lets you *Load* linetypes that are not already in the drawing. Just PICK the *Load* button to view and select from the list of available linetypes in the *Load or Reload Linetypes* dialog box (Fig. 12-13). The linetypes are loaded from the ACAD.LIN file. If you want to load all linetypes into the drawing, right-click to display a small cursor menu, then choose *Select All* to highlight all linetypes (see Fig. 12-12). You can also select multiple linetypes by holding down the Ctrl key or select a range by holding down the Shift key. The loaded linetypes are then available for assignment from both the *Linetype* tab and the *Layer* tab of the *Layer & Linetype Properties* dialog box.

AutoCAD supplies numerous linetypes as shown in Figure 12-14. The ACAD_ISO*n*W100 linetypes are intended to be used with metric drawings. For the ACAD_ISO*n*W100 linetypes, *Limits* should be set to metric sheet sizes to accommodate the relatively large linetype spacing. Avoid using both ACAD_ISO*n*W100 linetypes and other linetypes in one drawing due to the difficulty managing the two sets of linetype scales.

Delete

The *Delete* button is useful for deleting any unused linetypes from the drawing. Unused linetypes are those that have been loaded into the drawing but have not been assigned to layers or to objects. Freeing the drawing of unused linetypes should reduce file size slightly.

Typing *-Linetype* produces the command line format of *Linetype*. Although this format of the command does not offer the capabilities of the dialog box version, you can assign linetypes to objects or assign an object to use the layer's linetype by typing *-Linetype* or *-LA* (use the hyphen prefix). This action yields the following prompt:

```
Command: -linetype
?/Create/Load/Set:
```

You can list (*?*), *Load, Set,* or *Create* linetypes with the typed form of the command. Type a question mark (*?*) to display the list of available linetypes. Type *L* to *Load* any linetypes. The *Set* option of *Linetype* accomplishes the same action as selecting from the linetypes in the dialog box. That is, you can set a specific linetype for all subsequent objects regardless of the layer's linetype setting, or you can use *Set* to assign the new objects to use the layer's (*BYLAYER*) linetype.

You can create your own custom linetypes with the *Create* option or by using a text editor. Both simple linetypes (line, dash, and dot combinations) and complex linetypes (including text or other shapes) are possible.

Figure 12-13

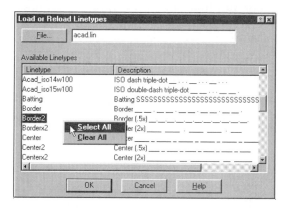

Figure 12-14

BORDER	
BORDER2	
BORDERX2	
CENTER	
CENTER2	
CENTERX2	
CONTINUOUS	
DASHDOT	
DASHDOT2	
DASHDOTX2	
DASHED	
DASHED2	
DASHEDX2	
DIVIDE	
DIVIDE2	
DIVIDEX2	
DOT	
DOT2	
DOTX2	
HIDDEN	
HIDDEN2	
HIDDENX2	
PHANTOM	
PHANTON2	
PHANTOMX2	
ACAD_ISO02W100	
ACAD_ISO03W100	
ACAD_ISO04W100	
ACAD_ISO05W100	
ACAD_ISO06W100	
ACAD_ISO07W100	
ACAD_ISO08W100	
ACAD_ISO09W100	
ACAD_ISO010W100	
ACAD_ISO011W100	
ACAD_ISO012W100	
ACAD_ISO013W100	
ACAD_ISO014W100	
ACAD_ISO015W100	

Linetype Control Drop-down List

The Object Properties toolbar contains a drop-down list for selecting linetypes (Fig. 12-15). Although this appears to be quick and easy, you can <u>assign linetypes only to objects</u> by this method unless *BYLAYER* is selected to use the layers' assigned linetypes. Any linetype you select from this list becomes the current object linetype. If you want to select linetypes for layers, make sure this list displays the *BYLAYER* setting, then use the *Layer & Linetype Properties* dialog box to select linetypes for layers.

Keep in mind that the Linetype Control drop-down list as well as the Layer Control and Color Control drop-down lists display the current settings for selected objects (objects selected when no commands are in use). When linetypes are assigned to layers rather than objects, the layer drop-down list reports a *BYLAYER* setting.

Figure 12-15 ─────────

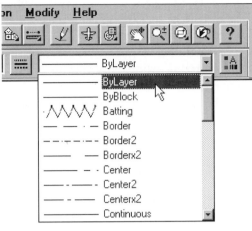

COLOR and *DDCOLOR*

Pull-down Menu	COMMAND (TYPE)	ALIAS (TYPE)	Short-cut	Screen (side) Menu	Tablet Menu
Format *Color*	*COLOR or* *DDCOLOR*	*COL*	...	*FORMAT* *Ddcolor*	*U,4*

Similar to linetypes, colors can be assigned to layers or to objects. Using the *Color* or *Ddcolor* command assigns a color for all newly created <u>objects</u>, regardless of the layer's color designation (unless the *BYLAYER* color is selected). This color setting <u>overrides</u> the layer color for any newly created objects so that all new objects are drawn with the specified color no matter what layers they are on. This type of color designation prohibits your ability to see which objects are on which layers by their color; however, for some applications object color setting may be desirable. Use the *Layer* tab of the *Layer &Linetype Properties* dialog box to set colors for layers.

Invoking this command by the menus or by typing *Ddcolor* presents the *Select Color* dialog box shown in Figure 12-16. This is essentially the same dialog box used for assigning colors to layers; however, the *BYLAYER* and *BYBLOCK* tiles are accessible. (The buttons are grayed out when this dialog is invoked from the *Layer* tab because, in that case, any setting is a *BYLAYER* setting.) *BYBLOCK* is discussed in Chapter 20.

If you type "*Color*," the command line format is displayed as follows:

Figure 12-16 ─────────

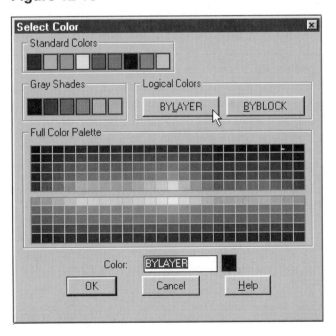

> Command: **color**
> New object color <BYLAYER>: **red** (Specify a color name, number, *BYLAYER* or *BYBLOCK* setting.)
> Command:

For the example, all new objects drawn will be *red* regardless of the layers' color settings until *Color* or *Ddcolor* is used again to set another color or *BYLAYER* setting.

The current color (whether *BYLAYER* or specific object color) is saved in the *CECOLOR* (Current Entity Color) variable. The current color can be set by changing the value in the *CECOLOR* variable directly at the command prompt (by typing "*CECOLOR*"), by using the *Color* command, *Select Color* dialog box, or by using the Color Control drop-down list in the Object Properties toolbar. The *CECOLOR* variable accepts a string value (such as "red" or "bylayer") or accepts the ACI number (0 through 255).

Color Control Drop-down List

Figure 12-17

When an object-specific color has been set, the top item in the Object Properties toolbar Color Control drop-down list displays the current color (Fig. 12-17). Beware—using this list to select a color assigns an <u>object-specific color unless *BYLAYER* is selected</u>. Any color you select from this list becomes the current object color. Selecting *Other. . .* from the bottom of the list invokes the *Select Color* dialog box (see Figure 12-16). If you want to assign colors to layers, make sure this list displays the *BYLAYER* setting, then use the *Layer* tab of the *Layer & Linetype Properties* dialog box to select colors for layers.

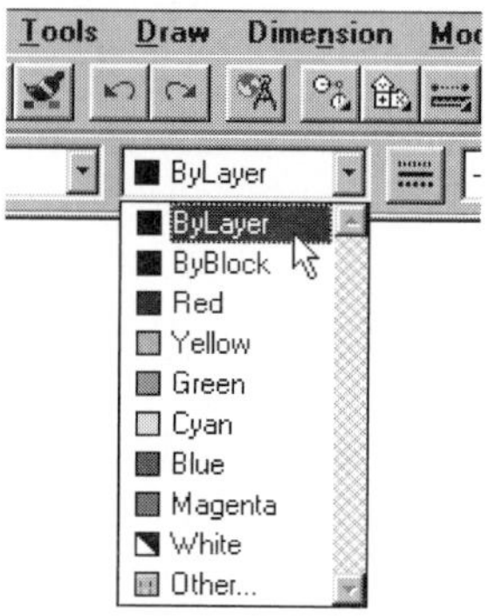

This drop-down list, as well as the others in the Object Properties toolbar, display the current settings for selected objects (objects selected when no commands are in use). A *BYLAYER* setting indicates that properties are assigned to layers rather than objects.

LTSCALE

	Pull-down Menu	COMMAND (TYPE)	ALIAS (TYPE)	Short-cut	Screen (side) Menu	Tablet Menu
	Format Linetype... Details >>	*LTSCALE*	*LTS*	...	*FORMAT Linetype Details >>*	...

Hidden, dashed, dotted, and other linetypes that have spaces are called <u>non-continuous</u> linetypes. When drawing objects that have non-continuous linetypes (either *BYLAYER* or object-specific linetype designations), the linetype's dashes or dots are automatically created and spaced. The *LTSCALE* (Linetype Scale) system variable controls the length and spacing of the dashes and/or dots. The value that is specified for *LTSCALE* affects the drawing <u>globally and retroactively</u>. That is, all existing non-continuous lines in the drawing as well as new lines are affected by *LTSCALE*. You can therefore adjust the drawing's linetype scale for all lines at any time with this one command.

If you choose to make the dashes of non-continuous lines smaller and closer together, reduce *LTSCALE*; if you desire larger dashes, increase *LTSCALE*. The *Hidden* linetype is shown in Figure 12-18 at various *LTSCALE* settings. Any positive value can be specified.

Figure 12-18

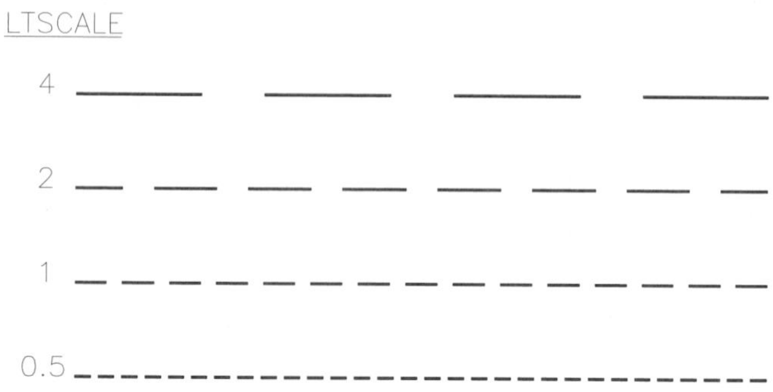

LTSCALE

4 ———— ———— ———— ————

2 — — — — — — — —

1 – – – – – – – – – – – – – –

0.5 –

LTSCALE can be set in the *Linetype* tab of the *Layer & Linetype Properties* dialog box or in command line format. In the *Linetype* tab, select the *Details* button to allow access to the *Global scale factor* edit box (Fig. 12-19). Changing the value in this edit box sets the LTSCALE variable. Changing the value in the *Current object scale* edit box sets the linetype scale for the current object only (see *CELTSCALE* below) but does not affect the global linetype scale (*LTSCALE*).

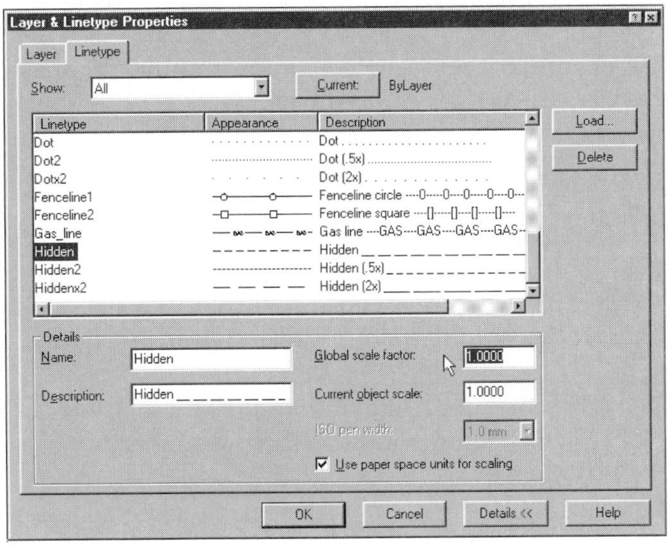

LTSCALE can also be used in the command line format.

Figure 12-19

```
Command: LTSCALE
New scale factor <1 or current value>:
(value) (Enter any positive value.)
Command:
```

The *LTSCALE* for the default template drawing (ACAD.DWT) is 1. This value represents an appropriate *LTSCALE* for objects drawn within the default *Limits* of 12 x 9.

As a general rule, you should change the *LTSCALE* proportionally when *Limits* are changed (more specifically, when *Limits* are changed to other than the intended plot sheet size). For example, if you increase the drawing area defined by *Limits* by a factor of 2 (to 24 x 18) from the default (12 x 9), you might also change *LTSCALE* proportionally to a value of 2. Since *LTSCALE* is retroactive, it can be changed at a later time or repeatedly adjusted to display the desired spacing of linetypes. (For more information on *LTSCALE*, see Chapters 13 and 14.)

The ACAD_ISO*n*W100 linetypes are intended to be used with metric drawings. Changing *Limits* to metric sheet sizes automatically displays these linetypes with appropriate linetype spacing. For these linetypes only, *LTSCALE* is changed automatically to the value selected in the *ISO Pen Width* box of the *Linetype* tab (see Fig. 12-19). Using both ACAD_ISO*n*W100 linetypes and other linetypes in one drawing is discouraged due to the difficulty managing two sets of linetype scales.

Even though you have some control over the size of spacing for non-continuous lines, you have almost no control over the placement of the dashes for non-continuous lines. For example, the short dashes of centerlines cannot always be controlled to intersect at the centers of a series of circles. The spacing can only be adjusted globally (all lines in the drawing) to reach a compromise. You can also adjust individual objects' linetype scale (*CELTSCALE*) to achieve the desired effect. See *CELTSCALE* below and Chapter 22, Multiview Drawing, for further discussion and suggestions on this subject.

CELTSCALE

	Pull-down Menu	COMMAND (TYPE)	ALIAS (TYPE)	Short-cut	Screen (side) Menu	Tablet Menu
	Format Linetype... Details>>	*CELTSCALE*	...	...	*FORMAT Linetype Details>>*	...

CELTSCALE stands for Current Entity Linetype Scale. This setting changes the object-specific linetype scale proportional to the *LTSCALE*. The *LTSCALE* value is global and retroactive, whereas *CELTSCALE* sets the linetype scale for all newly created objects and is not retroactive. *CELTSCALE* is object specific.

Using *CELTSCALE* to set an object linetype scale is similar to setting an object color and linetype in that the properties are <u>assigned to the specific object</u>.

For example, if you wanted all non-continuous lines (dashes and spaces) in the drawing to be two times the default size, set *LTSCALE* to 2 (*LTSCALE* is global and retroactive). If you then wanted only say, a select two or three lines to have smaller spacing, change *CELTSCALE* to .5, draw the new lines, and then change *CELTSCALE* back to 1.

You can also use the *Linetype* tab of the *Layer & Linetype Properties* dialog box to set the *CELTSCALE* (see Fig. 12-19). Set the desired *CELTSCALE* by listing the *Details* and changing the value in the *Current object scale* edit box. This setting affects all linetypes for all <u>newly created</u> objects. The linetype scale for individual objects can also be changed retroactively using the method explained in the following paragraph.

The <u>recommended</u> method for adjusting linetypes in the drawing to different scales is <u>not to use</u> *CELTSCALE* but rather the following strategy.

1. Set *LTSCALE* to an appropriate value for the drawing.
2. Create all objects in the drawing with the desired linetypes.
3. Adjust the *LTSCALE* again if necessary to globally alter the linetype scale.
4. Use *Ddmodify, Ddchprop,* or *Matchprop* to <u>retroactively</u> change the linetype scale (*CELTSCALE*) <u>for selected objects</u> (see *Ddmodify, Ddchprop,* and *Matchprop*).

CHANGING OBJECT PROPERTIES

Often it is desirable to change the properties of an object after it has been created. For example, an object's *Layer* property could be changed. This can be thought of as "moving" the object from one layer to another. When an object is changed from one layer to another, it assumes the new layer's color and linetype, provided the object was originally created with color and linetype assigned *BYLAYER*, as is generally the case. In other words, if an object was created on the wrong layer (possibly with the wrong linetype or color), it could be "moved" to the desired layer, therefore assuming the new layer's linetype and color.

Another example is to change an object's individual linetype scale. In some cases an individual object's linetype scale requires an adjustment to other than the global linetype scale (*LTSCALE*) setting. One of several methods can be used to adjust the object's individual linetype scale (*CELTSCALE*) to an appropriate value.

Several methods can be used to retroactively change properties of selected objects. The Object Properties toolbar, the *Change Properties (Ddchprop)* dialog box, the *Modify (Ddmodify)* dialog box, and the *Match Properties* command can be used to <u>retroactively</u> change the properties of individual objects. Properties that can be changed by all four methods are *Layer, Linetype, Color,* and object linetype scale (*CELTSCALE*).

Although some of the commands discussed in this section have additional capabilities, the discussion is limited to changing properties of objects. For this reason, these commands and features are also discussed in Chapter 16, Modify Commands III, and in other chapters.

Object Properties Toolbar

The three drop-down lists in the Object Properties toolbar (when not "dropped down") generally show the current layer, color, and linetype. However, if an object or set of objects is selected, the information in these boxes changes to display the current objects' settings. <u>You can change the selected objects' settings</u> by "dropping down" any of the lists and making another selection.

First, select (highlight) an object when no commands are in use. Use the pickbox (that appears on the crosshairs), Window, or Crossing Window to select the desired object or set of objects. The entries in the three boxes (Layer Control, Color Control, and Linetype Control) then change to display the settings for the <u>selected object or objects</u>. If several objects are selected that have different properties, the boxes display no information (go

Figure 12-20

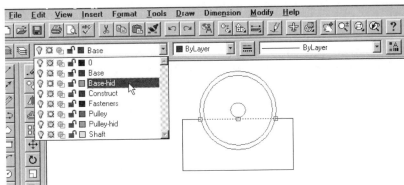

"blank"). Next, use any of the drop-down lists to make another selection (Fig. 12-20). The highlighted object's properties are changed to those selected in the lists. Press the Escape key to complete the process.

Remember that in most cases color and linetype settings are assigned *BYLAYER*. In this type of drawing scheme, to change the linetype or color properties of an object, you would change the object's <u>layer</u> (see Fig. 12-20). If you are using this type of drawing scheme, refrain from using the Color Control and Linetype Control drop-down lists for changing properties.

DDMODIFY

Pull-down Menu	COMMAND (TYPE)	ALIAS (TYPE)	Short-cut	Screen (side) Menu	Tablet Menu
Modify Properties...	DDMODIFY	MO	...	MODIFY1 Modify	Y,14

DDCHPROP

Pull-down Menu	COMMAND (TYPE)	ALIAS (TYPE)	Short-cut	Screen (side) Menu	Tablet Menu
Modify Properties...	DDCHPROP	CH	...	MODIFY1 Modify	Y,14

These dialog boxes can also be used to change an object's layer, linetype, or color designation retroactively. If you select the *Properties* icon button from the Object Properties toolbar, one of two dialog boxes appears depending on the number of objects selected. If one object is selected, the *Modify* dialog appears (Fig. 12-21), whereas <u>if two or more objects are selected</u>, the *Change Properties* dialog box appears (Fig. 12-22). Alternately, you can type *Ddmodify, Ddchprop*, or select *Properties. . .* from the *Modify* pull-down menu.

Figure 12-21

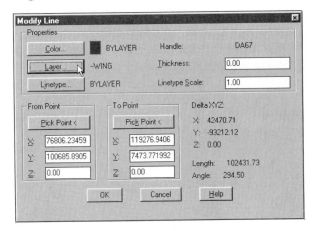

Either of these dialog boxes allows you to change the *Layer, Color, Linetype,* or *Linetype* scale (*CELTSCALE*) of the selected object(s). Selecting *Color* or *Linetype* other than *BYLAYER* sets the individual object's properties. For drawing schemes using the *BYLAYER* color and line-type settings, change the *Layer* to assume the desired linetype and color for the object.

NOTE: The *Linetype Scale* edit box in both of these dialog boxes affects the *CELTSCALE* variable and changes only the selected object's individual linetype scale. It does not affect the drawing *LTSCALE*. This is the recommended method for changing individual object linetype scale rather than setting *CELTSCALE* by command line format.

Figure 12-22

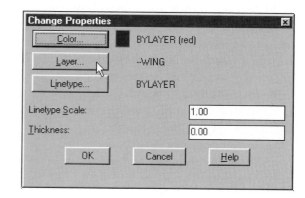

MATCHPROP

	Pull-down Menu	COMMAND (TYPE)	ALIAS (TYPE)	Short-cut	Screen (side) Menu	Tablet Menu
	Modify Match Properties	*MATCHPROP or PAINTER*	*MA*	...	*MODIFY1 Matchprp*	*Y,15*

Matchprop is used to "paint" the properties of one object to another. The process is simple. After invoking the command, select the object that has the desired properties (Source Object), then select the object you want to "paint" the properties to (Destination Object). The command prompt is as follows.

```
Command: matchprop
Select Source Object: PICK
Current active settings = color layer ltype ltscale thickness text dim hatch
Settings/<Select Destination Object(s)>: PICK
Settings/<Select Destination Object(s)>: Enter
Command:
```

Several "Destination Objects" can be selected. The "Destination Object(s)" assume all of the properties of the "Source Object" (listed as "Current active settings").

Use the *Settings* option to control which of eight possible properties and other settings are "painted" to the destination objects. At the "Settings/<Select Destination Object(s)>:" prompt, type *S* to display the *Property Settings* dialog box (Fig. 12-23). In the dialog box, designate which of the *Basic Properties* or *Special Properties* are to be painted to the "Destination Objects." The following *Basic Properties* correspond to properties discussed in this chapter:

Figure 12-23

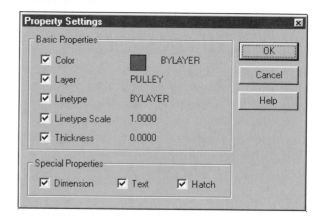

Color	paints the object or *BYLAYER* color
Layer	moves selected objects to Source Object layer
Linetype	paints the object or *BYLAYER* linetype
Linetype Scale	changes the individual object linetype scale (*CELTSCALE*) not global (*LTSCALE*)

The *Special Properties* of the *Property Settings* dialog box are discussed in Chapter 16, Modify Commands III. Also see Chapter 16 for a full explanation of *Change Properties* (*Ddchprop*) and *Modify* (*Ddmodify*).

R14

CHAPTER EXERCISES

1. **Layer tab of the Layer & Linetype Properties dialog box, Layer Control drop-down list, and Make Object's Layer Current**

   Open the **WILHOME.DWG** sample drawing located in the C:\Program Files\AutoCAD R14\Sample directory. Activate model space by typing *MS* at the command prompt or by double-clicking the word **PAPER** on the Status line. Next, use *SaveAs* to save the drawing in your working directory as **WILHOME2**.

   A. Invoke the *Layer* command in command line format by typing *-LAYER* or *-LA*. Use the *?* option to yield the list of layers in the drawing. Notice that all the layers are *On* except one. Which layer is *Off*?

   B. Use the **Object Properties** toolbar to indicate the layers of selected items. On the right side of the drawing, **PICK** the yellow lines around the floor plan. What layer are the dimensions on? Press the **Esc** key twice to cancel the selection. Next, **PICK** the cyan (light blue) text around the semi-circular room near the top of the floor plan on the right. What layer is the text on? Press the **Esc** key twice to cancel the selection.

   C. Invoke the *Layer* tab of the *Layer & Linetype Properties* dialog box by selecting the icon button or using the *Format* pull-down menu. Turn *Off* the **Dimensions** layer (click the light bulb icon). Select *OK* to return to the drawing and check the new setting. Are the dimensions displayed? Next, use the **Layer Control** drop-down list to turn the **Dimensions** layer back *On*.

   D. Type *Regen* and count the number of seconds it takes for the drawing to regenerate. Next, use the **Object Properties** toolbar to indicate the name of the layer of the green elevations on the left side of the drawing (**PICK** a green object and examine the layer name that appears). Use the **Layer Control** drop-down list to turn *Off* that layer (light bulb icon) and the layer with the cyan text. Now type *Regen* again and count the number of seconds it takes for the regeneration. Is there any change? Finally, *Freeze* both layers (snowflake/sun icon), then type *Regen* again and count the number of seconds it takes for the regeneration. Is there any difference between *Off* and *Freeze* for regenerations?

   E. Use the *Layer* tab of the *Layer & Linetype Properties* dialog box and **PICK** the *Select All* option (right-click for menu). *Freeze* all layers. Which layer will not *Freeze*? Notice the light bulb icon indicates the layers are still *On* although the layers do not appear in the drawing (*Freeze* overrides *On*). Use the same procedure to select all layers and *Thaw* them.

F. Use the *Layer* tab of the *Layer & Linetype Properties* dialog box again and drop down the *Show* list. Are there are any layers in the drawing that are unused (select *All unused*)? Use the *Show* list again and list *All* layers.

G. Now we want to *Freeze* all the *Yellow* layers. Using the *Layer* tab of the *Layer & Linetype Properties* dialog box again, sort the list by color. **PICK** the first layer name in the list of the four yellow ones. Now hold down the **Shift** key and **PICK** the last name in the list. All four layers should be selected. *Freeze* all four by selecting any one of the sun/snowflake icons. Select *OK* to return to the drawing.

H. Assuming you wanted to work only with the layers beginning with AR, you can set the filter to display only those layers. First, select the *Set Filter dialog* in the *Show* list of the *Layer* tab of the *Layer & Linetype Properties* dialog box. In the *Set Layer Filters* dialog box, enter **AR*** in the *Layer Names* edit box and make sure the checkbox at the bottom is set to apply the filter to the Object Properties toolbar. Select *OK*, then examine the list in the *Layer* tab. Now select *OK* in the *Layer & Linetype Properties* dialog box to return to the drawing. Examine the **Layer Control** drop-down list. Does it display only layer names beginning with AR? Return to the *Layer* tab and select *All* in the *Show* list. Select *OK* and examine the **Layer Control** drop-down list again. It should still have the filter applied even though the *Layer* tab shows *All* layers. Use the *Set Layer Filters* dialog box again and enter * in the *Layer Names* edit box to show all layer names in both lists.

I. Ensure all layers are *Thawed*. Use the *Layer* tab and sort the list alphabetically by *Name*. *Freeze* all layers beginning with **A** and **B**. Select *OK* and then use the drop-down list to display the names. Assuming you were working only with the thawed layers, it would be convenient to display <u>only</u> those names in the drop-down list. Use the *Set Layer Filter* dialog box to set a filter for *Thawed* layers. Ensure the checkbox at the bottom is checked. Return to the drawing and examine the drop-down list. Do only names of *Thawed* layers appear?

J. Next, you will "move" objects from one layer to another using the **Object Properties** toolbar. Make layer **0** the current layer by using the drop-down list and selecting the name. Use the *Layer* tab and right-click to *Select All* names. **PICK** one icon to *Freeze* all layers (except the current layer). Then select *Clear All* and *Thaw* layer **Gennote**. Select *OK* and return to the drawing. Use a Crossing Window to select all of the objects in the drawing (the small blue grips may appear). The **Object Properties** toolbar should indicate the layer on which the objects reside. Next, list the layers using the drop-down list and **PICK** layer **0**. Wait several seconds. Press **Esc** twice when the drawing regenerates. The objects should be red and on layer 0.

K. Let's assume you want to draw some additional objects on the same layer as the text and on the dimensions layer. First, *Zoom* in to the upper half of the floor plan on the right. Now select the *Make Object's Layer Current* icon button (on the far-right of the Object Properties toolbar). At the "Select object whose layer will become current" prompt, select a text object (cyan color). The **Gennote** layer name should appear in the Layer Control box as the current layer. Draw a *Line* and it should be on layer Gennote and cyan in color. Next, use *Make Object's Layer Current* again and select a dimension object. The **Dimensions** layer should appear as the current layer. Draw another *Line*. It should be on layer Dimensions and yellow in color. Keep in mind that this method of setting a layer current works especially well when you do not know the layer names or know what objects are on which layers.

L. Do not save changes to the drawing.

2. *Layer* and *Linetype* tabs of the *Layer & Linetype Properties* dialog box

   A. Begin a *New* drawing and use *Save* to assign the name **CH12EX2**. Set *Limits* to **11 x 8.5**. Use the *Linetype* tab of the *Layer & Linetype* dialog box to list the loaded linetypes. Are any linetypes already loaded? Right-click to *Select All* linetypes then **PICK** the *Load* button. Close the dialog box.

   B. Use the *Layer* tab to create 3 new layers named **OBJ, HID,** and **CEN**. Assign the following colors and line-types to the layers by clicking on the *Color* and *Linetype* columns:

OBJ	red	continuous
HID	yellow	hidden
CEN	green	center

Figure 12-24 ——————————

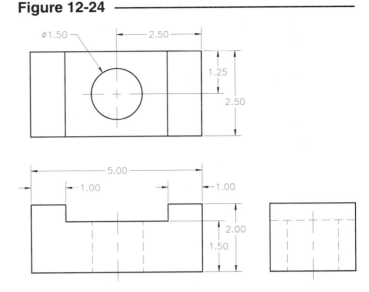

   C. Make the **OBJ** layer *Current* and PICK the *OK* tile. Verify the current layer by looking at the **Layer Control** drop-down list, then draw the visible object *Lines* and the *Circle* only (not the dimensions) as shown in Figure 12-24.

   D. When you are finished drawing the visible object lines, create the necessary hidden lines by making layer **HID** the *Current* layer, then drawing *Lines*. Notice that you only specify the *Line* endpoints as usual and AutoCAD creates the dashes.

   E. Next, create the centerlines for the holes by making layer **CEN** the *Current* layer and drawing *Lines*. Make sure the centerlines extend slightly past the *Circle* and past the horizontal *Lines* defining the hole. *Save* the drawing.

   F. Now create a *New* layer named **BORDER** and draw a border and title block of your design on that layer. The final drawing should appear like that in Figure 12-25.

   G. Open the *Layer & Linetype Properties* dialog box again and select the *Linetype* tab. Right-click to *Select All*, then select the *Delete* button. All unused linetypes should be removed from the drawing.

   H. *Save* the drawing.

Figure 12-25 ——————————

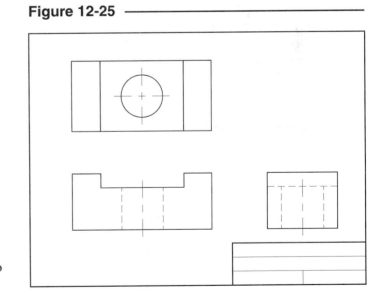

3. *LTSCALE, Change Properties dialog box, Matchprop*

Figure 12-26

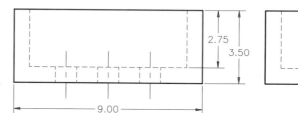

A. Begin a *New* drawing and use *Save* to assign the name **CH12EX3**. Create the same four *New* layers that you made for the previous exercise (**OBJ, HID, CEN,** and **BORDER**) and assign the same linetypes and colors. Set the *Limits* equal to a C size sheet, **22 x 17.**

B. Create the part shown in Figure 12-26. Draw on the appropriate layers to achieve the desired linetypes.

C. Notice that the *Hidden* and *Center* linetype dashes are very small. Use *LTSCALE* to adjust the scale of the non-continuous lines. Since you changed the *Limits* by a factor of slightly less than 2, try using **2** as the *LTSCALE* factor. Notice that all the lines are affected (globally) and that the new *LTSCALE* is retroactive for existing lines as well as for new lines. The dashes appear too long, however, because of the small hidden line segments, so *LTSCALE* should be adjusted. Remember that you cannot control where the multiple <u>short</u> centerline dashes appear for one line segment. Try to reach a compromise between the hidden and center line dashes.

D. The six short vertical hidden lines in the front view and two in the side view should be adjusted to a smaller linetype scale. Invoke the *Change Properties* dialog box by selecting *Properties* from the *Modify* pull-down menu or selecting the *Properties* icon button. Select <u>only the two short lines in the side view</u>. Use the dialog box to retroactively adjust the two individual objects' *Linetype Scale* to **0.6000.**

E. When the new *Linetype Scale* (actually, the *CELTSCALE*) for the two selected lines is adjusted correctly, use *Matchprop* to "paint" the *Linetype Scale* to the several other short lines in the front view. Select *Match Properties* from the *Modify* pull-down menu or select the "paint brush" icon. When prompted for the "Source Object," select one of the two short vertical lines in the side view. When prompted for the "Destination Object(s)," select <u>each</u> of the short lines in the front view. This action should "paint" the *Linetype Scale* to the lines in the front view.

F. When you have the desired linetype scales, *Exit* AutoCAD and *Save Changes*. Keep in mind that the drawing size may appear differently on the screen than it will on a print or plot. Both the plot scale and the paper size should be considered when you set *LTSCALE*. This topic will be discussed further in Chapter 13, Advanced Drawing Setup.

ADVANCED DRAWING SETUP

Chapter Objectives

After completing this chapter you should:

1. know the steps for setting up a drawing;

2. be able to determine an appropriate *Limits* setting for the drawing;

3. be able to calculate and apply the "drawing scale factor";

4. know the benefits and shortcomings of the *Quick Setup* and *Advanced Setup Wizards*;

5. be able to access existing and create new template drawings;

6. know what setup steps can be considered for creating template drawings.

CONCEPTS

When you begin a drawing, there are several steps that are typically performed in preparation for creating geometry, such as setting *Units, Limits,* and creating *Layers* with *linetypes* and *colors*. Some of these basic concepts were discussed in Chapters 6 and 12. This chapter discusses setting *Limits* for correct plotting as well as other procedures, such as layer creation and variables settings, that help prepare a drawing for geometry creation.

To correctly set up a drawing for printing or plotting to scale, <u>any two</u> of the following three variables must be known. The third can be determined from the other two.

> Drawing *Limits*
> Print or plot sheet (paper) size
> Print or plot scale

Initially, it would appear that using the automated setup "Wizards" available in the *Start Up* and *Create New Drawing* dialog boxes would be the fastest and easiest method for setting up a drawing. Surprisingly (unlike most other features in AutoCAD Release 14 that work so well), neither the *Quick Setup Wizard* nor the *Advanced Setup Wizard* correctly sets up a drawing for plotting to a standard scale. The reason is because the "wizards" consider only *Limits* and do not consider either sheet size or print/plot scale in their calculations. Therefore, you must have an understanding of the individual commands needed to correctly set up a drawing to scale. In addition, you should know when the setup wizards may and may not be of help to you.

Rather than performing the steps for drawing setup each time you begin, you can use "template drawings." Template drawings have many of the setup steps performed but contain no geometry. AutoCAD provides several template drawings and you can create your own template drawings. A typical engineering, architectural, design, or construction office generally produces drawings that are similar in format and can benefit from creation and use of individualized template drawings. The drawing similarities may be subject of the drawing (geometry), plot scale, sheet size, layering schemes, dimensioning styles, and/or text styles. Template drawing creation and use are also discussed in this chapter.

STEPS FOR DRAWING SETUP

Assuming that you have in mind the general dimensions and proportions of the drawing you want to create, and the drawing will involve using layers, dimensions, and text, the following steps are suggested for setting up a drawing:

1. Determine and set the *Units* that are to be used.
2. Determine and set the drawing *Limits;* then *Zoom All*.
3. Set an appropriate *Snap* value to be used if helpful.
4. Set an appropriate *Grid* value.
5. Change the *LTSCALE* value based on the new *Limits*.
6. Create the desired *Layers* and assign appropriate *linetype* and *color* settings.
7. Create desired *Text Styles* (optional, discussed in Chapter 18).
8. Create desired *Dimension Styles* (optional, discussed in Chapter 27).
9. Create or *Insert* the desired title block and border (optional).

Each of the steps for drawing setup is explained in detail here.

1. Set *Units*

> This task is accomplished by using the *Units* command, the *Units Control* dialog box (*Ddunits*), or the *Quick Setup* or *Advanced Setup Wizard*. Set the linear units and precision desired. Set angular units

and precision if needed. (See Chapter 6 for details on the *Units* command and see Setup Wizards in this chapter for setting *Units* using the wizards.)

2. Set *Limits*

Before beginning to create an AutoCAD drawing, determine the size of the drawing area needed for the intended geometry. Using the actual *Units*, appropriate *Limits* should be set in order to draw the object or geometry to the real-world size. *Limits* are set with the *Limits* command by specifying the lower-left and upper-right corners of the drawing area. Always *Zoom All* after changing *Limits*. *Limits* can also be set using the *Quick Setup* or *Advanced Setup Wizard*. (See Chapter 6 for details on the *Limits* command and see Setup Wizards in this chapter.)

If you are planning to plot the drawing to scale, *Limits* should be set to a proportion of the sheet size you plan to plot on. For example, if the sheet size is 11" x 8.5", set *Limits* to 11 x 8.5 if you want to plot full size (1"=1"). Setting *Limits* to 22 x 17 (2 times 11 x 8.5) provides 2 times the drawing area and allows plotting at 1/2 size (1/2"=1") on the 11" x 8.5" sheet. Simply stated, set *Limits* to a proportion of the paper size.

Setting *Limits* to the paper size allows plotting at 1=1 scale. Setting *Limits* to a proportion of the sheet size allows plotting at the reciprocal of that proportion. For example, setting *Limits* to 2 times an 11" x 8.5" sheet allows you to plot 1/2 size on that sheet. Or setting *Limits* to 4 times an 11" x 8.5" sheet allows you to plot 1/4 size on that sheet. (Standard paper sizes are given in Chapter 14.)

Drawing Scale Factor

The proportion of the *Limits* to the intended print or plot sheet size is the "drawing scale factor." This factor can be used as a general scale factor for other size-related drawing variables such as *LTSCALE, DIMSCALE* (dimensioning scale), and *Hatch* pattern scales. (Even if you plan to set up a drawing for plotting with Paper Space, the drawing scale factor should be calculated to determine system variable values and the *ZoomXP* scale factor.)

Most size-related AutoCAD drawing variables are set to 1 by default. This means that variables (such as *LTSCALE*) that control

Figure 13-1

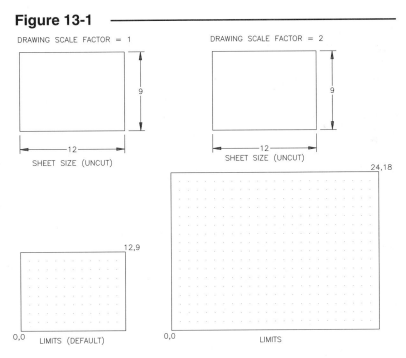

sizing and spacing of objects are set appropriately for creating a drawing plotted full size (1=1). Therefore, when you set *Limits* to the sheet size and print or plot at 1=1, the sizing and spacing of linetypes and other variable-controlled objects are correct. When *Limits* are changed to some proportion of the sheet size, the size-related variables should also be changed proportionally. For example, if you intend to plot on a 12 x 9 sheet and the default *Limits* (12 x 9) are changed by a factor of 2 (to 24 x 18), then 2 becomes the drawing scale factor. Then, as a general rule, the values of variables such as *LTSCALE, DIMSCALE,* and other scales should be multiplied by a factor of 2 (Fig. 13-1). When you then print the 24 x 18 area onto a 12" x 9" sheet, the plot scale would be 1/2, and all sized features would appear correct.

Limits should be set to the paper size or to a proportion of the paper size used for plotting or printing. In many cases, "cut" paper sizes are used based on the 11" x 8.5" module (as opposed to the 12" x 9" module called "uncut sizes"). Assume you plan to print on an 11" x 8.5" sheet. Setting *Limits* to the sheet size provides plotting or printing at 1=1 scale. Changing the *Limits* to a proportion of the sheet size <u>by some multiplier</u> makes that value the drawing scale factor. The <u>reciprocal</u> of the drawing scale factor is the scale for plotting on that sheet (Fig. 13-2).

Since the drawing scale factor (DSF) is the proportion of *Limits* to the sheet size, you can use this formula:

$$DSF = \frac{Limits}{Sheet\ size}$$

Because plot/print scale is the reciprocal of the drawing scale factor:

$$Plot\ scale = \frac{1}{DSF}$$

Figure 13-2

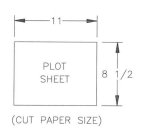

(CUT PAPER SIZE)

LIMITS	SCALE FACTOR	PLOT SCALE	
11 x 8.5	1	1=1	(FULL SIZE)
22 x 17	2	1=2	(HALF SIZE)
44 x 34	4	1=4	(1/4 SIZE)
110 x 85	10	1=10	(1/10 SIZE)

The term "drawing scale factor" is not a variable or command that can be found in AutoCAD or its official documentation. The setup wizards in Release 14 do not calculate or use a drawing scale factor based on sheet size. This concept has been developed by AutoCAD users to set system variables appropriately for printing and plotting drawings to standard scales.

3. Set *Snap*

Use the *Snap* command or *Drawing Aids* dialog box to set an appropriate *Snap* value if *Snap* is useful for your application. The value of *Snap* is dependent on the <u>interactive</u> drawing accuracy that is desired.

In some drawings where the *Limits* are relatively small and the dimensional increments are relatively large, *Snap* can be very useful. However, in other drawings with large *Limits* and small drawing increment lengths or with complex geometry of irregular interval lengths, *Snap* may not be useful and can be turned *Off*. (See Chapter 6 for details on the *Snap* command.)

The accuracy of the drawing and the size of the *Limits* should be considered when setting the *Snap* value. On one hand, to achieve accuracy and detail, you want the *Snap* to be as small of a dimensional increment as would be commonly used in the drawing. On the other hand, depending on the size of the *Limits*, the *Snap* value should be large enough to make interactive selection (PICKing) fast and easy. As a starting point for determining an appropriate *Snap* value, the default *Snap* value can be multiplied by the drawing scale factor.

If you use the *Quick Setup* or *Advanced Setup Wizard*, *Snap* is calculated for you. Because the wizards do not consider sheet size or scale, the resulting value is inappropriate for most applications. Examine the value and enter a new value based on the drawing scale factor.

4. Set *Grid*

The *Grid* value setting is usually set equal to or proportionally larger than that of the *Snap* value. *Grid* should be set to a proportion that is <u>easily visible</u> and to some value representing a regular increment (such as .5, 1, 2, 5, 10, or 20). Setting the value to a proportion of *Snap* gives visual indication of the *Snap* increment. For example, a *Grid* value of **1X, 2X,** or **5X** would give visual display of every 1, 2, or 5 *Snap* increments, respectively. If you are not using *Snap* because only extremely small or irregular interval lengths are needed, you may want to turn *Grid* off. (See Chapter 6 for details on the *Grid* command.)

If you use the *Quick Setup* or *Advanced Setup Wizard*, *Grid* is also calculated for you but may be inappropriately set. Examine the value and enter a new value based on the drawing scale factor.

R14

5. Set the *LTSCALE*

A change in *Limits* (then *Zoom All*) affects the display of non-continuous (hidden, dashed, dotted, etc.) lines. The *LTSCALE* variable controls the spacing of non-continuous lines. The drawing scale factor can be used to determine the *LTSCALE* setting. For example, if the default *Limits* (based on sheet size) have been changed by a factor of 2, the drawing scale factor=2; so set the *LTSCALE* to 2. In other words, if the DSF=2 and plot scale=1/2, you would want to double the linetype dashes to appear the correct size in the plot, so *LTSCALE*=2. If you prefer a *LTSCALE* setting of other than 1 for a drawing in the default *Limits*, multiply that value by the drawing scale factor.

American National Standards Institute (ANSI) requires that hidden lines be plotted with 1/8″ dashes. An AutoCAD drawing plotted full size (1 unit=1″) with the default *LTSCALE* value of 1 plots the *Hidden* linetype with 1/4″ dashes. Therefore, an *LTSCALE* of .5 is more appropriate for a drawing plotted full size. (Optionally, you can use the *Hidden2*, *Center2*, and other *2 linetypes with dash lengths of .5 times normal, then an *LTSCALE* of 1 is correct.) Multiply your *LTSCALE* value by the drawing scale factor when plotting to scale. If individual lines need to be adjusted for linetype scale, use *Modify Properties* (*Ddmodify* or *Ddchprop*) when the drawing is nearly complete. (See Chapter 12 for details on the *LTSCALE* and *CELTSCALE* variables.)

6. Create *Layers*; Assign *Linetypes* and *Colors*

Using the *Layer* command (*Layer* tab of the *Layer & Linetype Properties* dialog box), create the layers that you anticipate needing. Multiple layers can be created with the *New* option. You can type in several new names separated by commas. Assign a descriptive name for each layer, indicating its type of geometry, part name, or function. Include a *linetype* designator in the layer name if appropriate; for example, PART1-H and PART1-V indicate hidden and visible line layers for PART1.

Once the *Layers* have been created, assign a *Color* and *Linetype* to each layer. *Colors* can be used to give visual relationships among parts of the drawing. Geometry that is intended to be plotted with different pens (pen size or color) or printed with different line widths should be drawn in different screen colors. Use the *Select Linetype* dialog box to load the desired linetypes. You can load only linetypes that you anticipate using or load all linetypes and *Delete* the unused ones when the drawing is completed. (See Chapter 12 for details on creating *Layers*, *Linetypes*, and *Colors*.)

7. Create Text Styles

AutoCAD has only one text style as part of the standard template drawings (ACAD.DWT and ACADISO.DWT) and one or two text styles for most other template drawings. If you desire other *Text Styles*, they are created using the *Style* command or the *Text Style...* option from the *Data* pulldown menu. (See Chapter 18, Creating and Editing Text.) If you desire engineering standard text, create a *Text Style* using the ROMANS.SHX or .TTF font file.

8. Create Dimension Styles

If you plan to dimension your drawing, Dimension Styles can be created at this point; however, they are generally created during the dimensioning process. Dimension Styles are names given to groups of dimension variable settings. (See Chapter 27 for information on creating Dimension Styles.)

Although you do not have to create Dimension Styles until you are ready to dimension the geometry, it is helpful to create Dimension Styles as part of a template drawing. If you produce similar drawings repeatedly, your dimensioning techniques are probably similar. Much time can be saved by using a template drawing with previously created Dimension Styles. Several of the AutoCAD-supplied template drawings have prepared Dimension Styles.

9. Create a Title Block and Border

For 2D drawings, it is helpful to insert a title block and border early in the drawing process. This action gives a visual drawing boundary as well as reserves the space needed by the title block.

Since *Limits* are already set to facilitate drawing full size in the actual *Units,* creating a title block and border which will appear on the final plot in the appropriate size is not difficult. A simple method is to use the drawing scale factor to determine the title block size. Multiply the actual size of the title block and border by the drawing scale factor to determine their sizes for the drawing.

One common method is to use the *Insert* command to insert a title block and border as a *Block.* If the *Block* is the actual title block and border size needed for the plot sheet, simply use the drawing scale factor as the *X* and *Y scale factor* during the *Insert* command. (See Chapter 20 for information on *Block* creation and insertion.)

If you are preparing template drawings, a title block and border can be included as part of a template drawing. Most of the AutoCAD-supplied template drawings have a title block and border included, but in each case the title block and border is in Paper Space. Using Paper Space has many advantages and some disadvantages but is recommended for use only when you have had more experience with AutoCAD. See Chapter 19 for introductory information on Paper Space.

QUICK SETUP WIZARD AND *ADVANCED SETUP* WIZARD

Since the *Quick Setup* and *Advanced Setup Wizards* that are available in the *Start Up* and *Create New Drawing* dialog boxes do not allow input for either the plot/print sheet size or the intended plot scale, they cannot be used effectively for setting up a drawing correctly for plotting or printing to scale. The "Wizards" can be useful in some cases, however, such as setting up a drawing to print or plot on an A size sheet or for some of the initial steps to set up a metric drawing. This section closely examines the *Quick Setup Wizard* and the *Advanced Setup Wizard* to determine their shortcomings and usefulness. See Chapter 6 for coverage of the *Start from Scratch* option in the *Start Up* and *Create New Drawing* dialog boxes. Using template drawings is discussed later in this chapter.

Quick Setup Wizard

Select the *Use a Wizard* button in the *Start Up* or *Create New Drawing* dialog box to present a choice of *Quick Setup* or *Advanced Setup* (Fig. 13-3). The *Quick Setup Wizard* automates the first four steps listed under STEPS FOR DRAWING SETUP on the chapter's first page. Those functions, simply stated, are:

1. *Units*
2. *Limits*
3. *Snap*
4. *Grid*

Figure 13-3

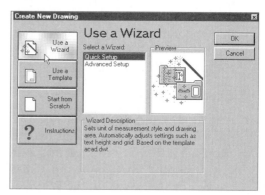

The Wizard prompts you for *Units* and *Area* (*Limits*) and sets the *Grid* and *Snap* values automatically, as well as setting many other size-related system variables, based on your input for the *Limits* values. The *Quick Setup Wizard* uses a multiplier to set the system variables similar to the concept of a drawing scale factor, except the multiplier is based only on the proportion of *Limits* to the default 12 x 9 *Limits* and not to the intended plot sheet size; therefore, the resulting setup is useful only for plotting to scale on an A size (11" x 8.5" or 12" x 9") sheet.

Choosing the *Quick Setup Wizard* invokes the *Quick Setup* dialog box (Fig. 13-4). There are two steps. To proceed through the steps, you can select the *Next* button or select each tab in succession.

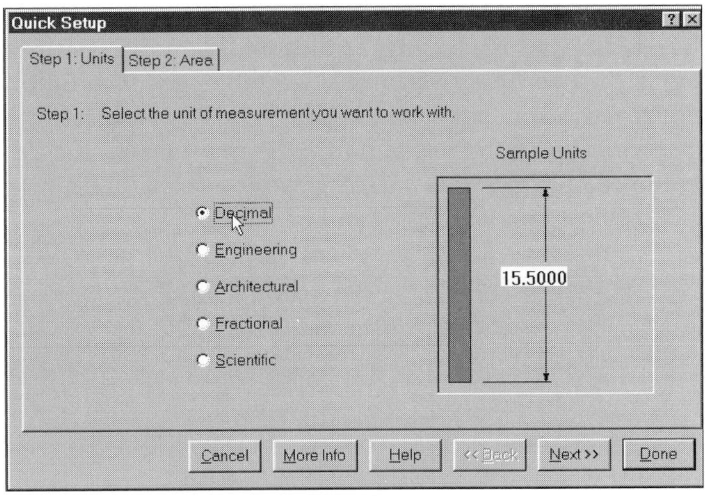

Step 1: Units
Press the desired radio button to display the units you want to use for the drawing. The options are:

Decimal	Use generic decimal units with a precision of 0.0000.
Engineering	Use feet and decimal inches with a precision of 0.0000.

Architectural   Use feet and fractional inches with a precision of 1/16 inch.
Fractional   Use generic fractional units with a precision of 1/16 units.
Scientific   Use generic decimal units showing a precision of 0.0000.

Use *Architectural* or *Engineering* units if you want to specify coordinate input using feet values with the ' (apostrophe) symbol. If you want to set additional parameters for units such as precision or system of angular measurement, use the *Units* or *Ddunits* command. Keep in mind the setting you select in this step changes only the display of units in the coordinate display area of the Status Line (*Coords*) and in dialog boxes but not necessarily for the dimension text format. Select the *Next* button or the *Area* tab after specifying *Units*.

Figure 13-4

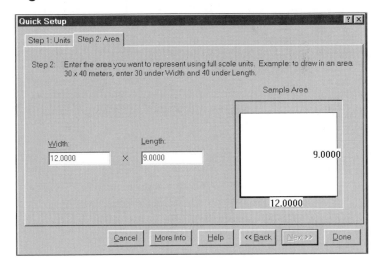

Step 2: Area
Enter two values that constitute the X and Y measurements of the area you want to work in (Fig. 13-5). These values set the *Limits* of the drawing. The first edit box labeled *Width* specifies the X value for *Limits*. The X value is usually the longer of the two measurements for your drawing area and represents the distance across the screen in the X direction or along the long axis of a sheet of paper. The second edit box labeled *Length* specifies the Y value for *Limits* (the Y distance of the drawing area). (Beware: The terms "width" and "length" are misleading since "length" is actually defined as the measurement of something along its greatest dimension. [*The American Heritage Dictionary, Office Edition*, Third Edition, 1994, Dell Publishing.] When setting AutoCAD *Limits*, the Y measurement is generally the shorter of the two measurements.) The two values together specify the upper-right corner of the drawing *Limits*. The Wizard automatically sets *Snap* and *Grid* values for you based on the values you specify for *Area* (*Limits*). The Wizard also calculates settings for most size-related system variables.

Figure 13-5

When you finish the two steps, *Quick Setup Wizard* automatically centers the drawing area (*Limits*) in the drawing editor. If you *Start from Scratch*, change *Limits*, then *Zoom All*, the left edge defined by *Limits* is always against the left edge of the screen.

Table of *Quick Setup Wizard* Settings

The table below gives a few examples (see Table of System Variables Affected by Wizards for full list) of the system variable settings (columns 3, 4, and 5) that result when specific *Area* (*Limits*) values are specified in the *Quick Setup Wizard*. (See Tables of *Limits* Settings, Chapter 14, for a complete list of standard plot scales, paper sizes, and appropriate *Limits* settings.) The area sizes that are used for these examples are proportional to the standard U.S. architectural (based on the 12 x 9 module) and engineering (based on the 11 x 8.5 module) paper sizes. The last column lists the plot scales that are appropriate for the variable settings *Quick Setup Wizard* generates.

Area (*Limits*)	Paper Proportion	*SNAP, GRID*	*LTSCALE, DIMSCALE*	*TEXTSIZE*	Plot Scale for "A" Size Sheet
12 x 9	1 x A architectural	.5	1	.2	1=1
18 x 12	1 x B architectural	.7	1.3333	.2667	3/4=1
24 x 18	1 x C architectural	1	2	.4	1/2=1
36 x 24	1 x D architectural	1	2.667	.5333	3/8=1
48 x 36	1 x E architectural	2	4	.8	1/4=1
12' x 9'	12 x A architectural	6"	12 (1')	2 3/8" (.2')	1"=1'
24' x 18'	24 x A architectural	12 (1')	24 (2')	4 13/16"(.4')	1/2"=1'
48' x 36'	48 x A architectural	24 (2')	48 (4')	9 5/8" (.8')	1/4"=1'
11 x 8.5	1 x A engineering	.5	.9444	.1889	1=1
17 x 11	1 x B engineering	.6	1.2222	.2444	3/4=1
22 x 17	1 x C engineering	.9	1.8889	.3778	1/2=1
34 x 22	1 x D engineering	1	2.4444	.4889	3/8=1
44 x 34	1 x E engineering	2	3.7778	.7556	1/4=1

Note that the *Quick Setup Wizard* generates the *DIMSCALE, LTSCALE,* and *TEXTSIZE* system variable settings (as well as other system variables listed in the Table of System Variables Affected by Wizards) based on the formula

$$N=D(Y/9)$$

where N is the new system variable setting,
D is the system variable default value,
Y is the *Area* (*Limits*) Y value, and
9 is the default *Area* (*Limits*) Y value.

The *Snap* and *Grid* values are rounded to the nearest 1/10 unit when N is under 1 unit and rounded to the nearest integer when N is over 1 unit.

Thus, the *Quick Setup Wizard* calculates appropriate settings (except *Snap* and *Grid* for some cases) for plotting to scale on A size sheets only. This is because the *Quick Setup Wizard* does not consider plot sheet size or plot scale in its calculation. Also notice the *Snap* and *Grid* values for 18 x 12, 17 x 11, and 22 x 17 are inappropriate and should be changed.

In summary, you can use the *Quick Setup Wizard* to automate setting up a drawing to plot to a standard scale on A size sheets. In some cases, you may want to change the *Snap* and *Grid* settings as well as other system variables.

Advanced Setup Wizard

The *Advanced Setup Wizard* performs the same tasks as the *Quick Setup Wizard* with the addition of allowing you to select units precision and other options (normally available in the *Units* dialog box), insert a title block, and enable Paper Space. Selecting *Advanced Setup* produces the dialog box series shown in Figures 13-6 through 13-12. Seven tabs, or "Steps," are involved in the series. To proceed through the steps, you can select the *Next* button or select each tab in succession.

The following list indicates the STEPS FOR DRAWING SETUP on the chapter's first page and the related tabs or "Steps" in the *Advanced Setup Wizard*:

1.	*Units*	*Step 1: Units*
		Step 2: Angle
		Step 3: Angle Measure
		Step 4: Angle Direction
2.	*Limits*	*Step 5: Area*
3.	*Snap*	(automatic)
4.	*Grid*	(automatic)
5.	*Insert Title Block*	*Step 6: Title Block*
		Step 7: Layout

Similar to the function of the *Quick Setup Wizard*, several system variables are changed from the default settings to other values based on your input for the *Limits* values.

Step 1: Units

You can select the units of measurement for the drawing as well as the unit's *Precision*. These are the same options available in the *Units* dialog box (see *Units*, Chapter 6). This is similar to *Step 1* in the *Quick Setup Wizard* but with the addition of *Precision*.

Similar to using the *Units* or *Ddunits* command, your choices in this and the next three dialog boxes determine the display of units for the coordinate display area of the Status line (*Coords*) and in dialog boxes. If you want to use feet units for coordinate input, select *Architectural* or *Engineering*.

Step 2: Angle

This tab provides for your input of the desired system of angular measurement. Select the drop-down list to select the angular *Precision*.

Figure 13-6

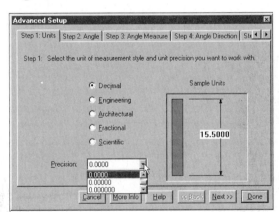

Figure 13-7

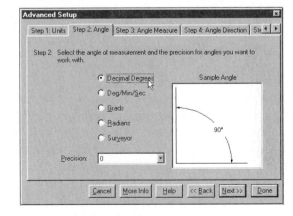

Step 3: Angle Measure

This step sets the direction for angle 0. East (X positive) is the AutoCAD default. This has the same function as selecting the *Angle 0 Direction* in the *Units Control* dialog box.

Figure 13-8

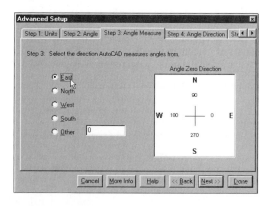

Step 4: Angle Direction

Select *Clockwise* if you want to change the AutoCAD default setting for <u>measuring</u> angles. This setting (identical to the *Units Control* dialog box option) affects the direction of positive and negative angles in commands such as *Rotate, Array Polar* and dimension commands that measure angles but does not affect the direction *Arcs* are drawn (always counter-clockwise).

Figure 13-9

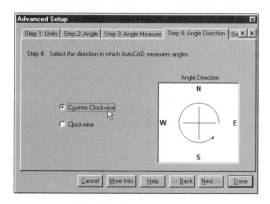

Step 5: Area

Enter values to define the upper-right corner for the *Limits* of the drawing. The *Width* refers to the X *Limits* component and the *Length* refers to the Y component. Generally, the *Width* edit box contains the larger of the two values unless you want to set up a vertically oriented drawing area. Even if you plan to insert a title block (*Step 6*) into Model Space (*Step 7*), do not skip this step just because *Limits* will be reset by the insertion. The values you input in this step are used to calculate other system variable settings (see Table of System Variables Affected by Wizards).

If you plan to print or plot to a standard scale, your input for *Area* should be based on the intended plot scale and sheet size. See the Tables of *Limits* Settings in Chapter 14 for appropriate values to use.

Figure 13-10

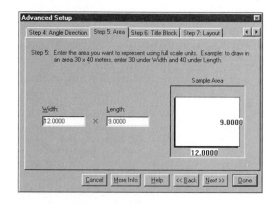

Step 6: Title Block

You can select from several AutoCAD-supplied title blocks listed next or select from custom (your own) title blocks by using the *Add* button to select a custom title block in DWG format. The files you select are added to the bottom of the drop-down list. The selected title block is inserted into the drawing as a *Block* object. A new layer is created called TITLE_BLOCK onto which the block is inserted. If you insert the title block into Model Space (*Step 7*), *Limits* are reset to match the title block size.

Figure 13-11

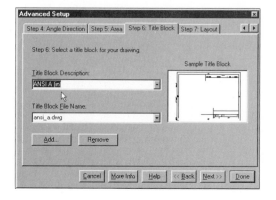

Inserting the AutoCAD-supplied title blocks into Model Space (*Step 7*) provides for plotting or printing a drawing at 1=1 scale. If you insert the title blocks into Model Space and intend to plot at other than 1=1, you will have to reset *Limits* and use *Scale* to rescale the title block to the appropriate size. If you insert the title blocks into Paper Space in *Step 7*, this procedure is not necessary.

Beginning a drawing with a predefined title block can also be accomplished by using a template drawing (including a title block). See Using and Creating Template Drawings.

Step 7: Layout
In this step you determine if you want to use Paper Space or not. Selecting *No* (not the default in this tab) allows you to work in Model Space, which is the space used for creating geometry and is generally the choice for simple 2D drawings. Select *Yes* if you want to use Paper Space. You should use Paper Space if you need <u>several views</u> (viewports) displaying one drawing or several drawings. The need to (1) plot or print several views of one drawing, (2) create a plot or print of several drawings, or (3) lay out different views of a 3D object are three good reasons for using Paper Space. Selecting the *Yes* option, however, creates only one Paper Space viewport. The viewport is automatically created on layer VIEWPORT. If you have selected a title block (*Step 6*), the title block is inserted into Paper Space.

Figure 13-12

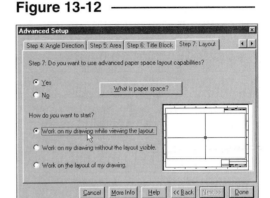

Below is a translation of the three choices for *How do you want to start?*

Option	Meaning
Work on my drawing while viewing the layout.	*Tilemode*=0, Model Space current
Work on my drawing without the layout visible.	*Tilemode*=1, Model Space current
Work on the layout of my drawing.	*Tilemode*=0, Paper Space current

See Chapter 19 for more information on Paper Space.

Table of System Variables Affected by Wizards

Your choice for *Area* (*Limits*) in both the *Quick Setup* and the *Advanced Setup Wizards* affects the following variables.

AFFECTED SYSTEM VARIABLE	DESCRIPTION
CHAMFERA	First *Chamfer* distance
CHAMFERB	Second *Chamfer* distance
CHAMFERC	*Chamfer* length
DIMSCALE	Overall dimensioning scale factor for size-related features
DIMSTYLE	Current dimensioning style
DONUTID	Default *Donut* inside diameter
DONUTOD	Default *Donut* outside diameter

AFFECTED SYSTEM VARIABLE	DESCRIPTION
FILLETRAD	Default *Fillet* radius
GRIDUNIT	*Grid* spacing for current viewport, X and Y
HPSCALE	Default *Hatch* pattern scale factor
HPSPACE	Default *Hatch* pattern line spacing for "U," user-defined simple patterns
LIMMAX	Upper-right drawing *Limits* for the current space
LTSCALE	Global linetype scale factor
OFFSETDIST	Default distance for *Offset*
SNAPUNIT	*Snap* spacing for current viewport, X and Y
TEXTSIZE	The default height for new *Text* objects drawn with the current text style
VIEWCTR	Center of view in current viewport
VIEWSIZE	Height of view in current viewport

All but one of these variables requires numeric input. The value used for each variable is calculated by the formula

$$N=D(Y/9)$$

where N is the new system variable setting,
       D is the system variable default value,
       Y is the *Area* (*Limits*) Y value, and
       9 is the default *Area* (*Limits*) Y value.

The *Snap* and *Grid* values are rounded to the nearest 1/10 unit when N is under 1 unit and are rounded to the nearest integer when N is over 1 unit.

The *DIMSTYLE* variable is set to "STANDARD_WIZARDSCALED" to provide dimensions that are scaled correctly, based on your input for *Area*.

Template Drawings
As an alternative to using *Start from Scratch* to begin a new drawing or using a *Setup Wizard* to set up your drawings, you should consider using template drawings. Template drawings can contain all the features provided by the wizards but can also include many other settings and features.

USING AND CREATING TEMPLATE DRAWINGS

Instead of going through the steps for setup each time you begin a new drawing, you can create one or more template drawings or use one of the AutoCAD-supplied template drawings. AutoCAD Release 14 template drawings are saved as a DWT file format. A template drawing is one which has the initial setup steps (*Units, Limits, Layers, Linetypes, Colors,* etc.) completed and saved, but no geometry has yet been created. Template drawings are used as a template or starting point each time you begin a new drawing. AutoCAD actually makes a copy of the template you select to begin the drawing. The creation and use of template drawings can save hours of preparation.

R14

Using Template Drawings

To use a template drawing, select the *Use a Template* option in the *Create New Drawing* dialog box (Fig. 13-13). This option allows you to select the desired template (DWT) drawing from a list including all templates in the TEMPLATE folder (usually in the C:\Program Files\AutoCAD R14\Template subdirectory). Any template drawings that you create using the *SaveAs* command (as described in the next section) appear in the list of available templates. Select the *More files...* option to browse other folders.

Figure 13-13

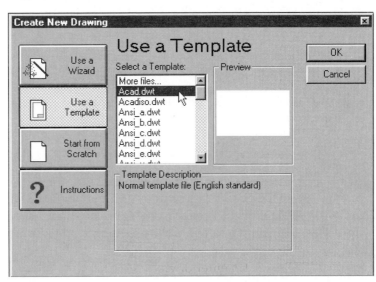

Several template drawings are supplied with AutoCAD Release 14 (see the selections in Fig. 13-13). You may notice similarities to the template drawing file names and the title block names that are available in the *Advanced Setup Wizard*. Generally, the template drawing files include the title blocks. In some cases, Dimension Styles have been created but other changes have not been made to the template drawings such as layer setups or loading linetypes.

See the Tables of Limits Settings in Chapter 14 for standard plotting scales and intended sheet sizes for the *Limits* settings in these templates.

Creating Template Drawings

To make a template drawing, begin a *New* drawing using the default template drawing (ACAD.DWT) or other template drawing (*.DWT), then make the initial drawing setups. Alternately, *Open* an existing drawing that is set up as you need (layers created, linetypes loaded, *Limits* and other settings for plotting or printing to scale, etc.), and then *Erase* all the geometry. Next, use *SaveAs* to save the drawing under a different descriptive name with a DWT file extension. Do this by selecting the *Drawing Template File (*.dwt)* option in the *Save Drawing As* dialog box (Fig. 13-14). AutoCAD automatically (by default) saves the drawing in the folder where other template (*.DWT) files are found (usually in the C:\Program Files\AutoCAD R14\Template subdirectory).

Figure 13-14

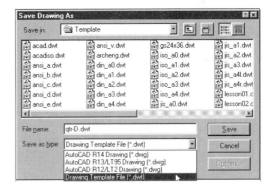

Multiple template drawings can be created, each for a particular situation. For example, you may want to create several templates, each having the setup steps completed but with different *Limits* and with the intention to plot or print each in a different scale or on a different size sheet. Another possibility is to create templates with different layering schemes. There are many possibilities, but the specific settings used depend on your applications (geometry), typical scales, or your plot/print devices.

Typical drawing steps that can be considered for developing template drawings are listed below:

> Set *Units*
> Set *Limits*
> Set *Snap*
> Set *Grid*
> Set *LTSCALE*
> Create *Layers* with color and linetypes assigned
> Create *Text Styles* (see Chapter 18)
> Create Dimension Styles (see Chapter 27)
> Create or *Insert* a standard title block and border

A popular practice is to create a template drawing for <u>each sheet size</u> used for plotting. *Limits* are set to the actual sheet size, and other settings and variables (*LTSCALE, DIMSCALE,* text sizes) are set to the actual desired size for the finished full-size plot. If you then want to create a new drawing and plot on a certain sheet size, start a new drawing using the appropriate template.

For laboratories or offices that use only one sheet size, create template drawings for <u>each plot scale</u> with the appropriate *Limits, Snap, Grid, Layers, LTSCALE, DIMSCALE,* etc.

You can also create templates for different plot scales <u>and</u> different sheet sizes. Make sure you assign appropriate and descriptive names to the template drawings.

Previous AutoCAD Releases

In previous releases of AutoCAD, any drawing file (*.DWG) could be used as a "prototype" drawing. This scheme is similar to the template concept in that a selected prototype drawing is used as a template, but because all prototypes have a .DWG file extension, there could be some confusion as to which files were to be used only as prototypes. The template (*.DWT) scheme in Release 14 provides a more secure method of file management and complies with similar template schemes in other software packages such as word processors and spreadsheets.

CHAPTER EXERCISES

For the first three exercises, you will set up several drawings that will be used in other chapters for creating geometry. Follow the typical steps for setting up a drawing given in this chapter.

1. **Start from Scratch**
    You are to create a drawing of a wrench. The drawing will be printed full size on an A size sheet (not on an A4 metric sheet), and the dimensions are in millimeters. Start a *New* drawing, select **Start from Scratch,** and select the **English** default settings. Follow these steps:

    A. Set **Units** to **Decimal** and **Precision** to **0.00.**
    B. Set **Limits** to a metric sheet size, **297 x 210;** then **Zoom All** (scale factor is approximately 25).
    C. Set **Snap** to **1.**
    D. Set **Grid** to **10.**
    E. Set **LTSCALE** to **12.5** (.5 times the scale factor of 25).
    F. **Load** the **Center Linetype.**

G. Create the following *Layers* and assign the *Colors* and *Linetypes* as shown:

WRENCH	*continuous*	*red*
CONSTR	*continuous*	*white*
CENTER	*center*	*green*
TITLE	*continuous*	*yellow*
DIM	*continuous*	*cyan*

H. *Save* the drawing and assign the name **WRENCH** for use in a later chapter exercise.

2. *Quick Setup Wizard*
    A drawing of a mechanical part is to be made and plotted full size (1"=1") on an 11" x 8.5" sheet. Use the *Quick Setup Wizard* to assist with the setup.

    A. Begin a *New* drawing. When the *Startup* or *Create new Drawing* dialog box appears, select *Use a Wizard*. Select *Quick Setup*.
    B. In the *Quick Setup* dialog box, select *Decimal* in the *Step 1: Units* tab.
    C. Press *Next* or select the *Step 2: Area* tab. Enter **11** and **8.5** in the two edit boxes so 11 appears in the *Sample Area* image tile along the bottom (X direction) and 8.5 appears along the right side (in the Y direction).
    D. Select the *Done* button.
    E. Check to ensure the *Limits* are set to 11,8.5 and *Grid* is set to .5.
    F. Change the *Snap* to **.250**.
    G. Set *LTSCALE* to **.5**.
    H. *Load* the *Center* and *Hidden Linetypes*.
    I. Create the following *Layers* and assign the given *Colors* and *Linetypes:*

VISIBLE	*continuous*	*red*
CONSTR	*continuous*	*white*
CENTER	*center*	*green*
HIDDEN	*hidden*	*yellow*
TITLE	*continuous*	*white*
DIM	*continuous*	*cyan*

    J. Save the drawing as **BARGUIDE** for use in another chapter exercise.

3. *Quick Setup Wizard*
    A floor plan of an apartment has been requested. Dimensions are in feet and inches. The drawing will be plotted at 1/2"=1' scale on an 24" x 18" sheet.

    A. Begin a *New* drawing and select the *Quick Setup Wizard*.
    B. Select *Architectural* in the *Step 1: Units* tab.
    C. In the *Step 2: Area* tab, set *Width* and *Length* to **48' x 36'** (576" x 432"), respectively (use the apostrophe symbol for feet). Select *Done*.
    D. Reset the *Snap* value to **1** (inch).
    E. Reset the *Grid* value to **12** (inches).
    F. Reset the *LTSCALE* to **12**.
    G. Create the following *Layers* and assign the *Colors*:

FLOORPLN	*red*
CONSTR	*white*
TEXT	*green*
TITLE	*yellow*
DIM	*cyan*

H. *Save* the drawing and assign the name **APARTMENT** to be used later.

4. *Quick Setup Wizard*

A drawing is to be created and is to be printed in 1/4"=1" scale on an A size (11" x 8.5") sheet. The *Quick Setup Wizard* can be used to automate the setup. (See Table of *Quick Setup Wizard* Settings, Plot Scale for A Size Sheet column.)

A. Begin a *New* drawing. When the *Startup* or *Create new Drawing* dialog box appears, select *Use a Wizard*. Select *Quick Setup*.
B. In the *Quick Setup* dialog box, select *Decimal* in the *Step 1: Units* tab.
C. Press *Next* or select the *Step 2: Area* tab. Enter **44** and **34** in the two edit boxes so 44 appears in the *Sample Area* image tile along the bottom (X direction) and 34 appears along the right side (in the Y direction).
D. Select the *Done* button.
E. Check to ensure the *Limits* are set to 44,34 and *Snap* and *Grid* are set to 1.
F. Save the drawing as **CH13EX4.**

5. *Advanced Setup Wizard*

In this exercise you will set up a drawing using a standard ANSI title block for printing on an A size (11" x 8.5") sheet for a mechanical engineering application.

A. Begin a *New* drawing. In the *Create New Drawing* dialog box, select *Use a Wizard,* then select *Advanced Setup*.
B. When the *Advanced Setup* dialog box appears, accept all of the default *Units* settings. Proceed to *Step 5: Area*.
C. In the *Area* tab, set the *Width* and *Length,* respectively, to **11** and **8.5**.
D. In the next tab, *Step 6: Title Block*, select the *ANSI A (in)* option.
E. In the *Step 7: Layout* tab, answer *No* to the *Do you want to use advanced paper space layout capabilities?* option.
F. Use *SaveAs* and assign the name **A_ANSI.**

6. *Advanced Setup Wizard*

Use the *Advanced Setup Wizard* to begin a drawing for an architectural application. The drawing is set up with a title block in Paper Space for a D size sheet and will be plotted in 1/4"=1' scale.

A. Begin a *New* drawing. In the *Create New Drawing* dialog box, select *Use a Wizard,* then select *Advanced Setup*.
B. When the *Advanced Setup* dialog box appears, select *Architectural* in the *Step 1: Units* tab. Select a *Precision* of *1/8"*.
C. Proceed to *Step 5: Area*. Enter *Width* and *Length* values of **144'** and **96'**, respectively. Make sure you use the ' (apostrophe) symbol or else inches are assumed.
D. In *Step 6: Title Block*, select the *Arch/Eng (in)* title block option.
E. Finally, in *Step 7: Layout,* select *Yes* to the option for using Paper Space. Also select *Work on my drawing while viewing the layout* to work in Model Space and see the title block in Paper Space. Select *Done*.
F. Reset *Snap* to **2'** (feet) and *Grid* to **4'** (feet).
G. Create the following *Layers* and assign the *Colors*:

FLOORPLN	*red*
HVAC	*cyan*
PLUMB	*yellow*
ELEC	*green*

| TEXT | *blue* |
| DIM | *magenta* |

H. Save the drawing in your working directory as **D_ARCH**.

7. **Create Template Drawings**
In this exercise, you will create several "generic" template drawings that can be used at a later time. Creating the templates now will save you time later when you begin new drawings.

A. Create a template drawing for use with decimal dimensions and using standard paper A size format. Begin a *New* drawing and select *Start from Scratch*. Select the *English* default settings. Include the following setups:

1. Set *Units* to *Decimal* and *Precision* to **0.00**.
2. Set *Limits* to **11** x **8.5**.
3. Set *Snap* to **.25**.
4. Set *Grid* to **1**.
5. Set *LTSCALE* to **.5**.
6. Create the following *Layers*, assign the *Linetypes* as shown, and assign your choice of *Colors*. Create any other layers you think you may need or any assigned by your instructor.

OBJECT	*continuous*
CONSTR	*continuous*
TEXT	*continuous*
TITLE	*continuous*
VPORTS	*continuous*
DIM	*continuous*
HIDDEN	*hidden*
CENTER	*center*
DASHED	*dashed*

7. Use *SaveAs* and name the template drawing **ASHEET**. Make sure you select *Drawing Template File (*.dwt)* from the *Save as Type:* drop-down list in the *Save Drawing As* dialog box. Also, from the *Save In:* drop-down list on top, select your working directory as the location to save the template. In the *Template Description* dialog box, enter "**A-size Engineering Sheet 11 × 8.5**."

B. Using the drawing in the previous exercise, create a template for a standard engineering B size sheet. First, use the *New* command and select the *Use a Template* option from the *Create New Drawing* dialog box. Select *More files...* from the list in center of the dialog box. When the *Select Template* dialog box appears, select the **ASHEET.DWT** from your working directory (not from the default folder with the AutoCAD supplied templates). When the drawing opens, set *Limits* to **17** x **11**. All other settings and layers are OK as they are. Use *SaveAs* and save the new drawing as a template (**DWT**) drawing in your working directory. Assign the name **BSHEET**. Add the appropriate description in the *Template Description* dialog box.

C. In this exercise, create a template for a standard engineering C size sheet. Use the *New* command and use the **ASHEET** template drawing (use the same methods explained in problem 7. B to use the template). Set *Limits* to **22** x **17**. Set *Snap* to **.5**. Keep all the other settings and layers as they are. Use *SaveAs*, assign the name **CSHEET**, and save the template (**DWT**) file in your working directory. Add the appropriate description in the *Template Description* dialog box.

D. Use the **ASHEET** template drawing (by the methods explained previously), only in this exercise create a template for a standard engineering D size sheet. Set *Limits* to **34** x **22**. All other settings and layers do not need to be changed. Assign the name **DSHEET** when you save it as a template drawing in your working directory. Add the appropriate description in the *Template Description* dialog box.

14

PRINTING AND PLOTTING

Chapter Objectives

After completing this chapter you should:

1. know the typical steps for printing or plotting;

2. be able to invoke and use the *Print/Plot Configuration* dialog box;

3. be able to select from available plotting devices and set the paper size and orientation;

4. be able to specify what area of the drawing you want to print or plot;

5. be able to preview the print/plot before creating a plotted drawing;

6. be able to specify a scale for printing or plotting a drawing;

7. know how to set up a drawing for plotting to a standard scale on a standard size sheet;

8. be able to use the tables of *Limits* Settings to determine *Limits,* scale, and paper size settings;

9. know how to configure plot and print devices in AutoCAD.

CONCEPTS

Plotting and printing are accomplished from within AutoCAD by invoking the *Plot* command. Using the *Plot* or *Print* command invokes the *Print/Plot Configuration* dialog box (Fig. 14-1). You have complete control of plotting and printing using the dialog box. In AutoCAD, the term "plotting" can refer to plotting on a pen plotter and/or printing with a printer.

Figure 14-1

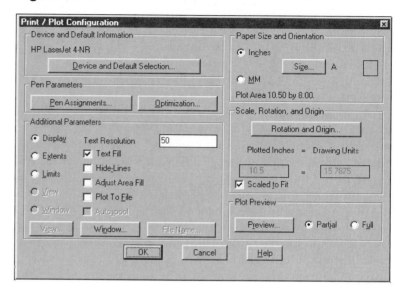

TYPICAL STEPS TO PLOTTING

Assuming the CAD system has been properly configured so the peripheral devices (plotters and/or printers) are functioning, the typical basic steps to printing and plotting using the *Print/Plot Configuration* dialog box are listed below:

1. Use *Save* to ensure the drawing has been saved in its most recent form before plotting (just in case some problem arises while plotting).

2. Make sure the plotter or printer is turned on, has paper and pens loaded, and is ready to accept the plot information from the computer.

3. Invoke the *Print/Plot Configuration* dialog box.

4. Check the upper-left corner of the dialog box to ensure that the intended device has been selected. If not, select the *Device and Default Selection* tile and make the desired choice.

5. Check the upper-right corner of the dialog box to ensure the desired paper size has been selected. If not, use the *Size...* tile to do so.

6. Only when necessary, change other options such as *Rotation and Origin* and *Pen Assignments*.

7. Determine and select which area of the drawing to plot: *Display, Extents, Limits, Window,* or *View*.

8. Enter the desired scale for the print or plot. If no standard scale is needed, toggle *Scaled to Fit* (so the check mark appears in the box).

9. Always *Preview* the plot to ensure the drawing will be printed or plotted as you expect. Select either a *Full* or *Partial* preview. If the preview does not display the plot as you intended, make the appropriate changes. Otherwise, needless time and media could be wasted.

10. If everything is OK, selecting the *OK* tile causes the drawing to be sent to the plotter or printer.

11. For additional plots or prints, you can use the *Preview* command to preview and plot the drawing based on the parameters previously set in the *Print/Plot Configuration* dialog box.

USING THE *PLOT* COMMAND

PLOT

	Pull-down Menu	COMMAND (TYPE)	ALIAS (TYPE)	Short-cut	Screen (side) Menu	Tablet Menu
	File *Print*	*PLOT* or *PRINT*	...	*Ctrl+P*	**FILE** *Plot*	*W,25*

Invoking *Plot* or *Print* normally invokes the *Print/Plot Configuration* dialog box (Fig. 14-1). (The *Print/Plot Configuration* dialog box can be suppressed by setting the *CMDDIA* system variable to **0**. This action displays a text screen instead and allows changing parameters by keyboard entry or plot script.)

Print/Plot Configuration Dialog Box

The *Print/Plot Configuration* dialog box allows you to change plotting parameters such as scale, paper size, pen assignments, rotation, and origin of the drawing on the sheet. Changes made to plotting options are saved (in the ACAD14.CFG file) so they do not have to be re-entered the next time.

Device and Default Selection

Many devices (printers or plotters) can be configured for use with AutoCAD, and all configurations are saved for your selection. For example, you can have both an A size and a D size plotter as well as a laser printer, any one of which could be used to plot the current file.

When you start AutoCAD for the first time, it automatically configures itself to use the Windows system printer. Additional devices for use specifically for AutoCAD can be configured within AutoCAD by using the *Printers* tab of the *Preferences* dialog box. You can also type *Config* to directly open the *Printers* tab of the *Preferences* dialog box. See Configuring Plotters and Printers near the end of this chapter.

Selecting the *Device and Default Selection* tile from the *Print/Plot Configuration* dialog box invokes the *Device and Default Selection* dialog box (Fig. 14-2). Highlighted near the top is the currently selected device along with the other configured choices.

Since multiple devices can be used and there may also be several typical sets of parameters frequently used for each device (scale, pen assignments, etc.), multiple settings of plotting parameters can be saved and retrieved by using the *Save...* and *Replace...* or *Save...* and *Merge...* tiles. The settings can be saved as a .PCP or .PC2 (plotter configuration file) and easily retrieved instead of having to make all of the selections each time you want to change parameters.

Figure 14-2

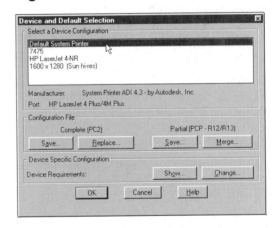

A .PCP (partial plot configuration) file keeps the settings you make in the *Pen Parameters* section of the dialog box (see below) but the information is device independent, so a .PCP file can be used for multiple print/plot devices. A .PC2 (complete plot configuration) file is a device-complete file. The file keeps the *Pen Parameter* information but also keeps the name and all other driver and configuration specifications for a <u>particular device</u>, so when a .PC2 file is loaded, it automatically <u>installs and configures the device and makes it the current plotter or printer</u>. The .PC2 files are useful for batch plotting because different printers and plotters can be controlled while AutoCAD is unattended.

Pen Parameters

Selecting the *Pen Assignments* tile produces the dialog box enabling you to specify the pens used for plotting based on screen colors (Fig. 14-3). The <u>screen color drives the plotting pen and/or printer line widths and colors</u>. Select one or several color numbers from the list (color 5 is selected in the example figure). The *Pen, Ltype, Speed,* and *Width* parameters can then be set on the right side of the dialog box.

Figure 14-3 ───────────────

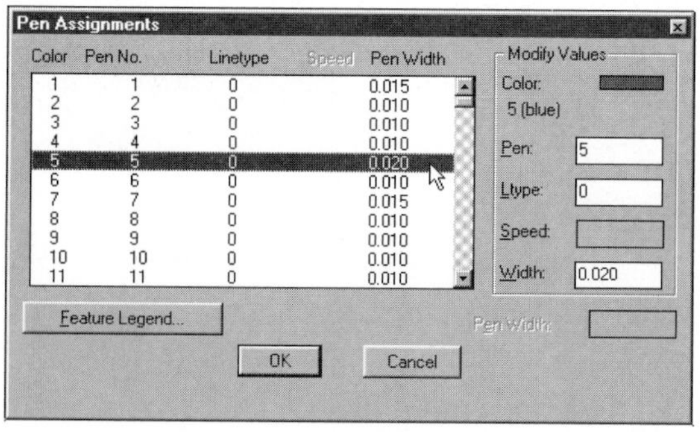

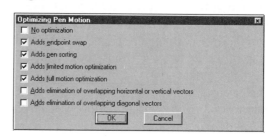

The parameters vary depending on the printer or plotter configured. Generally, multipen plotters and many printer drivers list the number of "pens" available. The *Linetype* column refers to <u>plotter or printer</u> linetypes, which are usually set to 0 (continuous), so AutoCAD's supplied drawing linetypes can be used instead (select *Feature Legend* to see what linetypes are available for your device). Pen plotters allow speed adjustment of pen motion (you may want to decrease *Speed* if the pens appear to skip).

Many printer drivers provide the capability to vary line width using the *Width* option (see the *Pen Width* column in Figure 14-3). This feature allows you to create excellent quality prints by varying the *Pen Widths* for different colors. Therefore, to achieve a print with multiple line widths, simply draw the geometry in the appropriate color (or on a layer with the appropriate color assigned) to determine the line width in the print. You should experiment with possibilities for your printing or plotting device.

The *Optimization* tile produces the dialog box (Fig. 14-4) for specifying up to seven levels of pen motion optimization (default levels are automatically set by most device drivers). Optimization is important for enabling pen plotters to draw lines on a sheet in the most efficient order.

Figure 14-4 ───────────────

Additional Parameters **(What to Plot)**

You specify what part of the drawing you want to plot by selecting the desired button from the lower-left corner of the dialog box (Fig. 14-1). The choices are listed below.

Display

This option plots the current display on the screen. If using viewports, it plots the current viewport display.

Extents

Plotting *Extents* is similar to *Zoom Extents*. This option plots the entire drawing (all objects), disregarding the *Limits*. Use the *Zoom Extents* command in the drawing to make sure the extents have been updated before using this plot option.

Limits

This selection plots the area defined by the *Limits* command (unless plotting a 3D object from other than the plan view).

View

With this option you can plot a view previously saved using the *View* command. Creating *Views* of specific areas of a drawing is very helpful when you plan to make repetitive plots of the same area or plots of multiple areas of one drawing.

Window

Window allows you to plot any portion of the drawing. You must specify the window by **PICK**ing or supplying coordinates for the lower-left and upper-right corners.

Text Resolution

This option allows you to control the resolution, or accuracy, for rounded sections of TrueType fonts such as the curve of the letter "o." The range is from 1 to 100. A low resolution of 5, for example, prints curved text with several very short, straight line segments, whereas a value of 100 produces smooth, rounded curves.

Text Fill

An AutoCAD drawing always displays TrueType fonts in solid filled format. You can, however, control if text is filled or not for the plot or print with this option. *Text Fill* off (no check mark) speeds plots of drawings with large text objects for pen plotters. This option sets the *TEXTFILL* system variable.

Hide Lines

This option removes hidden lines (edges obscured from view) if you are plotting a 3D surface or solid model.

Adjust Area Fill

Using this option forces AutoCAD to adjust for solid-filled areas (*Plines, Solids*, solid *Hatch* patterns) by pulling the pen inside the filled area by one-half the pen width. Use this option for critically accurate pen plots that have solid fills.

Plot to File

Choosing this option writes the plot to a file instead of making a plot. This action generally creates a .PLT file type. The format of the file (for example, PCL or HP/GL language) depends on the device brand and model that is configured. The plot file can be printed or plotted later <u>without</u> AutoCAD, assuming the correct interpreter for the device is available. Selecting this option enables the *File Name* tile for specifying the name and location for the plot file.

Paper Size and Orientation

Inches or *MM* should be checked to correspond to the units used in the drawing. Assuming the drawing was created using the correct units, a scale for plotting can be calculated without inch to millimeter conversion.

The *Size...* tile invokes the dialog box shown in Figure 14-5. You can select from the list of available paper sizes handled by the configured device. User sizes within the maximum plot area can also be defined.

Figure 14-5

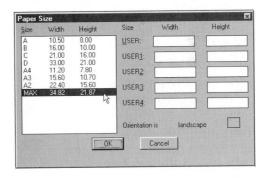

Scale, Rotation, and Origin

Selecting the *Rotation and Origin...* tile invokes the dialog box shown in Figure 14-6. The drawing can be rotated on the paper in 90 degree intervals. The origin specified indicates the location on the paper from point 0,0 (the plotter or printer's home position) from which the plot will be made. Home position for plotters is the <u>lower-left</u> corner (landscape orientation) and for printers it is the <u>upper-left</u> corner of the paper (portrait orientation).

Figure 14-6

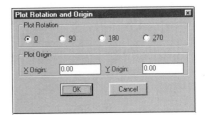

Scaled to Fit or **Plot to Scale**

Near the lower-right corner of the *Print/Plot Configuration* dialog box is the
area where you designate the plot scale (Fig. 14-7). You have two choices:
scale the drawing to automatically fit on the paper or calculate and indicate a
specific scale for the drawing to be plotted. By toggling the *Scaled to Fit* box,
the drawing is automatically sized to fit on the sheet, based on the specified
area to plot (*Display, Limits, Extents,* etc.). If you want to plot the drawing to
a specific scale, <u>remove the check</u> from the *Scaled to Fit* box and enter the
desired ratio in the boxes under *Plotted Inches* (or *Plotted MM*) and *Drawing
Units*. Decimals or fractions can be entered. For example, to prepare for a plot of one-half size
(1/2"=1"), the following ratios can be entered to achieve a plot at the desired scale: 1=2, 1/2=1, or .5=1.
For guidelines on plotting a drawing to scale, see Plotting to Scale in this chapter.

Figure 14-7

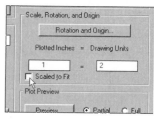

Plot Preview

Selecting the *Partial* tile (Fig. 14-1) displays the effective plotting area
as shown in Figure 14-8. Use the *Partial* option of *Plot Preview* to get
a quick check showing how the drawing will fit on the sheet. Two
rectangles and two sets of dimensions are displayed: the "Paper
size" is given and displayed in red and the "Effective area" is given
and displayed in blue. The "Effective area" is based on the selection
made in the *Additional Parameters* area of the plot dialog, that is,
Display, Extents, Limits, etc.

Figure 14-8

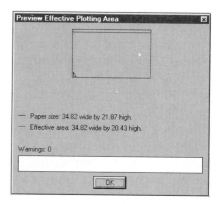

Selecting the *Full* option of *Plot Preview* displays the complete drawing as it will be plotted on the sheet
(Fig. 14-9). This function is particularly helpful to ensure the drawing will plot as you expect. When you
invoke a *Full Preview* in Release 14, the Drawing Editor displays a simulated sheet of paper with the
completed print or plot. The *Real Time Zoom* feature is on by default so you can check specific areas of
the drawing before
making the plot.
Right-clicking produces
the menu shown in
Figure 14-9, allowing you
to use *Real Time Pan* and
other *Zoom* options.
Changing the display
during a *Full Preview* <u>does
not change</u> the area of the
drawing to be plotted
or printed.

Figure 14-9

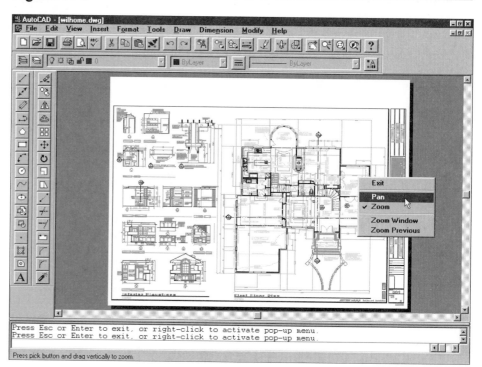

PREVIEW

	COMMAND (TYPE)	ALIAS (TYPE)	Short-cut	Screen (side) Menu	Tablet Menu
Pull-down Menu					
File *Print Preview*	*PREVIEW*	*PRE*	...	...	*X,24*

The *Preview* command accomplishes the same action as selecting a *Full Preview* in the *Print/Plot Configuration* dialog box. The advantage to using this command is being able to see a full print preview directly without having to invoke the *Print/Plot Configuration* dialog box first (see Figure 14-9). All functions of this preview (right-click for the *Real time Pan* and *Zoom* menu, etc.) operate the same as a full preview from the *Print/Plot Configuration* dialog box with the exception of one additional option—*Plot*.

You can print or plot during *Preview* directly from the right-click menu (Fig. 14-10). Using the *Plot* option creates a print or plot based on your settings in the *Print/Plot Configuration* dialog box. If you use this feature, ensure you first set the plotting parameters in the dialog box as you want.

Figure 14-10 ·

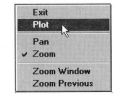

PLOTTING TO SCALE

When you create a <u>manual</u> drawing, a scale is determined before you can begin drawing. The scale is determined by the proportion between the size of the object on the paper and the actual size of the object. You then complete the drawing in that scale so that the actual object is proportionally reduced or enlarged to fit on the paper.

With a <u>CAD</u> drawing you are not restricted to a sheet of paper while drawing, so the geometry is created full size. Set *Limits* to provide an appropriate amount of drawing space; then the geometry can be drawn using the <u>actual dimensions</u> of the object. The resulting drawing on the CAD system is a virtual full-size dimensional replica of the actual object. Not until the CAD drawing is <u>plotted</u> on a fixed size sheet of paper, however, is it <u>scaled</u> to fit on the sheet.

Plotting an AutoCAD drawing to scale usually involves considering the intended sheet size and plot scale when specifying *Limits* during the initial drawing setup. The scale for plotting is based on the proportion of the paper size to the *Limits* (the reciprocal of the drawing scale factor).

For example, if you want to plot on an 11″ x 8.5″ sheet, you can set the *Limits* to 22 x 17 (2 x the sheet size). The value of **2** is then the drawing scale factor, and the plot scale (to enter in the *Print/Plot Configuration* dialog box) is **1/2** or **1=2** (the reciprocal of the drawing scale factor). If, in another case, the calculated drawing scale factor is **4**, then **1/4** or **1=4** would be the plot scale entered to achieve a drawing plotted at 1/4 actual size.

In order to calculate *Limits* and drawing scale factors correctly, you must know the standard paper sizes.

Standard Paper Sizes

Size	Engineering (")	Architectural (")
A	8.5 x 11	9 x 12
B	11 x 17	12 x 18
C	17 x 22	18 x 24
D	22 x 34	24 x 36
E	34 x 44	36 x 48

Size	Metric (mm)
A4	210 x 297
A3	297 x 420
A2	420 x 594
A1	594 x 841
A0	841 x 1189

Guidelines for Plotting to Scale

Plotting to scale preferably begins with the appropriate initial drawing setup. To correctly set up a drawing for printing or plotting to scale, <u>any two</u> of the following three variables must be known. The third variable can be determined from the other two.

> Drawing *Limits*
> Print or plot sheet (paper) size
> Print or plot scale

Your choice for *Limits* determines the drawing scale factor, which in turn is used to determine the plot scale or vice versa. Even though *Limits* can be changed and plot scale calculated at <u>any time</u> in the drawing process, it is usually done in the first few steps. Following are suggested steps for setting up a drawing for plotting to scale.

1. Set *Units* (*Decimal, Architectural, Engineering,* etc.) and *precision* to be used in the drawing.

2. Set *Limits* to a size, allowing geometry creation full size. *Limits* should be <u>set to the sheet size</u> used for plotting <u>times a factor</u>, if necessary, that provides enough area for drawing. This factor (proportion of the new *Limits* to the sheet size) becomes the <u>drawing scale factor</u>.

$$\text{DSF} = \frac{\text{Limits}}{\text{Sheet size}}$$

    You must also use a scale factor value that will yield a standard drawing scale (1/2"=1", 1/8"=1', 1:50, etc.), instead of a scale that is not a standard (1/3"=1", 3/5"=1', 1:23, etc.). See the Tables of *Limits* Settings for standard drawing scales.

3. The drawing scale factor is used as the scale factor (at least a starting point) for changing all size-related variables (*LTSCALE, DIMSCALE,* Hatch Pattern Scale, etc.). The <u>reciprocal</u> of the drawing scale factor is the plot scale to enter in the *Print/Plot Configuration* dialog box.

$$\text{Plot scale} = \frac{1}{\text{DSF}}$$

4. If the drawing is to be created using millimeter dimensions, set *Units* to *Decimal* and use metric values for the sheet size. In this way, the reciprocal of the drawing scale factor is the plot scale, the same as feet and inch drawings. However, multiply the drawing's scale factor by 25.4 (25.4mm = 1") to determine the factor for changing all size-related variables (*LTSCALE*, *DIM-SCALE*, etc.).

$$DSF(mm) = \frac{Limits}{Sheet\ size} \times 25.4$$

5. If drawing a border on the sheet, its maximum size cannot exceed the *Plot Area*. The *Plot Area* is given near the upper-right corner of the *Print/Plot Configuration* dialog box, but the values can be different for different devices. Since plotters or printers do not draw all the way to the edge of the paper, the border should not be drawn outside of the *Plot Area*. Generally, approximately 1/4" to 1/2" (6mm to 12mm) offset from each edge of the paper is required. To determine the distance in from the edge of the *Limits* after new *Limits* have been set, multiply 1/2 times the drawing scale factor. More precisely, calculate the distance based on the *Plot Area* listed for the current device and multiply by the drawing scale factor.

The **Tables of *Limits* Settings** (starting on the following page) help to illustrate this point. Sheet sizes are listed in the first column. The next series of columns from left to right indicate incremental changes in *Limits* and proportional changes in scale factor.

Simplifying the process of plotting to scale and calculating *Limits* and drawing scale factor can be accomplished by preparing template drawings. One method is to create a template for each sheet size that is used in the lab or office. In this way, the CAD operator begins the session by selecting the template drawing representing the sheet size and then multiples those *Limits* by some factor to achieve the desired *Limits* and drawing scale factor. Another method is to create a separate template drawing for each scale that you expect to use for plotting. In this way a template drawing can be selected with the final *Limits*, drawing scale factor, and plot scale already specified or calculated. This method requires creating a separate template for each expected scale and sheet size. (See Chapter 13 for help creating template drawings.)

Plotting to Scale Using Paper Space

If you are creating the drawing geometry in Model Space but plan to set up your title block and border in Paper Space, it is recommended that you calculate the drawing scale factor (DSF) and set Model Space *Limits* by these same methods. Although setting Model Space *Limits* is not critical in this case, calculating the DSF and *Limits* settings can be of assistance for plotting to scale.

To plot to scale using Paper Space, follow these steps:

1. Set Model Space *Limits* and create the geometry in Model Space as you would otherwise. Use the DSF to determine values for text size, dimension variables, hatch scales, etc.
2. Enable Paper Space (set TILEMODE to 0).
3. If you are using one of AutoCAD's template drawings or another template, the title block and border may already exist. If not, create, *Insert*, or *Xref* the title block and border.
4. Next, set Paper Space *Limits* to the plot or print sheet size.
5. In Model Space, use *Zoom* and enter an *XP* factor. The XP factor is the same as the plot scale (1/DSF).
6. Plot while Paper Space is enabled. Enter 1=1 as the plot scale in the *Print/Plot Configuration* dialog box.

See Chapter 19 for an introduction to using Paper Space.

TABLES OF *LIMITS* SETTINGS

Rather than making calculations of *Limits,* drawing scale factor, and print/plot scale for each drawing, the tables of limits settings on the following pages can be used to make calculating easier. There are five tables, one for each of the following applications:

Table

Mechanical Engineering
Architectural
Metric (using ISO standard metric sheets)
Metric (using U.S. standard engineering sheets)
Civil Engineering

To use the tables correctly, you must know any two of the following three variables. The third variable can be determined from the other two.

Approximate drawing *Limits*
Desired print or plot paper size
Desired print or plot scale

1. If you know the *Scale* and **paper size**:
    Assuming you know the scale you want to use, look along the top row to find the desired scale that you eventually want to print or plot. Find the desired paper size by looking down the left column. The intersection of the row and column yields the *Limits* settings to use to achieve the desired plot scale.

2. If you know the approximate *Limits* and **paper size**:
    Calculate how much space (minimum *Limits*) you require to create the drawing actual size. Look down the left column of the appropriate table to find the paper size you want to use for plotting. Look along that row to find the next larger *Limits* settings than your required area. Use these values to set the drawing *Limits*. The scale to use for the plot is located on top of that column.

3. If you know the *Scale* and approximate *Limits*:
    Calculate how much space (minimum *Limits*) you need to create the drawing objects actual size. Look along the top row of the appropriate table to find the desired scale that you eventually want to plot or print. Look down that column to find the next larger *Limits* than your required minimum area. Set the drawing *Limits* to these values. Look to the left end of that row to give the paper size you need to print or plot the *Limits* in the desired scale.

NOTE: Common scales are given in the tables of limits settings. If you need to create a print or plot in a scale that is not listed in the tables, you may be able to use the tables as a guide by finding the nearest table and standard scale to match your needs, then calculating proportional settings.

MECHANICAL TABLE OF *LIMITS* SETTINGS
(X axis x Y axis)

Paper Size (Inches)	Drawing Scale Factor 1 — Scale 1" = 1" — Plot 1 = 1	1.33 — 3/4" = 1" — 3 = 4	2 — 1/2" = 1" — 1 = 2	2.67 — 3/8" = 1" — 3 = 8	4 — 1/4" = 1" — 1 = 4	5.33 — 3/6" = 1" — 3 = 16	8 — 1/8" = 1" — 1 = 8
A 11 x 8.5 In.	11.0 x 8.5	14.7 x 11.3	22.0 x 17.0	29.3 x 22.7	44.0 x 34.0	58.7 x 45.3	88.0 x 68.0
B 17 x 11 In.	17.0 x 11.0	22.7 x 14.7	34.0 x 22.0	45.3 x 29.3	68.0 x 44.0	90.7 x 58.7	136.0 x 88.0
C 22 x 17 In.	22.0 x 17.0	29.3 x 22.7	44.0 x 34.0	58.7 x 45.3	88.0 x 68.0	117.0 x 90.7	176.0 x 136.0
D 34 x 22 In.	34.0 x 22.0	45.3 x 29.3	68.0 x 44.0	90.7 x 58.7	136.0 x 88.0	181.3 x 117.3	272.0 x 176.0
E 44 x 34 In.	44.0 x 34.0	58.7 x 45.3	88.0 x 68.0	117.3 x 90.7	176.0 x 136.0	235.7 x 181.3	352.0 x 272.0

ARCHITECTURAL TABLE OF *LIMITS* SETTINGS
(X axis x Y axis)

Paper Size (Inches)		Drawing Scale Factor 12 / Scale 1" = 1' / Plot 1 = 12	16 / 3/4" = 1' / 1 = 16	24 / 1/2" = 1' / 1 = 24	32 / 3/8" = 1' / 1 = 32	48 / 1/4" = 1' / 1 = 48	64 / 3/16" = 1' / 1 = 64	96 / 1/8" = 1' / 1 = 96
A 12 x 9	Ft	12 x 9	16 x 12	24 x 18	32 x 24	48 x 36	64 x 48	96 x 72
	In.	144 x 108	192 x 144	288 x 216	384 x 288	576 x 432	768 x 576	1152 x 864
B 18 x 12	Ft	18 x 12	24 x 16	36 x 24	48 x 32	72 x 48	96 x 64	144 x 96
	In.	216 x 144	288 x 192	432 x 288	576 x 384	864 x 576	1152 x 768	1728 x 1152
C 24 x 18	Ft	24 x 18	32 x 24	48 x 36	64 x 48	96 x 72	128 x 96	192 x 144
	In.	288 x 216	384 x 288	576 x 432	768 x 576	1152 x 864	1536 x 1152	2304 x 1728
D 36 x 24	Ft	36 x 24	48 x 32	72 x 48	96 x 64	144 x 96	192 x 128	288 x 192
	In.	432 x 288	576 x 384	864 x 576	1152 x 768	1728 x 1152	2304 x 1536	3456 x 2304
E 48 x 36	Ft	48 x 36	64 x 48	96 x 72	128 x 96	192 x 144	256 x 192	384 x 288
	In.	576 x 432	768 x 576	1152 x 864	1536 x 1152	2304 x 1728	3072 x 2304	4608 x 3456

METRIC TABLE OF *LIMITS* SETTINGS
FOR METRIC SHEET SIZES
(X axis x Y axis)

Paper Size (mm)		Drawing Scale Factor 25.4 / Scale 1:1 / Plot 1 = 1	50.8 / 1:2 / 1 = 2	127 / 1:5 / 1 = 5	254 / 1:10 / 1 = 10	508 / 1:20 / 1 = 20	1270 / 1:50 / 1 = 50	
A4 297 x 210	mm	297 x 210	594 x 420	1485 x 1050	2970 x 2100	5940 x 4200	14,850 x 10,500	29,700 x 21,000
	m	.297 x .210	.594 x .420	1.485 x 1.050	2.97 x 2.10	5.94 x 4.20	14.85 x 10.50	29.70 x 21.00
A3 420 x 297	mm	420 x 297	840 x 594	2100 x 1485	4200 x 2970	8400 x 5940	21,000 x 14,850	42,000 x 29,700
	m	.420 x .297	.840 x .594	2.100 x 1.485	4.20 x 2.97	8.40 x 5.94	21.00 x 14.85	42.00 x 29.70
A2 594 x 420	mm	594 x 420	1188 x 840	2970 x 2100	5940 x 4200	11,880 x 8400	29,700 x 21,000	59,400 x 42,000
	m	.594 x .420	1.188 x .840	2.97 x 2.10	5.94 x 4.20	11.88 x 8.40	29.70 x 21.00	59.40 x 42.00
A1 841 x 594	mm	841 x 594	1682 x 1188	4205 x 2970	8410 x 5940	16,820 x 11,880	42,050 x 29,700	84,100 x 59,400
	m	.841 x .594	1.682 x 1.188	4.205 x 2.970	8.41 x 5.94	16.82 x 11.88	42.05 x 29.70	84.10 x 59.40
A0 1189 x 841	mm	1189 x 841	2378 x 1682	5945 x 4205	11,890 x 8410	23,780 x 16,820	59,450 x 42,050	118,900 x 84,100
	m	1.189 x .841	2.378 x 1.682	5.945 x 4.205	11.89 x 8.41	23.78 x 16.82	59.45 x 42.05	118.90 x 84.10

METRIC TABLE OF *LIMITS* SETTINGS
FOR ENGINEERING (8.5 x 11 Format) SHEET SIZES
(X axis x Y axis)

Paper Size (mm)		Drawing Scale Factor 25.4 / Scale 1:1 / Plot 1=1	50.8 / 1:2 / 1=2	127 / 1:5 / 1=5	254 / 1:10 / 1=10	508 / 1:20 / 1=20	1270 / 1:50 / 1=50	2540 / 1:100 / 1=100
A 279.4 x 215.9	mm	279.4 x 215.9	558.8 x 431.8	1397 x 1079.5	2794 x 2159	5588 x 4318	13,970 x 10,795	27,940 x 21,590
	m	0.2794 x 0.2159	0.5588 x 0.4318	1.397 x 1.0795	2.794 x 2.159	5.588 x 4.318	13.97 x 10.795	27.94 x 21.59
B 431.8 x 279.4	mm	431.8 x 279.4	863.6 x 558.8	2159 x 1397	4318 x 2794	8636 x 5588	21,590 x 13,970	43,180 x 27,940
	m	0.4318 x 0.2794	0.8636 x 0.5588	2.159 x 1.397	4.318 x 2.794	8.636 x 5.588	21.59 x 13.97	43.18 x 27.94
C 558.8 x 431.8	mm	558.8 x 431.8	1117.6 x 863.6	2794 x 2159	5588 x 4318	11,176 x 8636	27,940 x 21,590	55,880 x 43,180
	m	0.5588 x 0.4318	1.1176 x 0.8636	2.794 x 2.159	5.588 x 4.318	11.176 x 8.636	27.94 x 21.59	55.88 x 43.18
D 863.6 x 558.8	mm	863.6 x 558.8	1727.2 x 1117.6	4318 x 2794	8636 x 5588	17,272 x 11,176	43,180 x 27,940	86,360 x 55,880
	m	0.8636 x 0.5588	1.7272 x 1.1176	4.318 x 2.794	8.636 x 5.588	17.272 x 11.176	43.18 x 27.94	86.36 x 55.88
E 1117.6 x 863.6	mm	1117.6 x 863.6	2235.2 x 1727.2	5588 x 4318	11,176 x 8636	22,352 x 17,272	55,880 x 43,180	111,760 x 86,360
	m	1.1176 x 0.8636	2.2352 x 1.7272	5.588 x 4.318	11.176 x 8.636	22.352 x 17.272	55.88 x 43.18	111.76 x 86.36

CIVIL TABLE OF *LIMITS* SETTINGS
(For Engineering Units)
(X axis x Y axis)

Paper Size (Inches)	Drawing Scale Factor 120 1" = 10' 1 = 120	240 1" = 20' 1 = 240	360 1" = 30' 1 = 360	480 1" = 40' 1 = 480	600 1" = 50' 1 = 600
A. 11 x 8.5 In.	1320 x 1020	2640 x 2040	3960 x 3060	5280 x 4080	6600 x 5100
Ft	110 x 85	220 x 170	330 x 255	440 x 340	550 x 425
B. 17 x 11 In.	2040 x 1320	4080 x 2640	6120 x 3960	8160 x 5280	10,200 x 6600
Ft	170 x 110	340 x 220	510 x 330	680 x 440	850 x 550
C. 22 x 17 In.	2640 x 2040	5280 x 4080	7920 x 6120	10,560 x 8160	13,200 x 10,200
Ft	220 x 170	440 x 340	660 x 510	880 x 680	1100 x 850
D. 34 x 22 In.	4080 x 2640	8160 x 5280	12,240 x 7920	16,320 x 10,560	20,400 x 13,200
Ft	340 x 220	680 x 440	1020 x 660	1360 x 880	1700 x 1100
E. 44 x 34 In.	5280 x 4080	10,560 x 8160	15,840 x 12,240	21,120 x 16,320	26,400 x 20,400
Ft	440 x 340	880 x 680	1320 x 1020	1760 x 1360	2200 x 1700

CONFIGURING PLOTTERS AND PRINTERS

Figure 14-11

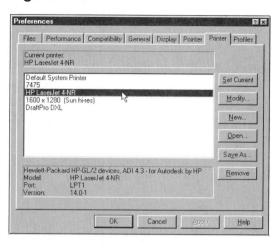

The *Printer* tab of the *Preferences* dialog box allows you to specify which devices you want to use in AutoCAD (Fig. 14-11). Access this dialog box by selecting *Preferences...* from the *Tools* pull-down menu or typing *Preferences*. You can also type *Config* to go directly to the *Printer* tab.

The central area of the tab lists the configured devices. When you begin AutoCAD Release 14 for the first time, the *Windows System Printer* is automatically configured (if you have a previously installed Windows printer). To make one device (from the list of configured devices) the current printer/plotter, highlight it and select the *Set Current* button. Setting a device current can also be accomplished from the *Print/Plot Configuration* dialog box. To configure other plotters or printers, select the *New...* tile.

New

Use this option to configure new print or plot devices. The list of available device drivers (manufacturers/device types) appears in the *Add a Printer* dialog box (Fig. 14-12). Selecting a manufacturer/device type, in turn, displays a text screen providing several similar devices to choose from and several questions to answer during the configuration process. You can also add your own description for the device. The device you select and the description then appear in the list in the *Printer* tab of the *Preferences* dialog box.

Figure 14-12

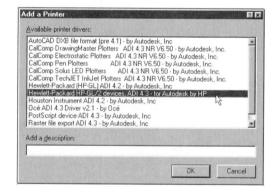

Modify

If you would like to change the description, change the settings for the particular printer or plotter, or configure a new printer or plotter from the same manufacturer/type list, use this option.

Open

For any single device, you can save the pen settings and optimization (see *Pen Parameters* under the *Plot* command) as well as the device-specific information. This option allows you to select a previously saved .PC2 file. A .PC2 file can be saved from this dialog box or from the *Device and Default Selection* dialog box (from the *Print/Plot Configuration* dialog box). Opening a .PC2 file loads all the information needed to "install" and configure a device as well as making it the current plotter/printer.

SaveAs

Use this option to save a .PC2 file for the selected device. (See *Device and Default Selection* earlier in this chapter.)

Remove

Selecting this button removes the selected printer or plotter from the list of configured devices. A warning message appears before deleting the device driver. If a .PC2 file was saved for the device before using *Remove*, the device can automatically be "reinstalled" and configured by opening the .PC2 file.

Using a Configured Raster File Converter as a Plotter

Using the *Add a Printer* dialog box (Fig. 14-12), you can select *Raster file export* to configure a driver for converting AutoCAD drawings to a variety of raster file formats.

CHAPTER EXERCISES

1. *Open* the **GASKETA** drawing that you created in Chapter 9 Exercises. What are the drawing *Limits*?

   A. *Plot* the drawing *Extents* on an 11″ x 8.5″ sheet. Select the *Scale to Fit* box.
   B. Next, *Plot* the drawing *Limits* on the same size sheet. *Scale to Fit*.
   C. Now, *Plot* the drawing *Limits* as before but plot the drawing at **1=1**. Measure the drawing and compare the accuracy with the dimensions given in the exercise in Chapter 9.
   D. Compare the three plots. What are the differences and why did they occur? (When you finish, there is no need to *Save* the changes.)

2. This exercise requires facilities for plotting an engineering C size sheet. *Open* drawing **CH12EX3**. Check to ensure the *Limits* are set at **22 x 17**. *Plot* the *Limits* at **1=1**. Measure the plot for accuracy by comparing with the dimensions given for the exercise in Chapter 12. To refresh your memory, this exercise involved adjusting the *LTSCALE* factor. Does the *LTSCALE* in the drawing yield hidden line dashes of 1/8″ (for the long lines) on the plot? If not, make the *LTSCALE* adjustment and plot again.

3. For this exercise, you will use a previously created template drawing to set up a drawing for plotting to scale. (The drawing will be used in later chapters for creating geometry.) If you know the final plot scale and sheet size, you can plan for the plot during the drawing setup. The *Limits* and other scale-related factors for plotting can be set at this phase. Follow the steps below.

   A drawing of a hammer is needed. The overall dimensions of the hammer are 12 1/2″ x about 5″. The drawing will be plotted on an 11″ x 8.5″ sheet. Begin a *New* drawing and use the template **ASHEET** you created in Chapter 13. Use *Save* and assign the name **HAMMER.DWG**. Follow these steps.

   A. Set *Units* to *Decimal* and *Precision* to **.000**.
   B. Set *Snap* to **.125**.
   C. Set *Grid* to **.5**.
   D. Since the geometry will not fit within *Limits* of 11 x 8.5, plotting at full size (1=1) will not be possible. In order to make a plot to a standard scale and show the geometry at the largest possible size on an 11″ x 8.5″ sheet, consult the Mechanical Table of *Limits* Settings to determine *Limits* settings for the drawing and a scale for plotting. Find the values for setting *Limits* for a plot at **3/4=1** scale. Set *Limits* appropriately; then *Zoom All*.
   E. An *LTSCALE* of .5 (previously set) creates hidden lines with the 1/8″ standard dashes when plotted at 1=1. Therefore, multiply the *LTSCALE* of .5 times the scale factor indicated in the table.
   F. *Save* the drawing.

4. Create a new template for architectural applications to plot at 1/8″=1′ scale on a D size sheet. Begin a *New* drawing using the template **DSHEET** and use *Save* to create a new template (DWT) named **D-8-AR** in <u>your working folder</u> (<u>not</u> where the AutoCAD-supplied templates are stored). Set *Units* to *Architectural*. Use the Architectural Table of *Limits* Settings to determine and set the new *Limits* for a D size sheet to plot at **1/8″=1′**. Multiply the existing *LTSCALE* (.5) times the scale factor shown in the table. Turn *Snap* and *Grid* off. *Save* as a template (DWT) drawing in your working folder.

5. Create a new template for metric applications to plot at 1:2 scale (1/2 size) on an A size sheet. Begin a *New* drawing using the template **ASHEET**, then use *Save* to create a new template drawing named **A-METRIC-2** in your working folder (not where the AutoCAD-supplied templates are stored). Use the Metric Table of *Limit* Settings to determine and set the new *Limits* for an **A** size sheet to plot at **1:2**. Multiply the existing *LTSCALE, Snap*, and *Grid* times the scale factor shown. *Save* as a template (DWT) drawing in your working folder.

6. Create a new template for civil engineering applications to plot at 1"=20′ scale on a C size sheet. Begin a *New* drawing using the template **CSHEET** and use *Save* to create a new template (DWT) named **C-CIVIL-20** in your working folder (not where the AutoCAD-supplied templates are stored). Set *Units* to *Engineering*. Use the Civil Table of *Limits* Settings to determine and set the new *Limits* for a **C** size sheet to plot at **1"=20′**. Multiply the existing *LTSCALE* (.5) times the scale factor shown. Turn *Snap* and *Grid* off. *Save* the drawing as a template (DWT) drawing in your working folder.

7. *Open* the **MANFCELL** drawing from Chapter 10. Check to make sure that the *Limits* settings are at **0,0** and **40′,30′**. Make two plots of this drawing on your plotter, according to the instructions.

    A. Make one plot of the drawing using a standard architectural scale. The scale you use is your choice and should be based on the sheet sizes available as well as the existing geometry size. *Plot* the drawing *Limits*. You may have to alter the *Limits* in order to plot the *Limits* to a standard scale. Use the Architectural Table of *Limits* Settings for guidance.
    B. Make one plot of the drawing using a standard civil engineering scale. Again, the choice of scale is yours and should be based on your available sheet sizes as well as the existing geometry. *Plot* the drawing *Limits*. You may have to alter the *Limits* for this plot also in order to plot the *Limits* to a standard scale. Use the Civil Table of *Limits* Settings for guidance.

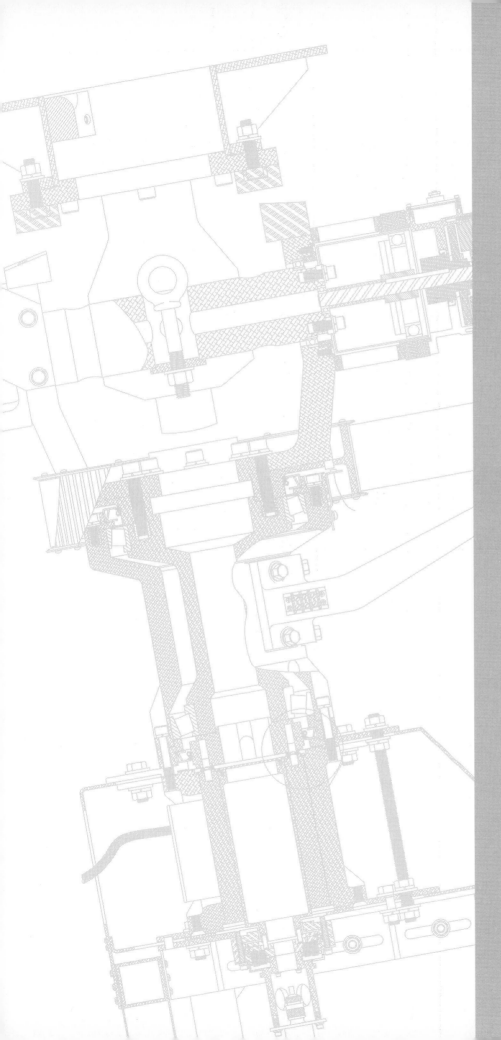

15

DRAW
COMMANDS II

Chapter Objectives

After completing this chapter you should:

1. be able to create construction lines using the *Xline* and *Ray* commands;
2. be able to create *Polygons* by the *Circumscribe*, *Inscribe*, and *Edge* methods;
3. be able to create *Rectangles* by PICKing two corners;
4. be able to use *Donut* to create circles with width;
5. be able to create *Spline* curves passing exactly through the selected points;
6. be able to create *Ellipses* using the *Axis End* method, the *Center* method, and the *Arc* method;
7. be able to use *Divide* to add points at equal parts of an object and *Measure* to add points at specified segment lengths along an object.

CONCEPTS

Remember that *Draw* commands create AutoCAD objects. The draw commands addressed in this chapter create more complex objects than those discussed in Chapter 8, Draw Commands I. The draw commands covered previously (*Line, Circle, Arc, Point,* and *Pline*) create simple objects composed of one object. The shapes created by the commands covered in this chapter are more complex. Most of the objects <u>appear</u> to be composed of several simple objects, but each shape is actually treated by AutoCAD as <u>one object</u>. Only the *Divide* and *Measure* commands create multiple objects. The following commands are explained in this chapter.

Xline, Ray, Polygon, Rectangle, Donut, Spline, Ellipse, Divide and *Measure*

These draw commands can be accessed using any of the command entry methods, including the menus and icon buttons, as illustrated in Chapter 8. Only the icon buttons that appear in the default *Draw* toolbar are given in this chapter by each command, although tools (icons) are available for the other draw commands if you want to customize your own toolbar.

COMMANDS

XLINE

Pull-down Menu	COMMAND (TYPE)	ALIAS (TYPE)	Short-cut	Screen (side) Menu	Tablet Menu
Draw *Construction Line*	XLINE	XL	...	*DRAW 1* *Xline*	L,10

When you draft with pencil and paper, light "construction" lines are used to lay out a drawing. These construction lines are not intended to be part of the finished object lines but are helpful for preliminary layout such as locating intersections, center points, and projecting between views.

There are two types of construction lines—*Xline* and *Ray*. An *Xline* is a line with infinite length, therefore having no endpoints. A *Ray* has one "anchored" endpoint and the other end extends to infinity. Even though these lines extend to infinity, they do not affect the drawing *Limits* or *Extents* or change the display or plot area in any way.

An *Xline* has no endpoints (*Endpoint Osnap* cannot be used) but does have a <u>root</u>, which is the theoretical <u>midpoint</u> (*Midpoint Osnap* can be used). If *Trim* or *Break* is used with *Xlines* or *Rays* such that two endpoints are created, the construction lines become *Line* objects. *Xlines* and *Rays* are drawn on the current layer and assume the current linetype and color (object-specific or *BYLAYER*).

There are many ways that you can create *Xlines* as shown by the options appearing at the command prompt. All options of *Xline* automatically repeat so you can easily draw multiple lines.

Command: **xline**
Hor/Ver/Ang/Bisect/Offset/<From point:>

From point

The default option only requires that you specify two points to construct the *Xline* (Fig. 15-1). The first point ("From point") becomes the root and anchors the line for the next point specification. The second point, or "Through point," can be PICKed at any location and can pass through any point (*Osnaps* can be used). If horizontal or vertical *Xlines* are needed, *ORTHO* can be used in conjunction with the "From point:" option.

    Command: *xline*
    Hor/Ver/Ang/Bisect/Offset/<From point:> **PICK**
    Through point:

Figure 15-1

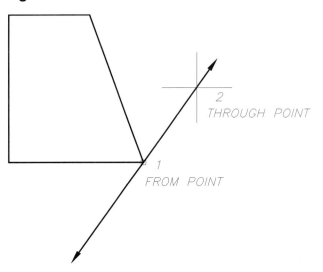

Hor

This option creates a horizontal construction line. Type the letter "H" at the "From point:" prompt. You only specify one point, the "Through point" or root (Fig. 15-2).

Figure 15-2

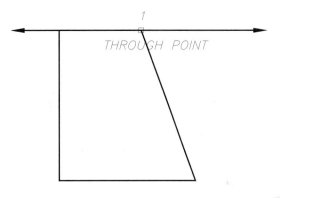

Ver

Ver creates a vertical construction line. You only specify one point, the "Through point" (root) (Fig. 15-3).

Figure 15-3

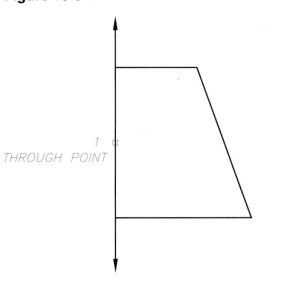

Ang

The *Ang* option provides two ways to specify the desired angle. You can (1) *Enter angle* or (2) select a *Reference* line (*Line*, *Xline*, *Ray*, or *Pline*) as the starting angle, then specify an angle from the selected line (in a counter-clockwise direction) for the *Xline* to be drawn (Fig. 15-4).

    Command: **XLINE**
    Hor/Ver/Ang/Bisect/Offset/<From point>: **a**
    Reference/<Enter angle (0)>:

Figure 15-4

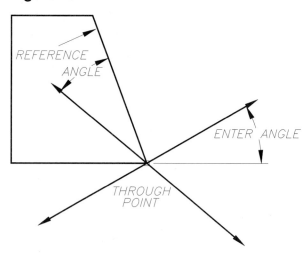

Bisect

This option draws the *Xline* at an angle between two selected points. First, select the angle vertex, then two points to define the angle (Fig. 15-5).

    Command: **XLINE**
    Hor/Ver/Ang/Bisect/Offset/<From point>: **b**
    Angle vertex point: **PICK**
    Angle start point: **PICK**
    Angle end point: **PICK**
    Angle end point: **Enter**
    Command:

Figure 15-5

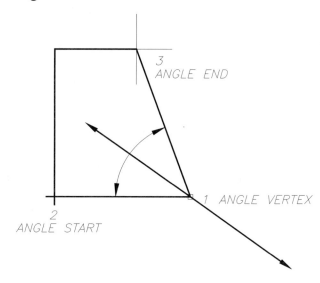

Offset

Offset creates an *Xline* parallel to another line. This option operates similarly to the *Offset* command. You can (1) specify a *Distance* from the selected line or (2) PICK a point to create the *Xline Through*. With the *Distance* option, enter the distance value, select a line (*Line*, *Xline*, *Ray*, or *Pline*), and specify on which side to create the offset *Xline*.

    Command: **XLINE**
    Hor/Ver/Ang/Bisect/Offset/<From point>: **o**
    Offset distance or Through <1.0000>:
    (value) or PICK
    Select a line object: **PICK**
    Side to offset?
    Select a line object: **Enter**
    Command:

Figure 15-6

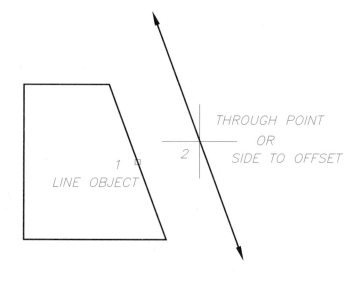

Using the *Through* option, select a line (*Line, Xline, Ray,* or *Pline*); then specify a point for the *Xline* to pass through. In each case, the anchor point of the *Xline* is the "root." (See *Offset,* Chapter 10.)

```
Command: XLINE
Hor/Ver/Ang/Bisect/Offset/<From point>: o
Offset distance or Through <1.0000>: T
Select a line object: PICK
Through point: PICK
Select a line object: Enter
Command:
```

RAY

Pull-down Menu	COMMAND (TYPE)	ALIAS (TYPE)	Short-cut	Screen (side) Menu	Tablet Menu
Draw Ray	RAY	...	...	DRAW 1 Ray	K,10

A *Ray* is also a construction line (see *Xline*), but it extends to infinity in only <u>one direction</u> and has one "anchored" endpoint. Like an *Xline*, a *Ray* extends past the drawing area but does not affect the drawing *Limits* or *Extents*. The construction process for a *Ray* is simpler than for an *Xline*, only requiring you to establish a "From point" (endpoint) and a "Through point" (Fig. 15-7). Multiple *Rays* can be created in one command.

Figure 15-7

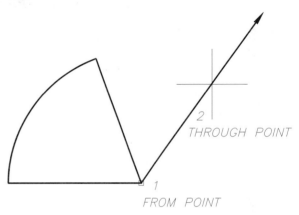

```
Command: ray
From point: PICK or (coordinates)
Through point: PICK or (coordinates)
Through point: PICK or (coordinates)
Through point: Enter to complete the command
Command:
```

Rays are especially helpful for construction of geometry about a central reference point or for construction of angular features. In each case, the geometry is usually constructed in only one direction from the center or vertex. If horizontal or vertical *Rays* are needed, just toggle on *ORTHO*. To draw a *Ray* at a specific angle, use relative polar coordinates at the "through point:" prompt. *Endpoint* and other appropriate *Osnaps* can be used with *Rays*. A *Ray* has one *Endpoint* but no *Midpoint*.

POLYGON

Pull-down Menu	COMMAND (TYPE)	ALIAS (TYPE)	Short-cut	Screen (side) Menu	Tablet Menu
Draw Polygon	POLYGON	POL	...	DRAW 1 Polygon	P,10

The *Polygon* command creates a regular polygon (all angles are equal and all sides have equal length). A *Polygon* object appears to be several individual objects but, like a *Pline*, is actually <u>one</u> object. In fact, AutoCAD uses *Pline* to create a *Polygon*. There are two basic options for creating *Polygons*: you can specify an *Edge* (length of one side) or specify the size of an imaginary circle for the *Polygon* to *Inscribe* or *Circumscribe*.

Inscribe/Circumscribe

The command sequence for this default method follows:

> Command: **polygon**
> Number of sides: (Enter a value for the number of sides.)
> Edge/<Center of polygon>: **PICK** or (**coordinates**) to specify the center
> Inscribed in circle/Circumscribed about circle: **I** or **C**
> Radius of circle: **PICK** or (**value**) or (**coordinates**)

The orientation of the *Polygon* and the imaginary circle are shown in Figure 15-8. Note that the *Inscribed* option allows control of one-half of the distance <u>across the corners,</u> and the *circumscribed* option allows control of one-half of the distance <u>across the flats.</u>

Using *ORTHO ON* with specification of the *radius of circle* forces the *Polygon* to a 90 degree orientation.

Figure 15-8

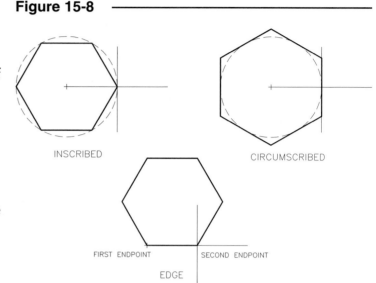

INSCRIBED   CIRCUMSCRIBED

FIRST ENDPOINT   SECOND ENDPOINT

EDGE

Edge

The *Edge* option only requires you to indicate the number of sides desired and to specify the two endpoints of one edge (Fig. 15-8).

> Command: **polygon**
> Number of sides: (**value**) (Enter a value for the number of sides.)
> Edge/<Center of polygon>: **E** (Invokes the edge option.)
> First endpoint of edge: **PICK** or (**coordinates**) (Interactively select or enter coordinates in any format.)
> Second endpoint of edge: **PICK** or (**coordinates**) (Interactively select or enter coordinates in any format.)

Because *Polygons* are created as *Plines*, *Pedit* can be used to change the line width or edit the shape in some way (see *Pedit*, Chapter 16). *Polygons* can also be *Exploded* into individual objects similar to the way other *Plines* can be broken down into component objects (see *Explode*, Chapter 16).

RECTANG

	COMMAND (TYPE)	ALIAS (TYPE)	Short-cut	Screen (side) Menu	Tablet Menu
Pull-down Menu					
Draw Rectangle	*RECTANG*	*REC*	...	*DRAW 1 Rectang*	*Q,10*

The *Rectang* command only requires the specification of two diagonal corners for construction of a rectangle, identical to making a selection window (Fig. 15-9). The corners can be PICKed or coordinates can be entered. The rectangle can be any proportion,

Figure 15-9

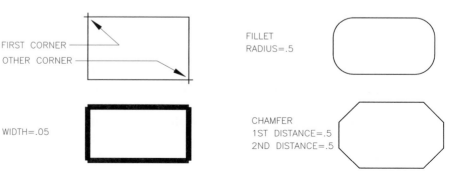

FIRST CORNER
OTHER CORNER

FILLET
RADIUS=.5

WIDTH=.05

CHAMFER
1ST DISTANCE=.5
2ND DISTANCE=.5

but the sides are always horizontal and vertical. The completed rectangle is <u>one AutoCAD object</u>, not four separate objects.

    Command: ***rectangle***
    Chamfer/Elevation/Fillet/Thickness/Width/<First corner>: **PICK** or (**coordinates**)
    Other corner: **PICK** or (**coordinates**)
    Command:

Rectangle uses the *Pline* command to construct the shape; therefore, several options of the *Rectangle* command affect the shape as if it were a *Pline* (see *Pline* and *Pedit*). For example, a *Rectangle* can have *width:*

    Chamfer/Elevation/Fillet/Thickness/Width/<First corner>: ***w***
    Width for rectangles <0.0000>: (**value**)

The *Fillet* and *Chamfer* options allow you to specify values for the fillet radius or chamfer distances:

    Chamfer/Elevation/Fillet/Thickness/Width/<First corner>: ***f***
    Fillet radius for rectangles <0.0000>: (**value**)

    Chamfer/Elevation/Fillet/Thickness/Width/<First corner>: ***c***
    First chamfer distance for rectangles <0.0000>: (**value**)
    Second chamfer distance for rectangles <0.0000>: (**value**)

The *Thickness* and *Elevation* options are 3D properties.

DONUT

Pull-down Menu	COMMAND (TYPE)	ALIAS (TYPE)	Short-cut	Screen (side) Menu	Tablet Menu
Draw *Donut*	*DONUT*	*DO*	...	*DRAW 1* *Donut*	*K,9*

A *Donut* is a circle with width (Fig. 15-10). Invoking the command allows changing the inside and outside diameters and creating multiple *Donuts*:

    Command: ***donut***
    Inside diameter <current>: (**value**) or **Enter**
    Outside diameter <current>: (**value**) or **Enter**
    Center of doughnut: **PICK** or (**coordinates**)
    Center of doughnut: **PICK** or (**coordinates**) or **Enter**

Donuts are actually solid filled circular *Plines* with width. The solid fill for *Donuts*, *Plines*, and other "solid" objects can be turned off with the *Fill* command or *FILLMODE* system variable.

Figure 15-10

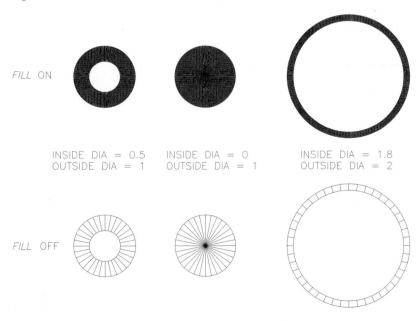

FILL ON

INSIDE DIA = 0.5
OUTSIDE DIA = 1

INSIDE DIA = 0
OUTSIDE DIA = 1

INSIDE DIA = 1.8
OUTSIDE DIA = 2

FILL OFF

SPLINE

Pull-down Menu	COMMAND (TYPE)	ALIAS (TYPE)	Short-cut	Screen (side) Menu	Tablet Menu
Draw *Spline*	*SPLINE*	*SPL*	...	DRAW 1 *Spline*	*L,9*

The *Spline* command creates a NURBS (non-uniform rational bezier spline) curve. The non-uniform feature allows irregular spacing of selected points to achieve sharp corners, for example. A *Spline* can also be used to create regular (rational) shapes such as arcs, circles, and ellipses. Irregular shapes can be combined with regular curves, all in one spline curve definition.

Spline is the newer and more functional version of a *Spline*-fit *Pline* (see *Pedit*, Chapter 16). The main difference between a *Spline*-fit *Pline* and a *Spline* is that a *Spline* curve passes through the points selected, while the points selected for construction of a *Spline*-fit *Pline* only have a "pull" on the curve. Therefore, *Splines* are more suited to accurate design because the curve passes exactly through the points used to define the curve (data points) (Fig. 15-11).

Figure 15-11

SPLINE
(DEFAULT TANGENTS)

The construction process involves specifying points that the curve will pass through and determining tangent directions for the two ends (for non-closed *Splines*) (Fig. 15-12).

Figure 15-12

END TANGENT

The *Close* option allows creation of closed *Splines* (these can be regular curves if the selected points are symmetrically arranged) (Fig. 15-13).

```
Command: spline
Object/<Enter first point>:PICK or (coordinates)
Enter point: PICK or (coordinates)
Close/Fit Tolerance/<Enter point>: PICK or (coordinates)
Close/Fit Tolerance/<Enter point>: PICK or (coordinates)
Close/Fit Tolerance/<Enter point>: Enter
Enter start tangent: PICK or Enter (Select direction for
tangent or Enter for default tangent.)
Enter end tangent: PICK or Enter (Select direction for
tangent.)
Command:
```

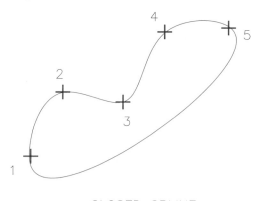

Figure 15-13

CLOSED SPLINE

The *Object* option allows you to convert *Spline*-fit *Plines* into NURBS *Splines*. Only *Spline*-fit *Plines* can be converted (see *Pedit*, Chapter 16).

```
Command: spline
Object/<Enter first point>:  o
Select objects to convert to splines.
Select Objects: PICK
Select Objects: Enter
Command:
```

A *Fit Tolerance* applied to the *Spline* "loosens" the fit of the curve. A tolerance of 0 (default) causes the *Spline* to pass exactly through the data points. Entering a positive value allows the curve to fall away from the points to form a smoother curve (Fig. 15-14).

Figure 15-14 ────────────

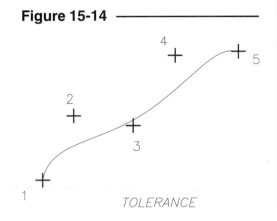

TOLERANCE

```
Close/Fit Tolerance/<Enter point>:  F
Enter Fit tolerance <0.0000>: (Enter a positive value.)
```

ELLIPSE

Pull-down Menu	COMMAND (TYPE)	ALIAS (TYPE)	Short-cut	Screen (side) Menu	Tablet Menu
Draw *Ellipse*	*ELLIPSE*	*EL*	...	*DRAW 1* *Ellipse*	*M,9*

An *Ellipse* is one object. AutoCAD *Ellipses* are (by default) NURBS curves (see *Spline*). There are three methods of creating *Ellipses* in AutoCAD: (1) specify one <u>axis</u> and the <u>end</u> of the second, (2) specify the <u>center</u> and the ends of each axis, and (3) create an elliptical <u>arc</u>. Each option also permits supplying a rotation angle rather than the second axis length.

```
Command: ellipse
Arc/Center/<Axis endpoint 1>:PICK or (coordinates) (This is the first endpoint of either the major or
minor axis.)
Axis endpoint 2: PICK or (coordinates) (Select a point for the other endpoint of the first axis.)
<Other axis distance>/Rotation: PICK or (coordinates) (This distance is measured perpendicularly
from the established axis.)
```

Axis End
This default option requires PICKing three points as indicated in the command sequence above (Fig. 15-15).

Figure 15-15 ────────────

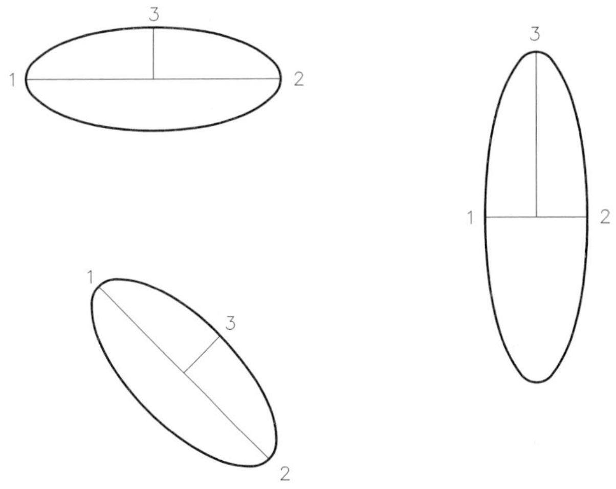

Rotation

If the *Rotation* option is used with the *Axis End* method, the following syntax is used:

> <Other axis distance>/Rotation: **R**
> Rotation around major axis: **PICK** or **(value)**

The specified angle is the number of degrees the shape is rotated <u>from the circular position</u> (Fig. 15-16).

Figure 15-16

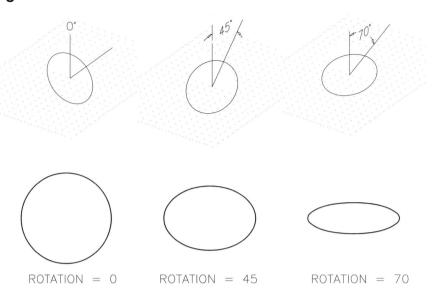

ROTATION = 0   ROTATION = 45   ROTATION = 70

Center

With many practical applications, the center point of the ellipse is known, and therefore the *Center* option should be used (Fig. 15-17):

> Command: **ellipse**
> Arc/Center/<Axis endpoint 1>: **C**
> Center of ellipse: **PICK** or **(coordinates)**
> Axis endpoint: **PICK** or **(coordinates)**
> <Other axis distance>/Rotation: **PICK** or **(coordinates)** (This distance is measured perpendicularly from the established axis.)

The *Rotation* option appears and can be invoked after specifying the *Center* and first *Axis endpoint*.

Figure 15-17

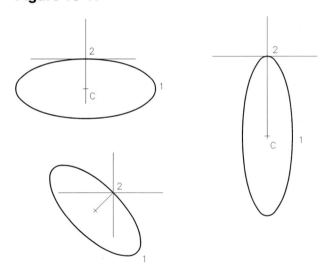

Arc

Use this option to construct an elliptical arc (partial ellipse). The procedure is identical to the *Center* option with the addition of specifying the start- and endpoints for the arc (Fig. 15-18):

> Command: **ellipse**
> Arc/Center/<Axis endpoint 1>: **a**
> <Axis endpoint 1>/Center: **PICK** or **(coordinates)**
> Axis endpoint 2: **PICK** or **(coordinates)**
> <Other axis distance>/Rotation: **PICK** or **(coordinates)**
> Parameter/<start angle>: **PICK** or **(angular value)**
> Parameter/Included/<end angle>: **PICK** or **(angular value)**
> Command:

Figure 15-18

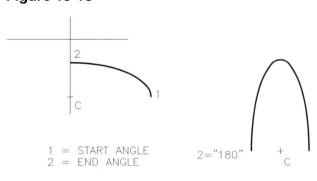

1 = START ANGLE
2 = END ANGLE

2="180"

The *Parameter* option allows you to specify the start point and endpoint for the elliptical arc. The parameters are based on the parametric vector equation: p(u)=c+a*cos(u)+b*sin(u), where c is the center of the ellipse and a and b are the major and minor axes.

DIVIDE

Pull-down Menu	COMMAND (TYPE)	ALIAS (TYPE)	Short-cut	Screen (side) Menu	Tablet Menu
Draw *Point >* *Divide*	*DIVIDE*	*DIV*	...	DRAW 2 *Divide*	V,13

The *Divide* and *Measure* commands add *Point* objects to existing objects. Both commands are found in the *Draw* pull-down menu under *Point* because they create *Point* objects.

The *Divide* command finds equal intervals along an object such as a *Line, Pline, Spline,* or *Arc* and adds a *Point* object at each interval. The object being divided is <u>not</u> actually broken into parts—it remains as <u>one</u> object. *Point* objects are automatically added to display the "divisions."

Figure 15-19

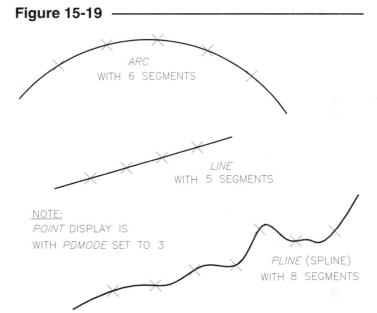

The point objects that are added to the object can be used for subsequent construction by allowing you to *OSNAP* to equally spaced intervals (*Nodes*).

The command sequence for the *Divide* command is as follows:

> Command: **divide**
> Select object to divide: **PICK** (Only one object can be selected.)
> <Number of segments>/Block: (**value**)
> (Enter a value for the number of segments.)
> Command:

Point objects are added to divide the object selected into the desired number of parts. Therefore, there is <u>one less</u> *Point* added than the number of segments specified.

After using the *Divide* command, the *Point* objects may not be visible unless the point style is changed with the *Point Style...* dialog box (*Format* pull-down menu) or by changing the *PDMODE* variable by command line format (see Chapter 8). A *Regen* must be invoked before the new point style will be displayed. Figure 15-19 shows *Points* displayed using a *PDMODE* of 3.

You can request that *Block*s be inserted rather than *Point* objects along equal divisions of the selected object. Figure 15-20 displays a generic rectangular-shaped block inserted with *Divide*, both aligned and not aligned with a *Line*, *Arc*, and *Pline*. In order to insert a *Block* using the *Divide* command, the name of an <u>existing</u> *Block* must be given. (See Chapter 20, Blocks.)

Figure 15-20 ───────────────

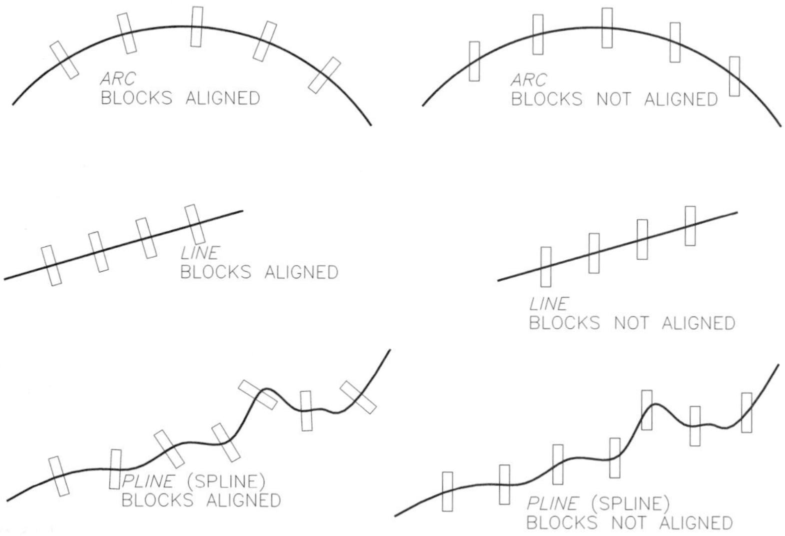

ARC BLOCKS ALIGNED

ARC BLOCKS NOT ALIGNED

LINE BLOCKS ALIGNED

LINE BLOCKS NOT ALIGNED

PLINE (SPLINE) BLOCKS ALIGNED

PLINE (SPLINE) BLOCKS NOT ALIGNED

MEASURE

Pull-down Menu	COMMAND (TYPE)	ALIAS (TYPE)	Short-cut	Screen (side) Menu	Tablet Menu
Draw Point > Measure	*MEASURE*	*ME*	...	DRAW 2 *Measure*	*V,12*

The *Measure* command is similar to the *Divide* command in that *Point* objects (or *Block*s) are inserted along the selected object. The *Measure* command, however, allows you to designate the <u>length</u> of segments rather than the <u>number</u> of segments as with the *Divide* command.

    Command: **measure**
    Select object to measure: **PICK** (Only one
    object can be selected.)
    <Segments length>/Block: (**value**) (Enter a
    value for length of one segment.)
    Command:

Point objects are added to the selected object at the designated intervals (lengths). One *Point* is added for each interval <u>beginning at the end nearest</u> the end used for object selection (Fig. 15-21). The intervals are of equal length except possibly the last segment, which is whatever length is remaining.

Figure 15-21 ───────────────

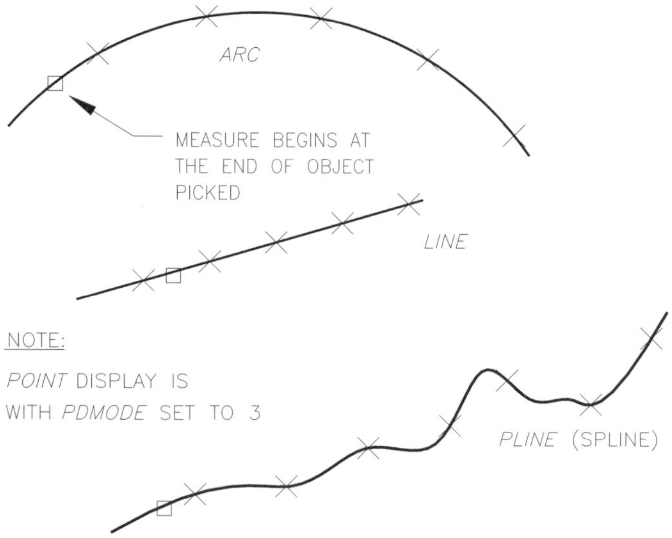

ARC

MEASURE BEGINS AT THE END OF OBJECT PICKED

LINE

<u>NOTE:</u>

POINT DISPLAY IS WITH *PDMODE* SET TO 3

PLINE (SPLINE)

You can request that *Block*s be inserted rather than *Point* objects at the designated intervals of the selected object. Inserting *Block*s with *Measure* requires that an <u>existing</u> *Block* be used.

TRACKING

Pull-down Menu	COMMAND (TYPE)	ALIAS (TYPE)	Short-cut	Screen (side) Menu	Tablet Menu	Cursor Menu (button 3 or Shift+button 2)
...	*(during command)* TRACKING	TK	...	****(asterisks) *Tracking*	T,15	*Tracking*

Tracking is not a command. *Tracking* is a method of specifying points that align orthogonally to other points. *Tracking* is used during a command, usually a draw command, similar to the way *OSNAPs* are used to locate points on objects. For example, when a command prompts you to PICK a point or enter coordinate values, you can use *Tracking* to find a point <u>orthogonal</u> to a *Midpoint, Endpoint,* or other point. You can also "build" the desired point by using *Tracking* in a progressive manner, for example, specify a point that aligns (orthogonally) with one *Endpoint* and aligns with another *Midpoint.* Any *OSNAPs* can be used in conjunction with *Tracking.*

Tracking can be invoked by several methods. Because *Tracking* is generally used in conjunction with *OSNAPs*, it is located in the *Osnaps* flyout group on the Standard toolbar (Fig. 15-22) and on the separate *Osnap* toolbar that can be activated from the *Toolbars* dialog box.

Figure 15-22 ——

When *Tracking* is turned on, the "rubberband" line extends only in vertical or horizontal directions, similar to the effect of turning on *ORTHO*. Each *Tracking* point specifies only one direction—horizontal or vertical. To indicate which direction you want to track (horizontal or vertical), move the cursor to the first point of tracking, then in either a vertical or horizontal direction. For example, Figure 15-23 illustrates the possible directions for *Tracking* from the *Endpoint* of an existing *Line.*

Figure 15-23 ——————

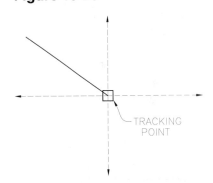

Tracking is invoked like any other *OSNAP*, but other *OSNAPs* can be used <u>after activating tracking</u>. For example, examine this illustration of creating a *Circle* of 2 units radius whose center aligns orthogonally with the *Midpoint* of an adjacent *Line* (Fig. 15-24):

> Command: **CIRCLE**
> 3P/2P/TTR/<Center point>: **tracking** (select the *Tracking* icon or menu option)
> First tracking point: **mid of** (select the *Midpoint* icon or menu option, then) **PICK** (see P1)
> Next point (Press ENTER to end tracking): (move cursor with *ORTHO* on in the desired direction and enter a value) **2.75**
> Next point (Press ENTER to end tracking): **Enter**
> Diameter/<Radius> <1.5000>: **1**
> Command:

Figure 15-24 —————————

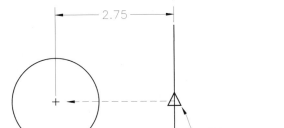

A classic example of using *Tracking* effectively is drawing a *Circle* with its center located at the center of an existing rectangle (Figs. 15-25 and 15-26). The coordinate for the <u>center of the rectangle</u> is aligned orthogonally with the midpoint of a horizontal (bottom or top) line and also with the midpoint of a vertical (side) line.

The command syntax for the operation in Figures 15-25 and 15-26 is as follows:

> Command: **CIRCLE**
> 3P/2P/TTR/<Center point>: **tracking**
> First tracking point: **mid of** **PICK** (P1, Fig. 15-25, then move vertically)
> Next point (Press ENTER to end tracking): **mid of** **PICK** (P2, Fig. 15-25)
> Next point (Press ENTER to end tracking): **Enter** (to end *Tracking*)
> Diameter/<Radius>: **1.5** (Fig. 15-26)
> Command:

NOTE: When *Tracking* two points, <u>it is critical that you first move in the direction you intend from the tracking point</u>. If you are not paying attention to this detail, you can easily become confused.

Another example is given in Figure 15-27. Two *Lines* exist. Two other *Lines* are drawn to complete a foursided shape such that the new corner aligns orthogonally with the endpoints of the two existing *Lines*:

> Command: **LINE**
> From point: **End of** **PICK** (P1)
> To point: **tracking**
> First tracking point: **End of** **PICK** (P1, then move vertically)
> Next point (Press ENTER to end tracking): **End of** **PICK** (P2) (creates vertical *Line*)
> Next point (Press ENTER to end tracking): **Enter**
> To point: **PICK** (P2) (creates horizontal *Line*)
> To point: **Enter**
> Command:

In summary, *Tracking* can be used for a variety of applications whenever you need to specify points that align orthogonally with other geometry.

Figure 15-25

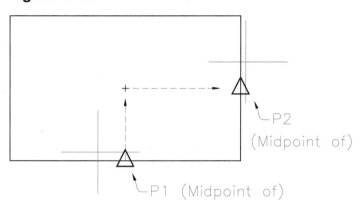

Figure 15-26

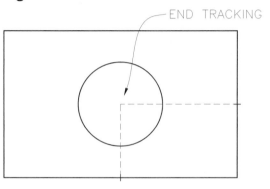

Figure 15-27

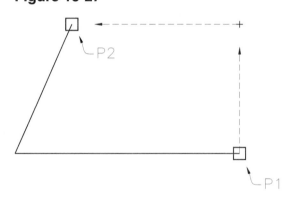

CHAPTER EXERCISES

1. *Polygon*

   Open the **PLINE1** drawing you
   created in Chapter 8 Exercises. Use
   Polygon to construct the polygon
   located at the *Center* of the existing
   arc, as shown in Figure 15-28.
   When finished, *SaveAs*
   POLYGON1.

Figure 15-28 ───────────────────────────

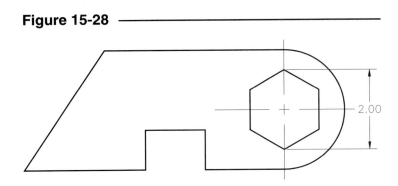

2. *Xline, Ellipse, Rectangle*

   Open the **APARTMENT** drawing that you created in Chapter 13. Draw the floor plan of the efficiency apartment on layer FLOORPLAN as shown in Figure 15-29. Use *SaveAs* to assign the
   name **EFF-APT**. Use *Xline* and *Line* to construct the floor plan with 8″ width for the exterior
   walls and 5″ for interior walls. Use the *Ellipse* and *Rectangle* commands to design and construct
   the kitchen sink, tub, wash basin, and toilet. *Save* the drawing but do not exit AutoCAD.
   Continue to Exercise 3.

Figure 15-29 ───────────────────────────

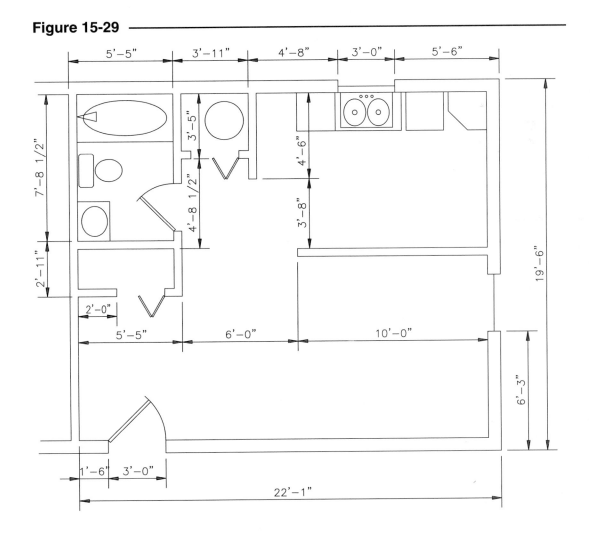

3. *Polygon, Sketch*

Create a plant for the efficiency apartment as shown in Figure 15-30. Locate the plant near the entry. The plant is in a hexagonal pot (use *Polygon*) measuring 18" across the corners. Use *Sketch* to create 2 leaves as shown in figure A. Create a *Polar Array* to develop the other leaves similar to figure B. *Save* the **EFF-APT** drawing.

Figure 15-30 ———————————

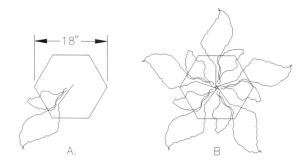

A.

B

4. *Donut*

Open **SCHEMATIC1** drawing you created in Chapter 10 Exercises. Use the *Point Style* dialog box (*Format* pull-down menu) to change the style of the *Point* objects to dots instead of small circles. Type *Regen* to display the new point style. Turn on the *Node Running Osnap* mode. Use the *Donut* command and set the *Inside diameter* and the *Outside diameter* to an appropriate value for your drawing such that the *Inside diameter* is ½ the value of the *Outside diameter*. Create a *Donut* at each existing *Node*. *SaveAs* **SCHEMATIC1B**. Your finished drawing should look like that in Figure 15-31.

Figure 15-31 ———————————

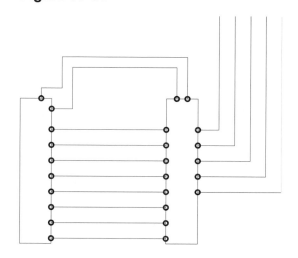

5. *Ellipse, Polygon*

Open the drawing that you set up in Chapter 13 named **WRENCH**. Complete the construction of the wrench shown in Figure 15-32. Center the drawing within the *Limits*. Use *Line, Circle, Arc, Ellipse,* and *Polygon* to create the shape. For the "break lines," create one connected series of jagged *Lines*, then *Copy* for the matching set. Utilize *Trim, Rotate,* and other edit commands where necessary. HINT: Draw the wrench head in an orthogonal position; then rotate the entire shape 15 degrees. Notice the head is composed of ½ of a *Circle* (R25) and ½ of an *Ellipse*. *Save* the drawing when completed.

Figure 15-32 ———————————

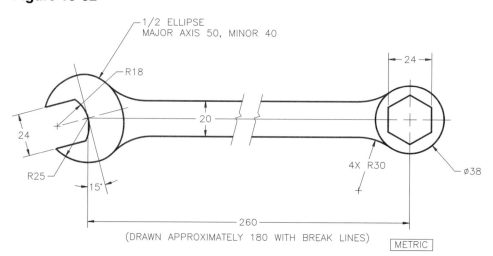

(DRAWN APPROXIMATELY 180 WITH BREAK LINES)

METRIC

6. *Xline, Spline*

Create the handle shown in Figure 15-33. Construct multiple *Horizontal* and *Vertical* *Xlines* at the given distances on a separate layer (shown as dashed lines in Figure 15-33). Construct two *Splines* that make up the handle sides by PICKing the points indicated at the *Xline* intersections. Note the *End Tangent* for one *Spline*. Connect the *Splines* at each end with horizontal *Lines*. *Freeze* the construction (*Xline*) layer. *Save* the drawing as **HANDLE**.

Figure 15-33

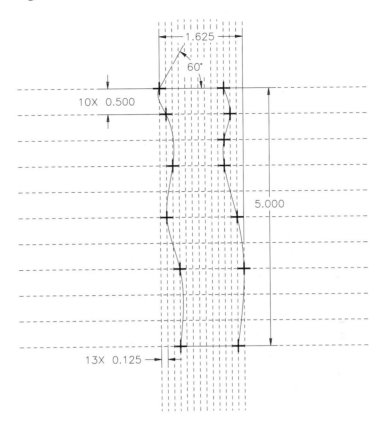

7. *Divide, Measure*

Use the **ASHEET** drawing as a template and assign the name **BILL-MATL**. Create the table in Figure 15-34 to be used as a bill of materials. Draw the bottom *Line* (as dimensioned) and a vertical *Line*. Use *Divide* along the bottom *Line* and *Measure* along the vertical *Line* to locate *Points* as desired. Create *Offsets Through* the *Points* using *Node OSNAP*. (*ORTHO* and *Trim* may be of help.)

Figure 15-34

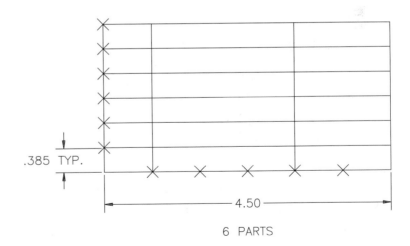

6 PARTS

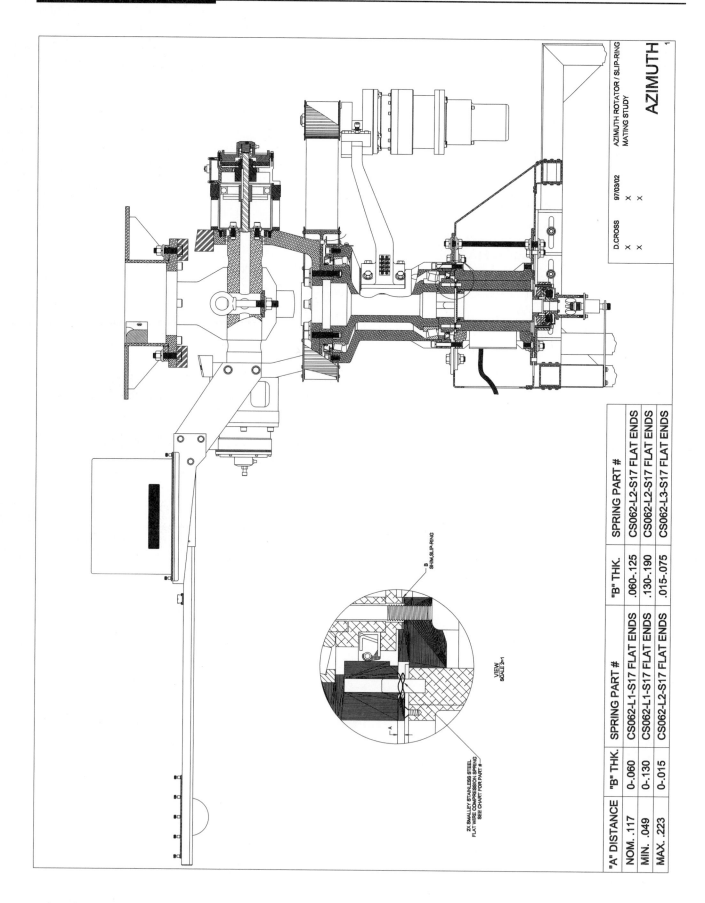

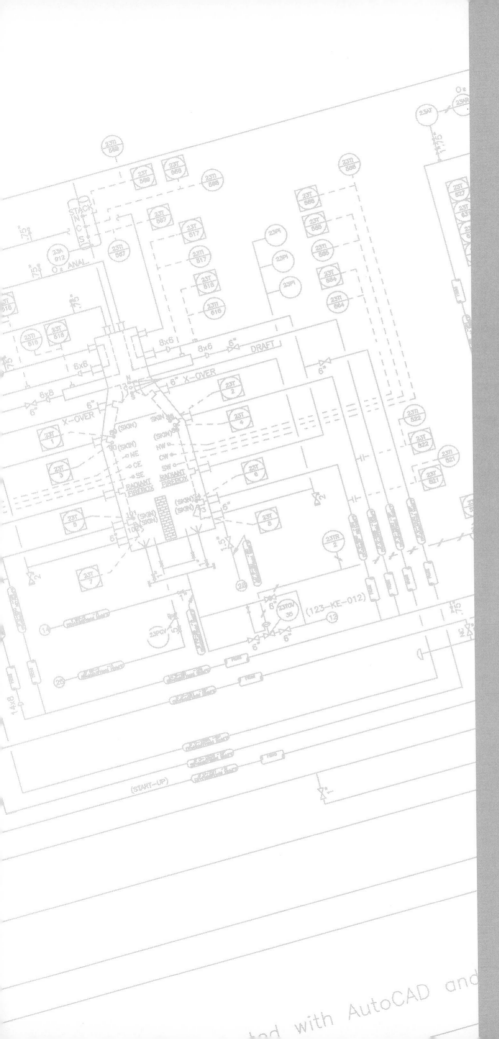

16

MODIFY COMMANDS III

Chapter Objectives

After completing this chapter you should:

1. be able to use *Ddchprop* and *Chprop* to change an object's layer, color, and linetype properties;

2. know how to use *Change* to change points or properties;

3. be able to use *Ddmodify* to modify any type of object;

4. be able to use *Matchprop* and the *Object Properties* toolbar to change properties of an object;

5. be able to *Explode* a *Pline* into its component objects and know that *Explode* can be used with *Blocks* and other composite shapes;

6. be able to *Align* objects with other objects;

7. be able to use all the *Pedit* options to modify *Plines* and to convert *Lines* and *Arcs* to *Plines*;

8. be able to modify *Splines* with *Splinedit*.

CONCEPTS

This chapter examines commands that are similar to, but generally more advanced and powerful than, those discussed in Chapters 9 and 10, Modify Commands I and II. None of the commands in this chapter creates new duplicate objects from existing objects but instead modifies the properties of the objects or convert objects from one type to another. Several commands are used to modify specific types of objects such as *Pedit* (modifies *Plines*), *Splinedit* (modifies *Splines*), *Mledit* (modifies *Mlines*), and *Union, Subtract*, and *Intersect* (modify *Regions*). Only one command in this chapter, *Align*, does not modify object properties but combines *Move* and *Rotate* into one operation. Several of the commands and features discussed in this chapter were mentioned in Chapter 12 (*Ddchprop, Ddmodify, MatchProp*, and *Object Properties* toolbar) but are explained completely here.

Only the commands in this chapter that modify general object properties have icon buttons that appear in the AutoCAD Drawing Editor by default. For example, you can access *Ddmodify, Ddchprop*, and *Matchprop* from the *Object Properties* toolbar and *Explode* by using its icon from the *Modify* toolbar.

Other commands that modify specific objects such as *Pedit, Mledit, Splinedit*, and the Boolean operators (*Union, Subtract*, and *Intersect*) appear in the *Modify II* toolbar. Activate the *Modify II* toolbar by selecting *Toolbars...* from the *View* pull-down menu (Fig. 16-1).

Figure 16-1

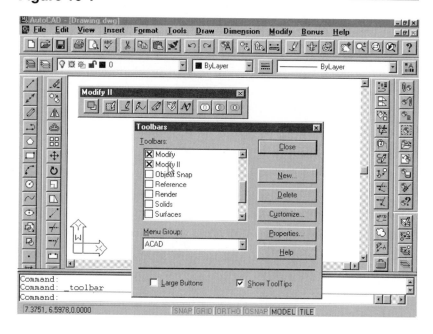

COMMANDS

You can use several methods to change properties of an object or of several objects. If you want to change an object's layer, color, or linetype only, the quickest method is by using the *Object Properties* toolbar (see *Object Properties* toolbar). If you want to change many properties of one or more objects (including layer, color, linetype, linetype scale, text style, dimension style, or hatch style) to match the properties of another existing object, use *Matchprop*. If you want to change any property (including coordinate data) for one object, use *Ddmodify*. You can use *Ddchprop* to change basic properties such as layer, linetype, and color for many objects at one time.

CHPROP and
DDCHPROP

	Pull-down Menu	COMMAND (TYPE)	ALIAS (TYPE)	Short-cut	Screen (side) Menu	Tablet Menu
	Modify *Properties...*	*CHPROP or* *DDCHPROP*	*CH*	...	*MODIFY1* *Modify*	*Y,14*

Typing *Chprop* (Change Properties) produces the command line format of this command. Typing *Ddchprop,* selecting the icon, or selecting *Properties* from the *Modify* pull-down menu invokes the *Change Properties* dialog box (Fig. 16-2).

Figure 16-2

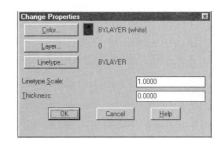

Ddchprop and *Chprop* allow you to change properties of several objects with one command. If you use the icon button or pull-down menu, you must <u>select more than one object</u> (in response to the "Select Objects:" prompt) if you want to use the *Change Properties* dialog box. If only one object is selected, the *Modify* dialog box appears instead (see *Ddmodify*).

Typing *Chprop* produces the command line format:

    Command: **chprop**
    Select Objects: **PICK**
    Select Objects: **Enter**
    Change what property (Color/LAyer/LType/ltScale/Thickness) ?

Layer
By changing an object's *Layer,* the object is effectively <u>moved</u> to the designated layer. In doing so, if the object's *Color* and *Linetype* are set to *BYLAYER,* the object assumes the color and linetype of the layer to which it is moved.

Color
It may be desirable in some cases to assign <u>explicit</u> *Color* to an <u>object</u> or objects independent of the layer on which they are drawn. The *Change Properties* command allows changing the color of an existing object from one object color to another or from *BYLAYER* assignment to an object color. <u>An object drawn with an object-specific color can also be changed to *BYLAYER* with this option.</u>

Linetype
An object can assume the *Linetype* assigned *BYLAYER* or can be assigned an <u>object-specific</u> *Linetype.* The *Linetype* option of *Change Properties* is used to change an object's *Linetype* to that of its layer (*BYLAYER*) or to any object-specific linetype that has been loaded into the current drawing. <u>An object drawn with an object-specific linetype can also be changed to *BYLAYER* with this option.</u>

Linetype Scale
An <u>object's individual linetype scale</u> can be changed with this option but <u>not the global</u> linetype scale (*LTSCALE*). <u>This is the recommended method to alter an individual object's linetype scale.</u> First, draw all the objects in the current global *LTSCALE.* *Ddchprop* or *Chprop* could then be used with this option to retroactively adjust the linetype scale of <u>specific</u> objects. The result would be similar to setting the *CELTSCALE* before drawing the specific objects. Using *Ddchprop* or *Chprop* to adjust a specific object's linetype scale <u>does not reset</u> the current *LTSCALE* or *CELTSCALE* variables.

Thickness

An object's *Thickness* can be changed by this option. *Thickness* is a three-dimensional quality (Z dimension) assigned to a two-dimensional object.

CHANGE

Pull-down Menu	COMMAND (TYPE)	ALIAS (TYPE)	Short-cut	Screen (side) Menu	Tablet Menu
...	*CHANGE*	-CH	...	...	...

The *Change* command allows changing three options: *Points, Properties,* or *Text*.

Point

This option allows changing the endpoint of an object or endpoints of several objects to one new position:

```
Command: change
Select Objects: PICK
Select Objects: Enter
Properties/<Change point>: PICK (Select a point to establish as new endpoint of all objects.
OSNAPs can be used.)
```

The endpoint(s) of the selected object(s) <u>nearest</u> the new point selected at the "Properties/<Change point>:" prompt is changed to the new point (Fig. 16-3).

Figure 16-3

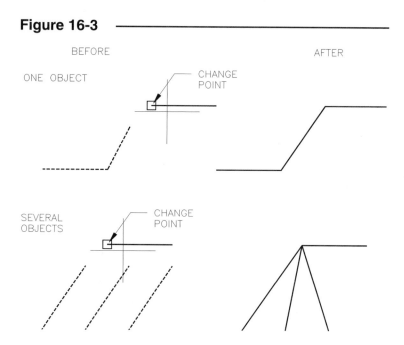

Properties

These options are discussed with the previous command. Selecting *Change* with the *Properties* option offers the same possibilities as using the *Ddchprop* or *Chprop* command. The *Elevation* property (a 3D property) of an object can also be changed with *Change* but not with *Ddchprop* or *Chprop*.

```
Properties/<Change point>: p
Change what property
(Color/Elev/LAyer/LType/ltScale/
Thickness) ?
```

Text

Although the word *"Text"* does <u>not</u> appear as an option, the *Change* command recognizes text created with *Text* or *Dtext* if selected. <u>*Change* does not change *Mtext*.</u> (See Chapter 18 for information on *Text, Dtext,* and *Mtext*.)

You can change the following characteristics of *Text* or *Dtext* objects:

    Text insertion point
    Text style
    Text height
    Text rotation angle
    Textual content

To change text, use the following command syntax:

Command: *change*
Select Objects: **PICK** (Select one or several lines of existing text.)
Select Objects: **Enter**
Properties/<Change point>: **Enter**
Enter text insertion point: **PICK** or **Enter** (PICK for a new insertion point or press Enter for no change.)
Text Style: ROMANS (Indicates the style of the selected text.)
New style or RETURN for no change: (**text style name**) or **Enter**
New height <0.2000>: (**value**) or **Enter** (Enter a value for new height or Enter if no change.)
New rotation angle: (**value**) or **Enter** (Enter value for new angle or Enter for no change.)
New text <FILLETS AND ROUNDS .125>: (**new text**) or **Enter** (Enter the new <u>complete line</u> of text or press Enter for no change.)

DDMODIFY

Pull-down Menu	COMMAND (TYPE)	ALIAS (TYPE)	Short-cut	Screen (side) Menu	Tablet Menu
Modify *Properties...*	*DDMODIFY*	*MO*	*...*	*MODIFY1* *Modify*	*Y,14*

Ddmodify allows you to <u>change any property of one object</u>. If using the icon or menus, <u>PICK only one object to invoke the *Modify (object type)* dialog box</u>. If more than one object is selected, the *Change Properties* dialog box appears instead.

The *Ddmodify* command always invokes a dialog box. The power of this feature is apparent because the configuration of the dialog box that appears is <u>specific</u> to the <u>type</u> of object that you select. For example, if you select a *Line*, a dialog box appears, allowing changes to any properties that a *Line* possesses; or, if you select a *Circle* or some *Text*, a dialog box appears specific to *Circle* or *Text* properties.

Figure 16-4 displays the dialog box after selecting a *Line*. Notice the dialog box title. Any aspect of the *Line* can be modified.

The *Properties* section always appears at the top of the *Modify (object type)* dialog box. This section has the same capabilities as the *Change Properties* dialog box. <u>All properties that an object possesses</u>, including position in the drawing, can be modified with the respective *Modify* dialog box.

Figure 16-4

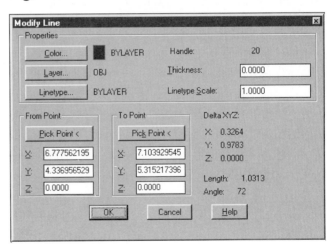

Figure 16-5 displays the dialog box after a *Pline* selection. Properties or individual vertices can be changed. Features of the *Pline* that can be modified here are similar to the capabilities of the *Pedit* command (see *Pedit*, this chapter).

Figure 16-5

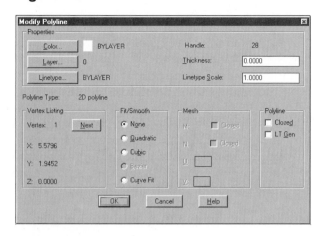

Each *Modify (object type)* dialog box offers a different set of changeable properties and features, depending on the object selected. Notice that the *Modify Circle* dialog box (Fig. 16-6) allows changing the radius and coordinates of the center as well as the typical properties (layer, linetype, etc.). Additional information such as diameter, circumference, and area is given.

Figure 16-6

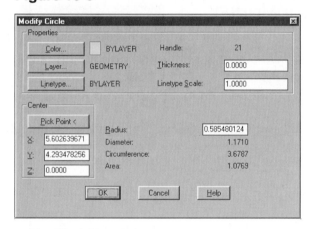

In some cases, the *Modify (object type)* dialog box enables editing by invoking another dialog box. If an *Mtext* object is selected, you can select the *Full editor...* tile (Fig. 16-7), which produces the same dialog box that was originally used to create the text.

Figure 16-7

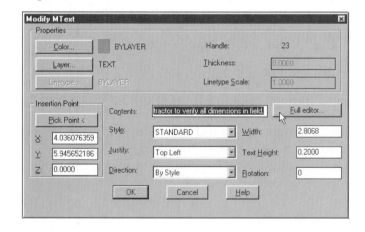

Selecting the *Full editor...* tile produces the *Multiline Text Editor* (Fig. 16-8), enabling you to change any properties that text possesses. (See Chapter 18 for information on *Mtext* and other text objects.)

Figure 16-8

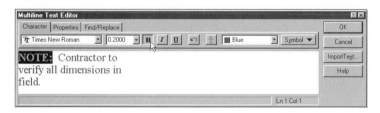

If hatch lines (created with the *Bhatch* command) are selected, the *Modify Hatch* dialog appears. Selecting the *Hatch Edit...* tile (Fig. 16-9) invokes the *Hatchedit* dialog box, providing access to parameters that were used to create the hatched area. (See Chapter 24 for information on hatching.)

Figure 16-9

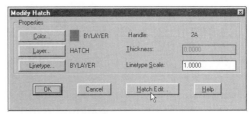

MATCHPROP

Pull-down Menu	COMMAND (TYPE)	ALIAS (TYPE)	Short-cut	Screen (side) Menu	Tablet Menu
Modify *Match Properties*	*MATCHPROP or PAINTER*	*MA*	...	*MODIFY1* *Matchprp*	*Y,15*

Matchprop is explained briefly in Chapter 12 but is explained again in this chapter with the full details of the *Special Properties*.

Matchprop is used to "paint" the properties of one object to another. Simply invoke the command, select the object that has the desired properties ("Source Object"), then select the object you want to "paint" the properties to ("Destination Object"). The command prompt is as follows:

Command: **matchprop**
Select Source Object: **PICK**
Current active settings = color layer ltype ltscale thickness text dim hatch
Settings/<Select Destination Object(s)>: **PICK**
Settings/<Select Destination Object(s)>: **Enter**
Command:

Figure 16-10

You can select Several "Destination Objects." The "Destination Object(s)" assume all of the "Current active settings" of the "Source Object."

The *Property Settings* dialog box can be used to set which of the *Basic Properties* and *Special Properties* are to be painted to the "Destination Objects" (Fig. 16-10). At the "Settings/<Select Destination Object(s)>:" prompt, type *S* to display the dialog box. You can control the following *Basic Properties*. Only the checked properties are "painted" to the destination objects.

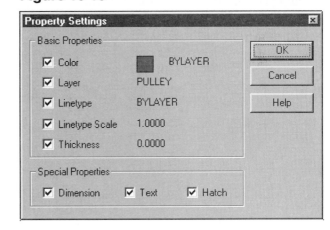

Color	This option paints the object-specific or *BYLAYER* color.
Layer	Move selected objects to the Source Object layer with this option checked.
Linetype	Paint the object-specific or *BYLAYER* linetype of the "Source Object" to the "Destination Object."
Linetype Scale	This option changes the individual object's linetype scale (*CELTSCALE*) not global linetype scale (*LTSCALE*).
Thickness	Thickness is a 3-dimensional quality.

The *Special Properties* section allows you to specify features of dimensions, text, and hatch patterns to match, as explained:

Dimension	This setting paints the *Dimension Style*. A *Dimension Style* defines the appearance of a dimension such as text style, size of arrows and text, tolerances if used, and many other features. (See Chapter 27.)
Text	This setting paints the "Source Object's" text *Style*. The text *Style* defines the text font and many other parameters that affect the appearance of the text. (See Chapter 18.)
Hatch	Checking this box paints the hatch properties of the Source Object to the "Destination Object(s)." The properties can include the hatch *Pattern, Angle, Scale,* and other characteristics. (See Chapter 24.)

Matchprop is a simple and powerful command to use, especially if you want to convert dimensions, text or hatch patterns to look like other objects in the drawing. This method works only when you have existing objects in the drawing you want to "match." If you want to convert only layer, linetype, and color properties without changing the other properties or if you do not have existing objects to "match," you can use the *Object Properties* toolbar.

Object Properties **Toolbar**

The three drop-down lists in the Object Properties toolbar (when not "dropped down") generally show the <u>current</u> layer, color, and linetype. However, if an object or set of objects is selected, the information in these boxes changes to display the current <u>object's</u> settings. You can change an object's settings by picking an object (when no commands are in use), then "dropping down" any of the three lists and selecting a different layer, linetype, or color.

Make sure you select (highlight) an object when no commands are in use. Use the pickbox (that appears on the cursor), Window, or Crossing Window to select the desired object or set of objects. The entries in the three boxes (Layer Control, Color Control, and Linetype Control) then change to display the settings for the <u>selected object or objects</u>. If several objects are selected that have different properties, the boxes display no information (go "blank"). Next, use any of the drop-down lists to make another selection (Fig. 16-11). The highlighted object's properties are changed to those selected in the lists. Press the Escape key to complete the process.

Figure 16-11 ——————

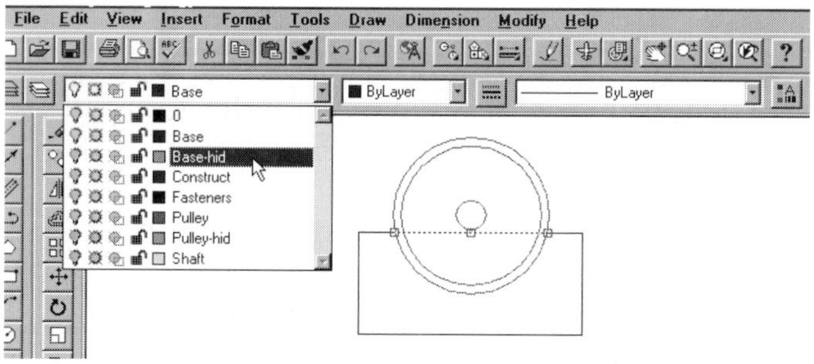

Remember that in most cases color and linetype settings are assigned *BYLAYER*. In this type of drawing scheme, to change the linetype or color properties of an object, you would change only the object's <u>layer</u> (see Figure 16-11). If you are using this type of drawing scheme, refrain from using the Color Control and Linetype Control drop-down lists for changing properties.

This method for changing object properties is about as quick and easy as *Matchprop*; however, only the layer, linetype, or color can be changed with the *Object Properties* toolbar. The *Object Properties* toolbar method works well if you do not have other existing objects to "match" or if you want to change <u>only</u> layer, linetype, and color <u>without</u> matching text, dimension, and hatch styles.

EXPLODE

Pull-down Menu	COMMAND (TYPE)	ALIAS (TYPE)	Short-cut	Screen (side) Menu	Tablet Menu
Modify *Explode*	*EXPLODE*	X	...	*MODIFY2* *Explode*	Y,22

Many graphical shapes can be created in AutoCAD that are made of several elements but are treated as one object, such as *Plines, Polygons, Blocks, Hatch* patterns, and dimensions. The *Explode* command provides you with a means of breaking down or "exploding" the complex shape from one object into its many component segments (Fig. 16-12). Generally, *Explode* is used to allow subsequent editing of one or more of the component objects of a *Pline, Polygon,* or *Block,* etc., which would otherwise be impossible while the complex shape is considered one object.

The *Explode* command has no options and is simple to use. You only need to select the objects to *Explode*.

Figure 16-12

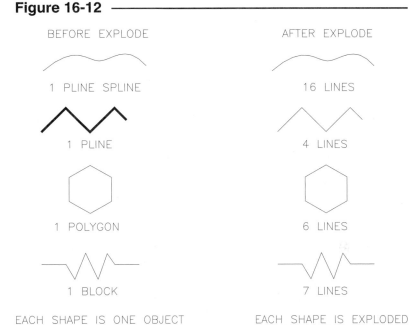

```
Command: explode
Select Objects: PICK (Select one or more Plines, Blocks, etc.)
Select Objects: Enter (Indicates selection of objects is complete.)
Command:
```

When *Plines, Polygons, Blocks,* or hatch patterns are *Exploded,* they are transformed into *Line, Arc,* and *Circle* objects. Beware, *Plines* having *width* lose their width information when *Exploded* since *Line, Arc,* and *Circle* objects cannot have width. *Exploding* objects can have other consequences such as losing "associativity" of dimensions and hatch objects and increasing file sizes by *Exploding Blocks.*

ALIGN

Pull-down Menu	COMMAND (TYPE)	ALIAS (TYPE)	Short-cut	Screen (side) Menu	Tablet Menu
Modify *3D Operations>* *Align*	*ALIGN*	AL	...	*MODIFY2* *Align*	X,14

Align provides a means of aligning one shape (a simple object, group of objects, *Pline, Boundary, Region, Block,* or a 3D object) with another shape. *Align* provides a complex motion, usually a combined translation (like *Move*) and rotation (like *Rotate*), in one command.

The alignment is accomplished by connecting source points (on the shape to be moved) to destination points (on the stationary shape). You should use *OSNAP* modes to select the source and destination points to assure accurate alignment. Either a 2D or 3D alignment can be accomplished with this command. The command syntax for alignment in a <u>2D alignment</u> is as follows:

```
Command: align
Select objects: PICK
Select objects: Enter
1st source point: PICK (with OSNAP)
1st destination point: PICK (with OSNAP)
2nd source point: PICK (with OSNAP)
2nd destination point: PICK (with OSNAP)
3rd source point: Enter
<2D> or 3D transformation: Enter
```

This command performs a translation and a rotation in one motion if needed to align the points as designated (Fig. 16-13).

Figure 16-13

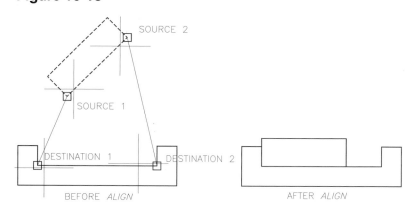

First, the 1st source point is connected to (actually touches) the 1st destination point (causing a translation). Next, the vector defined by the 1st and 2nd source points is aligned with the vector defined by the 1st and 2nd destination points (causing rotation). If no 3rd destination point is given (needed only for a 3D alignment), a 2D alignment is assumed and performed on the basis of the two sets of points.

PEDIT

Pull-down Menu	COMMAND (TYPE)	ALIAS (TYPE)	Short-cut	Screen (side) Menu	Tablet Menu
Modify Object > Polyline	*PEDIT*	*PE*	...	*MODIFY1 Pedit*	*Y,17*

This command provides numerous options for editing *Polylines* (*Plines*). As an alternative, *Ddmodify* can be used to change many of the *Pline*'s features in dialog box form (see Fig. 16-5).

The list of options below emphasizes the great flexibility possible with *Polylines*. The first step after invoking *Pedit* is to select the *Pline* to edit.

```
Command: pedit
Select polyline: PICK (Select the polyline for subsequent editing.)
Close or Open/Join/Width/Edit vertex/Fit/Spline/Decurve/ Ltype gen/Undo/eXit <X>: (option)
(Select the desired option from the screen menu or enter the capitalized letter for the desired option.)
```

Close

Close connects the last segment with the first segment of an existing "open" *Pline*, resulting in a "closed" *Pline* (Fig. 16-14). A closed *Pline* is one continuous object having no specific start or endpoint, as opposed to one closed by PICKing points. A *Closed Pline* reacts differently to the *Spline* option and to some commands such as *Fillet, Pline* option (see *Fillet*, Chapter 10).

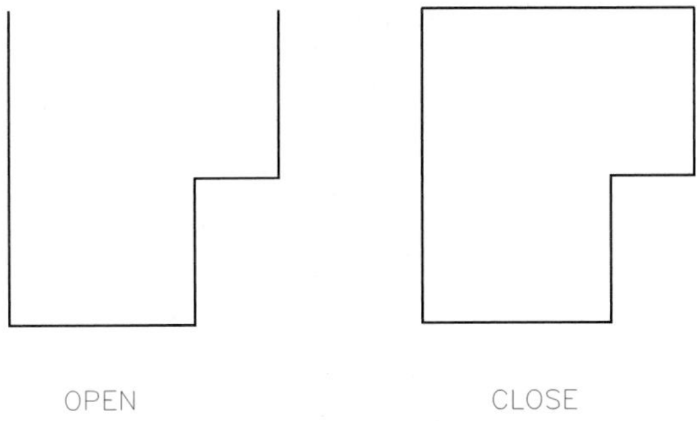

Open

Open removes the closing segment if the *Close* option was used previously (Fig. 16-14).

Figure 16-14 ——————————————

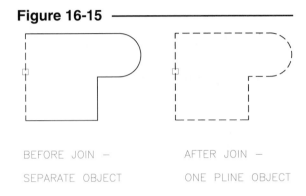

OPEN   CLOSE

Join

This option *Joins*, or connects, any *Plines, Lines,* or *Arcs* that have <u>exact</u> matching endpoints and adds them to the selected *Pline* (Fig. 16-15). Previously closed *Plines* cannot be *Joined*.

Figure 16-15 ——————————————

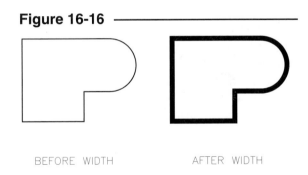

BEFORE JOIN —   AFTER JOIN —

SEPARATE OBJECT   ONE PLINE OBJECT

Width

Width allows specification of a uniform width for *Pline* segments (Fig. 16-16). Non-uniform width can be specified with the *Edit vertex* option.

Edit vertex

This option is covered in the next section.

Figure 16-16 ——————————————

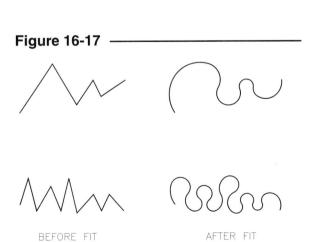

BEFORE WIDTH   AFTER WIDTH

Fit

This option converts the *Pline* from straight line segments to arcs. The curve consists of two arcs for each pair of vertices. The resulting curve can be radical if the original *Pline* consists of sharp angles. The resulting curve passes <u>through all</u> vertices (Fig. 16-17).

Figure 16-17 ——————————————

BEFORE FIT   AFTER FIT

Spline

This option converts the *Pline* to a B-spline (Bezier spline). The *Pline* vertices act as "control points" affecting the shape of the curve. The resulting curve passes through <u>only the end</u> vertices. A *Spline*-fit *Pline* is <u>not the same as a spline curve created with the *Spline* command</u>. This option produces a less versatile version of the newer *Spline* object.

Decurve

Decurve removes the *Spline* or *Fit* curve and returns the *Pline* to its original straight line segments state (Fig. 16-18).

When you use the *Spline* option of *Pedit*, the amount of "pull" can be affected by setting the *SPLINETYPE* system variable to either 5 or 6 <u>before</u> using the *Spline* option. *SPLINETYPE* applies either a quadratic (5=more pull) or cubic (6=less pull) B-spline function (Fig. 16-19).

Figure 16-18

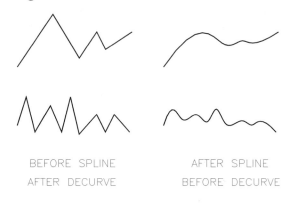

BEFORE SPLINE   AFTER SPLINE
AFTER DECURVE   BEFORE DECURVE

Figure 16-19

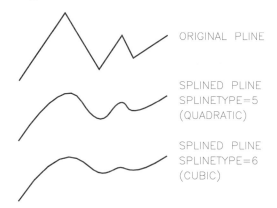

ORIGINAL PLINE

SPLINED PLINE
SPLINETYPE=5
(QUADRATIC)

SPLINED PLINE
SPLINETYPE=6
(CUBIC)

The *SPLINESEGS* system variable controls the number of line segments created when the *Spline* option is used. The variable should be set <u>before</u> using the option to any value (8=default): the higher the value, the more line segments. The actual number of segments in the resulting curve depends on the original number of *Pline* vertices and the value of the *SPLINETYPE* variable (Fig. 16-20).

Figure 16-20

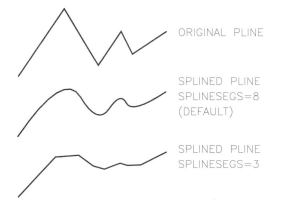

ORIGINAL PLINE

SPLINED PLINE
SPLINESEGS=8
(DEFAULT)

SPLINED PLINE
SPLINESEGS=3

Changing the *SPLFRAME* variable to 1 causes the *Pline* frame (the original straight segments) to be displayed for *Splined* or *Fit Plines*. *Regen* must be used after changing the variable to display the original *Pline* "frame" (Fig. 16-21).

Figure 16-21

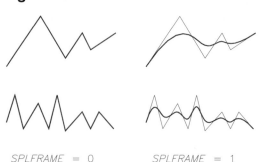

SPLFRAME = 0   SPLFRAME = 1

Ltype gen

This setting controls the generation of non-continuous linetypes for *Plines*. If *Off*, non-continuous linetype dashes start and stop at each vertex, as if the *Pline* segments were individual *Line* segments. For dashed linetypes, each line segment begins and ends with a full dashed segment (Fig. 16-22). If *On*, linetypes are drawn in a consistent pattern, disregarding vertices. In this case, it is possible for a vertex to have a space rather than a dash. Using the *Ltype gen* option <u>retroactively</u> changes *Plines* that have already been drawn. *Ltype gen* affects objects composed of *Plines* such as *Polygons*, *Rectangles*, and *Boundaries*.

Figure 16-22

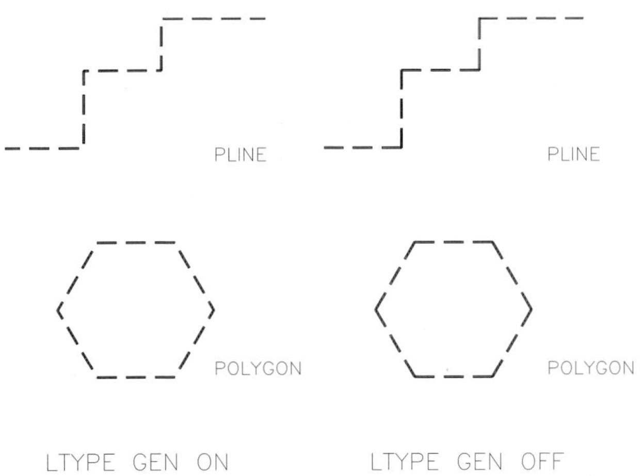

Similarly, the *PLINEGEN* system variable controls how <u>new</u> non-continuous linetypes are drawn for *Plines*. A setting of 1 creates a consistent linetype pattern, disregarding vertices (like *Ltype gen On*). A *PLINEGEN* setting of 0 creates linetypes stopping and starting at each vertex (like *Ltype gen Off*). However, *PLINEGEN* is <u>not retroactive</u>—it affects only <u>new</u> *Plines*.

Undo

Undo reverses the most recent *Pedit* operation.

eXit

This option exits the *Pedit* options, keeps the changes, and returns to the Command: prompt.

Vertex Editing

Upon selecting the *Edit Vertex* option from the *Pedit* options list, the group of suboptions is displayed on the screen menu and command line:

    Command: **pedit**
    Select polyline: **PICK** (Select the polyline for subsequent editing.)
    Close or Open/Join/Width/Edit vertex/Fit/Spline/Decurve/Ltype gen/Undo/eXit <X>: **E** (Invokes the Edit vertex suboptions.)
    Next/Previous/Break/Insert/Move/Regen/Straighten/Tangent/ Width/eXit <N>:

Next

AutoCAD places an **X** marker at the first endpoint of the *Pline*. The *Next* and *Previous* options allow you to sequence the marker to the desired vertex (Fig. 16-23).

Previous

See *Next* above.

Figure 16-23

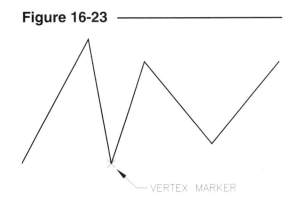

VERTEX MARKER

PREVIOUS AND NEXT OPTIONS WILL MOVE THE MARKER TO THE DESIRED VERTEX

Break

This selection causes a break between the marked vertex and another vertex you then select using the *Next* or *Previous* option (Fig. 16-24).

Next/Previous/Go/eXit <N>:

Selecting *Go* causes the break. An endpoint vertex cannot be selected.

Figure 16-24

BEFORE BREAK   AFTER BREAK

Insert

Insert allows you to insert a new vertex at any location <u>after</u> the vertex that is marked with the **X** (Fig. 16-25). Place the marker before the intended new vertex, use *Insert*, then PICK the new vertex location.

Figure 16-25

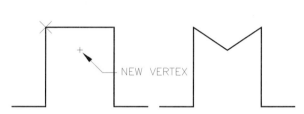

BEFORE INSERT   AFTER INSERT

NEW VERTEX

Move

You are prompted to indicate a new location to *Move* the marked vertex (Fig. 16-26).

Regen

Regen should be used after the *Width* option to display the new changes.

Figure 16-26

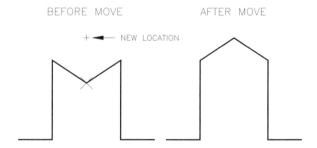

BEFORE MOVE   AFTER MOVE

NEW LOCATION

Straighten

You can *Straighten* the *Pline* segments between the current marker and the other marker that you then place by one of these options:

Next/Previous/Go/eXit <N>:

Selecting *Go* causes the straightening to occur (Fig. 16-27).

Figure 16-27

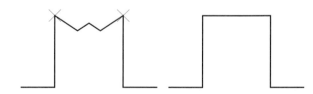

BEFORE STRAIGHTEN   AFTER STRAIGHTEN

Tangent

Tangent allows you to specify the direction of tangency of the current vertex for use with curve *Fitting*.

Width

This option allows changing the *Width* of the *Pline* segment immediately following the marker, thus achieving a specific width for one segment of the *Pline* (Fig. 16-28). *Width* can be specified with different starting and ending values. *Regen* must then be used to display the changes in width after using this option.

Figure 16-28

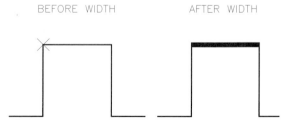

BEFORE WIDTH   AFTER WIDTH

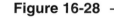

eXit

This option exits from vertex editing, saves changes, and returns to the main *Pedit* prompt.

Grips

Plines can also be edited easily using *Grips* (see Chapter 21). A *Grip* appears on each vertex of the *Pline*. Editing *Plines* with *Grips* is sometimes easier than using *Pedit* because *Grips* are more direct and less dependent on the command interface.

Converting *Lines* and *Arcs* to *Plines*

A very important and productive feature of *Pedit* is the ability to convert *Lines* and *Arcs* to *Plines* and closed *Pline* shapes. Potential uses of this option are converting a series of connected *Lines* and *Arcs* to a closed *Pline* for subsequent use with *Offset* or for inquiry of the area (*Area* command) or length (*List* command) of a single shape. The only requirement for conversion of *Lines* and *Arcs* to *Plines* is that the selected objects must have <u>exact</u> matching endpoints.

To accomplish the conversion of objects to *Plines*, simply select a *Line* or *Arc* object and request to turn it into one:

```
Command: pedit
Select polyline: PICK (Select one Line or Arc.)
Object selected is not a Pline
Do you want to turn it into one? <Y> Enter (Instructs AutoCAD to convert the selected object into a Pline.
Only one object can be selected.)
Close or Open/Join/Width/Edit vertex/Fit/Spline/Decurve/Ltype gen/Undo/eXit <X>: J
Select Objects: PICK  (Select the object or group of objects to join. Multiple objects can be selected.)
Select Objects: Enter (number) segments added to polyline
```

The resulting conversion is a closed *Polyline* shape.

SPLINEDIT

Pull-down Menu	COMMAND (TYPE)	ALIAS (TYPE)	Short-cut	Screen (side) Menu	Tablet Menu
Modify *Object >* *Splinedit*	*SPLINEDIT*	*SPE*	...	*MODIFY1* *Splinedit*	*Y,18*

Splinedit is an extremely powerful command for changing the configuration of existing *Splines*. You can use multiple methods to change *Splines*. All of the *Splinedit* methods fall under <u>two sets of options</u>.

The two groups of options that AutoCAD uses to edit *Splines* are based on two sets of points: <u>data points</u> and <u>control points</u>. Data points are the points that were specified when the *Spline* was created—the points that the *Spline* actually <u>passes through</u> (Fig. 16-29).

Figure 16-29

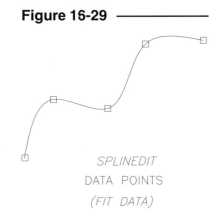

SPLINEDIT
DATA POINTS
(FIT DATA)

Control points are other points <u>outside of the path</u> of the *Spline* that only have a "<u>pull</u>" effect on the curve (Fig. 16-30).

Figure 16-30

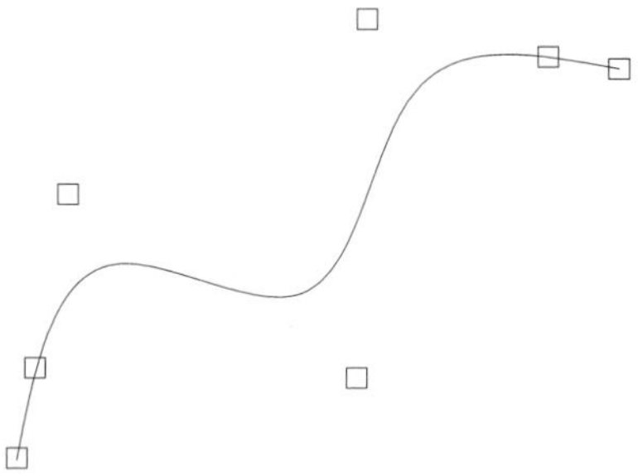

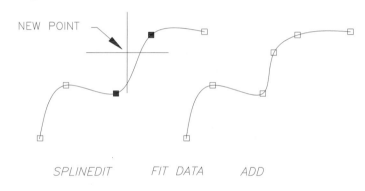

SPLINEDIT
CONTROL POINTS

Editing *Spline* <u>Data Points</u>

The command prompt displays several levels of options. The top level of options uses the control points method for editing. Select *Fit Data* to use data points for editing. The *Fit Data* methods are <u>recommended</u> for most applications. Since the curve passes directly through the data points, these options offer direct control of the curve path.

    Command: *splinedit*
    Select spline: **PICK**
    Fit Data/Close/Move Vertex/Refine/rEverse/Undo/eXit <X>: *f*
    Add/Close/Delete/Move/Purge/Tangents/toLerance/eXit <X>: (**option**)

Add

You can add points to the *Spline*. The *Spline* curve changes to pass through the new points. First, PICK an existing point on the curve. That point and the next one in sequence (in the order of creation) become highlighted. The new point will change the curve between those two highlighted data points (Fig. 16-31).

    Select point: **PICK** (Select an existing point on the curve before the intended new point.)
    Enter new point: **PICK** (PICK a new point location between the two marked points.)

Figure 16-31

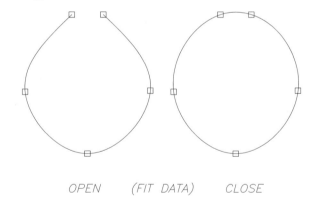

NEW POINT

SPLINEDIT   FIT DATA   ADD

Close/Open

The *Close* option appears only if the existing curve is open, and the *Open* prompt appears only if the curve is closed. Selecting either option automatically forces the opposite change. *Close* causes the two ends to become tangent, forming one smooth curve (Fig. 16-32). This <u>tangent continuity</u> is characteristic of *Closed Splines* only. *Splines* that have matching endpoints do not have tangent continuity unless the *Close* option of *Spline* or *Splinedit* is used.

Figure 16-32

OPEN   (FIT DATA)   CLOSE

Move

You can move any data point to a new location with this option (Fig. 16-33). The beginning endpoint (in the order of creation) becomes highlighted. Type *N* for next or *S* to select the desired data point to move; then PICK the new location.

    Next/Previous/Select Point/eXit/<Enter new location> <N>:

Purge

Purge <u>deletes all data points</u> and renders the *Fit Data* set of options unusable. You are returned to the control point options (top level). To reinstate the points, use *Undo*.

Figure 16-33

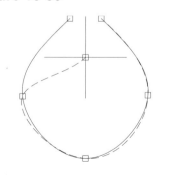

SPLINEDIT   FIT DATA   MOVE

Tangents

You can change the directions for the start and endpoint tangents with this option. This action gives the same control that exists with the "Enter start tangent" and "Enter end tangent" prompts of the *Spline* command used when the curves were created (see Fig. 15-12, *Spline, End Tangent*).

toLerance

Use *toLerance* to specify a value, or tolerance, for the curve to "fall" away from the data points. Specifying a tolerance causes the curve to smooth out, or fall, from the data points. The higher the value, the more the curve "loosens." The *toLerance* option of *Splinedit* is identical to the *Fit Tolerance* option available with *Spline* (see Fig. 15-14, *Spline, Tolerance*).

Editing *Spline* <u>Control Points</u>

Use the top level of command options (except *Fit Data*) to edit the *Spline's* control points. These options are similar to those used for editing the data points; however, the results are different since the curve does not pass through the control points.

```
Command: splinedit
Select spline: PICK
Fit Data/Close/Move Vertex/Refine/rEverse/Undo/eXit <X>:
```

Fit Data
Discussed previously.

Close/Open

These options operate similar to the *Fit Data* equivalents; however, the resulting curve falls away from the control points (Fig. 16-34; see also Fig. 16-32, *Fit Data, Close*).

Figure 16-34 ——————————————

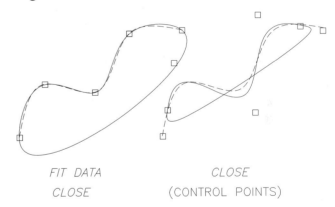

FIT DATA   CLOSE

CLOSE   (CONTROL POINTS)

Move Vertex

Move Vertex allows you to move the location of any control points. This is the control points' equivalent to the *Move* option of *Fit Data* (see Fig. 16-33, *Fit Data, Move*). The method of selecting points (*Next/Previous/Select point/eXit/*) is the same as that used for other options.

Refine

Selecting the *Refine* option reveals another level of options.

```
Add control point/Elevate Order/Weight/eXit <X>:
```

Add control points is the control points' equivalent to *Fit Data Add* (see Fig. 16-31). At the "Select a point on the Spline" prompt, simply PICK a point near the desired location for the new point to appear. Once the *Refine* option has been used, the *Fit Data* options are no longer available.

Elevate order allows you to <u>increase the number of control points</u> uniformly along the length of the *Spline*. Enter a value from *n* to 26, where *n* is the current number of points + one. Once a *Spline* is elevated, it cannot be reduced.

Weight is an option that you use to assign a value to the <u>amount of "pull"</u> that a <u>specific control point</u> has on the *Spline* curve. The higher the value, the more "pull," and the closer the curve moves toward the control point. The typical method of selecting points (*Next/Previous/Select point/eXit/*) is used.

rEverse

The *rEverse* option reverses the direction of the *Spline*. The first endpoint (when created) then becomes the last endpoint. Reversing the direction may be helpful for selection during the *Move* option.

Grips

Splines can also be edited easily using *Grips* (see Chapter 21). The *Grip* points that appear on the *Spline* are identical to the *Fit Data* points. Editing with *Grips* is a bit more direct and less dependent on the command interface.

CHAPTER EXERCISES

1. *Ddchprop* or *Chprop*

   Open the **PIVOTARM CH9** drawing that you worked on in Chapter 9 Exercises. *Load* the *Hidden* and *Center Linetypes*. Make two *New Layers* named **HID** and **CEN** and assign the matching linetypes and yellow and green colors, respectively. Check the *Limits* of the drawing; then calculate and set an appropriate *Ltscale* (usually .5 times the drawing scale factor). Use *Chprop* or *Ddchprop* to change the *Lines* representing the holes in the front view to the **HID** layer as shown in Figure 16-35. Use *SaveAs* and name the drawing **PIVOTARM CH16**.

   Figure 16-35

   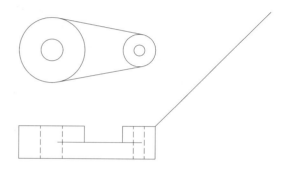

2. *Change*

   Open **CH8EX3**. *Erase* the *Arc* at the top of the object. *Erase* the *Points* with a window. Invoke the *Change* command. When prompted to *Select objects*, **PICK** all of the inclined *Lines* near the top. When prompted to <Change point>:, enter coordinate **6,8**. The object should appear as that in Figure 16-36. Use *SaveAs* and assign the name **CH16EX2**.

   Figure 16-36

   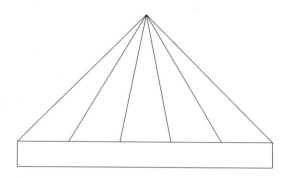

3. *Modify Circle*

   A design change is required for the bolt holes in **GASKETA** (from Chapter 9 Exercises). *Open* **GASKETA** and invoke the *Modify* dialog box (*Ddmodify*). Change each of the four bolt holes to **.375** diameter (Fig. 16-37). *Save* the drawing.

   Figure 16-37

   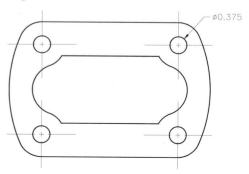

4. *Align*

   Open the **PLATES** drawing you
   created in Chapter 10. The three
   plates are to be stamped at one
   time on a single sheet of stock
   measuring 15″ x 12″. Place the
   three plates together to achieve
   optimum nesting on the sheet
   stock.

   A. Use *Align* to move the plate in
      the center (with 9 holes).
      Select the *1st source* and *des-
      tination points* (1S, 1D) and
      2nd source and *destination
      points* (2S, 2D) as shown in
      Figure 16-38.

   B. After the first alignment is
      complete, use *Align* to move
      the plate on the right (with 4
      diagonal holes). The *source*
      and *destination points* are
      indicated in Figure 16-39.

Figure 16-38

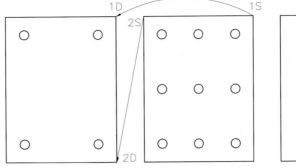

Figure 16-39

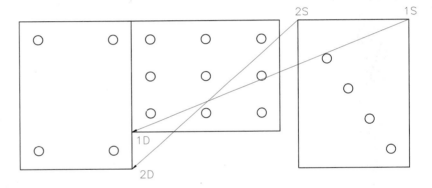

   C. Finally, draw the sheet stock outline (15″ x 12″)
      using *Line* as shown in Figure 16-40. The plates
      are ready for production. Use *SaveAs* and
      assign the name **PLATENEST**.

Figure 16-40

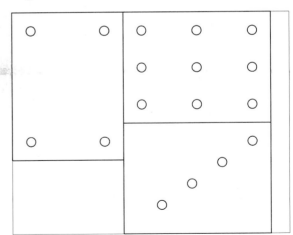

5. *Explode*

Figure 16-41

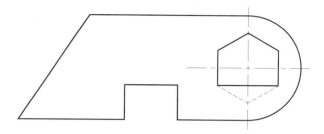

Open the **POLYGON1** drawing that you completed in Chapter 15 Exercises. You can quickly create the five-sided shape shown (continuous lines) in Figure 16-41 by *Exploding* the *Polygon*. First, *Explode* the *Polygon* and *Erase* the two *Lines* (highlighted). Draw the bottom *Line* from the two open *Endpoints*. Do not exit the drawing.

6. *Pedit*

Figure 16-42

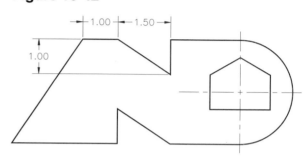

Use *Pedit* with the *Edit vertex* options to alter the shape as shown in Figure 16-42. For the bottom notch, use *Straighten*. For the top notch, use *Insert*. Use *SaveAs* and change the name to **PEDIT1**.

7. *Pline, Pedit*

Figure 16-43

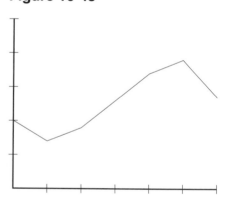

A. Create a line graph as shown in Figure 16-43 to illustrate the low temperatures for a week. The temperatures are as follows:

X axis	Y axis
Sunday	20
Monday	14
Tuesday	18
Wednesday	26
Thursday	34
Friday	38
Saturday	27

Use equal intervals along each axis. Use a *Pline* for the graph line. *Save* the drawing as **TEMP-A**. (You will label the graph at a later time.)

B. Use *Pedit* to change the *Pline* to a *Spline*. Note that the graph line is no longer 100% accurate because it does not pass through the original vertices (see Fig. 16-44). Use the *SPLFRAME* variable to display the original "frame" (*Regen* must be used after).

Figure 16-44

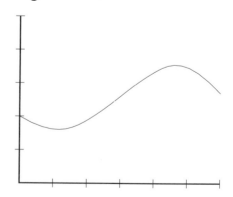

Use the *Modify Object* dialog box and try the *Cubic* and *Quadratic* options. Which option causes the vertices to have more pull? Find the most accurate option. Set *SPLFRAME* to **0** and *SaveAs* **TEMP-B**.

C. Use *Pedit* to change the curve from *Pline* to *Fit Curve*. Does the graph line pass through the vertices? *Saveas* **TEMP-C**.

D. *Open* drawing **TEMP-A**. *Erase* the *Splined Pline* and construct the graph using a *Spline* instead (Fig. 16-45, see Exercise 7A for data). *SaveAs* **TEMP-D**. Compare the *Spline* with the variations of *Plines*. Which of the four drawings (A, B, C, or D) is smoothest? Which is the most accurate?

Figure 16-45 ——————————————

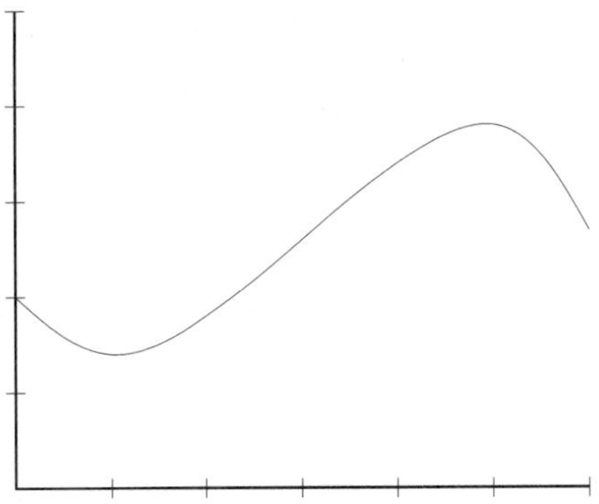

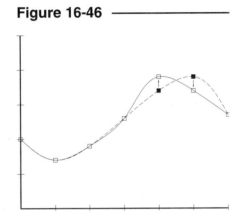

E. It was learned that there was a mistake in reporting the temperatures for that week. Thursday's low must be changed to 38 degrees and Friday's to 34 degrees. *Open* **TEMP-D** (if not already open) and use *Splinedit* to correct the mistake. Use the *Move* option of *Fit Data* so that the exact data points can be altered as shown in Figure 16-46. *SaveAs* **TEMP-E**.

Figure 16-46 ——————————————

8. **Converting** *Lines, Circles,* **and** *Arcs* **to** *Plines*

Figure 16-47 ——————————————————————————

A. Begin a *New* drawing and use **A-METRIC** as a template (that you created in Chapter 6 Exercises). Use *Save* and assign the name **GASKETC**. Change the *Limits* for plotting on an A sheet at 2:1 (refer to the Metric Table of *Limits* Settings and set *Limits* to 1/2 x *Limits* specified for 1:1 scale). Change the *LTSCALE* to **6**. Begin but do not complete the drawing of the gasket shown in Figure 16-47. First, draw only the <u>inside</u> shape using *Lines* and *Circles* (with *Trim*) or *Arcs*. Then convert the *Lines* and *Arcs* to one closed *Pline* using *Pedit*. Finally, locate and draw the 3 bolt holes. *Save* the drawing.

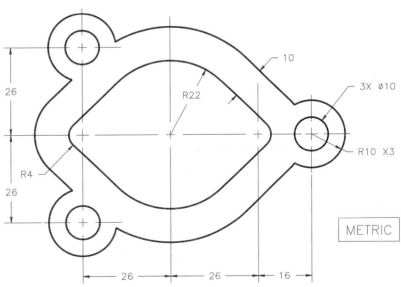

B. *Offset* the existing inside shape to create the outside shape. Use *Offset* to create concentric circles around the bolt holes. Use *Trim* to complete the gasket. *Save* the drawing and create a plot at 2:1 scale.

9. *Splinedit*   **Figure 16-48 —**   **Figure 16-49 —**

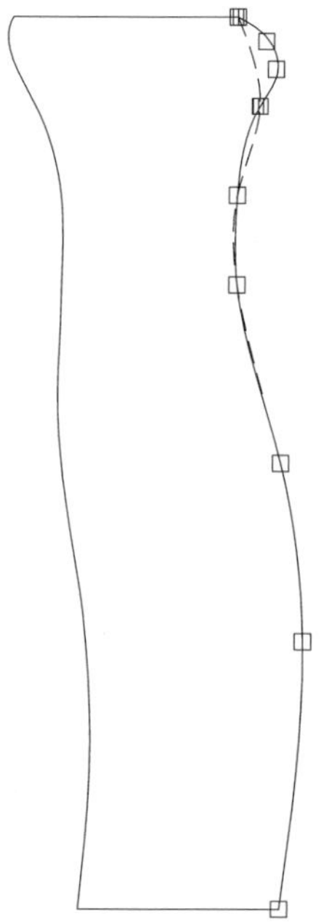

    Open the **HANDLE** drawing that you created in
    Chapter 15 Exercises. During the testing and analysis
    process, it was discovered that the shape of the handle
    should have a more ergonomic design. The finger side
    (left) should be flatter to accommodate varying sizes of
    hands, and the thumb side (right) should have more of a
    protrusion on top to prevent slippage.

    First, add more control points uniformly along the
    length of the left side with *Spinedit, Refine.* *Elevate* the
    Order from 4 to **6**, then use *Move Vertex* to align the
    control points as shown in Figure 16-48.

    On the right side of the handle, *Add* two points under
    the *Fit Data* option to create the protrusion shown in
    Figure 16-49. You may have to *Reverse* the direction of
    the *Spline* to add the new points between the two high-
    lighted ones. *SaveAs* **HANDLE2**.

10. **Hammer Drawing**

    Open the **HAMMER** drawing that you set up in Chapter 14 Exercises. Create the hammer shown
    in Figure 16-50. Make use of *Offset* for determining the center points for *Arc* and *Circle* radii (use
    the CONSTR layer for construction lines). Use *Fillet* wherever possible. When you finish, *Save*
    the drawing and create a plot to scale.

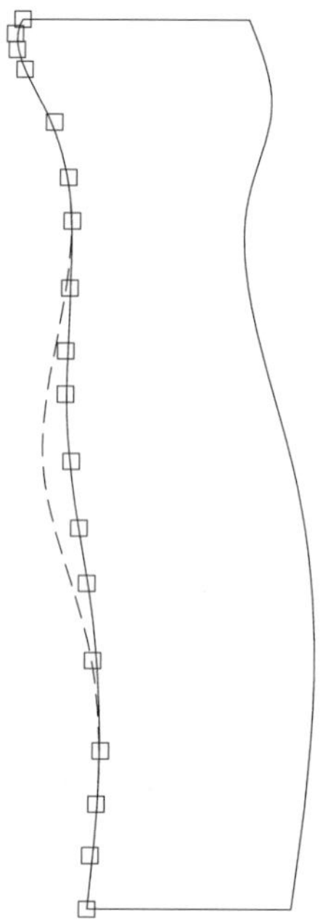

Figure 16-50 ───

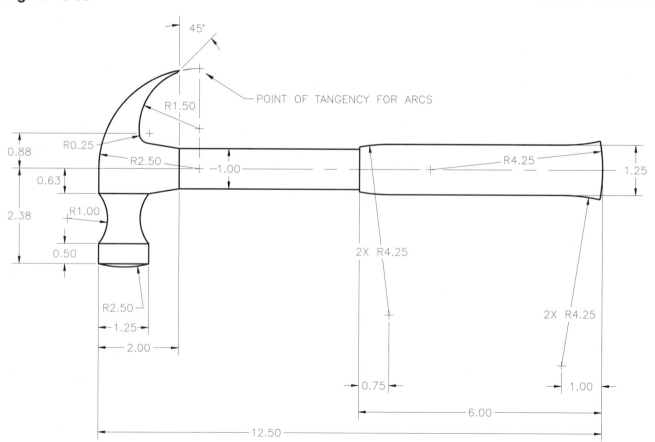

45°

R1.50

POINT OF TANGENCY FOR ARCS

0.88

R0.25

R2.50

-1.00

R4.25

1.25

0.63

2.38

R1.00

0.50

2X R4.25

R2.50

1.25

2.00

0.75

2X R4.25

1.00

6.00

12.50

17

INQUIRY COMMANDS

Chapter Objectives

After completing this chapter you should:

1. be able to list the *Status* of a drawing;

2. be able to *List* the AutoCAD database information about an object;

3. know how to list the entire database of all objects with *Dblist*;

4. be able to calculate the *Area* of a closed shape with and without "islands";

5. be able to find the *Distance* between two points;

6. be able to report the coordinate value of a selected point using the *ID* command;

7. know how to list the *Time* spent on a drawing or in the current drawing session;

8. be able to use *Setvar* to change system variable settings or list current settings.

CONCEPTS

AutoCAD provides several commands that allow you to find out information about the current drawing status and specific objects in a drawing. These commands as a group are known as "*Inquiry* commands" and are grouped together in the menu systems. The *Inquiry* commands are located on the Standard toolbar in a flyout fashion (Fig. 17-1). You can also use the *Tools* pull-down menu to access the *Inquiry* commands (Fig. 17-2).

Figure 17-1

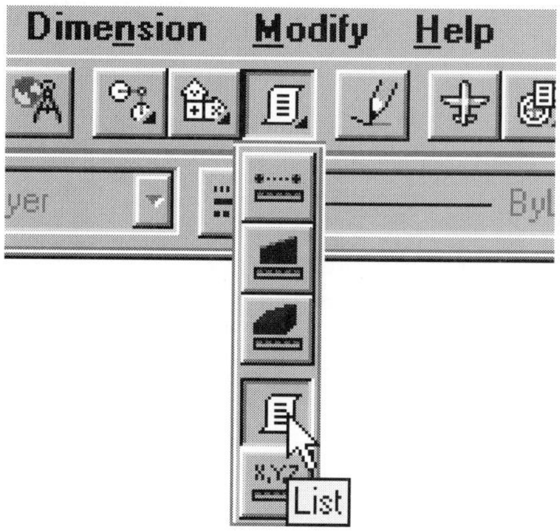

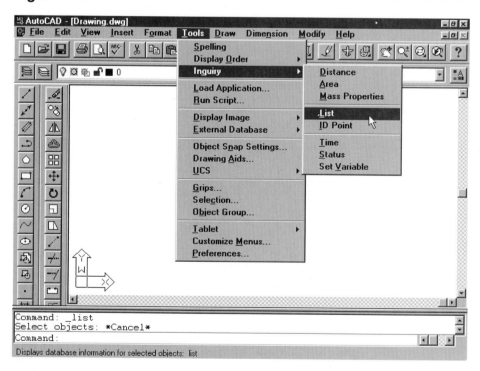

Figure 17-2

Using *Inquiry* commands, you can find out such information as the amount of time spent in the current drawing, the distance between two points, the area of a closed shape, the database listing of properties for specific objects (coordinates of endpoints, lengths, angles, etc.), and current settings for system variables as well as other information. The *Inquiry* commands are:

Status, List, Dblist, Area, Distance, ID, Time, and *Setvar*

COMMANDS

STATUS

Pull-down Menu	COMMAND (TYPE)	ALIAS (TYPE)	Short-cut	Screen (side) Menu	Tablet Menu
Tools *Inquiry >* *Status*	*STATUS*	...	...	*TOOLS 1* *Status*	...

The *Status* command gives many pieces of information related to the current drawing. Typing or **PICK**ing the command from the icon or one of the menus causes a text screen to appear similar to that shown in Figure 17-3. The information items are:

Figure 17-3

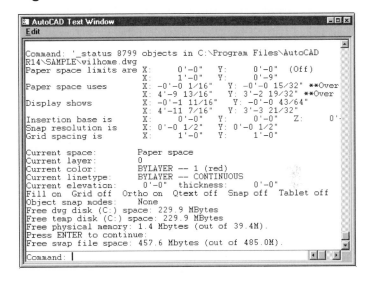

Total number of objects in the current drawing
Paper space limits: values set by the *Limits* command in Paper Space
Paper space uses: area used by the objects (drawing extents) in Paper Space
Model space limits: values set by the *Limits* command in Model Space
Model space uses: area used by the objects (drawing extents) in Model Space
Display shows: current display or windowed area
Insertion basepoint: point specified by the *Base* command or default (0,0)
Snap resolution: value specified by the *Snap* command
Grid spacing: value specified by the *Grid* command
Current space: Paper Space or Model Space
Current layer: name
Current color: current color assignment
Current linetype: current linetype assignment
Current elevation, thickness: 3D properties – current height above the XY plane and Z dimension
On or off status: *FILL, GRID, ORTHO, QTEXT, SNAP, TABLET*
Object Snap Modes: current *Running OSNAP* modes
Free dwg disk: space on the current drawing hard disk drive
Free temp disk: space on the current temporary files hard disk drive
Free physical memory: amount of free RAM (total RAM)
Free swap file space: amount of free swap file space (total allocated swap file)

LIST

Pull-down Menu	COMMAND (TYPE)	ALIAS (TYPE)	Short-cut	Screen (side) Menu	Tablet Menu
Tools Inquiry > List	LIST	LS or LI	...	TOOLS 1 List	U,8

The *List* command displays the database list of information in text window format for one or more specified objects. The information displayed depends on the type of object selected. Invoking the *List* command causes a prompt for you to select objects. AutoCAD then displays the list for the selected objects (see Figs. 17-4 and 17-5).

A *List* of a *Line* and an *Arc* is given in Figure 17-4.

For a *Line*, coordinates for the endpoints, line length and angle, current layer, and other information are given.

For an *Arc*, the center coordinate, radius, start and end angles, and length are given.

The *List* for a *Pline* is shown in Figure 17-5. The location of each vertex is given, as well as the length and perimeter of the entire *Pline*.

Note that in Release 14, *Plines* are created and listed as *Lwpolylines*, or "lightweight polylines." In previous releases, complete data for each *Pline* vertex (starting width, ending width, color, etc.) was stored along with the coordinate values of the vertex, then repeated for each vertex. In Release 14, the data common to all vertices is stored only once and only the coordinate data is stored for each vertex. Because this data structure saves file space, the new *Plines* are known as light weight *Plines*.

Figure 17-4

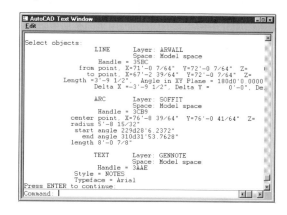

Figure 17-5

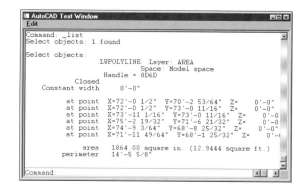

R14

DBLIST

Pull-down Menu	COMMAND (TYPE)	ALIAS (TYPE)	Short-cut	Screen (side) Menu	Tablet Menu
...	DBLIST	...	...	...	...

The *Dblist* command is similar to the *List* command in that it displays the database listing of objects; however, *Dblist* gives information for every object in the current drawing! This command is generally used when you desire to send the list to a printer or when only a few objects are in the drawing. If you use this command in a complex drawing, be prepared to page through many screens of information. Press Escape to cancel *Dblist* and return to the Command: prompt. Press F2 to close the text window and return to the Drawing Editor.

AREA

Pull-down Menu	COMMAND (TYPE)	ALIAS (TYPE)	Short-cut	Screen (side) Menu	Tablet Menu
Tools *Inquiry >* *Area*	*AREA*	*AA*	...	TOOLS 1 Area	T,7

The *Area* command is helpful for many applications. With this command AutoCAD calculates the area and the perimeter of any enclosed shape in a matter of milliseconds. You specify the area (shape) to consider for calculation by PICKing the *Object* (if it is a closed *Pline, Polygon, Circle, Boundary, Region* or other closed object) or by PICKing points (corners of the outline) to define the shape. The options are given below.

First Point

The command sequence for specifying the area by PICKing points is shown below. This method should be used only for shapes with <u>straight</u> sides. An example of the *Point* method (PICKing points to define the area) is shown in Figure 17-6.

Figure 17-6

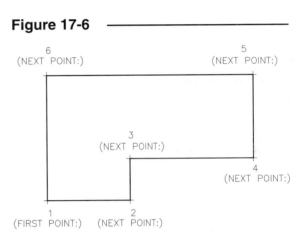

```
Command: area
<First point>/Object/Add/Subtract: PICK (Locate the first point to designate the shape.)
Next point: PICK (Select the second point to define the shape.)
Next point: PICK (Select the third point.)
Next point: PICK (Continue selecting points until all corners have been selected to completely define
the shape.)
Next point: Enter
Area = nn.nnn  perimeter = nn.nnn
Command:
```

Object

If the shape for which you want to find the area and perimeter is a *Circle, Polygon, Ellipse, Boundary, Region,* or closed *Pline,* the *Object* option of the *Area* command can be used. Select the shape with one PICK (since all of these shapes are considered as one object by AutoCAD).

The ability to find the area of a closed *Pline, Region,* or *Boundary* is extremely helpful. Remember that <u>any</u> closed shape, even if it includes *Arcs* and other curves, can be converted to a closed *Pline* with the *Pedit* command (as long as there are no gaps or overlaps) or can be used with the *Boundary* command. This method provides you with the ability to easily calculate the area of any shape, curved or straight. In short, convert the shape to a closed *Pline, Region,* or *Boundary* and find the *Area* with the *Object* option.

Add, Subtract

Add and *Subtract* provide you with the means to find the area of a closed shape that has islands, or negative spaces. For example, you may be required to find the surface area of a sheet of material that has several punched holes. In this case, the area of the holes is subtracted from the area defined by the perimeter shape. The *Add* and *Subtract* options are used specifically for that purpose. The following command sequence displays the process of calculating an area and subtracting the area occupied by the holes.

> Command: **area**
> <First point>/Object/Add/Subtract: **a** (Defining the perimeter shape is prefaced by the *Add* option.)
> <First point>/Object/Subtract: **o** (Assuming the perimeter shape is a *Boundary* or closed object.)
> (Add mode) Select object: **PICK** (Select the closed object.)
>
> Area = nn.nnn, Perimeter = nn.nnn
> Total area = nn.nnn
>
> (Add mode) Select object: **Enter** (Indicates completion of the *Add* mode.)
> <First point>/Object/Subtract: **s** (Change to the *Subtract* mode.)
> <First point>/Object/Add: **o** (Assuming the holes are *Circle* objects.)
> (Subtract mode) Select object: **PICK** (Select the first *Circle* to subtract.)
>
> Area = nn.nnn, Perimeter = nn.nnn
> Total area = nn.nnn
>
> (Subtract mode) Select object: **PICK** (Select the second *Circle* to subtract.)
>
> Area = nn.nnn, Perimeter = nn.nnn
> Total area = nn.nnn
>
> (Subtract mode) Select object: **Enter** (Completion of *Subtract* mode.)
> <First point>/Object/Add: **Enter** (Completion of *Area* command.)
> Command:

Make sure that you press Enter between the *Add* and *Subtract* modes.

An example of the last command sequence used to find the area of a shape minus the holes is shown in Figure 17-7. Notice that the object selected in the first step is a closed *Pline* shape, including an *Arc*.

Figure 17-7

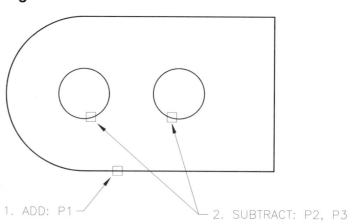

1. ADD: P1   2. SUBTRACT: P2, P3

DISTANCE

Pull-down Menu	COMMAND (TYPE)	ALIAS (TYPE)	Short-cut	Screen (side) Menu	Tablet Menu
Tools Inquiry > Distance	DIST	DI	...	TOOLS 1 Dist	T,8

The *Distance* command reports the distance between any two points you specify. *OSNAPs* can be used to "snap" to the existing points. This command is helpful in many engineering or architectural applications, such as finding the clearance between two mechanical parts, finding the distance between columns in a building, or finding the size of an opening in a part or doorway. The command is easy to use.

```
Command: dist
First point: PICK (Use OSNAPs if needed.)
Second point: PICK
Distance = nn.nnn, Angle in XY Plane = nn, Angle from XY Plane = nn
Delta X = nn.nnn, Delta Y = nn.nnn, Delta Z = nn.nnn
```

AutoCAD reports the absolute and relative distances as well as the angle of the line between the points.

ID

Pull-down Menu	COMMAND (TYPE)	ALIAS (TYPE)	Short-cut	Screen (side) Menu	Tablet Menu
Tools Inquiry > ID Point	ID	...	...	TOOLS 1 ID	U,9

The *ID* command reports the coordinate value of any point you select with the cursor. If you require the location associated with a specific object, an *OSNAP* mode (*Endpoint, Midpoint, Center,* etc.) can be used.

```
Command: ID
Point: PICK or (coordinate) (Select a point or enter a coordinate. OSNAPs can be used.)
X = nn.nnn, Y = nn.nnn, Z = nn.nnn
Command:
```

NOTE: *ID* also sets AutoCAD's "last point." The last point can be referenced in commands by using the @ (at) symbol with relative rectangular or relative polar coordinates. *ID* creates a "blip" at the location of the coordinates you enter or PICK.

TIME

Pull-down Menu	COMMAND (TYPE)	ALIAS (TYPE)	Short-cut	Screen (side) Menu	Tablet Menu
Tools Inquiry > Time	TIME	...	...	TOOLS 1 Time	...

This command is useful for keeping track of the time spent in the current drawing session or total time spent on a particular drawing. Knowing how much time is spent on a drawing can be useful in an office situation for bidding or billing jobs. The *Time* command reports the information shown in Figure 17-8.

The *Total editing time* is automatically kept, starting from when the drawing was first created until the current time. Plotting and printing time is not included in this total, nor is the time spent in a session when changes are discarded.

Figure 17-8

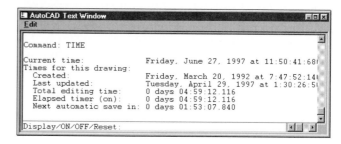

Display
The *Display* option causes *Time* to repeat the display with the updated times.

ON/OFF/Reset
The Elapsed timer is a separate compilation of time controlled by the user. The *Elapsed timer* can be turned *ON* or *OFF* or can be *Reset*.

Time also reports when the next automatic save will be made. The time interval of the Automatic Save feature is controlled by the *SAVETIME* system variable. To set the interval between automatic saves, type *SAVETIME* at the command line and specify a value for time (in minutes). The default interval is 120 minutes (see also *SAVETIME*, Chapter 2).

SETVAR

Pull-down Menu	COMMAND (TYPE)	ALIAS (TYPE)	Short-cut	Screen (side) Menu	Tablet Menu
Tools Inquiry > Set Variable	SETVAR	SET	...	TOOLS 1 Setvar	U,10

The settings (values or on/off status) that you make for many commands, such as *Limits*, *Grid*, *Snap*, *Running Osnaps*, *Fillet* values, *Pline* width, etc. are saved in system variables. In AutoCAD Release 14 there are 286 system variables. The variables store the settings that are used to create and edit the drawing. *Setvar* ("set variable") gives you access to the system variables. (See Appendix A for a complete list of the system variables, including an explanation, default setting, and possible settings for each.)

Setvar allows you to perform two functions: (1) change the setting for any system variable and (2) display the current setting for one or all system variables. To change a setting for a system variable using *Setvar*, just use the command and enter the name of the variable. For example, the following syntax lists values for the *GRID* setting and current *Fillet* radius:

Command: *setvar*
Variable name or ?: *gridunit*
New value for GRIDUNIT <0.5000,0.5000>:

Command: *setvar*
Variable name or ?: *filletrad*
New value for FILLETRAD <0.5000>:

With recent releases of AutoCAD, <u>*Setvar* is not needed to set system variables</u>. You can enter the variable name directly at the Command: prompt without using *Setvar* first.

Command: *filletrad*
New value for FILLETRAD <0.5000>:

To list the <u>current settings for all system variables</u>, use *Setvar* with the *?* (question mark) option. The complete list of system variables is given in a text window with the current setting for each variable (Fig. 17-9).

You can also use *Help* to list the system variables with a short explanation for each. In the *Help Topics* dialog box, select the *Contents* tab, select *Command Reference*, then *System Variables*.

Figure 17-9

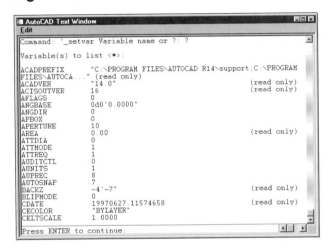

MASSPROP

The *Massprop* command is used to give mass and volumetric properties for AutoCAD solids. See Chapter 32, Advanced Solids Features, for use of *Massprop*.

CHAPTER EXERCISES

1. *List*

   A. *Open* the **PLATENEST** drawing from Chapter 16 Exercises. Assume that a laser will be used to cut the plates and holes from the stock, and you must program the coordinates. Use the *List* command to give information on the *Lines* and *Circles* for the one plate with four holes in a diagonal orientation. Determine and write down the coordinates for the 4 corners of the plate and the centers of the 4 holes.

   B. *Open* the **EFF-APT** drawing from Chapter 15 Exercises. Use *List* to determine the area of the inside of the tub. If the tub were filled with 10" of water, what would be the volume of water in the tub?

2. *Area*

   A. *Open* the **EFF-APT** drawing. The entry room is to be carpeted at a cost of $12.50 per square yard. Use the *Area* command (with the PICK points option) to determine the cost for carpeting the room, not including the closet.

   B. *Open* the **PLATENEST** drawing from Chapter 16 Exercises. Use *SaveAs* to create a file named **PLATE-AREA**. Using the *Area* command, calculate the wasted material (the two pieces of stock remaining after the 3 plates have been cut or stamped). HINT: Use *Boundary* to create objects from the waste areas for determining the *Area*.

   C. Create a *Boundary* (with islands) of the plate with four holes arranged diagonally. *Move* the new boundary objects **10** units to the right. Using the *Add* and *Subtract* options of *Area*, calculate the surface area for painting 100 pieces, both sides. (Remember to press Enter between the *Add* and *Subtract* operations.) *Save* the drawing.

3. *Dist*

   A. *Open* the **EFF-APT** drawing. Use the *Dist* command to determine the best location for installing a wall-mounted telephone in the apartment. Where should the telephone be located in order to provide the most equal access from all corners of the apartment? What is the farthest distance that you would have to walk to answer the phone?

   B. Using the *Dist* command, determine what length of pipe would be required to connect the kitchen sink drain (use the center of the far sink) to the tub drain (assume the drain is at the far end of the tub). Calculate only the direct distance (under the floor).

4. *ID*

   Open the **PLATENEST** drawing once again. You have now been assigned to program the laser to cut the plate with 4 holes in the corners. Use the *ID* command (with *OSNAP*s) to determine the coordinates for the 4 corners and the hole centers.

5. *Time*

   Using the *Time* command, what is the total amount of editing time you spent with the **PLATENEST** drawing? How much time have you spent in this session? How much time until the next automatic save?

6. *Setvar*

   Using the **PLATENEST** drawing again, use *Setvar* to list system variables. What are the current settings for *Fillet* radius (**FILLETRAD**), the last point used (**LASTPOINT**), *Linetype Scale* (**LTSCALE**), automatic file save interval (**SAVETIME**), and text *Style* (**TEXTSTYLE**).

18

CREATING AND EDITING TEXT

Chapter Objectives

After completing this chapter you should:

1. be able to create lines of text in a drawing using *Text* and *Dtext*;

2. be able to create and format <u>paragraph</u> text using *Mtext*;

3. be able to *Justify* text using each of the methods;

4. be able to <u>create</u> text styles with the *Style* <u>command</u>;

5. know that *Ddedit* can be used to edit the content of existing text and *Ddmodify* can be used to modify any property of existing text;

6. be able use *Spell* to check spelling and be able to *Find* and *Replace* text in a drawing.

CONCEPTS

The *Dtext, Mtext,* and *Text* commands provide you with a means of creating text in an AutoCAD drawing. "Text" in CAD drawings usually refers to sentences, words, or notes created from alphabetical or numerical characters that appear in the drawing. The numeric values that are part of specific dimensions are generally <u>not</u> considered "text," since dimensional values are a component of the dimension created automatically with the use of dimensioning commands.

Text in technical drawings is typically in the form of notes concerning information or descriptions of the objects contained in the drawing. For example, an architectural drawing might have written descriptions of rooms or spaces, special instructions for construction, or notes concerning materials or furnishings (Fig. 18-1). An engineering drawing may contain, in addition to the dimensions, manufacturing notes, bill of materials, schedules, or tables (Fig. 18-2). Technical illustrations may contain part numbers or assembly notes. Title blocks also contain text.

Figure 18-1

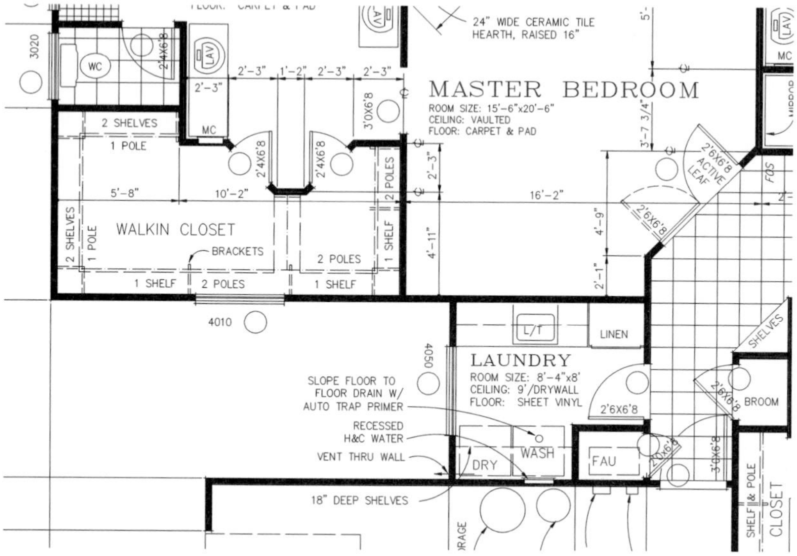

A line of text or paragraph of text in an AutoCAD drawing is treated as an object, just like a *Line* or a *Circle*. Each text object can be *Erased, Moved, Rotated,* or otherwise edited as any other graphical object. The letters themselves can be changed individually with special text editing commands. A spell checker is available by using the *Spell* command. Since text is treated as a graphical element, the use of many lines of text in a drawing can slow regeneration time and increase plotting time significantly.

Figure 18-2

ROUTE	POINT	STATION	COORDINATES	
			NORTH	EAST
℄ BASE LINE	P.I.	502+67.63	452555.90	2855805.56
℄ BASE LINE	P.C.	505+67.16	453207.20	2855642.55
℄ BASE LINE	P.I.	506+86.85	453307.61	2855577.41
℄ BASE LINE	P.T.	508+00.90	453427.17	2855572.01
℄ BASE LINE	P.T.	510+00.00	453626.07	2855563.02
℄ BASE LINE	P.T.	516+50.00	454275.41	2855533.69
℄ BASE LINE	P.O.T.	519+00.00	454264.13	2855783.44
DETOUR @ STA 551+80				
℄ DETOUR	P.O.T.	794+20.74	453044.48	2855659.62
℄ DETOUR	P.C.	795+67.79	453103.09	2855524.75
℄ DETOUR	P.I	796+47.79	453134.98	2855451.38
℄ DETOUR	P.T.	797+09.41	453214.98	2855451.38
℄ DETOUR	P.C.	806+94.29	454199.87	2855451.38
℄ DETOUR	P.I.	807+69.29	454274.87	2855451.38
℄ DETOUR	P.T.	808+41.38	454341.30	2855486.12
℄ DETOUR	P.C.	808+53.49	454352.03	2855491.73
℄ DETOUR	P.I.	809+28.49	454418.50	2855526.48
℄ DETOUR	P.C.	810+00.00	454493.42	2855523.09

R13

The *Dtext, Mtext,* and *Text* commands perform basically the same function; they create text in a drawing. *Mtext* is the newest and most sophisticated method of text entry. With *Mtext* (multiline text) you can create a paragraph of text that "wraps" within a text boundary (rectangle) that you specify. An *Mtext* paragraph is treated as one AutoCAD object. *Dtext* (dynamic text) displays each character in the drawing as it is typed and allows entry of multiple <u>lines</u> of text. *Text* allows only one line of text and requires an **Enter** or **Return** to display each line of text.

Many options for text justification are available. *Justification* is the method of aligning multiple lines of text. For example, if text is right justified, the right ends of the lines of text are aligned.

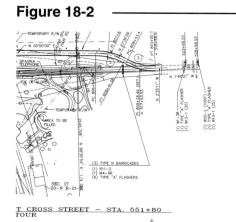

The form or shape of the individual letters is determined in AutoCAD by the text *Style*. Creating a *Style* begins with selecting a Windows standard TrueType or AutoCAD-supplied font file. Font files supplied with AutoCAD have file extensions of .TTF (TrueType) or .SHX (AutoCAD compiled shape files). The AutoCAD .SHX fonts are located in the C:\Program Files\AutoCAD R14\Fonts directory (by the default installation). The TrueType fonts are installed with the other Windows fonts in the C:\Windows\Fonts directory. Additional fonts can be purchased or may already be on your computer (supplied with Windows, word processors, or other software).

After a font for the *Style* is selected, other parameters (such as width and obliquing angle) can be specified to customize the *style* to your needs. Only one *style*, called *Standard*, has been created as part of the traditional default template drawing (ACAD.DWT) and uses the TXT.SHX font file. Other template drawings may have two or more created *styles*.

If any other style of text is desired, it must be created with the *Style* command. When a new *style* is created, it becomes the current one used by the *Dtext*, *Mtext*, or *Text* command. If several *styles* have been created in a drawing, a particular one can be recalled or made current by using the *style* <u>option</u> of the *Dtext*, *Mtext*, or *Text* commands.

In summary, the *Style* <u>command</u> allows you to design <u>new</u> styles with your choice of options, such as fonts, width factor, and obliquing angle, whereas the *style* <u>option</u> of *Dtext*, *Mtext*, and *Text* allows you to select from <u>existing</u> styles in the drawing that you previously created.

Commands related to creating or editing text in an AutoCAD drawing include:

Text	Places one line of text in a drawing, but the text is not visible until after pressing Enter.
Dtext	Places individual lines of text in a drawing and allows you to see each letter as it is typed.
Mtext	Places text in paragraph form (with word wrap) within a text boundary and allows many methods of formatting the appearance of the text.
Style	Creates text styles for use with any of the text creation commands. You can select from font files, specify other parameters to design the appearance of the letters, and assign a name for each style.
Spell	Checks the spelling of existing text in a drawing.
Ddedit	Invokes a dialog box for editing text. If you select *Text* or *Dtext* for editing, you can change the text (characters) only; if you select *Mtext* objects, you can edit individual characters and change the appearance of individual characters or the entire paragraph(s).
Ddmodify	A broad editing command for any type of text object (*Text*, *Dtext*, or *Mtext*), allowing revision of many aspects of text such as height, rotation angle, insertion point, style, or the text (characters).
Qtext	Short for quick-text, temporarily displays a line of text as a box instead of individual characters in order to speed up regeneration time and plotting time.

TEXT CREATION COMMANDS

The commands for creating text are formally named *Dtext*, *Mtext*, and *Text* (these are the commands used for typing). The *Draw* pull-down and screen (side) menus provide access to the two commonly used text commands, *Multiline Text...* (*Mtext*) and *Single-Line Text* (*Dtext*) (Fig. 18-3). Only the *Mtext* command has an icon button (by default) near the bottom of the *Draw* toolbar (Fig 18-4). The *Text* command must be typed at the command line.

Figure 18-3

Figure 18-4

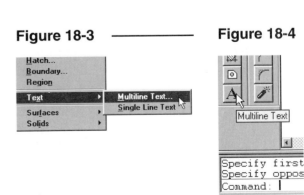

DTEXT

	Pull-down Menu	COMMAND (TYPE)	ALIAS (TYPE)	Short-cut	Screen (side) Menu	Tablet Menu
	Draw *Text >* *Single Line Text...*	DTEXT	DT	...	DRAW 2 *Dtext*	K,8

Dtext (dynamic text) lets you insert text into an AutoCAD drawing. *Dtext* displays each character in the drawing as it is typed. You can enter multiple lines of text without exiting the *Dtext* command. The lines of text do not "wrap." The options are presented below:

    Command: **dtext**
    Justify/Style/<Start point>:

Start Point
The *Start point* for a line of text is the <u>left end</u> of the baseline for the text (Fig. 18-5). *Height* is the distance from the baseline to the top of upper case letters. Additional lines of text are automatically spaced below and left justified. The *rotation angle* is the angle of the baseline (Fig. 18-6).

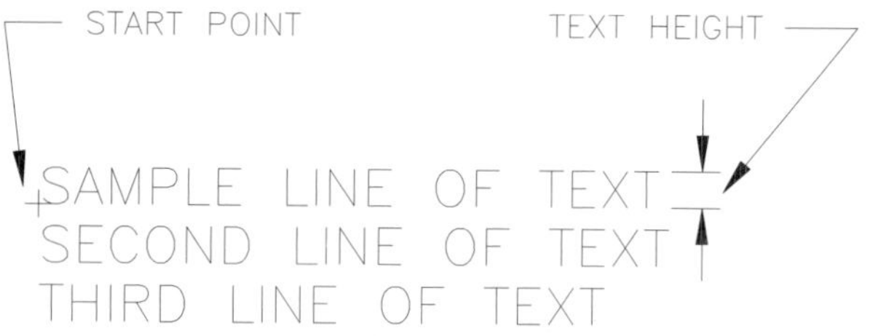

The command sequence for this option is:

    Command: **dtext**
    Justify/Style/<Start point>: **PICK** or (**coordinates**)
    Height <0.20>: **Enter** or (**value**)
    Rotation Angle <0>: **Enter** or (**value**)
    Text: (Type the desired line of text and press **Enter**.)
    Text: (Type another line of text and press **Enter**.)
    Text: **Enter**
    Command:

NOTE: When the "Text:" prompt appears, you can also PICK a new location for the next line of text anywhere in the drawing.

Justify
If you want to use one of the justification methods, invoking this option displays the choices at the prompt:

    Command: **dtext**
    Justify/Style/<Start point>: **J** (Invokes the justification options.)
    Align/Fit/Center/Middle/Right/TL/TC/TR/ML/MC/MR/BL/BC/BR: (**choice**) (Type capital letters.)

After specifying a justification option, you can enter the desired text in response to the "Text:" prompt. The text is not justified until <u>after</u> you press Enter.

Figure 18-5 ───────────────

Figure 18-6 ───────────────

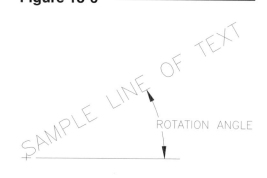

Align
Aligns the line of text between the two points specified (P1, P2). The text height is adjusted automatically (Fig. 18-7).

Fit
Fits (compresses or extends) the line of text between the two points specified (P1, P2). The text height does not change (Fig. 18-7).

Figure 18-7

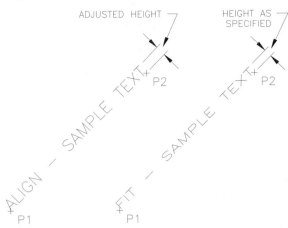

Center
Centers the baseline of the first line of text at the specified point. Additional lines of text are centered below the first (Fig 18-8).

Middle
Centers the first line of text both vertically and horizontally about the specified point. Additional lines of text are centered below it (Fig. 18-8).

Right
Creates text that is right justified from the specified point (Fig. 18-8).

Figure 18-8

LEFT JUSTIFIED TEXT
(START POINT)
SAMPLE

RIGHT JUSTIFIED TEXT
(R OPTION)
SAMPLE

CENTER JUSTIFIED TEXT
(C OPTION)
SAMPLE

MIDDLE JUSTIFIED TEXT
(M OPTION)
SAMPLE

TL
Top Left. Places the text in the drawing so the top line (of the first line of text) is at the point specified and additional lines of text are left justified below the point. The top line is defined by the upper case and tall lower case letters (Fig. 18-9).

Figure 18-9

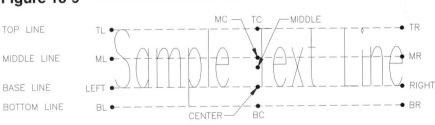

TC
Top Center. Places the text so the top line of text is at the point specified and the line(s) of text are centered below the point (Fig. 18-9).

TR
Top Right. Places the text so the top right corner of the text is at the point specified and additional lines of text are right justified below that point (Fig. 18-9).

ML
Middle Left. Places text so it is left justified and the middle line of the first line of text aligns with the point specified. The middle line is half way between the top line and the baseline, not considering the bottom (extender) line (Fig. 18-9).

MC
Middle Center. Centers the first line of text both vertically and horizontally about the midpoint of the middle line. Additional lines of text are centered below that point (Fig. 18-9).

MR
Middle Right. Justifies the first line of text at the right end of the middle line. Additional lines of text are right justified (Fig. 18-9).

BL
Bottom Left. Attaches the bottom (extender) line of the first line of text to the specified point. The bottom line is determined by the lowest point of lower case extended letters such as y, p, q, j, and g. If only upper-case letters are used, the letters appear to be located above the specified point. Additional lines of text are left justified (Fig. 18-9).

BC
Bottom Center. Centers the first line of text horizontally about the bottom (extender) line (Fig. 18-9).

BR
Bottom Right. Aligns the bottom (extender) line of the first line of text at the specified point. Additional lines of text are right justified (Fig. 18-9).

NOTE: Because there is a separate baseline and bottom (extender) line, the *MC* and *Middle* points do not coincide and the *BL, BC, BR* and *Left, Center, Right* options differ. Also, because of this feature, when all upper case letters are used, they ride above the bottom line. This can be helpful for placing text in a table because selecting a horizontal *Line* object for text alignment with *BL, BC,* or *BR* options automatically spaces the text visibly above the *Line.*

Style   (option of *Dtext, Mtext,* or *Text*)
The *style* <u>option</u> of the *Dtext, Mtext,* or *Text* command allows you to select from the <u>existing</u> text styles which have been previously created as part of the current drawing. The style selected from the list becomes the current style and is used when placing text with *Dtext, Mtext,* or *Text.*

Since only one text *style, Standard,* is available in the traditional (English) template drawing (ACAD.DWT) and the metric template drawing (ACADISO.DWT), other styles must be created before the *style* option of *Text* is of any use. Various text styles are created with the *Style* <u>command</u> (this topic is discussed later).

Use the *Style* option of *Dtext, Mtext,* or *Text* to list existing styles for the drawing. An example listing is shown below:

```
Command: dtext
Justify/Style/<Start point>: s
Style name (or ?) <STANDARD>: ?
Text style(s) to list <*>: Enter
Text styles:

Style name: ARCH-ITALIC   Font typeface: CityBlueprint
  Height: 0.0000  Width factor: 1.0000  Obliquing angle: 15
  Generation: Normal

Style name: ARCHITECT1   Font typeface: CityBlueprint
  Height: 0.0000  Width factor: 1.0000  Obliquing angle: 0
  Generation: Normal
```

Style name: COURIER   Font typeface: Courier New
  Height: 0.0000 Width factor: 1.0000 Obliquing angle: 15
  Generation: Normal

Style name: HELVETICA1   Font typeface: Arial
  Height: 0.0000 Width factor: 1.0000 Obliquing angle: 15
  Generation: Normal

Style name: ROMANSIMPLEX   Font files: romans.shx
  Height: 0.0000 Width factor: 1.0000 Obliquing angle: 0
  Generation: Normal

Style name: STANDARD   Font files: ISOCP
  Height: 0.0000 Width factor: 1.0000 Obliquing angle: 0
  Generation: Normal

Current text style: STANDARD
Justify/Style/<Start point>:

TEXT

Pull-down Menu	COMMAND (TYPE)	ALIAS (TYPE)	Short-cut	Screen (side) Menu	Tablet Menu
...	*TEXT*	...	...	...	...

Text is essentially the same as *Dtext* except that the text is not dynamically displayed one letter at a time as you type, but rather appears in the drawing only after pressing Enter. The other difference is that *Dtext* repeatedly displays the "Text:" prompt to allow entering multiple lines of text, whereas *Text* allows only one line. Otherwise, all the options and capabilities of *Text* are identical to *Dtext*.

```
Command: text
Justify/Style/<Start point>: PICK
Height <0.2000>: Enter or (value)
Rotation angle <0>: Enter or (value)
Text: Sample line of text. (The line of text appears in the drawing after pressing Enter.)
Command:
```

If you want to type another line of text below the previous line with the *Text* command, use *Text* again, but press Enter at the first prompt. The *Text* command then responds with the "Text:" prompt, at which time you can enter the next line of text. The new line is automatically spaced below and uses the same height, justification, and other options as the previous line.

MTEXT

Pull-down Menu	COMMAND (TYPE)	ALIAS (TYPE)	Short-cut	Screen (side) Menu	Tablet Menu
Draw *Text >* *Multiline Text...*	*MTEXT or* *-MTEXT*	*T, -T or* *MT*	...	*DRAW 2* *Mtext*	*J,8*

Multiline Text (*Mtext*) has more editing options than other text commands. You can apply underlining, color, bold, italic, font, and height changes to individual characters or words within a paragraph or multiple paragraphs of text.

Mtext allows you to create paragraph text defined by a text boundary. The <u>text boundary</u> is a reference rectangle that specifies the paragraph width. The *Mtext* object that you create can be a line, one paragraph, or several paragraphs. AutoCAD references *Mtext* (created with one use of the *Mtext* command) as one object, regardless of the amount of text supplied. Like *Text* and *Dtext*, several justification methods are possible.

```
Command: mtext
Current text style: STANDARD. Text height: 0.2000
Specify first corner: PICK
Specify opposite corner or [Height/Justify/Rotation/Style/Width]: PICK or (option)
```

You can PICK two corners to invoke the *Multiline Text Editor*, or enter the first letter of one of these options: *Height, Justify, Rotation, Style,* or *Width*. <u>All of the options</u> can also be accessed <u>within the Multiline Text Editor</u>.

Using the default option the *Mtext* command, you supply a "first corner" and "opposite corner" to define the diagonal corners of the text boundary (like a window). Although this boundary confines the text on two or three sides, one or two arrows indicate the direction text flows if it "spills" out of the boundary (Fig. 18-10). (See Text Flow and *Justification*.)

Figure 18-10 ————————————

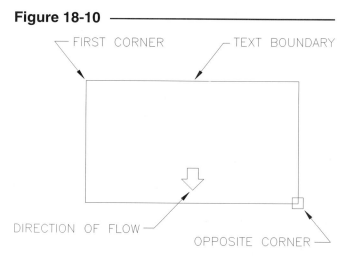

After you PICK the two points defining the text boundary, the *Multiline Text Editor* appears ready for you to enter the text (Fig. 18-11). Enter the desired text. The text wraps based on the width you defined for the text boundary. You can right-click for a menu allowing you to *Cut, Copy,* and *Paste* selected text. Select the *OK* button to have the text entered into the drawing within the text boundary.

Figure 18-11 ————————————

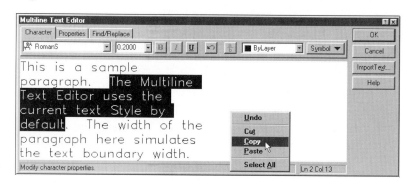

There are three tabs in the *Multiline Text Editor: Character* tab, *Properties* tab , and *Find and Replace* tab. Using the options in these tabs is <u>interactive</u>—text in the editor immediately reflects the changes made for most options in these tabs. There is also a button to *Import Text.*

Find/Replace **Tab**

See Editing Text later in this chapter.

Properties Tab

Use the properties tab to specify the format of the <u>entire paragraph</u>. Although this is the second tab, it is recommended that you <u>format the entire paragraph(s) here before editing individual characters</u> using the *Character* tab (Fig. 18-12). The following options are available.

Figure 18-12

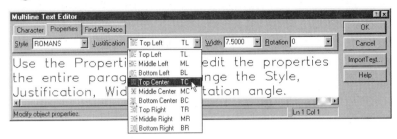

Style
Choose from a drop-down list of existing text styles (see *Style* command).

Justification
This property determines how the paragraph is located and direction of flow with respect to the text boundary (see Text Flow and *Justification*).

Width
Previous paragraph widths used are displayed in this drop-down list. You can enter a new value in the edit box to change the width of the existing text boundary. If a *Width* of 0 is entered or "no wrap" is selected, the lines of text will not "wrap" within the text boundary.

Rotation
The entire paragraph can be rotated to any angle. Changes made here are <u>not</u> reflected in the text appearing in the editor but only in the drawing itself. (See *Rotation*.)

You can type *MTPROP* at the command line to directly access the *Properties* tab of the *Multiline Text Editor*.

Character Tab

After formatting the entire paragraph, use the *Character* tab to alter individual characters in the paragraph(s) (Fig. 18-13). Using the options in this tab, first select (highlight) the desired characters or words, then set the desired options. The following options are available.

Figure 18-13

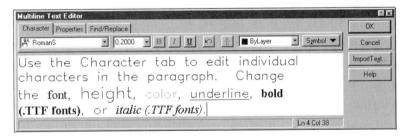

Font
Choose from any font in the drop-down list. Your selection here <u>overrides the text *Style*</u> used for the entire paragraph(s). Even though you can change the font for the entire *Mtext* object (paragraph), it is recommended to set the paragraph to the desired *Style* (in the *Properties* tab), rather than changing all characters to a different font here. See following NOTE.

Height
Select from the list or enter a new value for the height of selected words or letters. Your selection overrides the text *Height* used globally for the paragraph.

Bold, Italic, Underline
Select (highlight) the desired letters or words then PICK the desired button. Only authentic TrueType fonts (not the AutoCAD-supplied .SHX equivalents) can be bolded or italicized.

Stack/Unstack

If creating a stacked fraction, use a / (slash) between the numerator and denominator. If creating stacked text, place a ^ (caret) before the bottom text. Highlight the fraction or text, then use this option to stack or unstack the fraction or text.

Text Color

Select individual text, then use this drop-down list to select a color for the selected text. This selection overrides the layer color.

Symbol

Common symbols (plus/minus, diameter, degrees) can be inserted. Selecting *Other...* produces a character map to select symbols from (See *Other Symbols* below).

Other Symbols
The steps for inserting symbols from the *Character Map* dialog box (Fig. 18-14) are as follows:

Figure 18-14

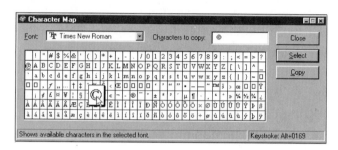

1. Highlight the symbol.
2. Double-click or PICK *Select* so the item appears in the *Characters to copy:* edit box.
3. Select the *Copy* button to copy the item(s) to the Windows Clipboard.
4. *Close* the dialog box.
5. In the *Multiline Text Editor*, move the cursor to the desired location to insert the symbol.
6. Finally, right-click and select *Paste* from the menu.

NOTE: The font, color, and height options in the *Character* tab override the properties of the entire paragraph. For example, changing the font in the *Character* tab overrides the text style's font used for the paragraph so it is possible to have one font used for the paragraph and others for individual characters within the paragraph. This is analogous to object-specific color and linetype assignment in that you can have a layer containing objects with different linetypes and colors than the linetype and color assigned to the layer. To avoid confusion, it is recommended to create text *Styles* to be used globally for the paragraphs, layer color to determine color for the paragraphs, and text *Height* for global paragraph formatting. Then if needed, use the *Character* tab to change fonts, colors, or height for <u>selected</u> text rather than for the entire paragraph.

Text Flow and *Justification*

When you PICK two corners to define the text boundary, you determine the width of the paragraph and the direction that the text flows. Unlike using *Mtext* in Release 13, you can draw the text boundary in any direction from the "first corner" to the "other corner." Text does not always fit within the boundary but is confined on two or three sides and "spills" out of the boundary (up, down, or both) in the direction of the arrow(s) displayed when you draw the boundary (see Figure 18-10).

Justification is the method of aligning the text with respect to the text boundary. The *Justification* options are *TL, TC, TR, ML, MC, MR, BL, BC,* and *BR* (*Top Left, Top Centered, Top Right, Middle Left, Bottom Left,* etc.) The text paragraph (*Mtext* object) is effectively "attached" to the boundary based on the *Justification* option selected. *Justification* can be specified by two methods: in command line format before specifying the text boundary or in the *Properties* tab of the *Multiline Text Editor*.

The *Justificaiton* methods are illustrated in Figure 18-15. The illustration shows the relationship among the *Justification* option, the text boundary, the direction of flow, and the resulting text paragraph.

Figure 18-15

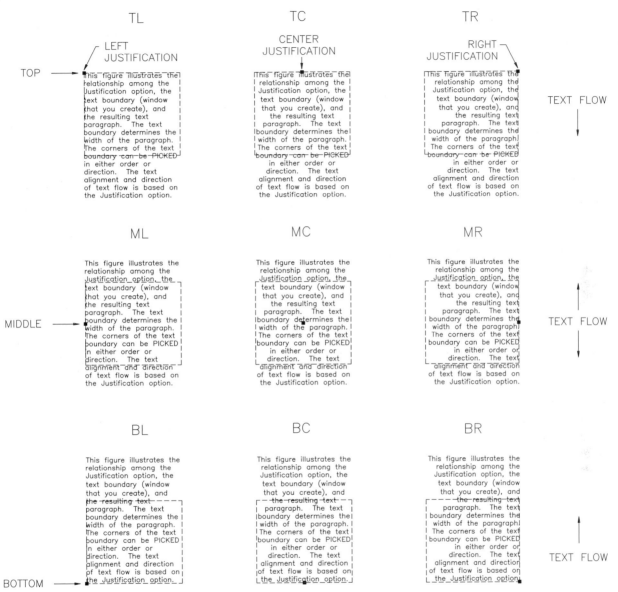

If you want to adjust the text boundary after creating the *Mtext* object, you can use *Ddedit* or *Ddmodify* (see Editing Text) or use Grips (Chapter 21) to stretch the text boundary. If you activate the Grips, four grips appear at the text boundary corners and one grip appears at the defined *Justification* point (Fig. 18-16).

Figure 18-16

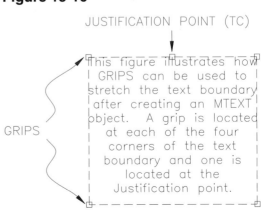

R14

Rotation

The *Rotation* option specifies the rotation angle of the <u>entire *Mtext* object (paragraph) including the text boundary.</u> *Rotation* can be specified by two methods: in command line format before specifying the text boundary or in the *Properties* tab of the *Multiline Text Editor.* It is recommended to use the command line method to specify a rotation angle so you can see the rotated text boundary as you specify the corners (Fig. 18-17). Using this method you can enter a value, or the angle can be specified by PICKing two points. If you use the *Rotation* option within the *Properties* tab of the *Multiline Text Editor,* the text boundary is rotated after the fact.

Figure 18-17

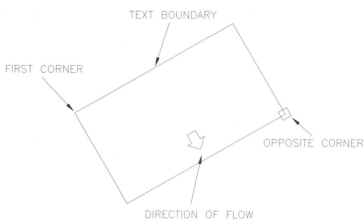

Calculating Text Height for Scaled Drawings

To achieve a specific height in a drawing intended to be plotted to scale, multiply the desired text height for the plot by the drawing scale factor. (See Chapter 14, Plotting.) For example, if the drawing scale factor is 48 and the desired text height on the plotted drawing is 1/8", enter **6** (1/8 x 48) in response to the "Height:" prompt of the *Dtext, Mtext,* or *Text* command.

If you know the plot scale (for example, ¼"= 1') but not the drawing scale factor (DSF), calculate the reciprocal of the plot scale to determine the DSF, then multiply the intended text height for the plot by the DSF, and enter the value in response to the "Height:" prompt. For example, if the plot scale is ¼"=1', then the DSF = 48 (reciprocal of 1/48).

If *Limits* have already been set and you do not know the drawing scale factor, use the following steps to calculate a text height to enter in response to the "Height:" prompt to achieve a specific plotted text height.

1. Determine the sheet size to be used for plotting (for example, 36" x 24").
2. Decide on the text height for the finished plot (for example, .125").
3. Check the *Limits* of the current drawing (for example, 144' x 96' or 1728" x 1152").
4. Divide the *Limits* by the sheet size to determine the drawing scale factor (1728"/36" = 48).
5. Multiply the desired text height by the drawing scale factor (.125 x 48 = 6).

TEXT STYLES AND FONTS

STYLE

Pull-down Menu	COMMAND (TYPE)	ALIAS (TYPE)	Short-cut	Screen (side) Menu	Tablet Menu
Format *Text Style...*	*STYLE or* *-STYLE*	*ST*	...	DRAW 2 *Dtext* *Style:*	U,2

Text styles can be created by using the *STYLE* command. A text *Style* is created by selecting a font file as a foundation and then specifying several other parameters to define the configuration of the letters.

Using the *Style* command invokes the *Text Style* dialog box (Fig 18-18). All options for creating and modifying text styles are accessible from this device. Selecting the font file is the initial step in creating a style. The font file selected then becomes a foundation for "designing" the new style based on your choices for the other parameters (*Effects*). Recommended steps for creating text *Styles* are as follows:

Figure 18-18 ────────────────────

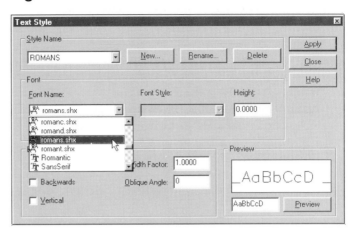

1. Use the *New* button to create a new text *Style*. This button opens the *New Text Style* dialog box (Fig. 18-19). By default, AutoCAD automatically assigns the name Style*n*, where *n* is a number that starts at 1. Enter a descriptive name into the edit box. Style names can be up to 31 characters long and can contain letters, numbers, and the special characters dollar sign ($), underscore (_), and hyphen (-).

Figure 18-19 ────────────────────

2. Select a font to use for the style from the *Font Name* drop-down list (see Fig. 18-18). TrueType equivalents of the traditional AutoCAD fonts files (.SHX) and authentic TrueType fonts are available.

3. Select parameters in the *Effects* section of the dialog box (Fig. 18-20, lower-left). Changes to the *Upside down*, *Backwards*, *Width Factor*, or *Oblique Angle* are displayed immediately in the *Preview* tile.

4. You can enter specific words or characters in the edit box in the lower-right, then press *Preview* to view the characters in the new style.

5. Select *Apply* to save the changes you made in the *Effects* section to the new style.

6. Create other new *Styles* using the same procedure listed in steps 1 through 5.

7. Select *Close*. The new (or last created) text style that is created automatically becomes the current style inserted when *Dtext*, *Mtext*, *Text*, or dimensioning is used.

Figure 18-20 ────────────────────

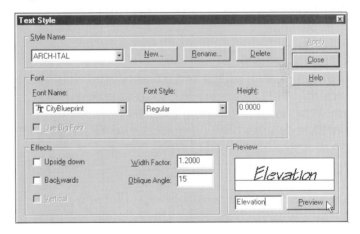

You can modify existing styles using this dialog box by selecting the existing *Style* from the list, making the changes, then selecting *Apply* to save the changes to the existing *Style*.

Rename
Select an existing *Style* from the drop-down list, then select *Rename*. Enter a new name in the edit box.

Delete
Select an existing *Style* from the drop-down list, then select *Delete*. You cannot delete the current *Style* or *Styles* that have been used for creating text in the drawing.

R14

Alternately, you can enter *-Style* (use the apostrophe prefix) at the command prompt to display the command line format of *Style* as shown below:

```
Command: -style
Text style name (or ?) <STANDARD>: name (Enter new Style name)
New style. Specify full font name or font filename <txt>: name (Enter desired font file)
Height <0.0000>: Enter or (value)
Width factor <1.0000>: Enter or (value)
Obliquing angle <0>: Enter or (value)
Backwards? <N> Enter or Y
Upside-down? <N> Enter or Y
Vertical? <N>  Enter or Y
(name) is now the current text style.
Command:
```

The options that appear in both the *Text Style* dialog box and in the command line format are described in detail here.

Height <0.000>

The <u>height should be 0.000</u> if you want to be prompted again for height each time the *Dtext, Mtext,* or *Text* command is used. In this way, the height is variable for the style each time you create text in the drawing. If you want the height to be constant, enter a value other than 0. Then, *Dtext, Mtext,* or *Text* will not prompt you for a height since it has already been specified. A specific height assignment with the *Style* command also overrides the *DIMTXT* setting (see Chapter 27).

Width factor <1.000>

A *width factor* of 1 keeps the characters proportioned normally. A value less than 1 compresses the width of the text (horizontal dimension) proportionally; a value of greater than 1 extends the text proportionally.

Obliquing angle <0>

An angle of 0 keeps the font file as vertical characters. Entering an angle of 15, for example, would slant the text forward from the existing position, or entering a negative angle would cause a back-slant on a vertically oriented font (Fig. 18-21).

Figure 18-21

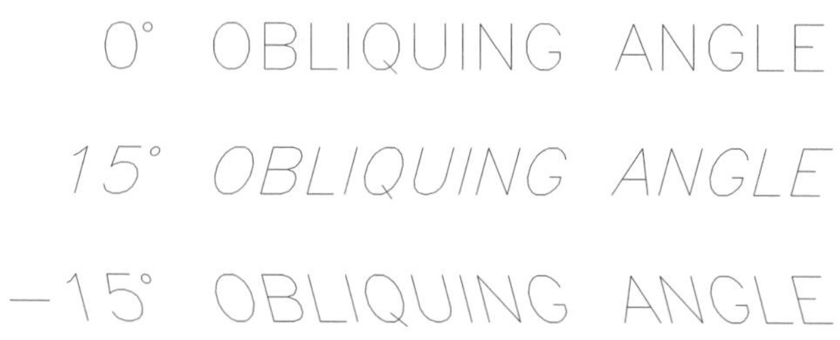

Backwards? <Y/N>

Backwards characters can be helpful for special applications, such as those in the printing industry.

Upside-down? <Y/N>

Each letter is created upside-down in the order as typed (Fig. 18-22). This is different than entering a rotation angle of 180 in the *Dtext* or *Text* commands. (Turn this book 180 degrees to read the figure.)

Figure 18-22

Vertical? <Y/N>

Vertical letters are shown in Figure 18-23. The normal rotation angle for vertical text when using *Dtext, Mtext,* or *Text* is 270. Only .SHX fonts can be used for this option. *Vertical* text does not display in the *Preview* image tile.

Figure 18-23 —

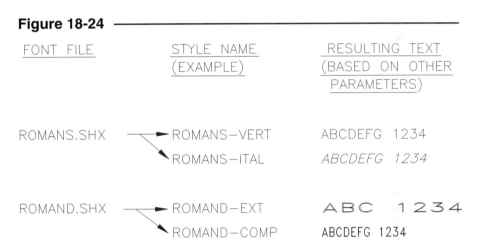

Since specification of a font file is an initial step in creating a style, it seems logical that different styles could be created using <u>one</u> font file but changing the other parameters. It is possible, and in many cases desirable, to do so. For example, a common practice is to create styles that reference the same font file but have different obliquing angles (often used for lettering on isometric planes). This can be done by using the *Style* command to assign the same font file for each style, but assign different parameters (obliquing angle, width factor, etc.) and a unique name to each style. The relationship between fonts, styles, and resulting text is shown in Figure 18-24.

Figure 18-24 ————

FONT FILE	STYLE NAME (EXAMPLE)	RESULTING TEXT (BASED ON OTHER PARAMETERS)
ROMANS.SHX	ROMANS—VERT	ABCDEFG 1234
	ROMANS—ITAL	*ABCDEFG 1234*
ROMAND.SHX	ROMAND—EXT	A B C 1 2 3 4
	ROMAND—COMP	ABCDEFG 1234

.SHX Fonts

The .SHX fonts are fonts created especially for AutoCAD drawings by Autodesk. They are generally smaller files and are efficient to use for AutoCAD drawings where file size and regeneration time is critical. These fonts, however, are composed of single line segments and are not solid-filled such as most TrueType fonts (use *Zoom* to closely examine and compare .SHX vs. TrueType fonts).

Because the *Multiline Text Editor* is a Windows-compliant dialog box, it can only display fonts that are recognized by Windows. Since AutoCAD .SHX fonts are not recognized by Windows, AutoCAD supplies a TrueType equivalent when you select an .SHX or any other non-TrueType font for display in the *Multiline Text Editor* only. When you then use a text creation command, AutoCAD creates the text in the drawing with the true .SHX font.

R14

Special Text Characters
Special characters that are often used in drawings can be entered easily in the internal *Multiline Text Editor* by selecting the *Symbol* button (see Fig. 18-13). In an external editor or with the *Text* and *Dtext* commands, you can code the text by using the "%%" symbols or by entering the Unicode values for high-ASCII (characters above the 128 range). The %% symbols can be used with *Text* and *Dtext* commands, but the Unicode values must be used for creating *Mtext* with an external text editor. The following codes are typed in response to the "Text:" prompt in the *Dtext* or *Text* command or entered in the external text editor.

External Editor	*Text* or *Dtext*	Result	Description
\U+2205	%%c	ø	diameter (metric)
\U+00b0	%%d	°	degrees
	%%o	‾	overscored text
	%%u	__	underscored text
\U+00b1	%%p	±	plus or minus
	%%*nnn*	varies	ASCII text character number
\U+*nnnn*		varies	Unicode text hexadecimal value

For example, entering "**Chamfer 45%%d**" at the "Text:" prompt of the *Dtext* command draws the following text: **Chamfer 45°**.

EDITING TEXT

SPELL

Pull-down Menu	COMMAND (TYPE)	ALIAS (TYPE)	Short-cut	Screen (side) Menu	Tablet Menu
Tools *Spelling*	*SPELL*	*SP*	...	*TOOLS 1* *Spell*	*T,10*

AutoCAD has an internal spell checker that can be used to spell check and correct existing text in a drawing after using *Dtext*, *Mtext*, or *Text*. The *Check Spelling* dialog box (Fig. 18-25) has options to *Ignore* the current word or *Change* to the suggested word. The *Ignore All* and *Change All* options treat every occurrence of the highlighted word.

AutoCAD matches the words in the drawing to the words in the current dictionary. If the speller indicates that a word is misspelled but it is a proper name or an acronym you use often, it can be added to a custom dictionary. Choose *Add* if you want to leave a word unchanged but add it to the current custom dictionary.

Figure 18-25

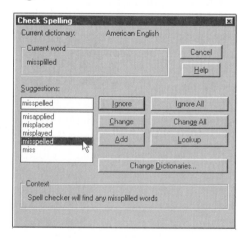

Selecting the *Change Dictionaries...* tile produces the dialog shown in Figure 18-26. You can select from other *Main Dictionaries* that are provided with your version of AutoCAD. The current main dictionary can also be changed in the *Files* tab of the *Preferences* dialog box and the name is stored in the *DCTMAIN* system variable.

The Release 14 default custom dictionary is SAMPLE.CUS (see Fig. 18-26). If you use the *Add* function (in the *Check Spelling* dialog box), the selected word is added to the current custom dictionary. You can also create a custom dictionary "on the fly" by entering any name in the *Custom dictionary* edit box. A file extension of .CUS should be used with the name (although other extensions will work). A custom dictionary name must be specified before you can add words. Words can be added by entering the desired word in the *Custom dictionary words* edit box or by using the *Add* tile in the *Check Spelling* dialog box. Custom dictionaries can be changed during a spell check. The custom dictionary can also be changed in the *Files* tab of the *Preferences* dialog box, and its name is stored in the *DCTCUST* system variable.

Figure 18-26

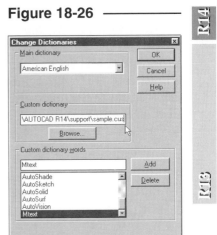

DDEDIT

Pull-down Menu	COMMAND (TYPE)	ALIAS (TYPE)	Short-cut	Screen (side) Menu	Tablet Menu
Modify *Object >* *Text...*	*DDEDIT*	*ED*	...	*Modify1* *Ddedit*	*Y,21*

Ddedit invokes a dialog box for editing existing text in a drawing. You can edit <u>individual characters</u> or the entire line or paragraph. If the selected text was created by the *Text* or *Dtext* command, the *Edit Text* dialog appears displaying one line of text (Fig. 18-27). If *Mtext* created the selected text, the *Multiline Text Editor* box appears (or current text editor).

Figure 18-27

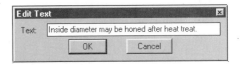

DDMODIFY

Pull-down Menu	COMMAND (TYPE)	ALIAS (TYPE)	Short-cut	Screen (side) Menu	Tablet Menu
Modify *Properties...*	*DDMODIFY*	*MO*	...	*MODIFY1* *Modify*	*Y,14*

Invoking this command prompts you to select objects and causes the *Modify Text* or *Modify Mtext* dialog box to appear. As described in Chapter 16, the dialog box that appears is <u>specific</u> to the type of object that is PICKed— kind of a "smart" dialog box. If a line of text (*Dtext* or *Text*) is PICKed in response to the "Select objects:" prompt, the dialog box shown in Figure 18-28 appears.

Modify Text allows <u>any</u> kind of editing to text, including an *Upside down, Backward, Width factor,* and *Obliquing angle* change.

Figure 18-28

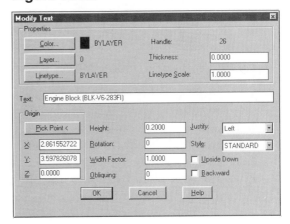

If a paragraph of text (created with *Mtext*) is selected, a similar *Modify Mtext* dialog box appears (Fig. 18-29). You can change the properties of the text such as *Style, Justification, Width,* or *Text Height.* Choose the *Full editor. . .* button to edit the text content in the *Multiline Text Editor* (or current text editor).

Figure 18-29

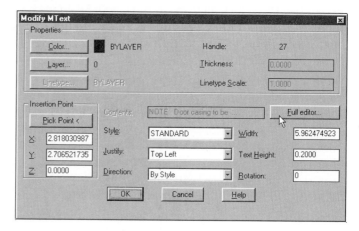

Find/Replace

The *Multiline Text Editor* includes a feature for finding and replacing text on a word-by-word basis. The *Find/Replace* tool replaces text based on content and case (upper, lower) but does not change character formatting or text properties. *Find/Replace* operates only for *Mtext* objects.

To *Find* Text

To *Find* text, follow these steps:

Figure 18-30

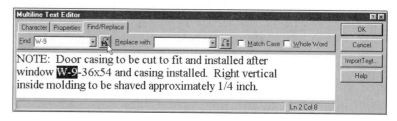

1. Use *Properties* from the *Modify* menu (or otherwise invoke *Ddmodify*) and select an *Mtext* object.
2. In the *Modify Mtext* dialog box (see Fig. 18-29), select *Full editor....*
3. When the *Multiline Text Editor* appears, select the *Find/Replace* tab.
4. Enter the word to find in the *Find* edit box (Fig. 18-30).
5. Check *Match Case* to locate only text that matches the case (upper and lower) of the *Find* box entry.
6. If you want AutoCAD to find the text entry as an entire word (not as part of another word), check the *Whole Word* box.
7. Choose the *Find* button.

To *Replace* Text

To *Replace* text, follow these steps:

1. Use *Properties* from the *Modify* menu (or otherwise invoke *Ddmodify*) and select an *Mtext* object.
2. In the *Modify Mtext* dialog box (see Fig. 18-29), select *Full editor. . ..*
3. When the *Multiline Text Editor* appears, select the *Find/Replace* tab.
4. Enter the word to be replaced (old word) in the *Find* edit box (Fig. 18-30).
5. Enter the new word (to replace the old new word), including the desired case, in the *Replace* box.
6. Choose the *Find* button.

If you want to replace a word with the same word but with the a change of a case (upper or lower) for one or more letters, AutoCAD makes the replacement correctly without using the *Match Case* option. The *Match Case* and *Whole Word* options are needed for the *Find* function only.

QTEXT

Pull-down Menu	COMMAND (TYPE)	ALIAS (TYPE)	Short-cut	Screen (side) Menu	Tablet Menu
...	QTEXT	...	...	...	...

Qtext (quick text) allows you to display a line of text <u>as a box</u> in order to speed up drawing and plotting times. Because text objects are treated as graphical elements, a drawing with much text can be relatively slower to regenerate and take considerably more time plotting than the same drawing with little or no text.

When *Qtext* is turned *ON* and the drawing is regenerated, each text line is displayed as a rectangular box (Fig. 18-31). Each box displayed represents one line of text and is approximately equal in size to the associated line of text.

For drawings with considerable amounts of text, *Qtext ON* noticeably reduces regeneration time.

Figure 18-31

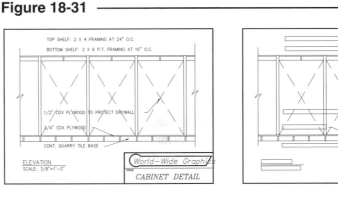

QTEXT OFF   QTEXT ON

For check plots (plots made during the drawing or design process used for checking progress), the drawing can be plotted with *Qtext ON*, requiring considerably less plotting time. *Qtext* is then turned *OFF* and the drawing must be *Regenerated* to make the final plot.

When *Qtext* is turned *ON*, the text remains in a readable state until a *Regen* is invoked or caused. When *Qtext* is turned *OFF*, the drawing must be regenerated to read the text again.

CHAPTER EXERCISES

1. *Dtext*

   Open the **EFF-APT** drawing. Make *Layer* **TEXT** *Current* and use *Dtext* to label the three rooms: **KITCHEN, LIVING ROOM,** and **BATH.** Use the *Standard style* and the *Start point* justification option. When prompted for the *Height:*, enter a value to yield letters of 3/16" on a 1/4"=1' plot (3/16 x the drawing scale factor = text height). *Save* the drawing.

2. *Style, Dtext, Ddedit*

   Open the drawing of the temperature graph you created as **TEMP-D**. Use *Style* to create two styles named **ROMANS** and **ROMANC** based on the *romans.shx* and *romanc.shx* font files (accept all defaults). Use *Dtext* with the *Center* justification option to label the days of the week and the temperatures (and degree symbols) with the **ROMANS** style as shown in Figure 18-32. Use a *Height* of **1.6**. Label the axes as shown using the **ROMANC** style. Use *Ddedit* for editing any mistakes. Use *SaveAs* and name the drawing **TEMPGRPH**.

Figure 18-32 —————

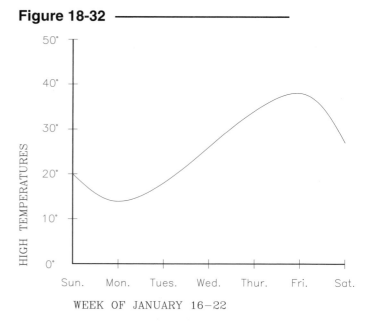

3. *Style, Dtext*

   Open the **BILLMATL** drawing created in the Chapter 15 Exercises. Use *Style* to create a new style using the *romans.shx* font. Use whatever justification methods you need to align the text information (not the titles) as shown in Figure 18-33. Next, type the *Style* command to create a new style that you name as **ROMANS-ITAL**. Use the *romans.shx* font file and specify a **15** degree *obliquing angle*. Use this style for the **NO.**, **PART NAME**, and **MATERIAL**. *SaveAs* **BILLMAT2**.

Figure 18-33 —————

NO.	PART NAME	MATERIAL
1	Base	Cast Iron
2	Centering Screw	N/A
3	Slide Bracket	Mild Steel
4	Swivel Plate	Mild Steel
5	Top Plate	Cast Iron

4. *Edit Text*

   Open the **EFF-APT** drawing. Create a new style named **ARCH1** using the *CityBlueprint* (.TTF) font file. Next, invoke the *Ddmodify* command. Use this dialog box to modify the text style of each of the existing room names to the new style as shown in Figure 18-34. *SaveAs* **EFF-APT2**.

Figure 18-34 —————

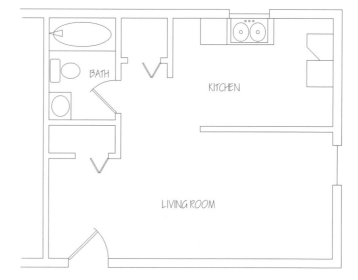

5. *Dtext, Mtext*

   Open the **CBRACKET** drawing from Chapter 10 Exercises. Using *romans.shx* font, use *Dtext* to place the part name and METRIC annotation (Fig 18-35). Use a *Height* of **5** and **4,** respectively, and the *Center Justification* option. For the notes, use *Mtext* to create the boundary as shown. Use the default *Justify* method (*TL*) and a *Height* of **3**. Use *Ddedit* or *Ddmodify* if necessary. *SaveAs* **CBRACTXT.**

Figure 18-35 ——————————

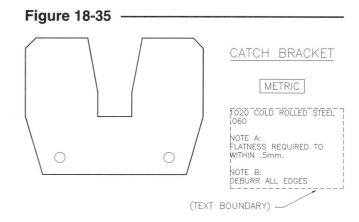

6. *Style*

   Create two new styles for each of your template drawings: **ASHEET, BSHEET,** and **CSHEET.** Use the *romans.shx* style with the default options for engineering applications or *CityBlueprint* (.TTF) for architectural applications. Next, design a style of your choosing to use for larger text as in title blocks or large notes.

Figure 18-36 ——————————

7. *Import Text, Ddedit, Ddmodify*

   Use a text editor such as Windows Wordpad or Notepad to create a text file containing words similar to "Temperatures were recorded at Sanderson Field by the National Weather Service." Then *Open* the **TEMPGRPH** drawing and use the *Import Text...* option of *Mtext* to bring the text into the drawing as a note in the graph as shown in Figure 18-36. Use *Ddedit* to edit the text if desired or use *Ddmodify* to change the text style or height.

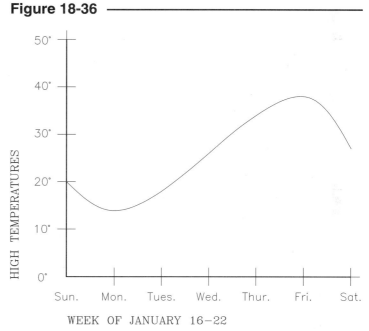

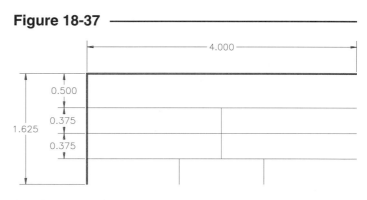

8. *Create a Title Block*

   A. Begin a *New* drawing and assign the name **TBLOCK.** Create the title block as shown in Figure 18-37 or design your own, allowing space for eight text entries. The dimensions are set for an A size sheet. Draw on *Layer* **0**. Use a *Pline* with **.02** *width* for the boundary and *Lines* for the interior divisions. (No *Lines* are needed on the right side and bottom because the title block will fit against the border lines.)

Figure 18-37 ——————————

B. Create two text *Styles* using *romans.shx* and *romanc.shx* font files. Insert text similar to that shown in Figure 18-38. Examples of the fields to create are:

        Company or School Name
        Part Name or Project Title
        Scale
        Designer Name
        Checker or Instructor Name
        Completion Date
        Check or Grade Date
        Project or Part Number

Choose text items relevant to your school or office. *Save* the drawing.

Figure 18-38

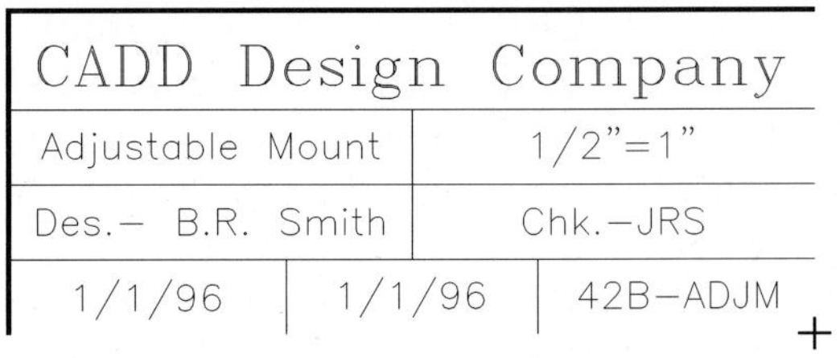

CADD Design Company		
Adjustable Mount	1/2"=1"	
Des.— B.R. Smith	Chk.—JRS	
1/1/96	1/1/96	42B—ADJM

19

TILED VIEWPORTS AND PAPER SPACE VIEWPORTS

Chapter Objectives

After completing this chapter you should:

1. know the difference between tiled viewports and paper space viewports and the purpose of the *TILEMODE* variable;

2. be able to create, *Save*, and *Restore* tiled viewport configurations using the *Vports* command and *Tiled Viewport Layout* dialog box;

3. be able to use the options of *Mview* to create viewports in paper space;

4. be able to use the *XP* option of *Zoom* to scale model space geometry to paper space units.

CONCEPTS

This chapter gives an introduction to the two types of viewports in AutoCAD (tiled viewports and paper space viewports) and the system variable (*TILEMODE*) that controls which type of viewport can be enabled. Tiled viewports are simple to understand and easy to use during construction of geometry. Paper space (or floating) viewports are more complex and are used mainly for arranging several views on a sheet for printing or plotting. Only an introduction to tiled and paper space viewports is given in this chapter.

Tiled viewports divide the screen into several areas that fit together with no space between, like tiles. Tiled viewports are created by the *Vports* command. Each tiled viewport can show a different area of the drawing by using display commands—especially helpful for 3D models. *Vports* divides the screen into multiple sections, allowing you to view different parts of a drawing on one screen. Tiled viewports can make construction and editing of complex drawings more efficient than repeatedly using display commands to view detailed areas. Tiled viewports were introduced with AutoCAD Release 10.

Paper space (floating) viewports are used for plotting several views of a drawing, or several drawings, on one sheet. Paper space viewports are used to set up a sheet of paper for plotting. The *Mview* command is used to create viewports in paper space. *Mview* allows several viewports to exist on the screen, but the viewports can be any size and rectangular proportion, unlike tiled viewports. Paper space viewports are generally used when the drawing is complete and you are ready to prepare a plot. Paper space viewports were introduced with AutoCAD Release 11.

AutoCAD allows you to work on your drawing in two separate spaces—model space and paper space. The space that you have been drawing in up to this time is called model space. The Drawing Editor begins a drawing in model space by default. The model geometry objects (which represent the subject of the drawing) are almost always drawn in model space. Paper space is normally used to set up the configuration of the plotting sheet.

Another fundamental difference between tiled viewports and paper space viewports is the space in which these viewports exist. Tiled viewports exist only when model space is active; therefore, they are sometimes referred to as "model space viewports" or "tiled model space." Paper space viewports do not exist in the same space as the model geometry. In order to create or use paper space viewports, paper space must first be activated. Release 14 of AutoCAD refers to viewports created in paper space as "floating viewports."

Figure 19-1

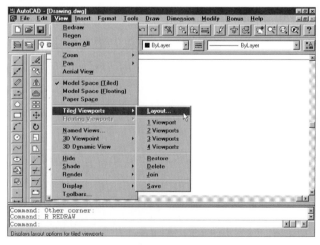

Although both tiled viewports and paper space viewports can be used in one drawing, they cannot be used at the same time. Therefore, the *TILEMODE* variable controls which type of viewport can be enabled at a specific point in time. Moreover, *TILEMODE* effectively controls which space is active—model space or paper space.

Commands related to creating and activating tiled and floating viewports are located in the *View* pull-down menu (Fig. 19-1). Note that only one of the two options in the menu (*Tiled Viewports* or *Floating Viewports*) is available at any one time, based on the setting of *TILEMODE* or whether *Model Space (Tiled)* or *Paper Space* is checked in the menu.

TILEMODE

Pull-down Menu	COMMAND (TYPE)	ALIAS (TYPE)	Short-cut	Screen (side) Menu	Tablet Menu
View *Model Space (Tiled)*	*TILEMODE*	*TM or TI*	...	*VIEW 1* *Tilemode*	*L,3*

When paper space viewports were introduced with Release 11, a new system variable called *TILEMODE* was also introduced. Since only <u>one</u> type of viewport can be used at a time, the *TILEMODE* system variable is used to enable one or the other type of viewports. More importantly, *TILEMODE* is a toggle that switches between using only model space or using paper space.

When *TILEMODE* is set to 1 (*TILEMODE* is *On*), only tiled viewports in model space can be used. The viewports themselves must be created with the *Vports* command. The variable can be typed, or you can select *Model Space (Tiled)* from the *View* pull-down menu. *TILEMODE* can also be toggled by double-clicking the word "TILE" on the Status Line. *TILEMODE On* is the default setting for the AutoCAD template drawing ACAD.DWT; so normally, model space is active and only tiled viewports can be used.

If *TILEMODE* is set to 0 (*TILEMODE* is *Off*), paper space is enabled and paper space-style viewports are available. Viewports in paper space must be created with the *Mview* command. Change *TILEMODE* to *Off* by typing the variable, selecting *Paper Space* from the *View* pull-down menu, or double-click the word "TILE" on the Status Line. Therefore, setting *TILEMODE* to 0 disables display of model space and tiled viewports and automatically enables paper space.

Whether or not viewports of any kind have been created with either the *Vports* or the *Mview* commands, the setting of *TILEMODE* displays the following:

TILEMODE setting	Resulting display
1	model space only—tiled (*Vports*) viewports can be used
0	paper space—paper space (*Mview*) viewports can be used

Switching the *TILEMODE* setting does not affect any objects you have drawn. The setting may affect the visibility of objects in the display.

This chapter discusses both tiled viewports and paper space viewports. Tiled viewports are discussed first because they are easier to use and understand.

TILED VIEWPORTS

All controls for tiled viewports are included in the *Vports* command. Tiled viewports are useful for displaying several views of one or more 2D drawings simultaneously on the screen. Using tiled viewports for displaying several views of a 3D drawing is discussed in Chapter 29.

VPORTS

Pull-down Menu	COMMAND (TYPE)	ALIAS (TYPE)	Short-cut	Screen (side) Menu	Tablet Menu
View *Tiled Viewports >*	*VPORTS*	...	...	*VIEW 1* *Vports*	*M,3*

The *Vports* command allows <u>tiled</u> viewports to be created on the screen. *Vports* divides the screen into several areas. *Vports* (tiled viewports) are available only when the *TILEMODE* variable is set to **1**. Tiled *Vports* affect <u>only the screen display</u>. The viewport configuration <u>cannot be plotted</u>. If the *Plot* command is used, only the <u>current</u> viewport display is plotted.

Figure 19-2 displays the AutoCAD drawing editor after the *Vports* command was used to divide the screen into tiled viewports. Tiled viewports always fit together like tiles with no space between. The shape and location of the viewports are not flexible as with paper space viewports.

Figure 19-2

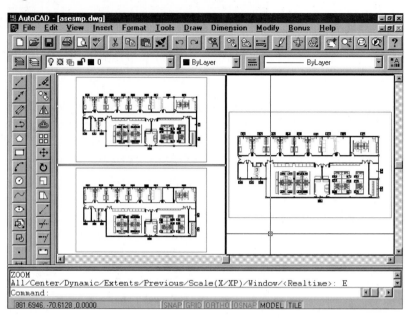

The *View* pull-down menu (Fig. 19-1) can be used to display the *Tiled Viewport Layout* dialog box (Fig. 19-3). This option in the pull-down menu is enabled only if *TILEMODE*=1. The dialog box allows you to PICK the image tile representing the layout that you want.

Figure 19-3

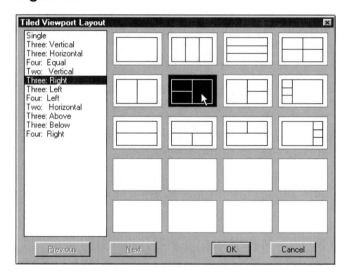

After you select the desired layout, the previous display appears in <u>each</u> of the viewports. For example, if a full view of the office layout was displayed when you use the *Vports* command or the related dialog box, the resulting display in <u>each</u> viewport would be the same full view of the office (Fig. 19-2). It is up to you then to use viewing commands (*Zoom*, *Pan*, etc.) in the active viewport to specify what areas of the drawing you want to see in each viewport. There is no automatic viewpoint configuration option with *Vports*.

A popular arrangement of view-points for construction and editing of 2D drawings is a combination of an overall view and one or two *Zoomed* views (Fig. 19-4). You cannot draw or project from one viewport to another. Keep in mind that there is only <u>one model</u> (drawing) but several views of it on the screen. Notice that the active or current viewport displays the cursor, while moving the pointing device to another viewport displays only the pointer (small arrow) in that view-port.

Figure 19-4

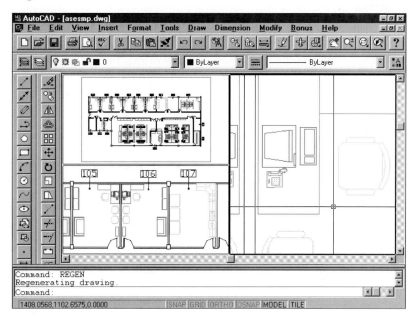

A viewport is made active by PICKing in it. Any display commands (*Zoom, Vpoint, Redraw,* etc.) and drawing aids (*SNAP, GRID, ORTHO*) used affect only the <u>current viewport</u>. Draw and edit commands that affect the model are potentially apparent in all viewports (for every display of the affected part of the model). *Redrawall* and *Regenall* can be used to redraw and regenerate all viewports.

You can begin a drawing command in one viewport and finish in another. In other words, you can toggle viewports within a command. For example, you can use the *Line* command to PICK the "From point:" in one viewport, then make another viewport current to PICK the "To point:".

Using the command line format, the syntax for *Vports* is as follows:

    Command: vports
    Save/Restore/Delete/Join/SIngle/?/2/<3>/4: Enter or (option)
    Horizontal/Vertical/Above/Below/Left/<Right>: Enter or (option)
    Regenerating drawing.
    Command:

Save
Allows you to assign a name and save the current viewport configuration. A viewport configuration is the particular arrangement of viewports and the display settings (*Zoom, Pan,* etc.) in each viewport. Up to 31 characters can be used when assigning a name. The configuration can be *Restored* at a later time.

Restore
Redisplays a previously *Saved* viewport configuration. AutoCAD prompts for the assigned name.

Delete
Deletes a named viewport configuration. AutoCAD prompts for the assigned name.

Join

This option allows you to combine (join) two adjacent viewports. The viewports to join must share a common edge the full length of each viewport. For example, if four equal viewports were displayed, two adjacent viewports could be joined to produce a total of three viewports. You must select a *dominant* viewport. The *dominant* viewport determines the display to be used for the new viewport.

SIngle

Changes back to a single screen display using the current viewport's display.

?

Displays the identification numbers and screen positions of named (saved) and active viewport configurations. The screen positions are relative to the lower-left corner of the screen (0,0) and the upper-right (1,1).

2, 3, 4

Use these options to create 2, 3, or 4 viewports. You can choose the configuration. The possibilities are illustrated if you use the *Tiled Viewport Layout* dialog box.

The *Vports* command can be used most effectively when constructing and editing drawings, whereas, paper space viewports are generally used when the model is complete and ready to prepare a plot. (See Chapter 29 for more information on using tiled viewports with 3D models.)

PAPER SPACE (FLOATING) VIEWPORTS

The two spaces that AutoCAD provides are model space and paper space. When you start AutoCAD and begin a drawing, model space is active by default. Objects that represent the subject of the drawing (model geometry) are normally drawn in model space. Dimensioning is also performed in model space because it is associative—directly associated to the model geometry. The model geometry is usually completed before using paper space.

Paper space represents the <u>paper that you plot or print on</u>. When you enter paper space for the first time, you see a blank "sheet." In order to see any model geometry, viewports in paper space must be created with *Mview* (like cutting rectangular holes) so you can "see" into model space. Any number or size of rectangular-shaped viewports can be created in paper space. Since there is only <u>one model space</u> in a drawing, you see the same model space geometry in each viewport. You can, however, control which layers are *Frozen* and *Thawed* and the scale of the geometry displayed <u>in each viewport</u>. Since paper space represents the actual paper used for plotting, plot from paper space at a scale of 1=1.

Consider this brief example to explain the basics of using paper space. In order to keep this example simple, only one viewport is created to set up a drawing for plotting.

First, the part geometry is created in model space as usual (Fig. 19-5). Associative dimensions are also created in model space. This step is the same method that you would have normally used to create a drawing.

Figure 19-5

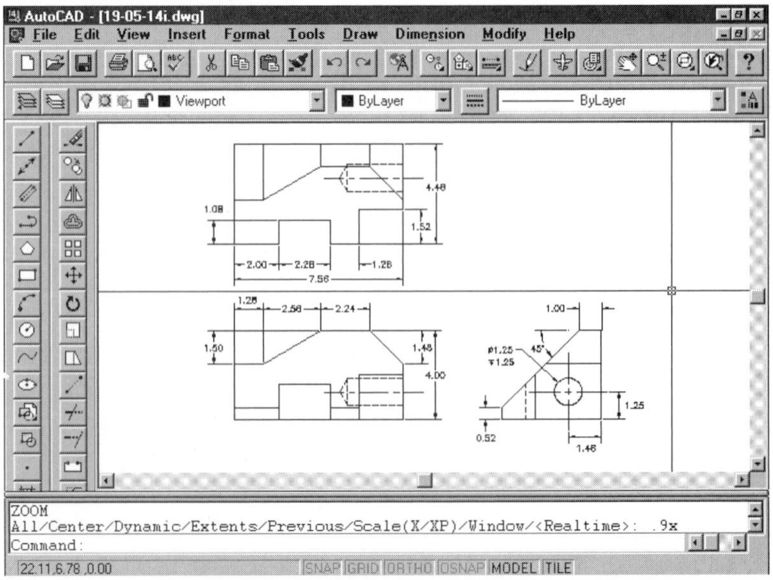

When the part geometry is complete, enable paper space. To do this, change the setting of the *TILEMODE* variable to 0. (See the previous *TILEMODE* command table for several possible methods of toggling the *TILEMODE* variable.)

By setting the *TILEMODE* variable to 0, you are automatically switched to paper space. When you enable paper space for the first time in a drawing, a "blank sheet" appears. The *Limits* command is used to set the paper space *Limits* to the sheet size intended for plotting. Objects such as a title block and border are created <u>in paper space</u> (Fig. 19-6). Normally, only objects that are <u>annotations</u> for the drawing (tile blocks, tables, border, company logo, etc.) are drawn in paper space.

A viewport must be created in the "paper" with the *Mview* command in order for you to "look" into model space. All of the model geometry in the drawing initially appears in the viewport. At this point, there is no specific scale relation between paper space units and the size of the geometry in the viewport (Fig. 19-7).

You can control the scale of model space to paper space and what part of model space geometry you see in a viewport. Two methods are used to control the geometry that is visible in a particular viewport:

Figure 19-6

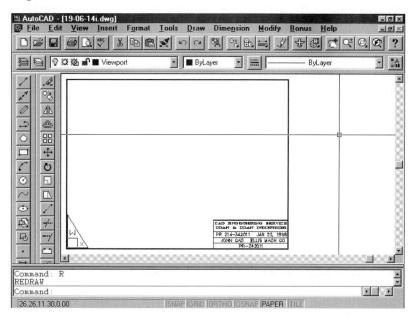

Figure 19-7

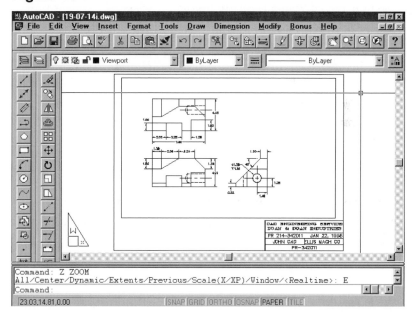

1. **Display commands**
   The *Zoom, Pan, View, Vpoint* and other display commands allow you to specify the part of model space you want to see in a viewport. There is even a *Zoom XP* option used to <u>scale</u> the geometry in model space units "times paper space" units. *Zoom XP* is the method used to define the <u>scale</u> of model geometry for the final plot.

2. **Viewport-specific layer visibility control**
   You can control what <u>layers</u> are visible in specific viewports. This function is often used for displaying different model space geometry in separate viewports. You can control which layers appear in which viewports by using the *Vplayer* command (command line format) or using the icons in the *Layer/Linetype Properties* dialog box or the *Layer Control* drop-down list (*Freeze/Thaw in current viewport* and *Freeze/Thaw in new viewport*).

For example, the *Zoom* command could be used in model space to display a detail of the model geometry in a viewport. In this case, the model geometry is scaled to 3/4 size by using *Zoom 3/4XP*.

While in a drawing session with paper space enabled, the *Mspace* (Model Space) and *Pspace* (Paper Space) commands allow you to switch between paper space (outside a viewport) and floating model space (inside a viewport) so you can draw or edit in either space. Commands that are used will affect the objects or display of the current space. An object <u>cannot be in both spaces</u>. You can, however, draw in paper space and <u>OSNAP to objects in model space</u>. When drawing is completed, activate paper space and plot at 1=1 since paper space *Limits* are set to the <u>actual paper size</u>.

When typing the *Pspace* and *Mspace* commands or selecting *Paper Space* or *Model Space (Floating)* from the *View* menu, the cursor control switches between paper space (outside the viewport) and model space (inside the viewport). The "crosshairs" can move completely across the screen when in paper space, but only within the viewport when using *Mspace* or *Model Space (Floating)*. The figures in this chapter show the cursor at 100% screen size to indicate this feature more clearly (see Figures 19-7 and 19-8). (Cursor size is variable using the *Pointer* tab of the *Preferences* dialog box.)

Figure 19-8 ──────────

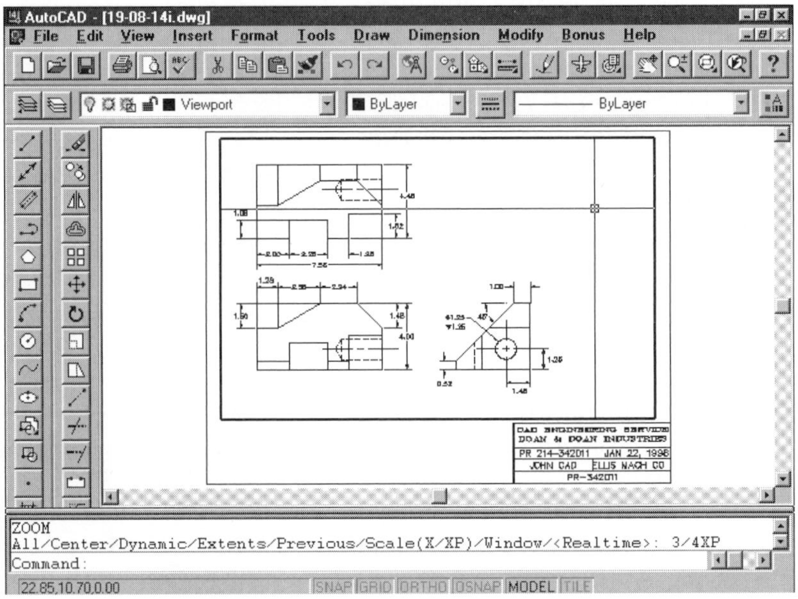

Reasons for Using Paper Space

Paper space is intended to be an aid to plotting. Since paper space represents a sheet of paper, it provides you with a means of preparing specific plotting sheet configurations.

In the previous example, paper space was used with the typical 2D drawing by creating the model geometry as usual but using paper space for the border, title block, tables, or other annotations. If you are creating a typical 2D drawing, you don't have to use paper space and probably shouldn't use paper space. The typical methods for creating drawings and plots all in model space are still valid and should be used except for special cases.

Paper space should be used if you want to prepare plots with <u>multiple views</u> or <u>multiple drawings</u> as listed here:

1. Use paper space to plot several views of the same drawing on one sheet, possibly in different scales, such as a main view and a detail. (Alternately, model space only can be used to accomplish the same result by using the *DIMLFAC* variable for dimensioning, for example, an enlarged detail.)

2. Use paper space to plot several drawings on one sheet. This action is accomplished by using *Xref* to bring several drawings into model space and by using layer visibility control to display individual drawings in separate viewports.

3. Use paper space to plot different views (e.g., top, front, side) of a 3D model on one sheet.

This chapter should supply you with only the information to create viewports to display a single drawing.

Guidelines for Using Paper Space

Although there are other alternatives, the typical steps using paper space to set up a drawing for plotting are listed here:

1. Create the part geometry in model space. Associative dimensions are also created in model space.

2. Switch to paper space by changing *TILEMODE* to 0 by any method.

3. Set up paper space using the following steps.

    A. Set *Limits* in paper space equal to the sheet size used for plotting. (Paper space has its own *Limits*, *Snap*, and *Grid* settings.)
    B. *Zoom All* in paper space.
    C. Set the desired *Snap* and *Grid* for paper space.
    D. Make a layer named BORDER or TITLE and *Set* that layer as current.
    E. Draw, *Insert*, or *Xref* a border and a title block.

4. Make viewports in paper space.

    A. Make a layer named VIEWPORT (or other descriptive name) and set it as the *Current* layer (it can be turned *Off* later if you do not want the viewport objects to appear in the plot).
    B. Use the *Mview* command to make viewports. Each viewport contains a view of model space geometry.

5. Control the display of model space graphics in each viewport. Complete these steps for each viewport.

    A. Use the *Mspace* (MS or *Floating Model Space*) command to "go into" model space (in the viewport). Move the cursor and PICK to activate the desired viewport.
    B. Use *Zoom XP* to scale the display of the model units to paper space units. This action dictates the plot scale for the model space graphics. The *Zoom XP* factor is the same as the plot scale factor that would otherwise be used (reciprocal of the "drawing scale factor"). For example: *Zoom .5XP* or *Zoom 1/2XP* would scale the model geometry at 1/2 times paper space units.
    C. Control the layer visibility for each viewport by using the *Vplayer* command or the icons in the *Layer/Linetype Properties* dialog box or the *Layer Control* drop-down list (*Freeze/Thaw in current viewport* and *Freeze/Thaw in new viewports*).

6. Plot from paper space at a scale of 1=1.

    A. Use the *Pspace* (PS) command to switch to paper space.
    B. If desired, turn *Off* the VIEWPORT layer so the viewport borders do not plot.
    C. Plot the drawing from paper space at a plot scale of 1=1 since paper space is set to the actual paper size. The paper space geometry will plot full size, and the resulting model space geometry will be plotted to the same <u>scale</u> as the *Zoom XP* factor.

PAPER SPACE COMMANDS AND VARIABLES

MVIEW

Pull-down Menu	COMMAND (TYPE)	ALIAS (TYPE)	Short-cut	Screen (side) Menu	Tablet Menu
View *Floating Viewports >*	MVIEW	MV	...	VIEW 1 *Mview*	M,4

Mview provides several options for creating paper space viewport objects (sometimes called "floating viewports"). *Mview* must be used in paper space (*TILEMODE*=0), so if you have switched to model space, invoking *Mview* (by typing or screen menu) automatically displays the reminder "** Command not allowed unless TILEMODE is set to 0 **." The *Floating Viewports* option in the *View* pull-down menu is enabled only when *TILEMODE*=0. There is no limit to the number of paper space viewports that can be created in a drawing. The command syntax to create one viewport (the default option) is as follows:

```
Command: mview
ON/OFF/Hideplot/Fit/2/3/4/Restore/<First Point>: PICK or (coordinates)
Other corner: PICK or (coordinates)
Regenerating drawing.
Command:
```

This action creates one rectangular viewport between the diagonal corners specified. The other options are described here briefly:

ON

On turns on the display of model space geometry in the selected viewport. Select the desired viewport to turn on.

OFF

Turn off the display of model space geometry in the selected viewport with this option. Select the desired viewport to turn off.

Hideplot

Hideplot causes hidden line removal in the selected viewport during a plot. This is used for 3D surface or solid models. Select the desired viewport.

Fit

Fit creates a new viewport and fits it to the size of the current display. If you are *Zoomed* in paper space, the resulting viewport is the size of the *Zoomed* area. The new viewport becomes the current viewport.

Restore

Use this option to create new viewports in the same relative pattern as a named tiled viewport configuration (see Tiled Viewports).

2/3/4

These options create a number of viewports within a rectangular area that you specify. The *2* option allows you to arrange 2 viewports either *Vertically* or *Horizontally*. The *4* option automatically divides the specified area into 4 equal viewports. Figure 19-9 displays the possible configurations using the *3* option. Using this option yields the following prompt:

Horizontal/Vertical/Above/Below/Left/<Right>:

Figure 19-9 ─────────────

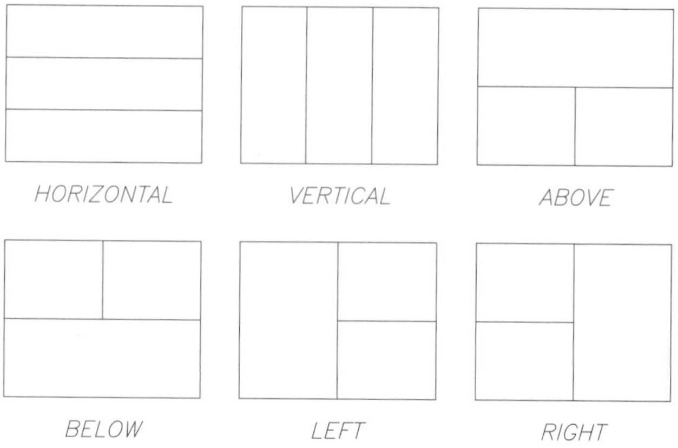

HORIZONTAL   VERTICAL   ABOVE

BELOW   LEFT   RIGHT

After making the desired selection, AutoCAD prompts: "Fit/<First Point>:". The *Fit* option automatically fills the current display with the specified number of viewport configurations (similar to the *Fit* option described earlier).

Paper space viewports created with *Mview* are treated by AutoCAD as <u>objects</u>. Like other objects, *Mview* viewports can be affected by most editing commands. For example, you could use *Mview* to create one viewport, then use *Copy* or *Array* to create other viewports. You could edit the size of viewports with *Stretch* or *Scale*. Additionally, you can use *Move* (Fig. 19-10) to relocate the position of viewports. Delete a viewport using *Erase*. You must be in paper space to PICK the viewport objects (borders). The viewports created by *Mview* or other editing methods are always rectangular and have a vertical and horizontal orientation (they cannot be *Rotated*).

Figure 19-10

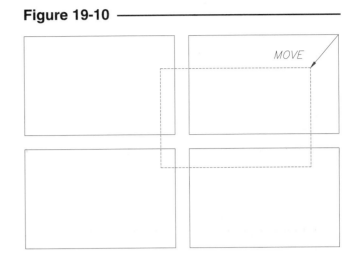

Floating viewports can also overlap (Fig. 19-10). This feature makes it possible for geometry appearing in different viewports to occupy the same area on the screen or on a plot.

NOTE: Avoid creating one viewport completely within another's border because visibility and selection problems may result.

Model Space, Paper Space, and Floating Model Space

Previously, it was stated that there are two spaces: paper space and model space. The *TILE-MODE* variable controls which of these two spaces is active. However, if paper space is enabled (*TILEMODE=0*) and viewports in paper space have been created, there are three possible spaces: model space, paper space, and floating model space. <u>Floating model space is model space inside a paper space viewport</u>. The *Pspace* and *Mspace* commands control whether the cursor is outside or inside a viewport (when *TILEMODE=0*) (Fig. 19-11).

Figure 19-11

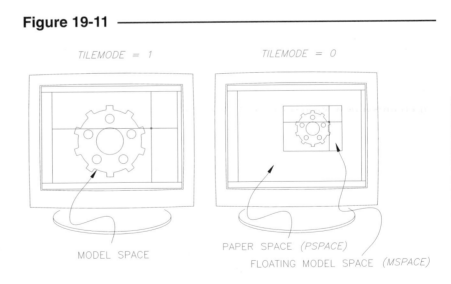

If you prefer typing, the correct combination of *TILEMODE*, *Pspace*, and *Mspace* must be used to work in one of the three spaces (normal model space, paper space, and floating model space). If you use the *View* pull-down menu, selecting one of the three options (*Model Space Tiled*, *Paper Space*, or *Model Space Floating*) automatically toggles the correct *TILEMODE* setting and invokes the *Pspace* or *Mspace* command.

PSPACE

Pull-down Menu	COMMAND (TYPE)	ALIAS (TYPE)	Short-cut	Screen (side) Menu	Tablet Menu
View *Paper Space*	*PSPACE*	*PS*	...	*VIEW 1* *Pspace*	*L,5*

The *Pspace* command switches from model space (inside a viewport) to paper space (outside a viewport). The cursor is displayed in paper space and can move <u>across the entire screen</u> when in paper space (see Fig. 19-7), while the cursor appears inside the current viewport only if the *Mspace* command is used (see Fig. 19-8). If you type this command or select it from the screen menu, paper space must be enabled by setting *TILEMODE* to 0. If *TILEMODE* is set to 1, the following message appears:

Command: **pspace**
** Command not allowed unless TILEMODE is set to 0 **
Command:

Selecting *Pspace* from the pull-down menu also automatically sets *TILEMODE* to 0 so paper space is enabled. Therefore, it is possible to use the menu selection to go from *TILEMODE=1* directly to paper space.

MSPACE

Pull-down Menu	COMMAND (TYPE)	ALIAS (TYPE)	Short-cut	Screen (side) Menu	Tablet Menu
View *Model Space (Floating)*	*MSPACE*	*MS*	...	*VIEW 1*	*L,4*

The *Mspace* command switches from paper space to model space <u>inside a viewport</u>. If floating viewports exist in the drawing when you use the command, you are switched to the last active viewport. The cursor appears only <u>in</u> that viewport. The current viewport displays a heavy border. You can switch to another viewport (make another current) by PICKing in it.

If you select *Model Space (Floating)* from the pull-down menu, *TILEMODE* is automatically set to 0. If no viewports exist, the *Mview* command is automatically invoked. If you type *MSpace* or use the screen menu, *TILEMODE* must previouldy be set to 0, there must be at least one paper space viewport, and it must be *On* (not turned *Off* by the *Mview* command) or AutoCAD issues a message and cancels the command.

ZOOM XP Factors

Paper space objects such as title block and border should correspond to the paper on a 1=1 scale. Paper space is intended to represent the plotting sheet. *Limits* in <u>paper space</u> should be set to the exact paper size, and the finished drawing is plotted <u>from paper space</u> to a scale of 1=1. The model space geometry, however, should be true scale in real-world units, and *Limits* in <u>model space</u> are generally set to accommodate that geometry.

When model space geometry appears in a viewport in paper space, the size of the <u>displayed</u> geometry can be controlled so that it appears and plots in the correct scale. This action is accomplished with the *XP* option of the *Zoom* command. *XP* means "times paper space." Thus, model space geometry is *Zoomed* to some factor "times paper space."

Since paper space is set to the actual size of the paper, the *Zoom XP* factor that should be used for a viewport is <u>equivalent to the plot scale</u> that would otherwise be used for plotting that geometry in model space. The *Zoom XP* factor is the reciprocal of the "drawing scale factor" (Chapter 13). *Zoom XP* <u>only</u> while you are "in" the desired model space viewport. Fractions or decimals are accepted.

For example, if the model space geometry would normally be plotted at 1/2"=1" or 1:2, the *Zoom* factor would be *.5XP* or *1/2XP*. If a drawing would normally be plotted at 1/4"=1', the *Zoom* factor would be *1/48XP*. Other examples are given below:

1:5	*Zoom .2XP* or 1/5 XP
1:10	*Zoom .1XP* or 1/10XP
1:20	*Zoom .05XP* or 1/20XP
1/2"=1"	*Zoom 1/2XP*
3/8"=1"	*Zoom 3/8XP*
1/4"=1"	*Zoom 1/4XP*
1/8"=1"	*Zoom 1/8XP*
3"=1'	*Zoom 1/4XP*
1"=1'	*Zoom 1/12XP*
3/4'=1'	*Zoom 1/16XP*
1/2"=1'	*Zoom 1/24XP*
3/8"=1'	*Zoom 1/32XP*
1/4"=1'	*Zoom 1/48XP*
1/8"=1'	*Zoom 1/96XP*

Refer to the Tables of Limits Settings, Chapter 14, for other plot scale factors.

Using the *Advanced Setup Wizard* and Template Drawings

The *Advanced Setup Wizard* available in the *Create New Drawing* dialog box can be used to create one paper space viewport when you begin a drawing. Alternately, several template drawings are available already set up with one paper space viewport and a title block and border.

Advanced Setup Wizard

When you use the *New* command, the *Create New Drawing* dialog box appears with an option for the *Advanced Setup Wizard* (see Chapters 6 and 13 for introductory information). In the steps for setting up a drawing using this utility, *Step 7: Layout* provides you with the option to use paper space and how to begin working in the drawing.

In *Step 7: Layout,* there are two decisions to make (Fig. 19-12). First, decide if you want to use paper space or not. The question, "Do you want to use advanced paper space capabilities?" means, "Do you want to use paper space?" (No special advanced capabilities are activated here—only the choice to use paper space or not.) Selecting *No* (not the default in this tab) allows you to work in model space (*TILEMODE*=1). Select *Yes* if you want to use paper space. Selecting the *Yes* option, however, creates only one paper space viewport. The viewport is automatically created on layer VIEWPORT. If you have selected a title block (*Step 6*), the title block is inserted into paper space. If you select *Yes*, the three options that follow are enabled.

Figure 19-12

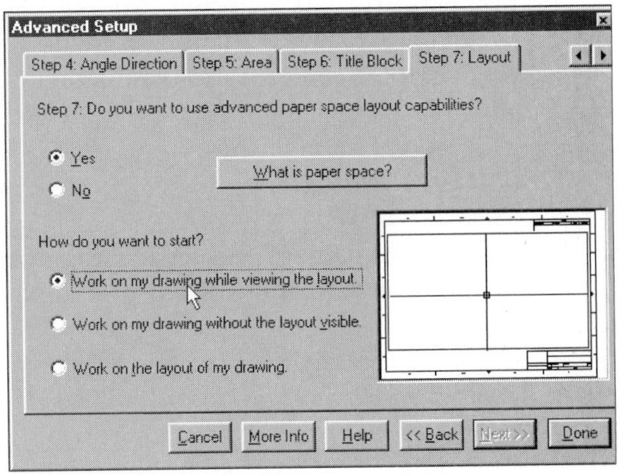

Next, make a decision on how you want to begin working. Below is a translation of the three choices for *How do you want to start?*

Option	Meaning
Work on my drawing while viewing the layout.	Tilemode=0, model space current
Work on my drawing without the layout visible.	Tilemode=1, model space current
Work on the layout of my drawing.	Tilemode=0, paper space current

Keep in mind that using the *Advanced Setup Wizard* automatically changes a number of system variables based on your input to *Step 5: Area*. The system variables may or (most likely) may not be the settings you need for printing or plotting a drawing to a specific scale, so some adjustments may be necessary. See Chapter 13, Advanced Drawing Setup, for more information on the *Setup Wizards*.

CHAPTER EXERCISES

1. *Vports* **(Tiled Viewports)**

   Open the **ASESMP** drawing. This drawing is available if you used the *Full Installation* option when AutoCAD was installed on your computer or network. The drawing is usually located in the C:\Program Files\AutoCAD R14\Sample directory (assuming the default installation location).

   A. Invoke **Tiled Viewports** by the pull-down menu; then select **Layout....** When the *Tiled Viewport Layout* dialog box appears, choose the **Three: Right** option (or image tile just left of center). The resulting viewport configuration should appear as Figure 19-2.

   B. Using **Zoom** and **Pan** in the individual viewports, produce a display with an overall view on the top and two detailed views as shown in Figure 19-4.

   C. Type the **Vports** command. Use the **Save** option to save the viewport configuration as **3R**. Click in the bottom left viewport to make it the current viewport. Use **Vports** again and change the display to a **Single** screen. Now use **SaveAs** to save the drawing as **ASESMP-2** and <u>reset the path to your working directory</u>.

   D. *Open* **ASESMP-2**. Use the *?* option of **Vports**; then **Restore 3R** to ensure the viewport configuration was saved as expected. Invoke a **Single** viewport again. Finally, use **Vports** or the **Tiled Viewport Layout** dialog box to create another viewport configuration of your choosing. *Save* the viewport configuration and assign an appropriate name. *Save* the **ASESMP-2** drawing.

2. *TILEMODE, Mview*

   In this exercise, you will use an existing drawing, enable paper space, draw a border and title block, create one viewport, and print or plot the drawing to scale.

   A. *Open* the **CBRACKET** drawing you worked on in Chapter 10 Exercises. If you have already created a title block and border, *Erase* them.

   B. Change **TILEMODE** to **0** by any method (type, double-click the word **TILE** on the Status Line, or select **Paper Space** from the **View** pull-down menu). You should be presented with a "blank sheet" and the paper space icon (drafting triangle) should appear.

C. Set *Limits* (in paper space) to an A size sheet, **11 x 8.5**. Set *Snap* to **.5**. Set layer **TITLE** *Current*. Draw a title block and border (in paper space). HINT: Use *Pline* with a *width* of .02, and draw the border with a .5 unit margin within the edge of the *Limits* (border size of 10 x 7.5). You can create a title block similar to the TBLOCK you created in Chapter 18 (Fig. 18-37) with a space for your school or company name, your name, part name, date, and scale. Enter a scale of "**1:1 METRIC**" in the title block. Save the drawing as **CBRACKET-PS**.

Figure 19-13 ───────────

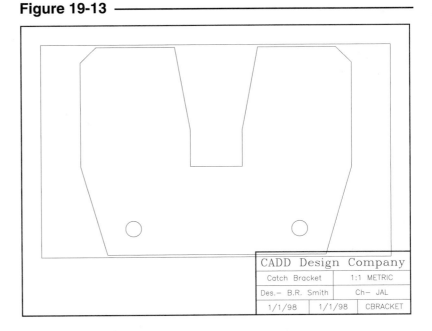

D. Create layer **VPORTS** and make it *Current*. Use *Mview* to create one viewport. Pick diagonal corners for the viewport at **1,2** and **10,7.5**. The saddle drawing should appear in the viewport at no particular scale, similar to Figure 19-13.

E. Next, use the *Mspace* command to bring the cursor into model space. Use *Zoom* with a *.03937XP* factor to scale the model space geometry to paper space. (The conversion from inches to milimeters is 25.4 and from milimeters to inches is .03937. Model space units are milimeters and paper space units are inches, therefore enter a *Zoom XP* factor of .03937). You may have to *Pan* (in model space) to center the drawing within the viewport.

F. Finally, use the *Pspace* command to activate paper space. Make layer 0 current and *Freeze* layer **VPORTS**. Your completed drawing should look like that in Figure 19-14. *Save* the drawing.

Figure 19-14 ───────────

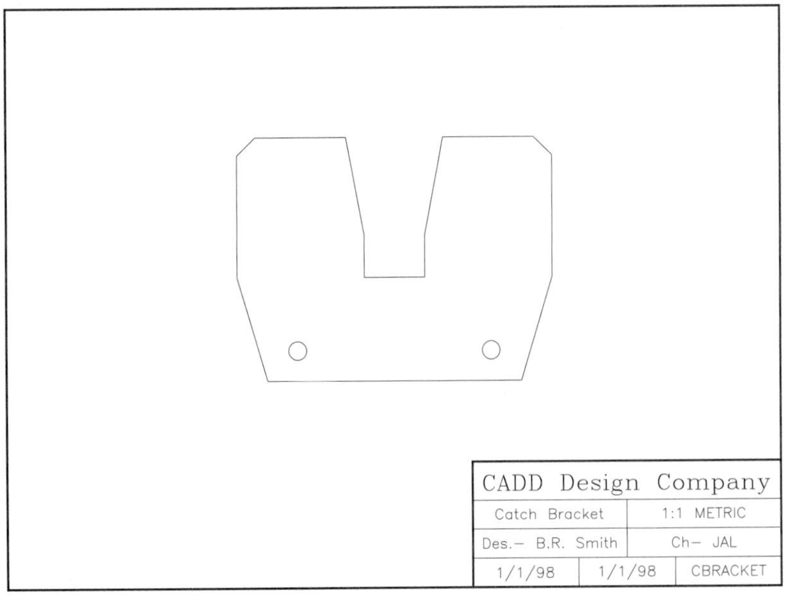

3. *TILEMODE, Mview*

This exercise gives you more experience using an existing drawing, creating one paper space viewport, and printing or plotting to scale.

A. *Open* the **EFF-APT2** drawing you worked with last in Chapter 18 Exercises. If you have already created a title block and border, *Erase* them.

B. Change *TILEMODE* to **0** by any method. You should be presented with a "blank sheet" and the paper space icon (drafting triangle) should appear.

C. Set *Limits* in paper space to a C size sheet, **24 x 18**, and *Zoom All*. Set *Snap* to **.25**. Make layer **TITLE** the *Current* layer. Draw a title block and border (in paper space). HINT: Use *Pline* with a *width* of .03, and draw the border with a .75 unit margin within the edge of the *Limits* (border size of 22.5 x 16.5). You can create a title block similar to the TBLOCK you created in Chapter 18 (Fig. 18-37) with a space for your school or company name, your name, part name, date, and scale. Enter a scale of **1/2"=1'** in the title block. *Save* the drawing as **EFF-APT-PS**.

D. Create a new layer named **VIEWPORT** and make it *Current*. Use *Mview* to create one viewport. Pick diagonal corners for the viewport at **1,1** and **23,17**. The apartment drawing should appear in the viewport at no particular scale.

E. Use the *Mspace* command to bring the cursor into model space. Use *Zoom* with a *1/24XP* factor. (The proportion of 1/2"=1' is the same as 1/2"=12", or 1=1/24). You may have to *Pan* (in model space) to center the drawing within the viewport.

F. Next, use the *Pspace* command to place your cursor in paper space. *Freeze* layer **VIEWPORT**. Your completed drawing should like that in Figure 19-15. *Save* the drawing.

Figure 19-15

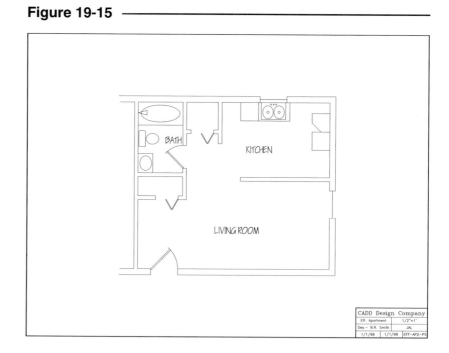

4. *Advanced Setup Wizard*

Figure 19-16

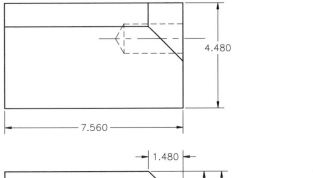
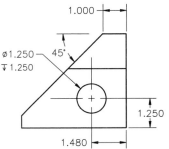
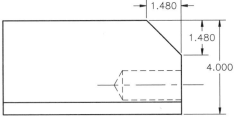

This exercise gives you practice using the *Advanced Setup Wizard* to set up a drawing to work in paper space and model space. You will construct the Wedge Block in Figure 19-16.

A. Begin a *New* drawing. In the *Create New Drawing* dialog box, select *Use a Wizard*, then select *Advanced Setup*. In the *Advanced Setup* dialog box that appears, take the following steps:

Step 1: Units	use *Decimal* with *.000 Precision*
Step 2: Angle	accept default setting
Step 3: Angle Measure	accept default setting
Step 4: Angle Direction	accept default setting
Step 5: Area	enter **18** for *Width* and **12** for *Length*
Step 6: Title Block	select the *Ansi_b* title block
Step 7: Layout	answer *Yes* to *"Do you want to use paper space?"* and select *Work on my drawing while viewing the layout.*

B. When the drawing appears, the title block is visible in paper space and model space is active (Fig. 19-17). The *Advanced Setup Wizard* sets many system variables based on your input to *Area*, but often these settings are not appropriate. Change the *Snap* to **.125** and *Grid* to **.5**. Change *LTSCALE* to **.5**. *Save* the drawing as **WEDGEBK**.

Figure 19-17

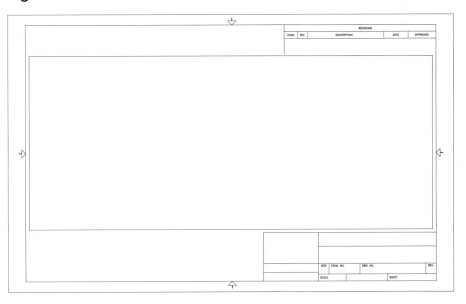

C. Use the *Mspace* command or select *Model Space (Floating)* from the *View* pull-down menu to activate model space and begin drawing. Alternately, you can change *TILEMODE* to **1** while you draw. Create layers for hidden lines, centerlines, and object lines. Create the drawing geometry as shown in Figure 19-16.

D. When you are finished with the views, ensure *TILEMODE* is set to **0** (title block is visible) and make model space active (*Mspace*). To scale the model space geometry to paper space, first *Zoom Extents*, then use *Zoom* with a *.5XP* factor. (The drawing is scaled to .5=1, or 1/2"=1" when plotted from paper space at 1=1.) You may also need to use *Pan* to move the geometry about within the viewport so it does not overlap with the title block.

E. Next, use *Pspace*. *Freeze* layer **Viewport**. Your completed drawing should look like that in Figure 19-18. Fill in the title block with appropriate information. *Save* the drawing. From paper space, print or plot the drawing on a B size sheet and enter a scale of **1=1**.

Figure 19-18 ───

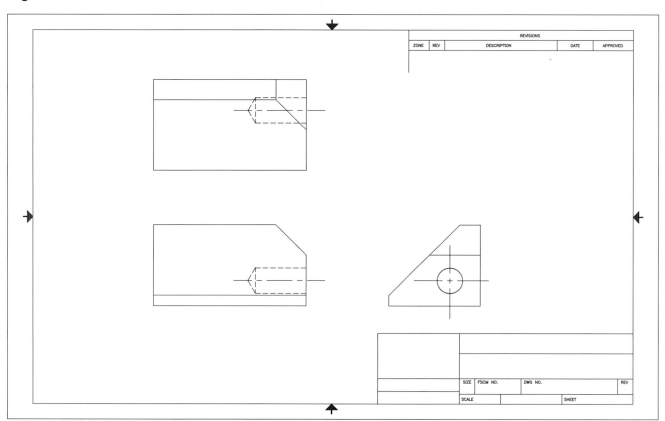

20

BLOCKS

Chapter Objectives

After completing this chapter you should:

1. understand the concept of creating and inserting symbols in AutoCAD drawings;

2. be able to use the *Block* command to transform a group of objects into one object that is stored in the current drawing's block definition table;

3. be able to use the *Insert* and *Minsert* commands to bring *Blocks* into drawings;

4. know that *color* and *linetype* of *Blocks* are based on conditions when the *Block* is made;

5. be able to convert *Blocks* to individual objects with *Explode*;

6. be able to use *Wblock* to prepare .DWG files for insertion into other drawings;

7. be able to redefine and globally change previously inserted *Blocks*;

8. be able to define an insertion base point for .DWG files using *Base*.

CONCEPTS

A *Block* is a group of objects that are combined into <u>one</u> object with the *Block* command. The typical application for *Blocks* is in the use of symbols. Many drawings contain symbols, such as doors and windows for architectural drawings, capacitors and resistors for electrical schematics, or pumps and valves for piping and instrumentation drawings. In AutoCAD, symbols are created first by constructing the desired geometry with objects like *Line*, *Arc*, and *Circle*, then transforming the set of objects comprising the symbol into a *Block*. A description of the objects comprising the *Block* is then stored in the drawing's "block definition table." The *Blocks* can then each be *Inserted* into a drawing many times and treated as a single object. Text can be attached to *Blocks* (called *Attributes*) and the text can be modified for each *Block* when inserted.

Figure 20-1 compares a shape composed of a set of objects and the same shape after it has been made into a *Block* and *Inserted* back into the drawing. Notice that the original set of objects is selected (highlighted) individually for editing, whereas, the *Block* is only one object.

Figure 20-1 ────────────────────────────

10 OBJECTS   1 OBJECT

Since an inserted *Block* is one object, it uses less file space than a set of objects that is copied with *Copy*. The *Copy* command creates a duplicate set of objects, so that if the original symbol were created with 10 objects, 3 copies would yield a total of 40 objects. If instead the original set of 10 were made into a *Block* and then *Inserted* 3 times, the total objects would be 13 (the original 10 + 3).

Upon *Inserting* a *Block*, its scale can be changed and rotational orientation specified without having to use the *Scale* or *Rotate* commands (Fig. 20-2). If a design change is desired in the *Blocks* that have already been *Inserted*, the original *Block* can be redefined and the previously inserted *Blocks* are automatically updated. *Blocks* can be made to have explicit *Linetype* and *Color*, regardless of the layer they are inserted onto, or they can be made to assume the *Color* and *Linetype* of the layer onto which they are *Inserted*.

Figure 20-2 ────────────────────────────

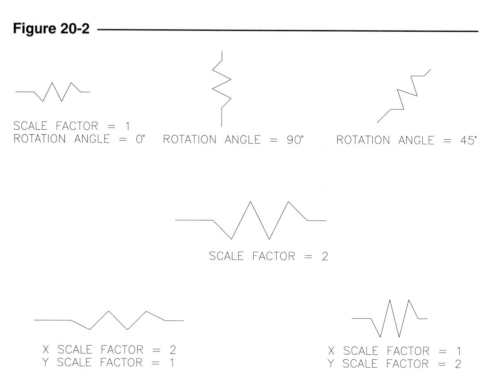

SCALE FACTOR = 1
ROTATION ANGLE = 0°   ROTATION ANGLE = 90°   ROTATION ANGLE = 45°

SCALE FACTOR = 2

X SCALE FACTOR = 2   X SCALE FACTOR = 1
Y SCALE FACTOR = 1   Y SCALE FACTOR = 2

Blocks can be <u>nested</u>; that is, one *Block* can reference another *Block*. Practically, this means that the definition of *Block* "C" can contain *Block* "A" so that when *Block* "C" is inserted, *Block* "A" is also inserted as part of *Block* "C" (Fig. 20-3).

Figure 20-3

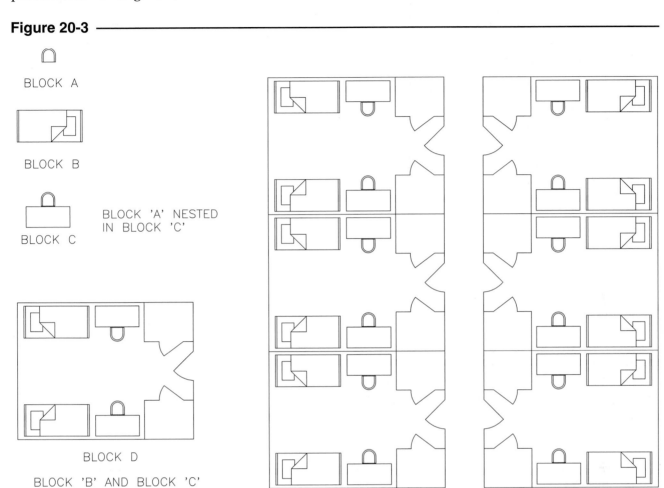

BLOCK A

BLOCK B

BLOCK 'A' NESTED
IN BLOCK 'C'

BLOCK C

BLOCK D

BLOCK 'B' AND BLOCK 'C'
NESTED IN BLOCK 'D'

Blocks created within the current drawing can be copied to disk as complete and separate drawing files (.DWG file) by using the *Wblock* command (Write Block). This action allows you to *Insert* the *Blocks* into other drawings. Specifically, when you use the *Insert* command, AutoCAD first searches for the supplied *Block* name in the current drawing's block definition table. If the designated *Block* is not located there, AutoCAD searches the directories for a .DWG file with the designated name.

Commands related to using *Blocks* are:

Bmake	Invokes a dialog box for creating *Blocks*
Block	Creates a *Block* from individual objects
Insert	Inserts a *Block* into a drawing
Ddinsert	Invokes a dialog box for inserting a *Block*
Minsert	Permits a multiple insert in a rectangular pattern
Explode	Breaks a *Block* into its original set of multiple objects
Wblock	Writes an existing *Block* or a set of objects to a file on disk
Base	Allows specification of an insertion base point
Purge	Deletes uninserted *Blocks* from the block definition table
Rename	Allows renaming *Blocks*

COMMANDS

BLOCK and
BMAKE

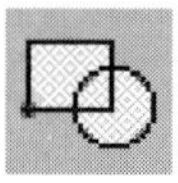

Pull-down Menu	COMMAND (TYPE)	ALIAS (TYPE)	Short-cut	Screen (side) Menu	Tablet Menu
Draw *Block >* *Make...*	*BLOCK* *or BMAKE*	*B or -B*	...	DRAW2 *Bmake*	N,9

Selecting the icon button, using the pull-down or screen menu, or typing *Bmake* produces the *Block Definition* dialog box shown in Figure 20-4. This dialog box provides the same functions as using the *Block* command (command line equivalent).

Figure 20-4 ——————

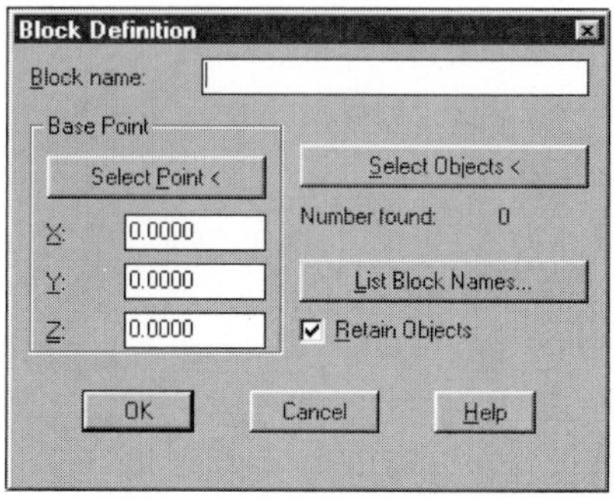

To make a *Block*, first create the *Lines, Circles, Arcs,* or other objects comprising the shapes to be combined into the *Block*. Next, use the *Bmake* or *Block* command to transform the objects into one object—a *Block*.

If you are using the *Block Definition* dialog box, enter the desired *Block* name in the *Block name* box. Then use the *Select Objects* tile to return to the drawing temporarily to select the objects you wish to comprise the *Block*. After selection of objects, the dialog box reappears. Use the *Select Point* button in the *Base Point* section of the dialog box if you want to use a point other than the default 0,0 as the "insertion point" when the *Block* is later inserted (usually select a point in the center or corner of the set of objects). When you select *OK*, the objects comprising the *Block* disappear and the new *Block* is defined and stored in the drawing's block definition table awaiting future insertions. A check appearing in the *Retain Objects* checkbox forces AutoCAD to retain the original "template" objects (similar to using *Oops* after the *Block* command), even though the definition of the *Block* remains in the table.

The *List Block Names* tile can be chosen to list all existing *Blocks* in the current drawing (Fig. 20-5).

Figure 20-5 ——————

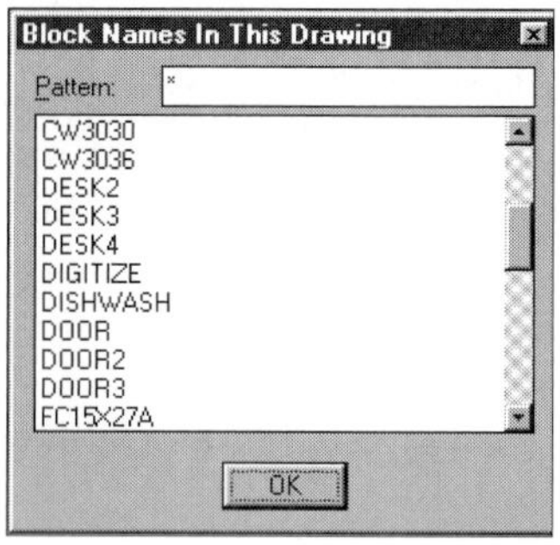

If you prefer to type, use *Block* to produce the command line equivalent of *Bmake*. The command syntax is as follows:

Command: **Block**
Block name (or ?): (**name**) (Enter a descriptive name for the *Block* up to 31 characters.)
Insertion base point: **PICK** or (**coordinates**) (Select a point to be used later for insertion.)
Select objects: **PICK**
Select objects: **PICK** (Continue selecting all desired objects.)
Select objects: **Enter**

The *Block* then <u>disappears</u> as it is stored in the current drawing's "block definition table." The *Oops* command can be used to restore the original set of "template" objects (they reappear), but the definition of the *Block* remains in the table. Using the *?* option of the *Block* command lists the *Blocks* stored in the block definition table.

Block Color and *Linetype* **Settings**

The <u>color</u> and <u>linetype</u> of an inserted *Block* are determined by one of the following settings when the *Block* is created:

1. When a *Block* is inserted, it is drawn on its original layer with its original *Color* and *Linetype* (when the objects were <u>created</u>), regardless of the layer or *color* and *linetype* settings that are current when the *Block* is inserted (unless conditions 2 or 3 exist).

2. If a *Block* is created on <u>Layer 0</u> (Layer 0 is current when the original objects comprising the *Block* are created), then the *Block* assumes the *color* and *linetype* of any layer that is current when it is inserted (Fig. 20-6).

3. If the *Block* is created with the special *BYBLOCK linetype* and *color* setting, the *Block* is inserted with the *Color* and *Linetype* settings that are <u>current during insertion</u>, whether the *BYLAYER* or explicit object *Color* and *Linetype* settings are current.

Figure 20-6

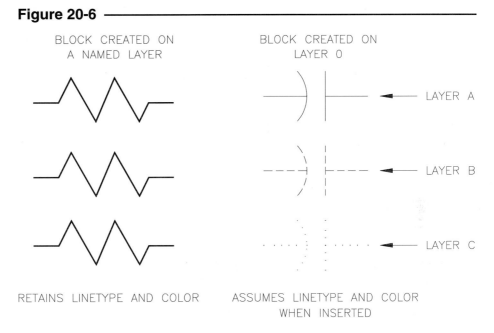

BLOCK CREATED ON
A NAMED LAYER

BLOCK CREATED ON
LAYER 0

◄—— LAYER A

◄—— LAYER B

◄—— LAYER C

RETAINS LINETYPE AND COLOR

ASSUMES LINETYPE AND COLOR
WHEN INSERTED

INSERT and
DDINSERT

Pull-down Menu	COMMAND (TYPE)	ALIAS (TYPE)	Short-cut	Screen (side) Menu	Tablet Menu
Insert *Block...*	*INSERT or* *DDINSERT*	*I or -I*	...	*INSERT* *Ddinsert*	*T,5*

Once the *Block* has been created, it is inserted back into the drawing at the desired location(s) with the *Insert* command. *Insert* also allows the *Blocks* to be scaled or rotated upon insertion. The command syntax is given here:

    Command: **insert**
    Block name (or ?): **name** (Type the name of an existing block or .DWG file to insert.)
    Insertion point: **PICK** or (**coordinates**) (Give the desired location of the *Block*.)
    X scale factor <1>/Corner/XYZ: **PICK** or (**value**) (Specifies the size of the *Block* in the X direction.)
    Y scale factor (default=X): (**value**) or **Enter** (Specifies the size in the Y direction.)
    Rotation angle: **PICK** or (**value**) (Enter an angle for *Block* rotation.)

Selecting the "*X scale factor*" with the cursor specifies both X and Y factors. The rotation angle can be forced to 90 degree increments by turning *ORTHO* (F8) *On*.

The *Insert* dialog box can be invoked by using the pull-down menu, icon buttons, or tablet menu or by typing *Ddinsert* (Fig. 20-7). Selecting the *Block* tile causes another box to pop up, listing the *Blocks* previously defined in the drawing's block definition table (Fig. 20-8). The desired *Block* is selected from the list. Selecting the *File* tile causes a box to pop up, allowing selection of any drawing (.DWG files) from any accessible drive and directory for insertion. The *Insert* dialog box provides explicit value entry of insertion point coordinates, scale, and rotation angle. *Explode* can also be toggled, which would insert the *Block* as multiple objects (see *INSERT* with *).

Figure 20-7 ——————————

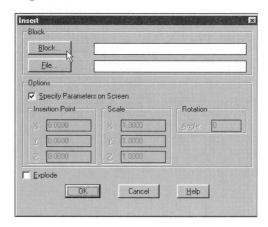

Figure 20-8 ——————

Insert Presets

Sometimes it is desirable to see the *Block* in the intended scale factor or rotation angle before you choose the insertion point. *Insert* presets allow you to specify a rotation angle or scale factor <u>before</u> you dynamically drag the *Block* to PICK the insertion point. (Normally, you have to select the insertion point before the prompts for scale factor and rotation angle appear.) Presets can be used with the *Insert* command or *Insert* dialog box. To do this, type in one of the following characters <u>at the "Insertion point:" prompt</u>:

R	rotation angle
S	scale factor (uniform)
X	X scale factor only
Y	Y scale factor only
Z	Z scale factor only

For example, to insert a *Block* at a 45 degree angle using the *Insert* dialog box, the command syntax reads as follows:

> Command: **ddinsert** (select *Block* name from the list)
> Insertion point: **R** (specifies *Rotation* preset)
> Rotation angle: **45** (rotates the *Block* to 45 degrees during dynamic insertion)
> Insertion point: X scale factor <1> / Corner / XYZ: **Enter** or (**value**)
> Y scale factor (default=X): **Enter** or (**value**)
> Command:

This action allows you to see the *Block* at the prescribed rotation angle as you dynamically drag it to PICK the insertion point.

MINSERT

Pull-down Menu	COMMAND (TYPE)	ALIAS (TYPE)	Short-cut	Screen (side) Menu	Tablet Menu
...	*MINSERT*	...	...	...	...

This command allows a <u>multiple insert</u> in a rectangular pattern (Fig. 20-9). *Minsert* is actually a combination of the *Insert* and the *Array Rectangular* commands. The *Blocks* inserted with *Minsert* are associated (the group is treated as one object) and cannot be edited independently (unless *Exploded*).

Examining the command syntax yields the similarity to a *Rectangular Array*.

Figure 20-9

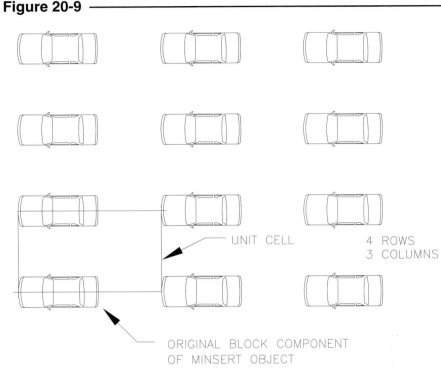

UNIT CELL   4 ROWS
                   3 COLUMNS

ORIGINAL BLOCK COMPONENT
OF MINSERT OBJECT

Command: **Minsert**
Block name (or ?): **name**
Insertion point: **PICK** or
  (**coordinates**)
X scale factor
  <1>/Corner/XYZ:
  (**value**) or **PICK**
Y scale factor (default=X): (**value**) or **Enter**
Rotation angle: (**value**) or **PICK**
Number of rows (---): (**value**)
Number of columns (III): (**value**)
Unit cell or distance between rows: (**value**) or **PICK** (Value specifies Y distance from *Block* corner
  to *Block* corner; PICK allows drawing a unit cell rectangle.)
Distance between columns: (**value**) or **PICK** (Specifies X distance between *Block* corners.)
Command:

EXPLODE

Pull-down Menu	COMMAND (TYPE)	ALIAS (TYPE)	Short-cut	Screen (side) Menu	Tablet Menu
Modify *Explode*	*Explode*	X	...	*MODIFY2* *Explode*	Y,22

Explode breaks a <u>previously</u> inserted *Block* back into its original set of objects (Fig. 20-10), which allows you to edit individual objects comprising the shape. *Blocks* that have been inserted with differing X and Y scale factors or *Blocks* that have been *Minsert*ed can be *Explode*d in Release 13 and 14. There are no options for this command.

Command: **explode**
Select objects: **PICK**
Select objects: **Enter**
Command:

Inserting with an * (asterisk) symbol accomplishes the same goal as using *Insert* normally, then *Explode*.

Figure 20-10

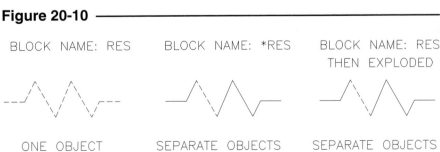

BLOCK NAME: RES   BLOCK NAME: *RES   BLOCK NAME: RES
                                                   THEN EXPLODED

ONE OBJECT   SEPARATE OBJECTS   SEPARATE OBJECTS

INSERT with *

Using the *Insert* command with the asterisk (*) allows you to insert a *Block,* not as one object, but as the original set of objects comprising the *Block.* In this way, you can edit individual objects in the *Block,* otherwise impossible if the *Block* is only one object (Fig. 20-10).

The normal *Insert* command is used; however, when the desired <u>Block name</u> is entered, it is prefaced by the asterisk (*) symbol:

Command: **insert**
Block name (or ?): ***** (**name**) (Type the * symbol, then the name of an existing block or .DWG file to insert.)
Command:

This action accomplishes the same goal as using *Insert;* then *Explode.*

WBLOCK

Pull-down Menu	COMMAND (TYPE)	ALIAS (TYPE)	Short-cut	Screen (side) Menu	Tablet Menu
File *Export...*	*WBLOCK*	*W*	...	FILE *Export*	*W,24*

The *Wblock* command writes a *Block* out to disk as a separate and complete drawing (.DWG) file. The *Block* used for writing to disk can exist in the current drawing's *Block* definition table or can be created by the *Wblock* command. Remember that the *Insert* command inserts *Blocks* (from the current drawing's block definition table) or finds and accepts .DWG files and treats them as *Blocks* upon insertion.

If you are using an existing *Block,* a copy of the *Block* is essentially transformed by the *Wblock* command to create a complete AutoCAD drawing (.DWG) file. The original block definition remains in the current drawing's block definition table. In this way, *Blocks* that were originally intended for insertion into the current drawing can be inserted into other drawings.

If you want to transform a set of objects to be used as a *Block* in other drawings but not in the current one, you can use *Wblock* to transform (a copy of) the objects in the current drawing into a separate .DWG file. This action does not create a *Block* in the current drawing.

As an alternative, if you want to create symbols specifically to be inserted into other drawings, each symbol could be created initially as a separate .DWG file. Figure 20-11 illustrates the relationship among a *Block,* the current drawing, and a *WBlock.*

To create *Wblocks* (.DWG files) <u>from existing Blocks,</u> follow this command syntax:

Command: **Wblock**
(At this point, the *Create Drawing File* dialog box appears, prompting you to supply a name for the .DWG file to be created. Typically, a new descriptive name would be typed in the edit box rather than selecting from the existing names.)
Block name: (**name**) (Enter the name of the desired existing *Block.* If the file name given in the previous step is the same as the existing *Block* name, an "=" symbol can be entered at this prompt.)
Command:

A copy of the existing *Block* is then created in the current or selected directory as a *Wblock* (.DWG file).

To create a *Wblock* (.DWG file) to be used as a *Block* in other drawings <u>but not in the current drawing,</u> follow the same steps as before, but when prompted for the "Block name:" press Enter or select *blank* from the screen menu. The next steps are like the *Block* command prompts:

Figure 20-11

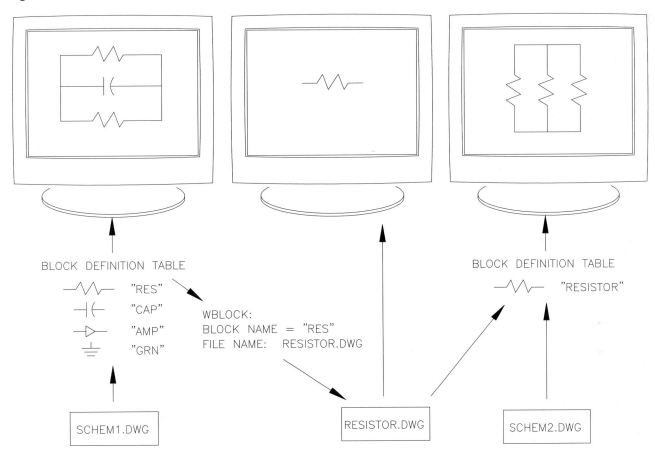

Command: **Wblock**
(The *Create Drawing File* dialog box appears, prompting you to supply a name for the .DWG file to be created)
Block name: (**Enter**) or (**blank**)
Insertion base point: **PICK** or (**coordinates**) (Pick a point to be used later for insertion.)
Select objects: **PICK**
Select objects: **Enter** (Press Enter to complete selection.)
Command:

An alternative method for creating a *Wblock* is using the *Export Data* dialog box accessed from the *File* pull-down menu. Make sure you select *Drawing (*.DWG)* as the type of file to export (bottom left of the box, Fig. 20-12). You can create a *Wblock* (.DWG file) from an existing *Block* or from objects that you select from the screen. After specifying the .DWG name you want to create and the dialog box disappears, you are presented with the same command syntax as above.

When *Wblocks* are *Inserted*, the *Color* and *Linetype* settings of the *Wblock* are determined by the

Figure 20-12

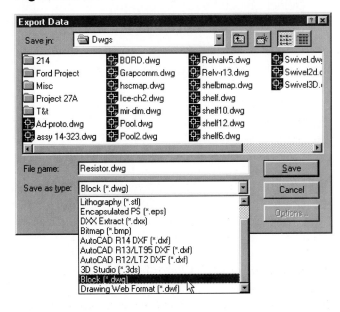

settings current when the original objects comprising the *Wblock* were created. The three possible settings are the same as those for *Blocks* (see the *Block* command, *Color* and *Linetype* Settings).

When a *Wblock* is *Inserted*, its parent (original) layer is also inserted into the current drawing. *Freezing* either the parent layer or the layer that was current during the insertion causes the *Wblock* to be frozen.

Redefining *Blocks*

If you want to change the configuration of a *Block*, even after it has been inserted, it can be accomplished by redefining the *Block*. In doing so, all of the previous *Block* insertions are automatically and globally updated (Fig. 20-13). AutoCAD stores two fundamental pieces of information for each *Block* insertion—the insertion point and the *Block* name. The actual block definition is stored in the block definition table. Redefining the *Block* involves changing that definition.

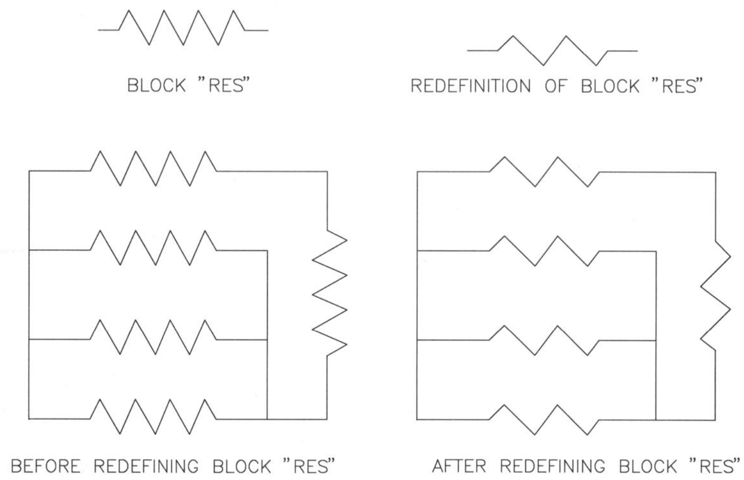

To redefine a *Block*, use the *Block* command. First, draw the new geometry or change the original "template" set of objects. (The change cannot be made using an inserted *Block* unless it is *Exploded* because a *Block* cannot reference itself.) Next, use the *Block* command and select the new or changed geometry. The old *Block* is redefined with the new geometry as long as the original Block name is used.

Figure 20-13

BLOCK "RES"   REDEFINITION OF BLOCK "RES"

BEFORE REDEFINING BLOCK "RES"   AFTER REDEFINING BLOCK "RES"

```
Command: block
Block name (or ?): name (Enter the original Block name.)
Block (name) already exists.
Redefine it? <N>: Yes (Answering Y or yes causes the redefinition.)
Insertion base point: PICK or (coordinates) (Select a point to be used later for insertion.)
Select objects: PICK
Select objects: PICK (Continue selecting all desired objects.)
Select objects: Enter
Command:
```

The *Block* is redefined and all insertions of the original *Block* display the new geometry.

The *Block* command can also be used to redefine *Wblocks* that have been inserted. In this case, enter the *Wblock* name (.DWG filename) at the "Block name:" prompt to redefine (actually replace) a previously inserted *Wblock*.

BASE

Pull-down Menu	COMMAND (TYPE)	ALIAS (TYPE)	Short-cut	Screen (side) Menu	Tablet Menu
Draw *Block* > *Base*	*BASE*	...	...	DRAW2 *Base*	...

The *Base* command allows you to specify an "insertion base point" (see the *Block* command) in the current drawing for subsequent insertions. If the *Insert* command is used to bring a .DWG file into another drawing, the insertion base point of the .DWG is 0,0 by default. The *Base* command permits you to specify another location as the insertion base point. The *Base* command is used in the symbol drawing, that is, used in the drawing <u>to be inserted</u>. For example, while creating

Figure 20-14

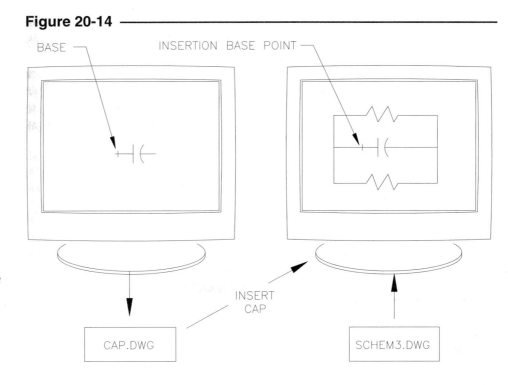

separate symbol drawings (.DWGs) for subsequent insertion into other drawings, the *Base* command is used to specify an appropriate point on the symbol geometry for the *Insert* command to use as a "handle" other than point 0,0 (see Fig. 20-14).

Blocks Search Path

In AutoCAD Release 14 you can use the *Preferences* dialog box to specify the search path used when the *Insert* command attempts to locate .DWG files. As you remember, when *Insert* is used and a *Block* name is entered, AutoCAD searches the drawing's block definition table and, if no *Block* by the specified name is found, searches the specified path for a .DWG file by the name. (If you use the *Insert* dialog box, you can select *File...* button to locate a .DWG file.) If the paths of the intended files are specified in the

"Support File Search Path," you can type *Insert* and the name of the file, and AutoCAD will locate and insert the file. Otherwise, you have to specify the name of the file prefaced by the path each time you *Insert* a new file.

The *Files* tab of the *Preferences* dialog box allows you to list and modify the path that AutoCAD searches for drawings to *Insert* (Fig. 20-15). The search path is also used for finding font files, menus, plug-ins, linetypes, and hatch patterns.

Suppose all of your symbols were stored in a folder called D:\Dwg\Symbols. Access the *Preferences* dialog box by typing *Preferences* or selecting from the *Tools* pull-down menu. In the *Files* tab, open the *Support File Search Path* section.

Figure 20-15

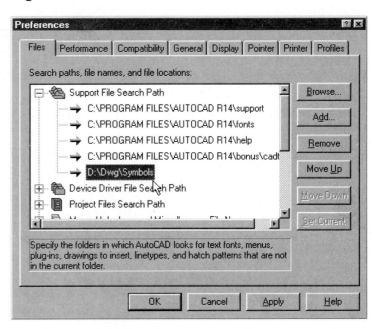

Select *Add..* or *Browse...* to enter the location of your symbols as shown in the figure. Use *Move Up* or *Move Down* to specify an order (priority) for searching. Then each time you type *"Insert"*, you could enter the name of a file without the path and AutoCAD can locate it.

PURGE

Pull-down Menu	COMMAND (TYPE)	ALIAS (TYPE)	Short-cut	Screen (side) Menu	Tablet Menu
File *Drawing Utilities >* *Purge >*	*PURGE*	*PU*	...	*FILE* *Purge*	X,25

Purge allows you to selectively delete a *Block* that is not referenced in the drawing. In other words, if the drawing has a *Block* defined but not appearing in the drawing, it can be deleted with *Purge*. In fact, *Purge* can selectively delete <u>any named object</u> that is not referenced in the drawing. Examples of unreferenced named objects are:

Blocks that have been defined but not *Inserted*;
Layers that exist without objects residing on the layers;
Dimstyles that have been defined, yet no dimensions are created in the style (see Chapter 27);
Linetypes that were loaded but not used;
Shapes that were loaded but not used;
Text Styles that have been defined, yet no text has been created in the *Style*;
Mlstyles (multiline styles) that have been defined, yet no *Mlines* have been drawn in the style.

Since these named objects occupy a small amount of space in the drawing, using *Purge* to delete unused named objects can reduce the file size. This may be helpful when drawings are created using template drawings that contain many unused named objects or when other drawings are *Inserted* that contain many unused *Layers* or *Blocks*, etc.

Purge is especially useful for getting rid of unused *Blocks* because unused Blocks can occupy a huge amount of file space compared to other named objects (*Blocks* are the only geometry-based named objects). Although some other named objects can be deleted by other methods (Windows 95/NT dialog boxes with right-click options), *Purge* is the only method for deleting *Blocks*.

Purge can be invoked by the methods shown in the command table. Using the *File* pull-down menu as shown in Figure 20-16 reveals all of the *Purge* options. The command options (when typing *Purge*) are listed next.

Figure 20-16

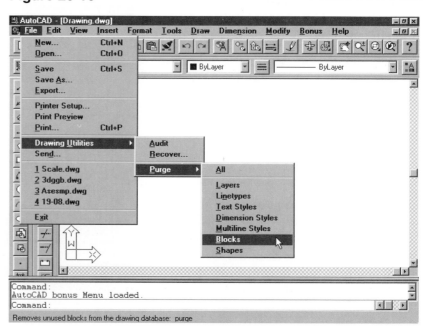

Command: *purge*
Purge unused Blocks/Dimstyles/LAyers/LTypes/SHapes/STyles/Mlinestyles/All: (**option**)
Names to purge <*>: **Enter** or (**name**)
Verify each name to be purged? <Y>

You can select one type of object to be *Purged* or select *All* to have all named objects listed. AutoCAD lists all named objects matching the type selected. For example, if you specified *Blocks* to purge, AutoCAD responds with a list of <u>unused</u> *Blocks* that may look something like this:

Purge block DESK? <N>
Purge block DESK2? <N>
Purge block CHAIR? <N>
Purge block TABLE? <N>

Answering with a "Y" causes the deletion of the *Blocks*. The other named object options operate similarly. Purge can be used <u>anytime</u> in a drawing session in Release 14 or Release 13.

RENAME and DDRENAME

Pull-down Menu	COMMAND (TYPE)	ALIAS (TYPE)	Short-cut	Screen (side) Menu	Tablet Menu
Format Rename...	*RENAME or DDRENAME*	*REN or -REN*	...	*FORMAT Ddrename*	*V,1*

These utility commands allow you to rename a *Block* or <u>any named object</u> that is part of the current drawing. *Rename* and the related dialog box accessed by the *Ddrename* command allow you to rename the named objects listed here:

Blocks, Dimension Styles, Layers, Linetypes, Text Styles, User Coordinate Systems, Views, and *Viewport* configurations

The *Rename* command can be used to rename objects one at a time. For example, the command sequence might be as follows:

Command: *rename*
Block/Dimstyle/LAyer/LType/Style/Ucs/VIew/VPort: **B** (use the *Block* option)
Old block name: **DESK**
New block name: **DESK-30**
Command:

Wildcard characters are not allowed with the *Rename* command but are allowed with the dialog box.

Optionally, if you prefer to use dialog boxes, you may type *Ddrename* or select *Rename...* from the *Format* pull-down menu to access the dialog box shown in Figure 20-17.

You may select from the *Named Objects* list to display the related objects existing in the drawing. Then select or type the old name so it appears in the *Old Name:* edit box. Specify the new name in the *Rename To:* edit box. <u>You must then PICK *the Rename To:* tile</u> and the new names will appear in the list. Then select the *OK* tile to confirm.

The *Rename* dialog box can be used with wildcard characters to specify a list of named objects for renaming.

Figure 20-17 ——————

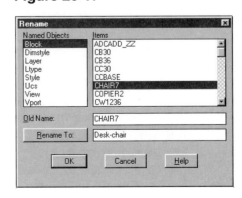

For example, if you desired to rename all the "DIM-*" layers so that the letters "DIM" were replaced with only the letter "D", the following sequence would be used.

1. In the *Old Name* edit box, enter "DIM-*" and press Enter. All of the layers beginning with "DIM-" are highlighted.

2. In the *Rename To:* edit box, enter "D-*." Next, the *Rename To:* tile must be PICKed. Finally, PICK the *OK* tile to confirm the change.

CHAPTER EXERCISES

1. ***Block, Insert***
    In the next several exercises, you will create an office floor plan, then create pieces of furniture as *Blocks* and *Insert* them into the office. All of the block-related commands are used.

    A. Start a *New* drawing. Select *Start from Scratch* and use the *English* defaults. Use *Save* and assign the name **OFF-ATT**. Set up the drawing as follows:

1. *Units*	*Architectural*	**1/2" Precision**
2. *Limits*	**48' x 36'**	(1/4"=1' scale on an A size sheet), drawing scale factor = 48
3. *Snap*	**3**	
4. *Grid*	**12**	
5. *Layers*	**FLOORPLAN**	**continuous** **colors of your choice, all different**
	FURNITURE	**continuous**
	ELEC-HDWR	**continuous**
	ELEC-LINES	**hidden**
	DIM-FLOOR	**continuous**
	DIM-ELEC	**continuous**
	TEXT	**continuous**
	TITLE	**continuous**
6. *Text Style*	*CityBlueprint*	**CityBlueprint (TrueType font)**
7. *Ltscale*	**24**	

    B. Create the floor plan shown in Figure 20-18. Center the geometry in the *Limits*. Draw on layer **FLOORPLAN**. Use any method you want for construction (e.g., *Line, Pline, Xline, Mline, Offset*).

    C. Create the furniture shown in Figure 20-19. Draw on layer **FURNITURE**. Locate the pieces anywhere for now. Do <u>not</u> make each piece a *Block*. *Save* the drawing as **OFF-ATT**.

    Now make each piece a *Block*. Use the *name* as indicated and the *insertion base point* as shown by the "blip." Next, use the *Block* command again but with the *?* option to list the block definition table. Use *SaveAs* and rename the drawing **OFFICE**.

    D. Use *Insert* to insert the furniture into the drawing, as shown in Figure 20-20. You may use your own arrangement for the furniture, but *Insert* the same number of each piece as shown. *Save* the drawing.

Figure 20-18

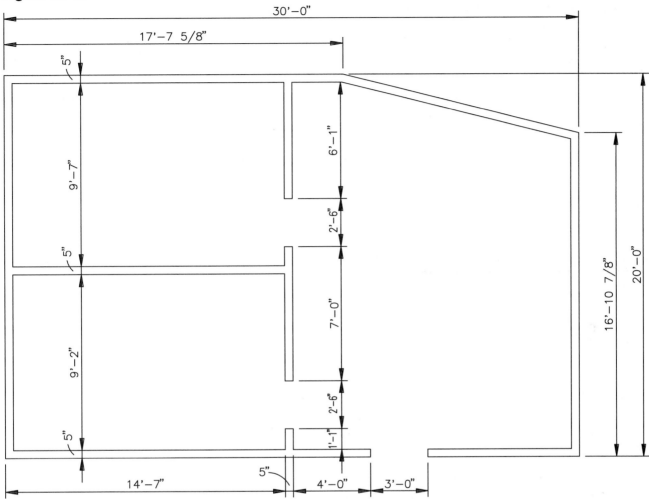

Figure 20-19

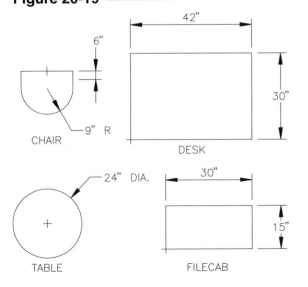

CHAIR

DESK

TABLE

FILECAB

Figure 20-20

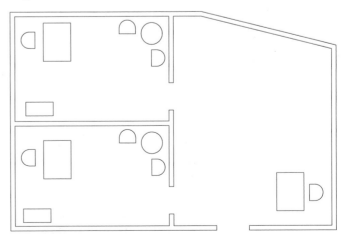

2. **Creating a .DWG file for** *Insertion, Base*

Begin a *New* drawing. Assign the name **CONFTABL**.
Create the table as shown in Figure 20-21 on *Layer* **0**.
Since this drawing is intended for insertion into the
OFFICE drawing, use the *Base* command to assign an
insertion base point at the lower-left corner of the table.

When you are finished, *Save* the drawing.

Figure 20-21

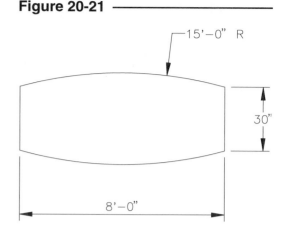

3. *Insert, Explode, Divide*

Figure 20-22

A. *Open* the **OFFICE**
    drawing. Ensure that
    layer **FURNITURE** is
    current. Use *Insert* to
    bring the **CONFTABL**
    drawing in as a *Block* in
    the placement shown in
    Figure 20-22.

    Notice that the CONF-
    TABL assumes the line-
    type and color of the
    current layer, since it was
    created on layer **0**.

B. *Explode* the CONFTABL.
    The *Exploded*
    CONFTABL
    returns to *Layer* **0**, so
    use *Chprop* to change it

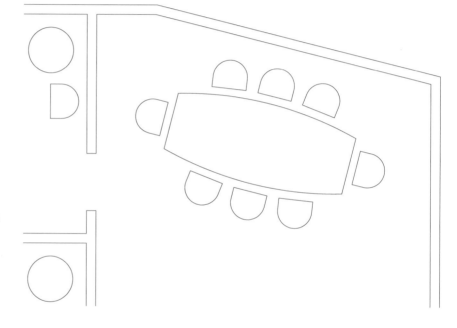

back to *Layer* **FURNITURE**. Then use the *Divide* command (with the *Block* option) to insert
the **CHAIR** block as shown in Figure 20-22. Also *Insert* a **CHAIR** at each end of the table.
Save the drawing.

4. *Wblock, BYBLOCK setting*

Figure 20-23

A. *Open* the **EFF-APT2** drawing you worked on in Chapter 18
    Exercises. Use the *Wblock* command to transform the plant
    into a .DWG file (Fig. 20-23). Use the name **PLANT** and
    specify the *Insertion base point* at the center. Do not save the
    EFF-APT2 drawing.

B. *Open* the **OFFICE** drawing and *Insert* the **PLANT** into one of
    the three rooms. The plant probably appears in a different
    color than the current layer. Why? Check the *Layer* listing to
    see if any new layers came in with the PLANT block. *Erase*
    the PLANT block.

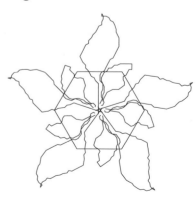

PLANT

C. *Open* the **PLANT** drawing. Change the *Color* and *Linetype* setting of the plant objects to **BYBLOCK**. *Save* the drawing.

D. *Open* the **OFFICE** drawing again and *Insert* the **PLANT** onto the **FURNITURE** layer. It should appear now in the current layer's *color* and *linetype*. *Insert* a **PLANT** into each of the 3 rooms. *Save* the drawing.

5. **Redefining a** *Block*

Figure 20-24 ────────────────

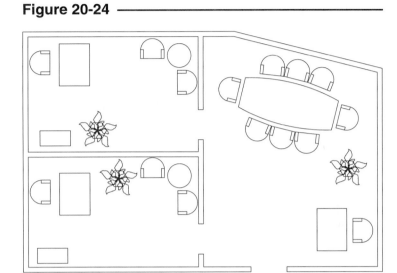

After a successful meeting, the client accepts the proposed office design with one small change. The client requests a slightly larger chair than that specified. *Explode* one of the **CHAIR** blocks. Use the *Scale* command to increase the size <u>slightly</u> or otherwise redesign the chair in some way. Use the *Block* command to redefine the **CHAIR** block. All previous insertions of the CHAIR should reflect the design change. *Save* the drawing. Your design should look similar to that shown in Figure 20-24. *Plot* to a standard scale based on your plotting capabilities.

6. *Rename*

Open the **OFFICE** drawing. Enter *Ddrename* (or select from the *Format* pull-down menu) to produce the *Rename* dialog box. Rename the *Blocks* as follows:

New Name	Old Name
DSK	DESK
CHR	CHAIR
TBL	TABLE
FLC	FILECAB

Next, use the *Rename* dialog box again to rename the *Layers* as indicated below:

New Name	Old Name
FLOORPLN-DIM	DIM-FLOOR
ELEC-DIM	DIM-ELEC

Use the *Rename* dialog box again to change the *Text Style* name as follows:

New Name	Old Name
ARCH-FONT	CITYBLUEPRINT

Use *SaveAs* to save and rename the drawing to **OFFICE-REN**.

7. *Purge*

Using the Windows Explorer, check the file size of **OFFICE-REN.DWG.** Now use *Purge*. Answer *Yes* to *Purge* any unreferenced named objects. *Exit AutoCAD* and *Save Changes*. Check the file size again. Is the file size slightly smaller?

8. Create the process flow diagram shown in Figure 20-25. Create symbols (*Blocks*) for each of the valves and gates. Use the names indicated (for the *Blocks*) and include the text in your drawing. *Save* the drawing as **PFD**.

Figure 20-25

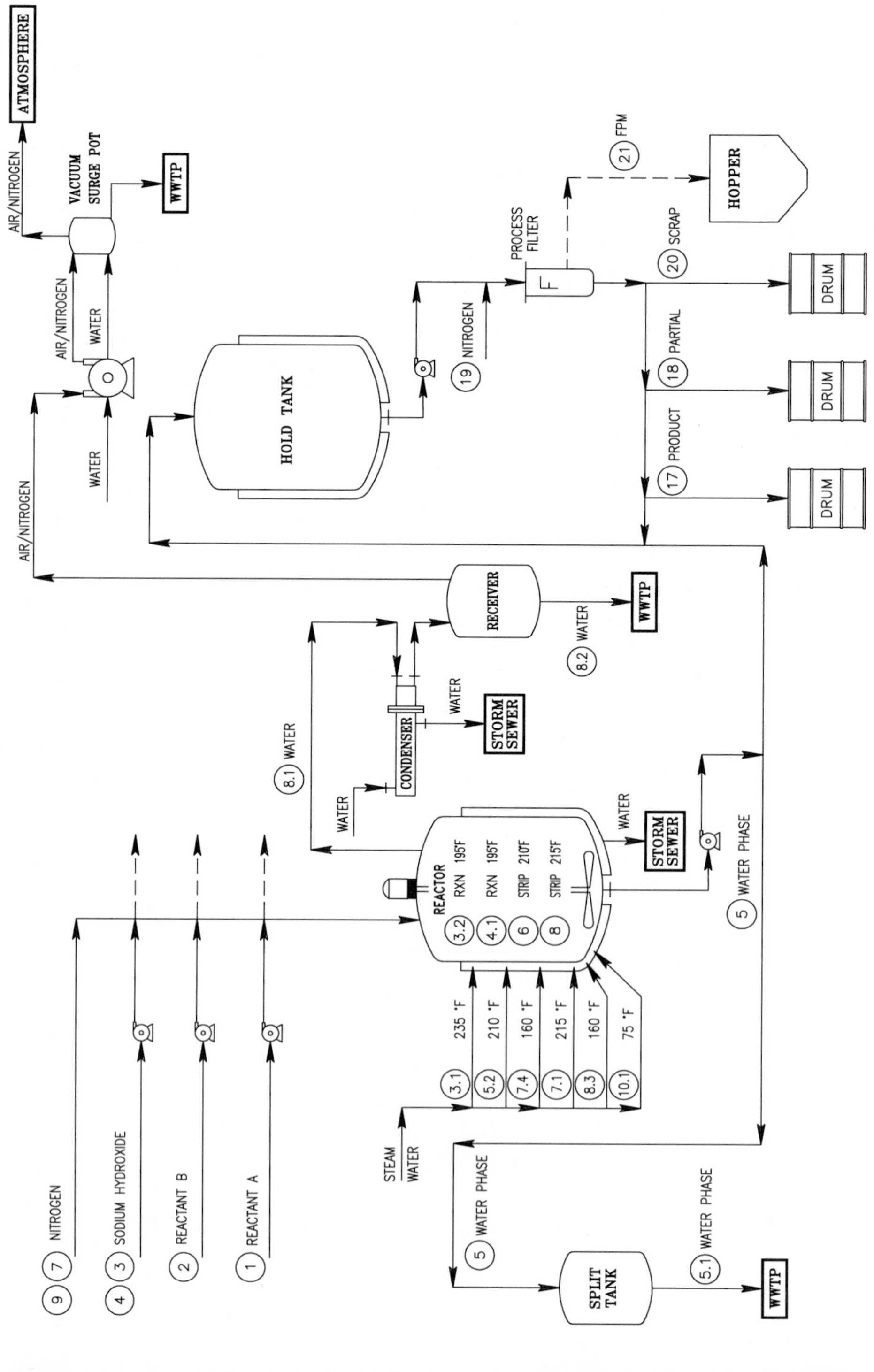

21

GRIP EDITING

Chapter Objectives

After completing this chapter you should:

1. be able to use the *GRIPS* variable to enable or disable object grips;

2. be able to activate the grips on any object;

3. be able to make an object's grips warm, hot, or cold;

4. be able to use each of the grip editing options, namely, STRETCH, MOVE, ROTATE, SCALE, AND MIRROR;

5. be able to use the Copy and Base suboptions;

6. be able to use the auxiliary grid that is automatically created when the Copy suboption is used;

7. be able to change grip variable settings using the *Grips* dialog box or the command line format.

CONCEPTS

Grips provide an alternative method of editing AutoCAD objects. The object grips are available for use by setting the *GRIPS* variable to **1**. Object grips are small squares appearing on selected objects at endpoints, midpoints, or centers, etc. The object grips are activated (made visible) by **PICK**ing objects with the cursor pickbox only <u>when no commands are in use</u> (at the open Command: prompt). Grips are like small, magnetic *OSNAPs* (*Endpoint, Midpoint, Center, Quadrant,* etc.) that can be used for snapping one object to another, for example. If the cursor is moved within the small square, it is automatically "snapped" to the grip. Grips can replace the use of *OSNAP* for many applications. The grip option allows you to STRETCH, MOVE, ROTATE, SCALE, MIRROR, or COPY objects without invoking the normal editing commands or *OSNAPs*.

As an example, the endpoint of a *Line* could be "snapped" to the endpoint of an *Arc* (shown in Fig. 21-1) by the following steps:

Figure 21-1

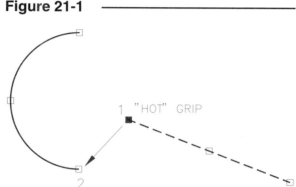

1. Activate the grips by selecting both objects. Selection is done when no commands are in use (during the open Command: prompt).
2. Select the grip at the endpoint of the *Line* (1). The grip turns **hot** (red).
3. The ** STRETCH ** option appears in place of the Command: prompt.
4. STRETCH the *Line* to the endpoint grip on the *Arc* (2). **PICK** when the cursor "snaps" to the grip.
5. The *Line* and the *Arc* should then have connecting endpoints. The Command: prompt reappears. Press Escape twice to cancel (deactivate) the grips.

GRIPS FEATURES

GRIPS and *DDGRIPS*

Pull-down Menu	COMMAND (TYPE)	ALIAS (TYPE)	Short-cut	Screen (side) Menu	Tablet Menu
Tools *Grips...*	*GRIPS or* *DDGRIPS*	*GR*	...	*TOOLS 2* *Ddgrips*	*X,10*

Grips are enabled or disabled by changing the setting of the system variable, *GRIPS*. A setting of 1 enables or turns *ON* *GRIPS* and a setting of **0** disables or turns *OFF GRIPS*. This variable can be typed at the Command: prompt, or the *Grips* dialog box can be invoked from the *Options* pull-down menu, or by typing *Ddgrips* (Fig. 21-2). Using the dialog box, toggle *Enable Grips* to turn *GRIPS ON*. The default setting in AutoCAD Release 14 for the *GRIPS* variable is **1** (*ON*). (See Grip Settings near the end of the chapter for explanations of the other options in the *Grips* dialog box.)

Figure 21-2

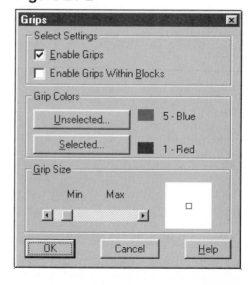

The *GRIPS* variable is saved in the system registry rather than in the current drawing as with most other system variables. Variables saved in the system registry are effective for any drawing session on that <u>particular computer</u>, no matter which drawing is current. The reasoning is that grip-related variables (and selection set-related variables) are a matter of personal preference and therefore should remain constant for a particular CAD station.

When *GRIPS* have been enabled, a small pickbox (3 pixels is the default size) appears at the center of the cursor crosshairs. (The pickbox also appears if the *PICKFIRST* system variable is set to **1**.) This pickbox operates in the same manner as the pickbox appearing during the "Select objects:" prompt. Only the pickbox, *AUTO window*, or *AUTO crossing window* methods can be used for selecting objects to activate the grips. (These three options are the only options available for Noun/Verb object selection as well.)

Activating Grips on Objects   **Figure 21-3**

The grips on objects are activated by selecting desired objects with the cursor pickbox, window, or crossing window. This action is done <u>when no commands are in use</u> (at the open Command: prompt). When an object has been selected, two things happen: the grips appear and the object is highlighted. The grips are the small blue (default color) boxes appearing at the endpoints, midpoint, center, quadrants, vertices, insertion point, or other locations depending on the object type (Fig. 21-3). Highlighting indicates that the object is included in the selection set.

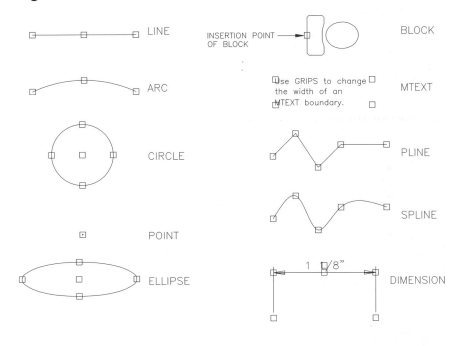

Warm, Hot, and Cold Grips

Grips can have three states: **warm, hot,** or **cold** (Fig. 21-4).

A grip is always **warm** first. When an object is selected, it is **warm**—its grips are displayed in blue (default color) and the object is highlighted. The grips can then be made **hot** or **cold**. **Cold** grips are also blue, but the object becomes <u>unhighlighted</u>.

Cold grips are created by <u>deselecting</u> a highlighted <u>object</u> that has **warm** grips. In other words, removing the **warm** grip object from the selection set with SHIFT+#1, or using the *Remove* mode, makes its grips **cold**. A **cold** grip can be used as a point to snap <u>to</u>. A **cold** grip's object is not in the selection set and therefore is not affected by the MOVE, ROTATE, or other editing action. Pressing Escape <u>once</u> changes all **warm** grips to **cold** and clears the selection set. Pressing Escape again deactivates all grips. (In short, press Escape <u>twice</u> to cancel grips entirely.)

Figure 21-4

COLD ▭──────▭──────▭ DESELECT OBJECT

WARM ▭ ─ ─ ─ ▭ ─ ─ ─ ▭ SELECT OBJECT

HOT ▭ ─ ─ ─ ▭ ─ ─ ─ ■ SELECT GRIP

R14

A **hot** grip is red (by default) and its object is almost always highlighted. Any grip can be changed to **hot** by selecting the grip itself. A <u>hot grip is the default base point</u> used for the editing action such as MOVE, ROTATE, SCALE, or MIRROR, or is the stretch point for STRETCH. When a **hot** grip exists, a new series of prompts appear in place of the Command: prompt that displays the various grip editing options. The grip editing options are also available from a right-click cursor menu (Fig. 21-5). <u>A grip must be changed to **hot** before the editing options appear.</u> A grip can be transformed from **cold** to **hot**, but the object is not highlighted (included in the selection set). Two or more grips can be made **hot** simultaneously by pressing SHIFT while selecting <u>each</u> grip.

Figure 21-5

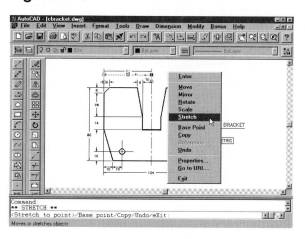

If you have made a *Grip* **hot** and want to deactivate it, possibly to make another *Grip* **hot** instead, press Escape once. This returns the object to a **warm** state. In effect, this is an undo only for the **hot** Grip. Pressing Escape twice makes the *Grips* **cold**, and pressing Escape again cancels all *Grips*.

The three states of grips can be summarized as follows:

Cold	blue grips unhighlighted object	The object is not part of the selection set, but cold grips can be used to "snap" to or used as an alternate base point.
Warm	blue grips highlighted object	The object is included in selection set and is affected by the editing action.
Hot	red grips highlighted object	The base point or control point for the editing action depending on which option is used. The object is included in selection set.

Pressing Escape demotes only the highest grips one level:

Grip State	Press Escape once	Press Escape twice	Press Escape three times
only **cold**	grips are deactivated		
warm, cold	**warm** demoted to **cold**	grips are deactivated	
hot, warm, cold	**hot** demoted to **warm**	**warm** demoted to **cold**	grips are deactivated

Grip Editing Options

When a **hot grip** has been activated, the grip editing options are available. The Command: prompt is replaced by the STRETCH, MOVE, ROTATE, SCALE, or MIRROR grip editing options. You can sequentially cycle through the options by pressing the Space bar or Enter. The editing options are displayed in Figures 21-6 through 21-10.

R14

Alternately, you can <u>right-click when a grip is **hot**</u> to activate the grip menu (see Fig. 21-5). This menu has the same options available in command line format with the addition of *Reference, Properties...,* and *Go to URL....* The options are described in the following figures.

** STRETCH **
<Stretch to point>/Base
point/Copy/Undo/eXit:

Figure 21-6

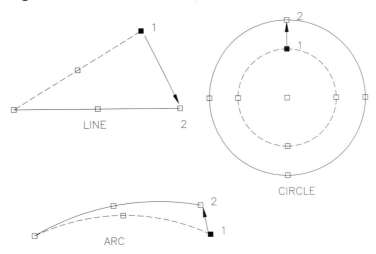

LINE

CIRCLE

ARC

** MOVE **
<Move to point>/Base point/Copy/Undo/eXit:

Figure 21-7

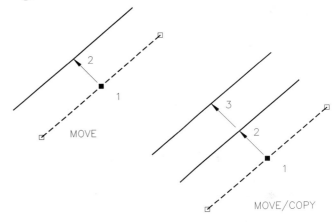

MOVE

MOVE/COPY

** ROTATE **
<Rotate angle>/Base
point/Copy/Undo/Reference/eXit:

Figure 21-8

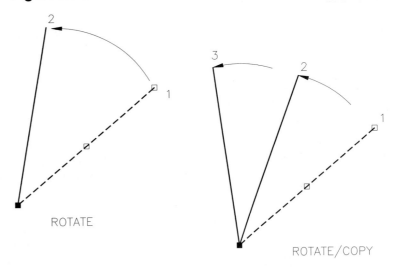

ROTATE

ROTATE/COPY

** SCALE **
<Scale factor>/Base point/Copy/Undo/Reference/eXit:

Figure 21-9 ────────────────────────────

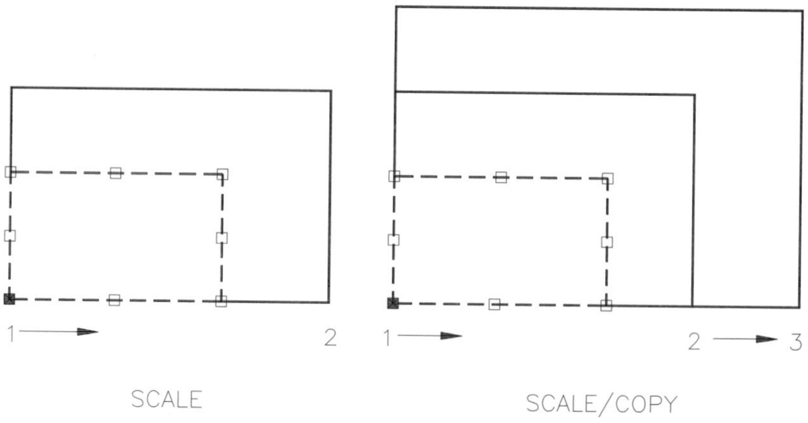

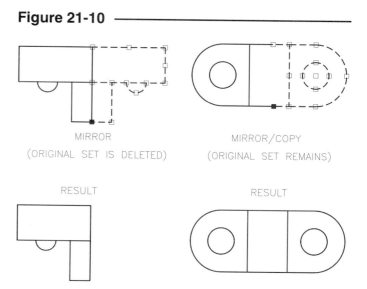

SCALE   SCALE/COPY

** MIRROR **
<Second point>/Base point/
Copy/Undo/eXit:

Figure 21-10 ────────────────────────────

MIRROR   MIRROR/COPY
(ORIGINAL SET IS DELETED)   (ORIGINAL SET REMAINS)

RESULT   RESULT

The *Grip* options are easy to understand and use. Each option operates like the full AutoCAD command by the same name. Generally, the editing option used (except for STRETCH) affects all highlighted objects. The **hot** grip is the base point for each operation. The suboptions, Base and Copy, are explained next.

NOTE: The STRETCH option differs from other options in that STRETCH affects only the object that is attached to the **hot** grip, rather than affecting all highlighted (**warm**) objects.

Base
The Base suboption appears with all of the main grip editing options (STRETCH, MOVE, etc.). Base allows using any other grip as the base point instead of the **hot** grip. Type the letter *B* or select from the right-click cursor menu to invoke this suboption.

Copy
Copy is a suboption of every main choice. Activating this suboption by typing the letter *C* or selecting from the right-click cursor menu invokes a Multiple copy mode, such that whatever set of objects is STRETCHed, MOVEd, ROTATEd, etc., becomes the first of an unlimited number of copies (see the previous five figures). The Multiple mode remains active until exiting back to the Command: prompt.

Undo
The Undo option, invoked by typing the letter *U* or selecting from the right-click cursor menu will undo the last Copy or the last Base point selection. Undo functions <u>only</u> after a Base or Copy operation.

Reference
This option operates similarly to the reference option of the *Scale* and *Rotate* commands. Use *Reference* to enter or PICK a new reference length (SCALE) or angle (ROTATE). (See *Scale* and *Rotate*, Chapter 9.) With grips, *Reference* is only enabled when the SCALE or ROTATE options are active.

Properties... **(Right-Click Menu Only)**
Selecting this option from the right-click grip menu (see Fig. 21-5) activates the *Modify (object)* dialog box. The object with the **hot** grip is the subject of the dialog box. Any property of the selected object can be changed with this dialog box. If more than one hot grip exists, the *Change Properties* dialog box appears. (See *Ddmodify* and *Ddchprop*, Chapter 16.)

Go to URL... **(Right-Click Menu Only)**
This option connects to the Internet URL (uniform resource locator) if one is attached to the object with the **hot** grip. A URL is an Internet address. Using the Internet tools, you can attach a URL to an AutoCAD object.

Figure 21-11

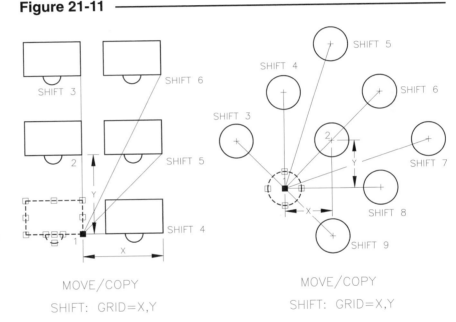

MOVE/COPY
SHIFT: GRID=X,Y

Auxiliary Grid

An auxiliary grid is <u>automatically</u> established on creating the first Copy (Fig. 21-11). The grid is activated by pressing Shift while placing the subsequent copies. The subsequent copies are then "snapped" to the grid in the same manner that *SNAP* functions. The spacing of this auxiliary grid is determined by the location of the first Copy, that is, the X and Y intervals between the base point and the second point.

For example, a "polar array" can be simulated with grips by using ROTATE with the Copy suboption (Fig. 21-12).

The "array" can be constructed by making one Copy, then using the auxiliary grid to achieve equal angular spacing. The steps for creating a "polar array" are as follows:

1. Select the object(s) to array.

2. Select a grip on the set of objects to use as the center of the array. Cycle to the ROTATE option by pressing Enter or selecting from the right-click cursor menu. Next, invoke the Copy suboption.

Figure 21-12

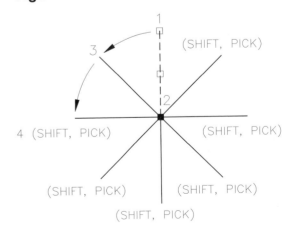

ROTATE/COPY
SHIFT – ANGLES EQUAL

3.   Make the first copy at any desired location.

4.   After making the first copy, activate the auxiliary angular grid by holding down SHIFT while making the other copies.

5.   Cancel the grips or select another command from the menus.

Editing Dimensions

Figure 21-13

One of the most effective applications of grips is as a dimension editor. Because grips exist at the dimension's extension line origins, arrowhead endpoints, and dimensional value, a dimension can be changed in many ways and still retain its associativity (Fig. 21-13). See Chapter 26, Dimensioning, for further information about dimensions, associativity, and editing dimensions with *Grips*.

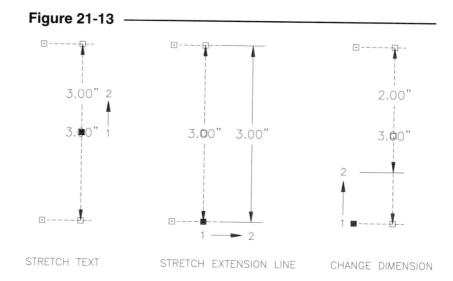

STRETCH TEXT   STRETCH EXTENSION LINE   CHANGE DIMENSION

GUIDELINES FOR USING GRIPS

Although there are many ways to use grips based on the existing objects and the desired application, a general set of guidelines for using grips is given here:

1.   Create the objects to edit.

2.   Select **warm** and **cold** grips first. This is usually accomplished by selecting <u>all</u> grips first (**warm** state), then <u>deselecting</u> objects to establish the desired **cold** grips.

3.   Select the desired **hot** grip(s). The *Grip* options should appear in place of the Command: line.

4.   Press Space or Enter to cycle to the desired editing option (STRETCH, MOVE, ROTATE, SCALE, MIRROR) or select from the right-click cursor menu.

5.   Select the desired suboptions, if any. If the *Copy* suboption is needed or the Base point needs to be re-specified, do so at this time. *Base* or *Copy* can be selected in either order.

6.   Make the desired STRETCH, MOVE, ROTATE, SCALE, or MIRROR.

Figure 21-14

7.   Cancel the grips by pressing Escape twice or selecting a command from a menu.

GRIPS SETTINGS

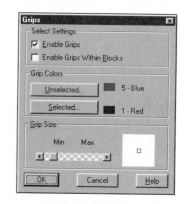

Several settings are available in the *Grips* dialog box (Fig. 21-14) that control the way grips appear or operate. The settings can also be changed by typing in the related variable name at the command prompt.

Enable Grips

If an "X" appears in the checkbox, *Grips* are enabled for the workstation. A check sets the *GRIPS* variable to 1 (on). Removing the check disables grips and sets *GRIPS=0* (off). The default setting is on.

Enable Grips Within Blocks

When no check appears in this box, only one grip at the block's insertion point is visible (Fig. 21-15). This allows you to work with a block as one object. The related variable is *GRIPBLOCK*, with a setting of 0=off. This is the default setting (disabled).

When the box is checked, *GRIP-BLOCK* is set to 1. All grips on all objects contained in the block are visible and functional (Fig. 21-15). This allows you to use any grip on any object within the block. This does not mean that the block is *Exploded*—the block retains its one-object integrity and individual entities in the block cannot be edited independently. This feature permits you to use the grips only on each of the block's components.

Figure 21-15

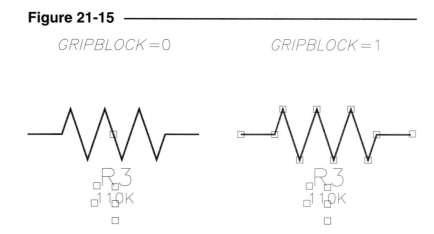

Notice in Figure 21-15 how the insertion point is not accessible when *GRIPBLOCK* is set to 1. Notice also that there are two grips on each of the *Normal* attributes and only one on the *Invisible* attribute and that these grips do not change with the two *GRIPBLOCK* settings. Making one of these grips hot allows you to "move" the attribute to any location and the attribute still retains its association with the block.

Grip Colors Unselected

This setting enables you to change the color of **warm** and **cold** grips. PICKing the *Unselected...* tile produces the *Select Color* dialog box (identical to that used with other color settings). The default setting is blue (ACI number 5). Alternately, the *GRIPCOLOR* variable can be typed and any ACI number from 0 to 255 can be entered.

Grip Colors Selected

The color of **hot** grips can be specified with this option. The *Selected...* tile produces the *Select Color* dialog box. You can also type *GRIPHOT* and enter any ACI number to make the change. The default color is red (1).

Grip Size

Use the slider bar to interactively increase or decrease the size of the grip "box." The *GRIPSIZE* variable could alternately be used. The default size is 3 pixels square (*GRIPSIZE=3*).

NOTE: All *Grip*-related variable settings are saved in the system registry rather than in the current drawing file. This generally means that changes in these variables remain with the computer, not the drawing being used.

Point Specification Precedence

When you select a point with the input device, AutoCAD uses a point specification precedence to determine which location on the drawing to find. The hierarchy is listed here:

1. Object Snap (*OSNAP*).
2. Explicit coordinate entry (absolute, relative, or polar coordinates).

3. *ORTHO.*
4. Point filters (XY filters).
5. *Grips* auxiliary grid (rectangular and circular).
6. *Grips* (on objects).
7. *SNAP* (F9 grid snap).
8. Digitizing a point location.

Practically, this means that *OSNAP* has priority over any other point selection mode. As far as *Grips* are concerned, *ORTHO* overrides *Grips*, so turn off *ORTHO* if you want to snap to *Grips*. Although *Grips* override *SNAP* (F9), it is suggested that *SNAP* be turned *Off* while using *Grips* to simplify PICKing.

More Grips

Because *Grips* have such a wide range of editing potential and because different AutoCAD objects react to *Grips* in different ways, discussion of grip editing is integrated into other chapters. For example, dimensions and surface models have special editing capabilities for *Grips*; therefore, those topics are covered in the related chapters.

CHAPTER EXERCISES

1. *Open* the **TEMP-D** drawing. Use the grips on the existing *Spline* to generate a new graph displaying the temperatures for the following week. **PICK** the *Spline* to activate **warm** grips. Use the **STRETCH** option to stretch the first data point grip to the value indicated below for Sunday's temperature. **Cancel** the grips; then repeat the steps for each data point on the graph. *Save* the drawing as **TEMP-F**.

Figure 21-16 ──────────────

X axis	Y axis
Sunday	29
Monday	32
Tuesday	40
Wednesday	33
Thursday	44
Friday	49
Saturday	45

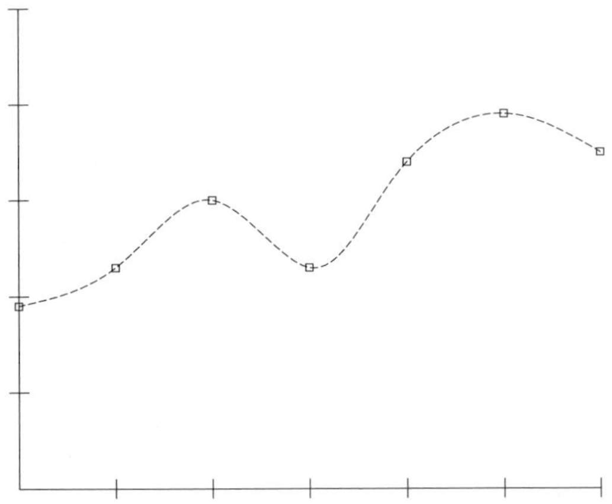

2. *Open* **CH16EX2** drawing. Activate **grips** on the *Line* to the far right. Make the top grip **hot**. Use the **STRETCH** option to stretch the top grip to the right to create a vertical *Line*. **Cancel** the grips.

    Next, activate the **grips** for all the *Lines* (including the vertical one on the right); then make the vertical *Line* grips **cold** as shown in Figure 21-17. **STRETCH** the top of all inclined *Lines* to the top of the vertical *Line* by making the common top grips **hot**, then stretching to the cold **grip** of the vertical *Line*. *Save* the drawing as **CH21EX2**.

Figure 21-17 ──────────────

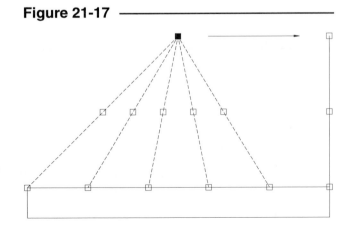

3. This exercise involves all the options of grip editing to create a Space Plate. Begin a *New* drawing or use the **ASHEET** *template* and use *SaveAs* to assign the name **SPACEPLT**.

Figure 21-18

A. Set the *Snap* value to **.125**.
Draw the geometry shown in Figure 21-18 using the *Line* and *Circle* commands.

B. Activate the **grips** on the *Circle*. Make the center grip **hot**. Cycle to the **MOVE** option. Enter *C* for the **Copy** option. You should then get the prompt for **MOVE (multiple)**. Make two copies as shown in Figure 21-19. The new *Circles* should be spaced evenly, with the one at the far right in the center of the rectangle. If the spacing is off, use **grips** with the **STRETCH** option to make the correction.

Figure 21-19

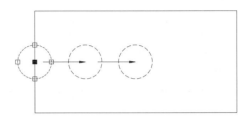

C. Activate the **grips** on the number 2 *Circle* (from the left) to make them **warm**. Also activate the **grips** on the bottom horizontal *Line*, but make them **cold**. Make the center *Circle* grip **hot** and cycle to the **MIRROR** option. Enter *C* for the **Copy** option. (You'll see the **MIRROR (multiple)** prompt.) Then enter *B* to specify a new base point as indicated in Figure 21-20. Turn *On ORTHO* and specify the mirror axis as shown to create the new *Circle* (shown in Fig. 21-20 in hidden linetype).

Figure 21-20

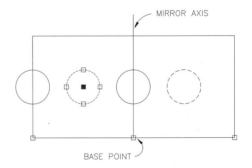

D. Use *Trim* to trim away the outer half of the *Circle* and the interior portion of the vertical *Line* on the left side of the Space Plate. Activate the **grips** on the two vertical *Lines* and the new *Arc*. Make the common grip **hot** (on the *Arc* and *Line* as shown in Fig. 21-21) and **STRETCH** it downward .5 units. (Note how you can affect multiple objects by selecting a common grip.) **Stretch** the upper end of the *Arc* upward .5 units.

Figure 21-21

E. *Erase* the *Line* on the right side of the Space Plate. Use the same method that you used in step C to **MIRROR** the *Lines* and *Arc* to the right side of the plate (as shown in Fig. 21-22).

(REMINDER: Make sure that you make the grips on the bottom *Line* **cold**. After you select the **hot** grip, use the **Copy** option and the **Base** point option. Use *ORTHO*.)

F. In this step, you will **STRETCH** the top edge

Figure 21-22

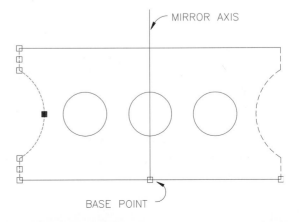

Figure 21-23

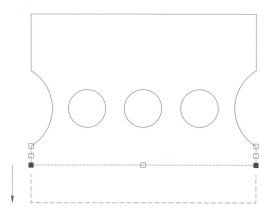

Figure 21-24

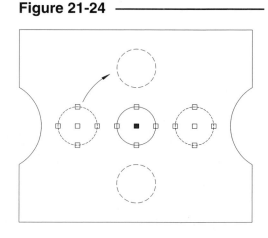

Figure 21-25

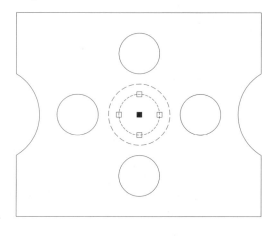

upward one unit and the bottom edge downward one unit by selecting multiple **hot** grips.

Select the desired horizontal and attached vertical *Lines*. Hold down Shift while selecting each of the endpoint grips, as shown in Figure 21-23. Although they appear **hot**, you must select one of the two again to activate the **STRETCH** option.

G. In this step, two more *Circles* are created by the **ROTATE** option (see Fig. 21-24). Select the three existing *Circles* to make the grips **warm**. Deselect the center *Circle* so that it will not be copied. PICK the center grip to make it **hot**. Cycle to the **ROTATE** option. Enter *C* for the **Copy** option. Make sure *ORTHO* is *On* and create the new *Circles*.

H. Select the center *Circle* and make the center grip **hot** (Fig. 21-25). Cycle to the **SCALE** option. The scale factor is **1.5**. Since the *Circle* is a 1 unit diameter, it can be interactively scaled (watch the *Coords* display), or you can enter the value. The drawing is complete. *Save* the drawing as **SPACEPLT**.

Figure 21-26

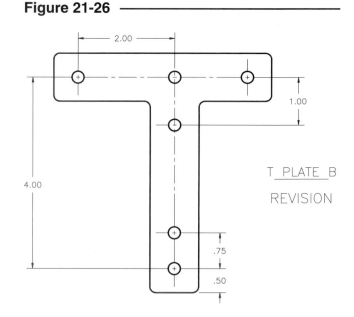

4. *Open* the **T-PLATE** drawing that you created in Chapter 10. An order has arrived for a modified version of the part. The new plate requires two new holes along the top, a 1" increase in the height, and a .5" increase from the vertical center to the hole on the left (Fig. 21-26). Use grips to **STRETCH** and **Copy** the necessary components of the existing part. Use *SaveAs* to rename the part to **TPLATEB**.

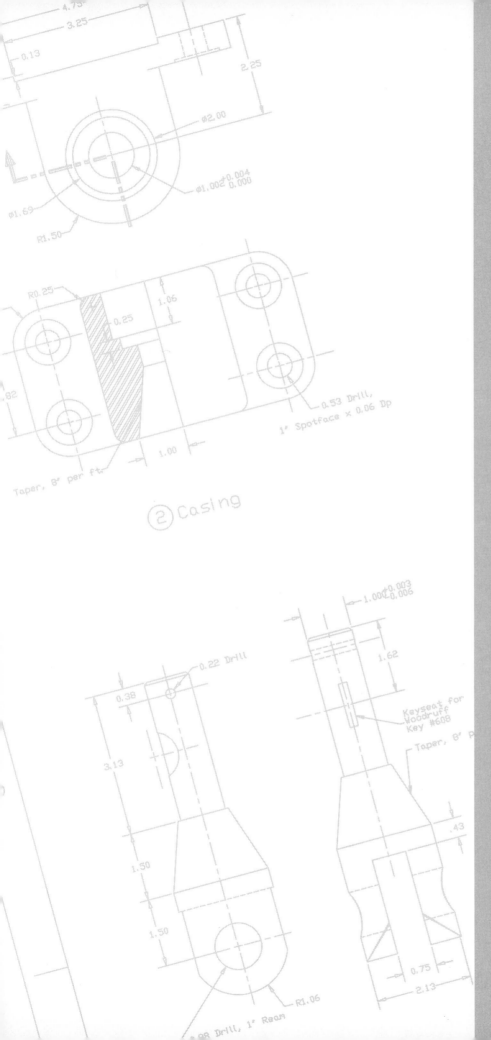

22

MULTIVIEW DRAWING

Chapter Objectives

After completing this chapter you should:

1. be able to draw projection lines using *ORTHO* and *SNAP*;

2. be able to use *Xline* and *Ray* to create construction lines;

3. know how to use *Offset* for construction of views;

4. be able to use point filters and *Tracking* for alignment of lines and views;

5. be able to use construction layers for managing construction lines and notes;

6. be able to use linetypes and layers to draw and manage ANSI standard linetypes;

7. know how to create fillets, rounds, and runouts;

8. know the typical guidelines for creating a three-view multiview drawing.

CONCEPTS

Multiview drawings are used to represent 3D objects on 2D media. The standards and conventions related to multiview drawings have been developed over years of using and optimizing a system of representing real objects on paper. Now that our technology has developed to a point that we can create 3D models, some of the methods we use to generate multiview drawings have changed, but the standards and conventions have been retained so that we can continue to have a universally understood method of communication.

This chapter illustrates methods of creating 2D multiview drawings with AutoCAD (without a 3D model) while complying with industry standards. Many techniques can be used to construct multiview drawings with AutoCAD because of its versatility. The methods shown in this chapter are the more common methods because they are derived from traditional manual techniques. Other methods are possible.

PROJECTION AND ALIGNMENT OF VIEWS

Projection theory and the conventions of multiview drawing dictate that the views be aligned with each other and oriented in a particular relationship. AutoCAD has particular features, such as *SNAP*, *ORTHO*, construction lines (*Xline, Ray*), Object Snap, *Tracking*, and point filters, that can be used effectively for facilitating projection and alignment of views.

Using *ORTHO* and *OSNAP* to Draw Projection Lines

ORTHO (F8) can be used effectively in concert with *OSNAP* to draw projection lines during construction of multiview drawings. For example, drawing a *Line* interactively with *ORTHO ON* forces the *Line* to be drawn in either a horizontal or vertical direction.

Figure 22-1 simulates this feature while drawing projection *Lines* from the top view over to a 45 degree miter line (intended for transfer of dimensions to the side view). The "From point:" of the *Line* originated from the *Endpoint* of the *Line* on the top view.

Figure 22-1 ———————————————

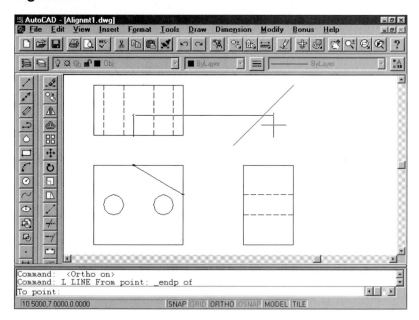

Figure 22-2 illustrates the next step. The vertical projection *Line* is drawn from the *Intersection* of the 45 degree line and the last projection line. *ORTHO* forces the *Line* to the correct vertical alignment with the side view.

Remember that any draw or edit command that requires PICKing is a candidate for *ORTHO* and/or *OSNAP*.

NOTE: *OSNAP* overrides *ORTHO*. If *ORTHO* is *ON* and you are using an *OSNAP* mode to PICK the "To point:" of a *Line*, the *OSNAP* mode has priority; and, therefore, the construction may not result in an orthogonal *Line*.

Figure 22-2

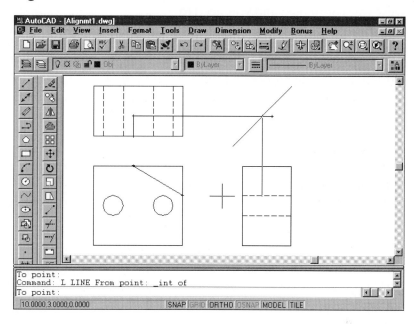

Using *Xline* and *Ray* for Construction Lines

Another strategy for constructing multiview drawings is to make use of the AutoCAD construction line commands *Xline* and *Ray*.

Xlines can be created to "box in" the views and ensure proper alignment. *Ray* is suited for creating the 45 degree miter line for projection between the top and side views. The advantage to using this method is that horizontal and vertical *Xlines* can be created quickly.

These lines can be *Trimmed* to become part of the finished view, or other lines could be drawn "on top of" the construction lines to create the final lines of the views. In either case, <u>construction lines should be drawn on a separate layer</u> so that the layer can be frozen before plotting. If you intend to *Trim* the construction lines so that they become part of the final geometry, draw them originally on the view layers.

Figure 22-3

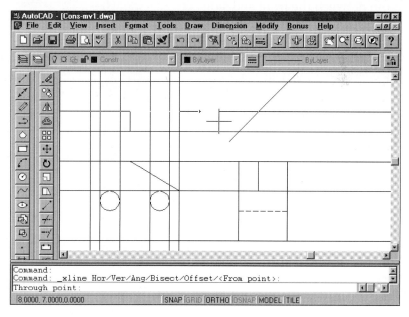

Using *Offset* for Construction of Views

An alternative to using the traditional miter line method for construction of a view by projection, the *Offset* command can be used to transfer distances from one view and to construct another. The *Distance* option of *Offset* provides this alternative.

For example, assume that the side view was completed and you need to construct a top view (Fig. 22-4). First, create a horizontal line as the inner edge of the top view (shown highlighted) by *Offset* or other method. To create the outer edge of the top view (shown in phantom linetype), use *Offset* and PICK points (1) and (2) to specify the *distance*. Select the existing line (3) as the "*Object to Offset:*", then PICK the "*Side to offset?*" at the current cursor position.

Figure 22-4

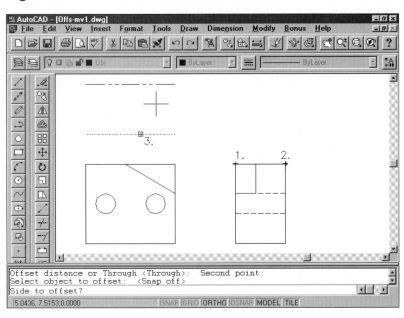

Realignment of Views Using *ORTHO* and *SNAP*

Another application of *ORTHO* and *OSNAP* is the use of *Move* to change the location of an entire view.

For example, assume that the views of the multiview in Figure 22-5 are complete and ready for dimensioning; however, there is not enough room between the front and side views. You can invoke the *Move* command, select the entire view with a *window* or other option, and "slide" the entire view outward. *ORTHO* assures the proper alignment. Make sure *SNAP* is *ON* during the *Move* (if *SNAP* was used in the original construction). This action assures that the final position of the view is on a *SNAP* point and that the objects retain their orientation with respect to the *SNAP*.

An alternative to *Mov*ing the view, interactively is use of coordinate specification (absolute, relative rectangular, or relative polar or direct distance entry). (See Chapter 9, Fig. 9-5, for an example of moving a view with direct distance entry.)

Using Point Filters for Alignment

If you want to draw a *Line* so that it ends or begins in alignment with another object or view, X or Y point filters can be used. For example, Figure 22-6

Figure 22-5

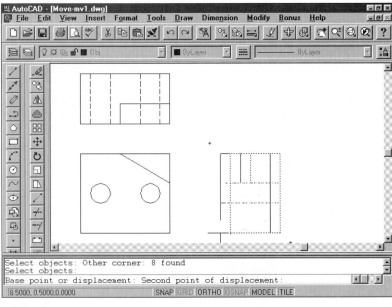

Figure 22-6

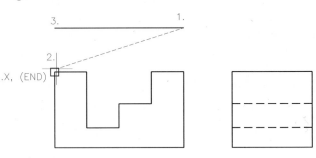

shows the construction of a *Line* in the top view. It is desired to end the *Line* in alignment with the X position directly above the left side of the front view. An *.X* point filter is used to accomplish this action. <u>*ORTHO* is *On*</u>. The following command syntax is used for the construction:

Figure 22-7

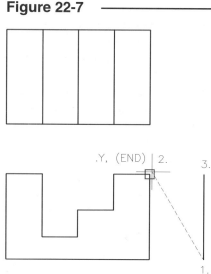

> Command: **line**
> From point: **PICK** (1.)
> To point: **.x** (Type ".X"; then press **Enter**.)
> **of Endpoint of PICK** (2.) (AutoCAD uses only the X value of the point selected.)
> (need YZ): **PICK** (anywhere near 3. AutoCAD uses the Y and Z points selected.)
> To point: **Enter**
> Command:

Another possibility is to use a .Y point filter for construction in alignment with the Y value (height) of existing objects (Fig. 22-7). In this case, enter a **.Y** in response to the "To point:" prompt and press Enter. The point you PICK (2) supplies only the Y value. AutoCAD responds with "(need XZ)". PICK again near (3) to supply the X value. Make sure *ORTHO* is *On*.

Using *Tracking* for Alignment

Tracking can also be used for drawing or moving objects into alignment with other objects in much the same way as point filters are used. *Tracking*, a new Release 14 feature, is a method of specifying points that align orthogonally to other points. *Tracking* is used during a command similar to the way *OSNAPs* are used to locate points on objects. For example, when a command prompts you to PICK a point or enter coordinate values, you can use *Tracking* to find a point <u>orthogonal</u> to a *Midpoint, Endpoint,* or other point. See *Tracking* in Chapter 15.

As an example, you can use *Tracking* to move a misplaced view into alignment with another view (Fig. 22-8). In this example, *Tracking* is used in conjunction with *OSNAPs* during the *Move* command to align the top view orthogonally with the front view. The command sequence is as follows:

Figure 22-8

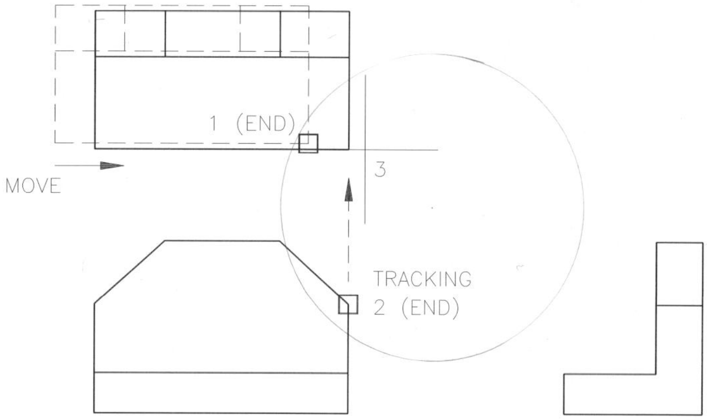

> Command: **move**
> Select objects: **PICK** (top view objects)
> Select objects: **PICK** (top view objects)
> Select objects: **Enter**
> Base point or displacement: **endp of**
> **PICK** (P1)
> Second point of displacement: **tracking**
> First tracking point: **endp of PICK** (P2)
> Next point (Press ENTER to end tracking): **PICK** (P3)
> Next point (Press ENTER to end tracking): **Enter**
> Command:

Using *Tracking* before specifying the second point (P2) ensures that the <u>next point</u> will be aligned with the *Endpoint* of the front view. The next point (P3) can be PICKed anywhere above (P2) and it will remain in vertical alignment with (P2).

The use of layers for isolating construction lines can make drawing and editing faster and easier. The creation of multiview drawings can involve construction lines, reference points, or notes that are not intended for the final plot. Rather than *Erasing* these construction lines, points, or notes before plotting, they can be created on a separate layer and turned *Off* or made *Frozen* before running the final plot. If design changes are required, as they often are, the construction layers can be turned *On*, rather than having to recreate the construction.

There are two strategies for creating construction objects on separate layers:

1.   Use Layer 0 for construction lines, reference points, and notes. This method can be used for fast, simple drawings.

2.   Create a new layer for construction lines, reference points, and notes. Use this method for more complex drawings or drawings involving use of *Blocks* on Layer 0.

Figure 22-9

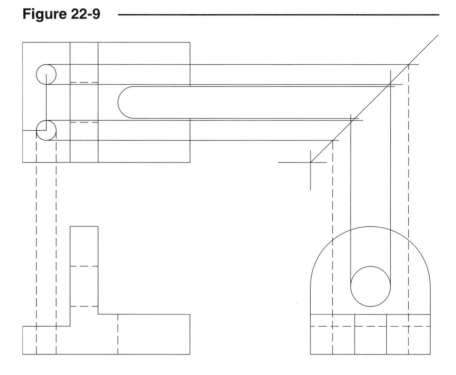

For example, consider the drawing during construction in Figure 22-9. A separate layer has been created for the construction lines, notes, and reference points.

In Figure 22-10, the same drawing is shown ready for making the final plot. Notice that the construction Layer has been *Frozen*.

Figure 22-10

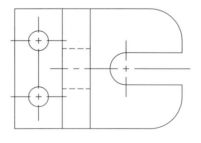

If you are plotting the *Limits*, and the construction layer has objects <u>outside</u> the *Limits*, the construction layer should be *Frozen*, rather than being turned *Off*, unless only *Xlines* and *Rays* exist on the layer.

USING LINETYPES

Different types of lines are used to represent different features of

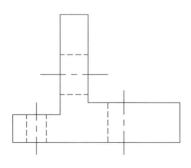

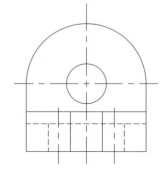

USING LINETYPES

Different types of lines are used to represent different features of a multiview drawing. Linetypes in AutoCAD are accessed by the *Linetype* command or through the *Linetype* tab of the *Layer & Linetype Properties* dialog box. *Linetypes* can be changed retroactively by the *Change Properties* or the *Modify* dialog boxes. *Linetypes* can be assigned to individual objects specifically or to layers (*BYLAYER*). See Chapter 12 for a full discussion on this topic.

AutoCAD complies with the ANSI and ISO standards for linetypes. The principal AutoCAD linetypes used in multiview drawings and the associated names are shown in Figure 22-11.

Many other linetypes are provided in AutoCAD. Refer to Chapter 12 for the full list and illustration of the linetypes.

Other standard lines are created by AutoCAD automatically. For example, dimension lines can be automatically drawn when using dimensioning commands (Chapter 26), and section lines can be automatically drawn when using the *Hatch* command (Chapter 24).

Figure 22-11

CONTINUOUS

HIDDEN

CENTER

PHANTOM

DASHED

AutoCAD linetypes do not have a specified thickness, but this can be accomplished by either creating *Plines* and specifying *Width* or by plotting with specific width pens. (See Chapters 15 and 16 for information on *Plines* and Chapter 14 for information on plotting.)

Drawing Hidden and Center Lines

Figure 22-12 illustrates a typical application of AutoCAD *Hidden* and *Center* linetypes. Notice that the horizontal center line in the front view does not automatically locate the short dashes correctly, and the hidden lines in the right side view incorrectly intersect the center vertical line.

Although AutoCAD supplies ANSI standard linetypes, the application of those linetypes does not always follow ANSI standards. For example, you do <u>not</u> have control over the

Figure 22-12

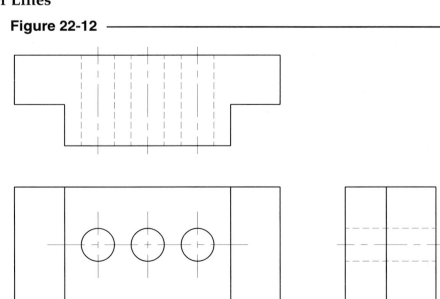

placement of the individual dashes of center lines and hidden lines. (You have control of only the endpoints of the lines and the *Ltscale*.) Therefore, the short dashes of center lines may not cross exactly at the circle centers, or the dashes of hidden lines may not always intersect as desired.

You do, however, have control of the endpoints of the lines. Draw lines with the *Center* linetype such that the endpoints are symmetric about the circle or group of circles. This action assures that the short dash occurs at the center of the circle (if an odd number of dashes are generated). Figure 22-13 illustrates correct and incorrect technique.

Figure 22-13

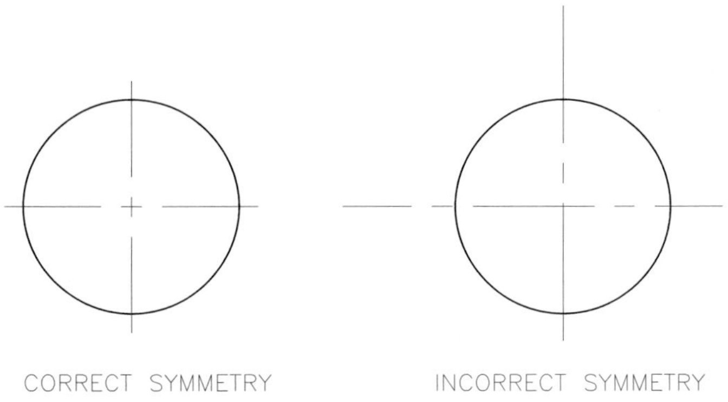

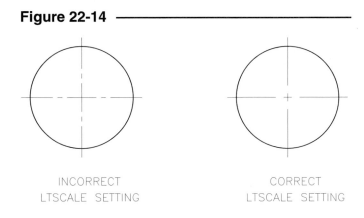

CORRECT SYMMETRY   INCORRECT SYMMETRY

You can also control the relative size of non-continuous linetypes with the *Ltscale* variable. In some cases, the variable can be adjusted to achieve the desired results.

For example, Figure 22-14 demonstrates the use of *Ltscale* to adjust the center line dashes to the correct spacing. Remember that *Ltscale* adjusts linetypes globally (all linetypes across the drawing).

Figure 22-14

INCORRECT   CORRECT
LTSCALE SETTING   LTSCALE SETTING

When the *Ltscale* has been optimally adjusted for the drawing globally, use the *Ddmodify* dialog box to adjust the linetype scale of individual objects. In this way, the drawing lines can originally be created to the global linetype scale without regard to the *Celtscale*. The finished drawing linetype scale can be adjusted with *Ltscale* globally; then only those objects that need further adjusting can be fine-tuned retroactively with *Ddmodify* or *Ddchprop*.

The *Center* command (a dimensioning command) can be used to draw center lines automatically with correct symmetry and spacing (see Chapter 26, Dimensioning).

Managing Linetypes and Colors

There are two strategies for assigning linetypes and colors: *BYLAYER* and object-specific assignment. In either case, thoughtful layer utilization for linetypes will make your drawings more flexible and efficient.

BYLAYER Linetypes and Colors

The *BYLAYER* linetype and color settings are recommended when you want the most control over linetype visibility and plotting. This is accomplished by creating layers with the *Layer Control* dialog box and using *Set Color* and *Set Ltype* for each layer. After you assign *Linetypes* to specific layers, you simply set the layer (with the desired linetype) as the *Current* layer and draw on that layer in order to draw objects in a specific linetype.

A template drawing for creating typical multiview drawings could be set up with the layer and linetype assignments similar to that shown in Figure 22-15. Notice the layer names, associated colors, and assigned *Ltypes*.

Figure 22-15 ────────────────────

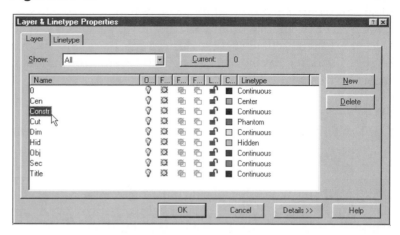

Using this strategy (*BYLAYER Linetype* and *Color* assignment) gives you flexibility. You can control the linetype visibility by controlling the layer visibility. You can control the associated plotting pen for each linetype (in the *Plot Control* dialog box) if each linetype is assigned a specific color. You can also <u>retroactively</u> change the linetype and color of an existing object by changing the object's *Layer* property with the *Change Properties* or the *Modify* dialog box. Objects changed to a different layer assume the *color* and *linetype* of the new layer.

Another strategy for multiview drawings involving several parts, such as an assembly, is to create layers for each linetype <u>specific to each part</u>, as shown in Figure 22-16. With this strategy, each part has the complete set of linetypes, but only one color per part in order to distinguish the part from others in the display.

Figure 22-16 ────────────────────

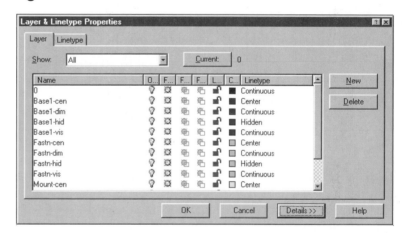

Related layer <u>groups</u> can be selected within the dialog box using the *Filters* dialog box. If the *Layer* command is used instead, wildcards can be typed for layer selection. For example, entering "?????-HID" selects all of the layers with hidden lines, or entering "MOUNT*" would select all of the layers associated with the "MOUNT" part.

Object *Linetypes and Colors*
Although this method can be complex, object-specific linetype and color assignment can also be managed easier by skillful utilization of layers. One method is to create one layer for each part or each group of related geometry (Fig. 22-17). The *BYLAYER* color and linetype settings should be left to the defaults. Then object-specific linetype settings can be assigned using the *Linetype* command or through the *Linetype* tab of the *Layer & Linetype Properties* dialog box, and

Figure 22-17 ────────────────────

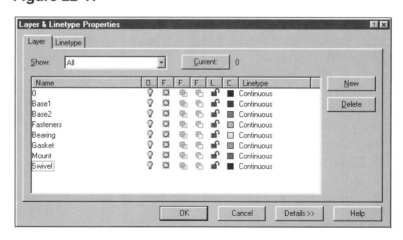

color settings can be assigned using the *Color* command or *Ddcolor* dialog box. (Colors provide the means for controlling plotted line weights). For assemblies, all lines related to one part would be drawn on one layer. Remember that you can draw everything with one linetype and color setting, then use *Ddchprop* to retroactively set the desired color and linetype for each set of objects. Visibility of parts can be controlled by *Freezing* or *Thawing* part layers. You cannot isolate and control visibility of linetypes or colors by this method.

CREATING FILLETS, ROUNDS, AND RUNOUTS

Many mechanical metal or plastic parts manufactured from a molding process have slightly rounded corners. The otherwise sharp corners are rounded because of limitations in the molding process or for safety. A convex corner is called a round and a concave corner is called a fillet. These fillets and rounds are created easily in AutoCAD by using the *Fillet* command.

The example in Figure 22-18 shows a multiview drawing of a part with sharp corners before the fillets and rounds are drawn.

Figure 22-18 ――――――――

The corners are rounded using the *Fillet* command. First, use *Fillet* to specify the *Radius*. Once the *Radius* is specified, just select the desired lines to *Fillet* near the end to round.

If the *Fillet* is in the middle portion of a *Line* instead of the end, *Extend* can be used to reconnect the part of the *Line* automatically trimmed by *Fillet*, or *Fillet* can be used in the *Notrim* mode.

The finish marks (V-shaped symbols) indicate machined surfaces. Finished surfaces have sharp corners.

Figure 22-19 ――――――――

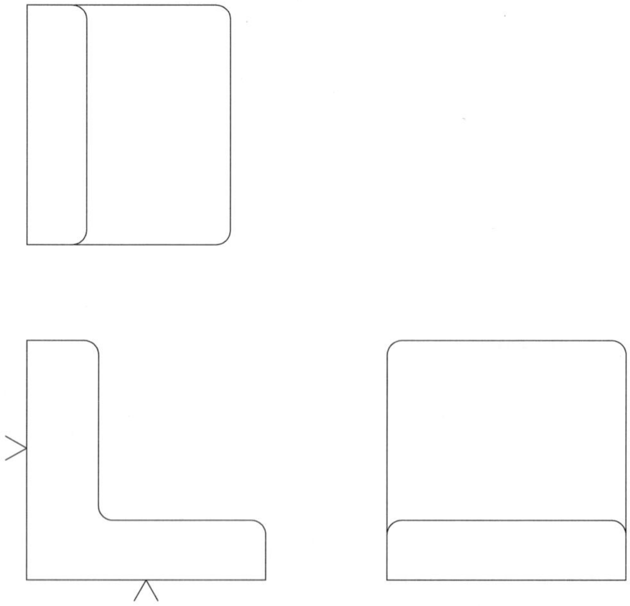

although other options can be used. Alternately, the *Circle TTR* option can be used with *Trim* to achieve the desired effect. As a general rule, use the same radius or slightly larger than that given for the fillets and rounds, but draw it less than 90 degrees.

Figure 22-20

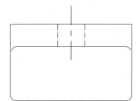

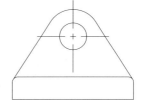

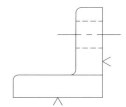

GUIDELINES FOR CREATING A TYPICAL THREE-VIEW DRAWING

Following are some guidelines for creating the three-view drawing in Figure 22-21. This object is used only as an example. The steps or particular construction procedure may vary, depending on the specific object drawn. Dimensions are shown in the figure so you can create the multiview drawing as an exercise.

1. Drawing Setup

    Units are set to *Decimal* with 3 places of *Precision*. *Limits* of .22 x 17 are set to allow enough drawing space for both views. The finished drawing can be plotted on a B size sheet at a scale of 1"=1" or on an A size sheet at a scale of 1/2"=1". *Snap* is set to an increment of .125. *Grid* is set to

Figure 22-21

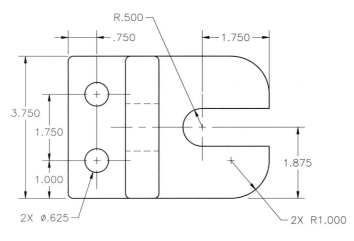

FILLETS AND ROUNDS .125

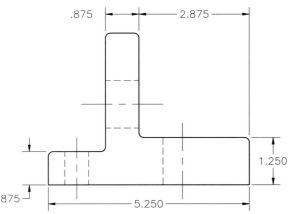

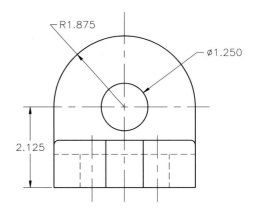

an increment of .5. *Ltscale* is not changed from the default of 1. Layers are created (OBJ, HID, CEN, DIM, BORDER, and CONSTR) with appropriate *Ltypes* and *Colors* assigned.

2.   An outline of each view is "blocked in" by drawing the appropriate *Lines* and *Circles* on the OBJ layer similar to that shown in Figure 22-22. Ensure that *SNAP* is *ON*. *ORTHO* should be turned *ON* when appropriate. Use the cursor to ensure that the views align horizontally and vertically. Note that the top edge of the front view was determined by projecting from the *Circle* in the right side view.

Figure 22-22 ——————————

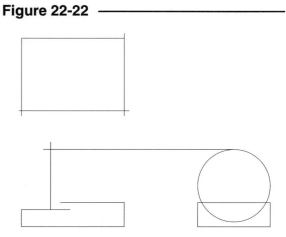

Another method for construction of a multiview drawing is shown in Figure 22-23. This method uses the *Xline* and *Ray* commands to create construction and projection lines. Here all the views are blocked in and some of the object lines have been formed. The construction lines should be kept on a separate layer, except in the case where *Xlines* and *Rays* can be trimmed and converted to

the object lines. (The following illustrations do not display this method because of the difficulty in seeing which are object and construction lines.)

Figure 22-23 ——————————

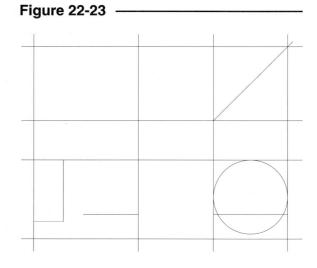

3.   This drawing requires some projection between the top and side views (Fig. 22-24). The CONSTR layer is set as *Current*. Two *Lines* are drawn from the inside edges of the two views (using *OSNAP* and *ORTHO* for alignment). A 45 degree miter line is constructed for the projection lines to "make the turn." A *Ray* is suited for this purpose.

4.   Details are added to the front and top views (Fig. 22-25). The projection line from the side view to the front view (previous figure) is *Trimmed*. A *Circle* representing the hole is drawn in the side view and projected up and over to the top view and to the front view. The object lines are drawn on layer OBJ and some projection lines are drawn on layer CONSTR. The horizontal projection lines from the 45 degree miter line

Figure 22-24 ——————————

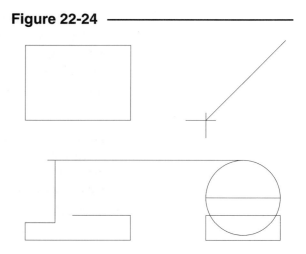

4. Details are added to the front and top views
    (Fig. 22-25). The projection line from the side view to
    the front view (previous figure) is *Trimmed*. A *Circle*
    representing the hole is drawn in the side view and
    projected up and over to the top view and to the front
    view. The object lines are drawn on layer OBJ and
    some projection lines are drawn on layer CONSTR.
    The horizontal projection lines from the 45 degree
    miter line are drawn on layer HID awaiting *Trimming*.
    Alternately, those two projection lines could be drawn
    on layer CONSTR and changed to the appropriate
    layer with *Change Properties* after *Trimming*.

Figure 22-25 ───────────

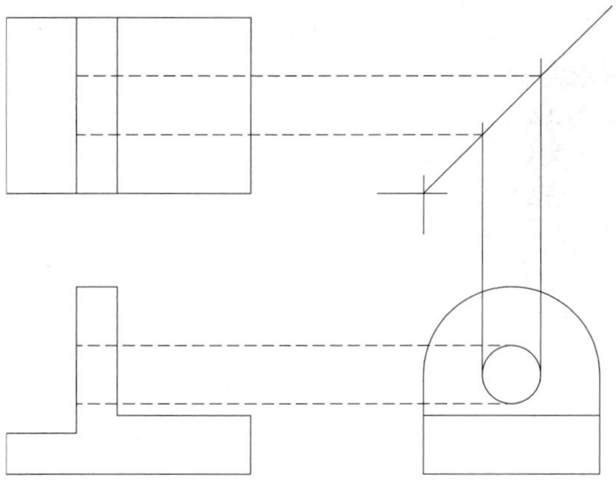

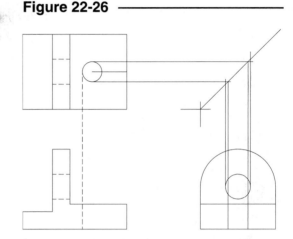

5. The hidden lines used for projection to the top view
    and front view (previous figure) are *Trimmed*. The slot
    is created in the top view with a *Circle* and projected to
    the side and front views. It is usually faster and easier
    to draw round object features in their circular view
    first, then project to the other views. Make sure you
    use the correct layers (OBJ, CONSTR, HID) for the
    appropriate features. If you do not, *Ddmodify* can be
    used retroactively.

Figure 22-26 ───────────

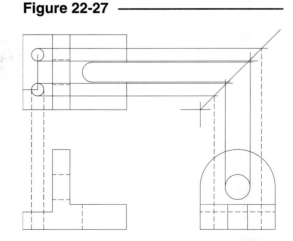

6. The lines shown in the previous figure as projection
    lines or construction lines for the slot are *Trimmed*
    or *Erased*. The holes in the top view are drawn on
    layer OBJ and projected to the other views. The
    projection lines and hidden lines are drawn on
    their respective layers.

Figure 22-27 ───────────

7. *Trim* the appropriate hidden lines. *Freeze* layer CONSTR. On layer OBJ, use *Fillet* to create the rounded corners in the top view. Draw the correct center lines for the holes on layer CEN. The value for *Ltscale* should be adjusted to achieve the optimum center line spacing. *Ddmodify* can be used to adjust individual object linetype scale.

Figure 22-28 ————

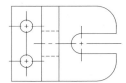

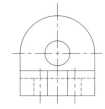

8. Fillets and rounds are added using the *Fillet* command. The runouts are created by drawing a *3point Arc* and *Trimming* or *Extending* the *Line* ends as necessary. Use *Zoom* for this detail work.

Figure 22-29 ————

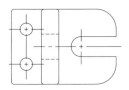
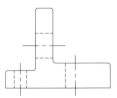
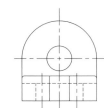

9. Add a border and a title block using *Pline*. Include the part name, company, draftsperson, scale, date, and drawing file name in the title block. The drawing is ready for dimensioning and man-ufacturing notes.

Figure 22-30 ————

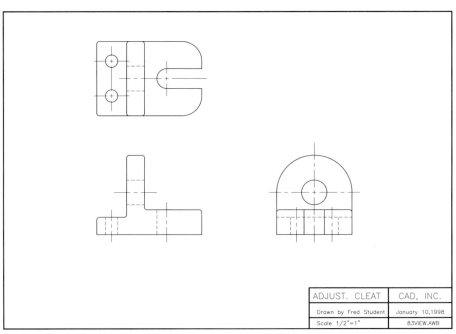

ADJUST. CLEAT	CAD, INC.
Drawn by Fred Student	January 10,1998
Scale 1/2"=1"	83VIEW.AWB

CHAPTER EXERCISES

Figure 22-31 ———————————

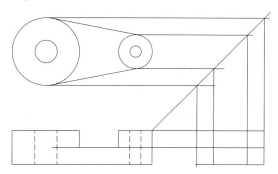

1. *Open* the **PIVOTARM CH16** drawing.

   A. Create the right side view. Use *OSNAP* and *ORTHO* to create *Lines* or *Rays* to the miter line and down to the right side view as shown in Figure 22-31. *Offset* may be used effectively for this purpose instead. Use *Extend, Offset,* or *Ray* to create the projection lines from the front view to the right side view.

   B. *Trim* or *Erase* the unwanted projection lines, as shown in Figure 22-32. Draw a *Line* or *Ray* from the *Endpoint* of the diagonal *Line* in the top view down to the front to supply the boundary edge for *Trimming* the horizontal *Line* in the front view as shown.

Figure 22-32 ———————————

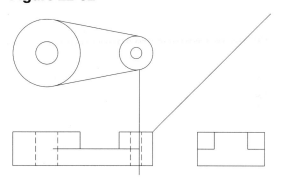

   C. Next, create the hidden lines for the holes by the same fashion as before. Use previously created *Layers* to achieve the desired *Linetypes*. Complete the side view by adding the horizontal hidden *Line* in the center of the view.

Figure 22-33 ———————————

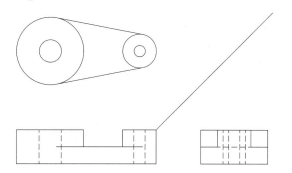

   D. Another hole for a set screw must be added to the small end of the Pivot Arm. Construct a *Circle* of 4mm diameter with its center located 8mm from the top edge in the side view as shown in Figure 22-34. Project the set screw hole to the other views.

Figure 22-34 ———————————

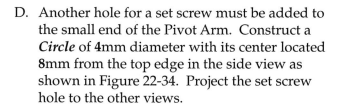

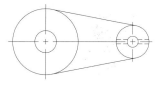

E.   Make new layers **CONSTR, OBJ,** and **TITLE** and change objects to the appropriate layers with *Change Properties*. *Freeze* layer **CONSTR**. Add centerlines on the **CEN** layer as shown in Figure 22-34. Change the *Ltscale* to **18**. To complete the PIVOTARM drawing, draw a *Pline* border (*width* **.02** x scale factor) and *Insert* the **TBLOCK** drawing that you created in Chapter 20 Exercises. *SaveAs* **PIVOTARM CH22**.

For exercises 2 through 5, construct and plot the multiview drawings as instructed. Use an appropriate *template* drawing for each exercise unless instructed otherwise. Use conventional practices for *layers* and *linetypes*. Draw a *Pline* border with the correct *width* and *Insert* your **TBLOCK** drawing.

Figure 22-35 ———————————————————— **Figure 22-36** ————————————————

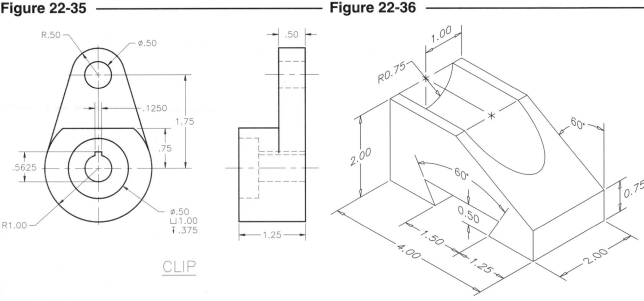

CLIP

2.   Make a two-view multiview drawing of the Clip (Fig. 22-35). *Plot* the drawing full size (**1=1**). Use the **ASHEET** template drawing to achieve the desired plot scale. *Save* the drawing as **CLIP**.

3.   Make a three-view multiview drawing of the Bar Guide (Fig. 22-36) and *Plot* it **1=1**. Use the **BAR-GUIDE** drawing you set up in Chapter 13 Exercises. Note that a partial *Ellipse* will appear in one of the views.

4.   Construct a multiview drawing of the Saddle (Fig. 22-37). Three views are needed. The channel along the bottom of the part intersects with the saddle on top to create a slotted hole visible in the top view. *Plot* the drawing at **1=1**. *Save* as **SADDLE**.

Figure 22-37 ————————————————————

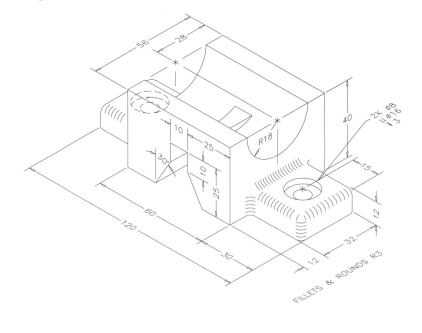

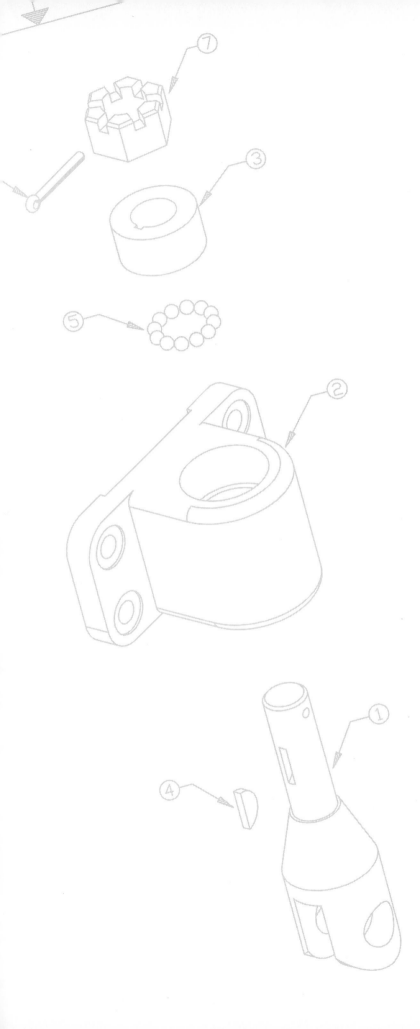

23

PICTORIAL DRAWINGS

Chapter Objectives

After completing this chapter you should be able to:

1. activate the *Isometric Style* of *Snap* for creating isometric drawings;

2. draw on the three isometric planes by toggling *Isoplane* using Ctrl+E;

3. create isometric ellipses with the *Iso-circle* option of *Ellipse*;

4. construct an isometric drawing in AutoCAD;

5. create Oblique Cavalier and Cabinet drawings in AutoCAD.

CONCEPTS

Isometric drawings and oblique drawings are pictorial drawings. Pictorial drawings show three principal faces of the object in one view. A pictorial drawing is a drawing of a 3D object as if you were positioned to see (typically) some of the front, some of the top, and some of the side of the object. All three dimensions of the object (width, height, and depth) are visible in a pictorial drawing.

Multiview drawings differ from pictorial drawings because a multiview only shows two dimensions in each view, so two or more views are needed to see all three dimensions of the object. A pictorial drawing shows all dimensions in the one view. Pictorial drawings depict the object similar to the way you are accustomed to viewing objects in everyday life, that is, seeing all three dimensions. Figure 23-1 and Figure 23-2 show the same object in multiview and in pictorial representation, respectively. Notice that multiview drawings use hidden lines to indicate features that are normally obstructed from view, whereas <u>hidden lines are normally omitted</u> in isometric drawings (unless certain hidden features must be indicated for a particular function or purpose).

Types of Pictorial Drawings

Pictorial drawings are classified as follows:

1. Axonometric drawings
   a. Isometric drawings
   b. Dimetric drawings
   c. Trimetric drawings

2. Oblique drawings

Axonometric drawings are characterized by how the angle of the edges or axes (axon-) are measured (-metric) with respect to each other.

Isometric drawings are drawn so that each of the axes have equal angular measurement. ("Isometric" means equal measurement.) The isometric axes are always drawn at 120 degree increments (Fig. 23-3). All rectilinear lines on the object (representing horizontal and vertical edges—not inclined or oblique) are drawn on the isometric axes.

A 3D object seen "in isometric" is thought of as being oriented so that each of three perpendicular faces (such as the top, front, and side) are seen equally. In other words, the angles formed between the line of sight and each of the principal faces are equal.

Figure 23-1

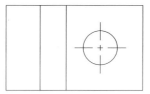

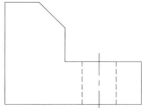

Figure 23-2

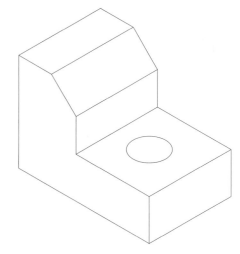

Figure 23-3

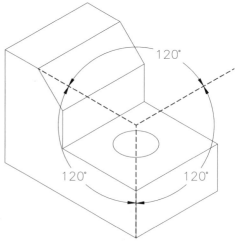

Dimetric drawings are constructed so that the angle between any two of the three axes is equal. There are many possibilities for dimetric axes. A common orientation for dimetric drawings is shown in Figure 23-4. For 3D objects seen from a dimetric viewpoint, the angles formed between the line of sight and each of two principal faces are equal.

Trimetric drawings have three unequal angles between the axes. Numerous possibilities exist. A common orientation for trimetric drawings is shown in Figure 23-5.

Oblique drawings are characterized by a vertical axis and horizontal axis for the two dimensions of the front face and a third (receding) axis of either 30, 45, or 60 degrees (Fig. 23-6). Oblique drawings depict the true size and shape of the front face, but add the depth to what would otherwise be a typical 2D view. This technique simplifies construction of drawings for objects that have contours in the profile view (front face) but relatively few features along the depth. Viewing a 3D object from an oblique viewpoint is not possible.

This chapter will explain the construction of isometric and oblique drawings in AutoCAD.

Pictorial Drawings Are 2D Drawings

Isometric, dimetric, trimetric, and oblique drawings are 2D drawings, whether created with AutoCAD or otherwise. Pictorial drawing was invented before the existence of CAD and therefore was intended to simulate a 3D object on a 2D plane (the plane of the paper). If AutoCAD is used to create the pictorial, the geometry lies on a 2D plane—the XY plane. All coordinates defining objects have X and Y values with a Z value of 0. When the *Isometric* style of *Snap* is activated, an isometrically structured *SNAP* and *GRID* appear on the XY plane.

Figure 23-7 illustrates the 2D nature of an isometric drawing created in AutoCAD. Isometric lines are created on the XY plane. The *Isometric SNAP* and *GRID* are also on the 2D plane. (The *Vpoint* command was used to give other than a *Plan* view of the drawing in this figure.)

Although pictorial drawings are based on the theory of projecting 3D objects onto 2D planes, it is physically possible to achieve an axonometric (isometric, dimetric, or trimetric) viewpoint of a 3D object using a 3D CAD system. In AutoCAD, the *Vpoint* command is used to specify the observer's position in 3D space with respect to a 3D model. Chapter 29 discusses the specific commands and values needed to attain axonometric viewpoints of a 3D model.

Figure 23-4

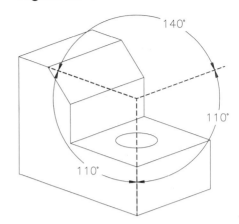

Figure 23-5

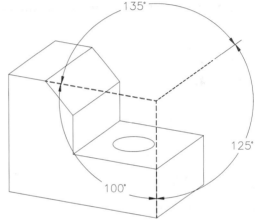

Figure 23-6

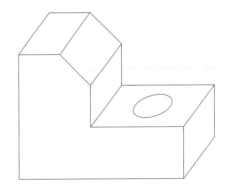

Figure 23-7

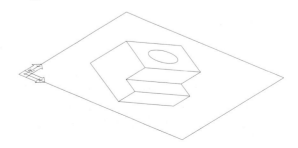

ISOMETRIC DRAWING IN AutoCAD

AutoCAD provides the capability to construct isometric drawings. An isometric *SNAP* and *GRID* are available, as well as a utility for creation of isometrically correct ellipses. Isometric lines are created with the *Line* command. There are no special options of *Line* for isometric drawing, but isometric *SNAP* and *GRID* can be used to force *Lines* to an isometric orientation. Begin creating an isometric drawing in AutoCAD by activating the *Isometric Style* option of the *Snap* command. This action can be done using any of the options listed in the following command table.

SNAP

Pull-down Menu	COMMAND (TYPE)	ALIAS (TYPE)	Short-cut	Screen (side) Menu	Tablet Menu
Tools *Drawing Aids...*	*SNAP*	*SN*	*F9 or* *Ctrl+B*	*TOOLS 2* *Ddrmodes*	*W,10*

```
Command: snap
Snap spacing or ON/OFF/Aspect/Rotate/Style <1.0000>: s
Standard/Isometric <S>: I
Vertical spacing <1.0000>: Enter
Command:
```

Alternately, toggling the indicated checkbox in the lower-right corner of the *Drawing Aids* dialog box activates the *Isometric SNAP* and *GRID* (Fig. 23-8).

Figure 23-9 illustrates the effect of setting the *Isometric SNAP* and *GRID*. Notice the new position of the cursor.

Using Ctrl+E (pressing the Ctrl key and the letter "E" simultaneously) toggles the cursor to one of three possible *Isoplanes* (AutoCAD's term for the three faces of the isometric pictorial). If *ORTHO* is *ON*, only isometric lines are drawn; that is, you can only draw *Lines* aligned with the isometric axes. *Lines* can be drawn on only two axes for each isoplane. Ctrl+E allows drawing on the two axes aligned with another face of the object. *ORTHO* is *OFF* in order to draw inclined or oblique lines (not on the isometric axis). The functions of *GRID* (F7) and *SNAP* (F9) remain unchanged.

With *SNAP ON*, toggle *Coords* (F6) several times and examine the read-out as you move the cursor. The absolute coordinate format is of no particular assistance while drawing

Figure 23-8

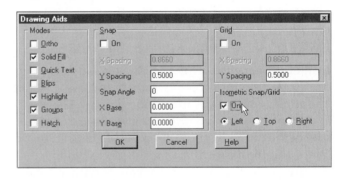

Figure 23-9

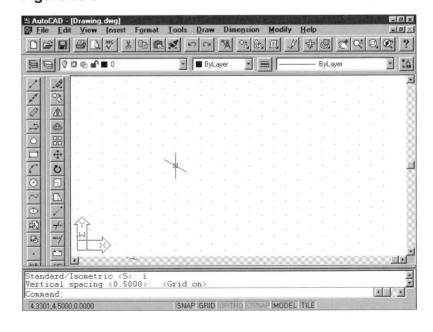

in isometric because of the configuration of the *GRID*. The <u>relative</u> <u>polar</u> format, however, is <u>very</u> <u>helpful</u>. Use relative polar format for *Coords* while drawing in isometric (see Fig. 23-10).

The effects of changing the *Isoplane* are shown in the following figures. Press Ctrl+E to change *Isoplane*.

With *ORTHO ON*, drawing a *Line* is limited to the two axes of the current *Isoplane*. Only one side of a cube, for example, can be drawn on the current *Isoplane*. Watch *Coords* (in a polar format) to give the length of the current *Line* as you draw (lower-left corner of the screen).

Toggling Ctrl+E switches the cursor and the effect of *ORTHO* to another *Isoplane*. One other side of a cube can be constructed on this *Isoplane* (Fig. 23-11).

Toggling Ctrl+E again forces the cursor and the effect of *ORTHO* to the third *Isoplane*. The third side of the cube can be constructed similarly.

Figure 23-10

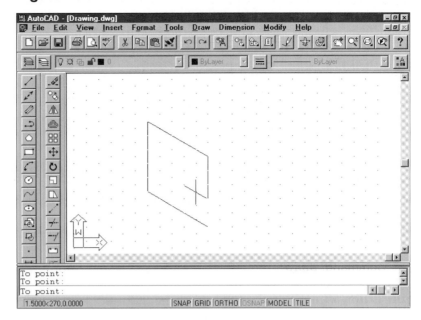

Figure 23-11

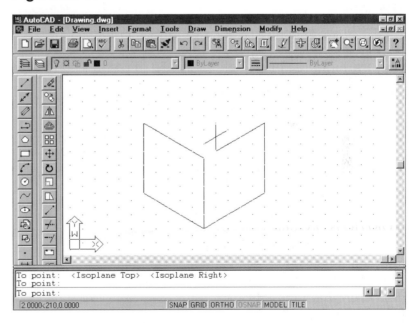

Isometric Ellipses

Isometric ellipses are easily drawn in AutoCAD by using the *Isocircle* option of the *Ellipse* command. This option appears <u>only</u> when the isometric *SNAP* is *ON*.

ELLIPSE

Pull-down Menu	COMMAND (TYPE)	ALIAS (TYPE)	Short-cut	Screen (side) Menu	Tablet Menu
Draw *Ellipse*	*ELLIPSE*	*EL*	...	*DRAW 1* *Ellipse*	*M,9*

Although the *Isocircle* option does not appear in the pull-down or digitizing tablet menus, it can be invoked as an option of the *Ellipse* command. You <u>must type "I"</u> to use the *Isocircle* option. The command syntax is as follows:

Command: *ellipse*
Arc/Center/Isocircle/<Axis endpoint 1>: *I*
Center of circle: **PICK** or (**coordinates**)
<Circle radius>/Diameter: **PICK** or (**coordinates**)
Command:

After selecting the center point of the *Isocircle*, the isometrically correct ellipse appears on the screen on the current *Isoplane*. Use Ctrl+E to toggle the ellipse to the correct orientation. When defining the radius interactively, use *ORTHO* to force the rubberband line to an isometric axis (Fig. 23-12).

Since isometric angles are equal, all isometric ellipses have the same proportion (major to minor axis). The only differences in isometric ellipses are the size and the orientation (*Isoplane*).

Figure 23-12

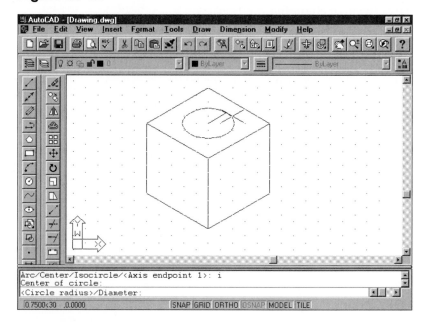

Figure 23-13 shows three ellipses correctly oriented on their respective faces. Use Ctrl+E to toggle the correct *Isoplane* orientation: *Isoplane Top, Isoplane Left,* or *Isoplane Right.*

When defining the radius or diameter of an ellipse, it should always be measured in an isometric direction. In other words, an isometric ellipse is always measured on the two isometric axes (or center lines) parallel with the plane of the ellipse.

If you define the radius or diameter interactively, use *ORTHO ON*. If you enter a value, AutoCAD automatically applies the value to the correct isometric axes.

Figure 23-13

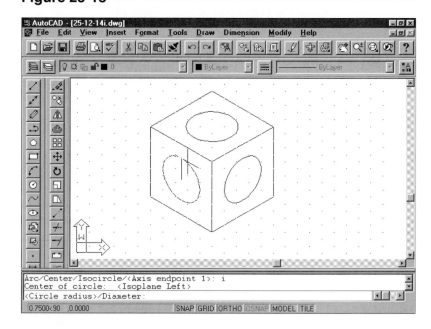

Creating an Isometric Drawing

Figure 23-14

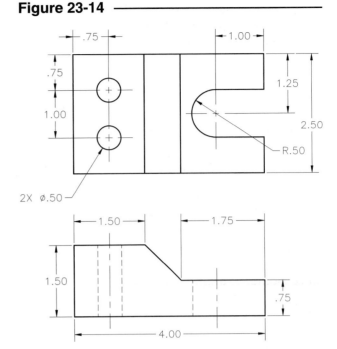

In this exercise, the object in Figure 23-14 is drawn in isometric.

The initial steps to create an isometric drawing begin with the typical setup (see Chapter 6, Drawing Setup):

1. Set the desired *Units*.

2. Set appropriate *Limits*.

3. Set the *Isometric Style* of *Snap* and specify an appropriate value for spacing.

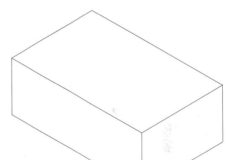

4. The next step involves creating an isometric framework of the desired object. In other words, draw an isometric box equal to the overall dimensions of the object. Using the dimensions given in Figure 23-14, create the encompassing isometric box with the *Line* command (Fig. 23-15).

   Use *ORTHO* to force isometric *Lines*. Watch the *Coords* display (in a relative polar format) to give the current lengths as you draw or use direct distance entry.

Figure 23-15

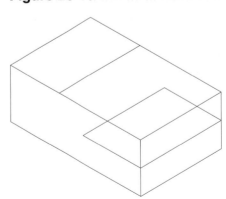

5. Add the lines defining the lower surface. Define the needed edge of the upper isometric surface as shown.

Figure 23-16

6. The <u>inclined</u> edges of the inclined surface can be drawn (with *Line*) only when *ORTHO* is *OFF*. <u>Inclined</u> lines in isometric cannot be drawn by transferring the lengths of the lines, but only by defining the <u>ends</u> of the inclined lines on <u>isometric</u> lines, then connecting the endpoints. Next, *Trim* or *Erase* the necessary *Lines*.

Figure 23-17 ──────────

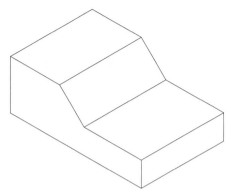

7. Draw the slot by constructing an *Ellipse* with the *Isocircle* option. Draw the two *Lines* connecting the circle to the right edge. *Trim* the unwanted part of the *Ellipse* (highlighted) using the *Lines* as cutting edges.

Figure 23-18 ──────────

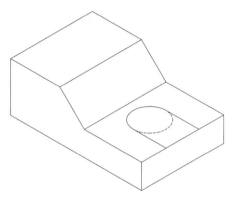

8. *Copy* the far *Line* and the *Ellipse* down to the bottom surface. Add two vertical *Lines* at the end of the slot.

Figure 23-19 ──────────

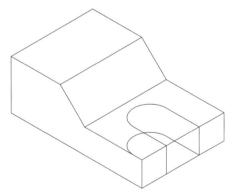

9. Use *Trim* to remove the part of the *Ellipse* that would normally be hidden from view. *Trim* the *Lines* along the right edge at the opening of the slot.

Figure 23-20 ──────────

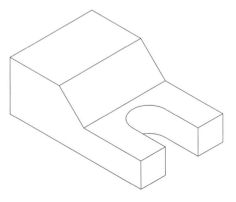

10. Add the two holes on the top with *Ellipse, Isocircle* option. Use *ORTHO ON* when defining the radius. *Copy* can also be used to create the second *Ellipse* from the first.

Figure 23-21

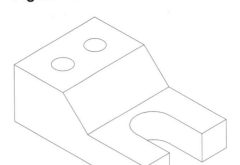

Dimensioning Isometric Drawings in AutoCAD

Refer to Chapter 26, Dimensioning, for details on how to dimension isometric drawings.

OBLIQUE DRAWING IN AutoCAD

Figure 23-22

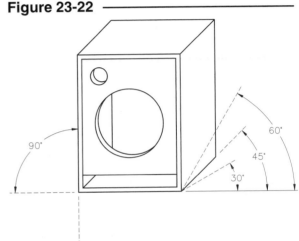

Oblique drawings are characterized by having two axes at a 90 degree orientation. Typically, you should locate the <u>front face</u> of the object along these two axes. Since the object's characteristic shape is seen in the front view, an oblique drawing allows you to create all shapes parallel to the front face true size and shape as you would in a multiview drawing. Circles on or parallel to the front face can be drawn as circles. The third axis, the receding axis, can be drawn at a choice of angles, 30, 45, or 60 degrees, depending on whether you want to show more of the top or the side of the object.

Figure 23-22 illustrates the axes orientation of an oblique drawing, including the choice of angles for the receding axis.

Another option allowed with oblique drawings is the measurement used for the receding axis. Using the full depth of the object along the receding axis is called <u>Cavalier</u> oblique drawing. This method depicts the object (a cube with a hole in this case) as having an elongated depth (Fig. 23-23).

Figure 23-23

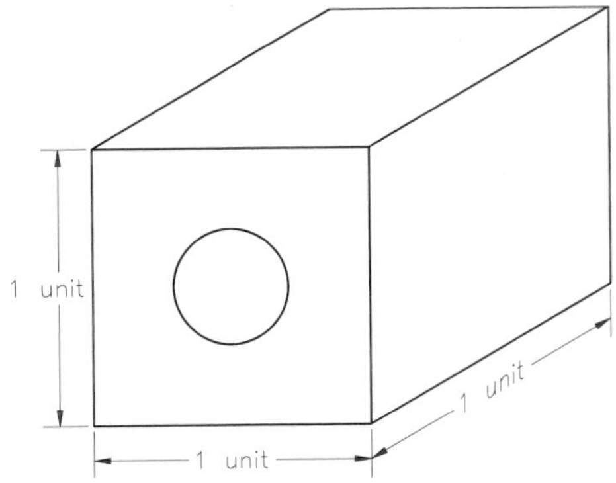

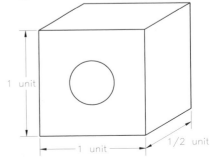

Using 1/2 or 3/4 of the true depth along the receding axis gives a more realistic pictorial representation of the object. This is called a <u>Cabinet</u> oblique (Fig. 23-24).

No functions or commands in AutoCAD are intended specifically for oblique drawing. However, *SNAP* and *GRID* can be aligned with the receding axis using the *Rotate* option of the *Snap* command to simplify the construction of edges of the object along that axis. The steps for creating a typical oblique drawing are given next.

Figure 23-24

The object in Figure 23-25 is used for the example. From the dimensions given in the multiview, create a cabinet oblique with the receding axis at 45 degrees.

Figure 23-25 ———————

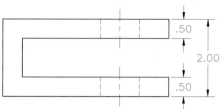

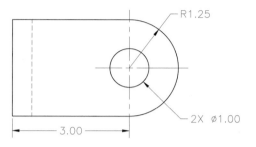

1. Create the characteristic shape of the front face of the object as shown in the front view.

Figure 23-26 ———————

2. Use *Copy* with the *Multiple* option to copy the front face back on the receding axis. Polar coordinates can be entered as the "second point of displacement." For example, the first *Copy* should be located at **@.25<45** relative to the "Base point." (Remember to calculate 1/2 of the actual depth.) As an alternative, a *Line* can be drawn at 45 degrees and *Divided* to place *Points* at .25 increments. *Multiple Copies* of the front face can be *OSNAP*ed with the *Node* option.

Figure 23-27 ———————

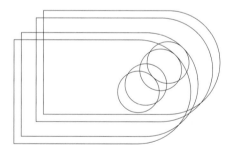

3. Draw the *Line* representing the edge on the upper-left of the object along the receding axis. Use *Endpoint OSNAP* to connect the *Lines*. Make a *Copy* of the *Line* or draw another *Line* .5 units to the right. Drop a vertical *Line* from the *Intersection* as shown.

Figure 23-28 ———————

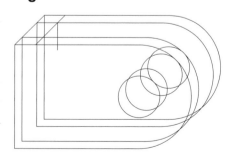

4. Use *Trim* and *Erase* to remove the unwanted parts of the *Lines* and *Circles* (those edges that are normally obscured).

Figure 23-29

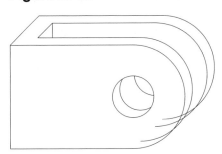

5. *Zoom* with a *window* to the lower right corner of the drawing. Draw a *Line Tangent* to the edges of the arcs to define the limiting elements along the receding axis. *Trim* the unwanted segments of the arcs.

Figure 23-30

The resulting cabinet oblique drawing should appear like that in Figure 23-31.

Figure 23-31

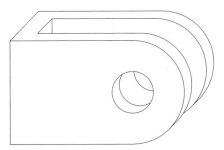

CHAPTER EXERCISES

Isometric Drawing

For exercises 1, 2, and 3 create isometric drawings as instructed. To begin, use an appropriate template drawing and draw a *Pline* border and insert the **TBLOCK**.

1. Create an isometric drawing of the cylinder shown in Figure 23-32. *Save* the drawing as **CYLINDER** and *Plot* so the drawing is *Scaled to Fit* on an A size sheet.

Figure 23-32

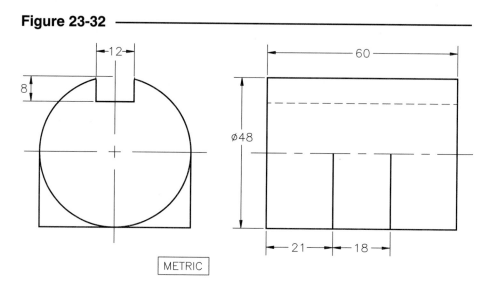

METRIC

2. Make an isometric drawing of the Corner Brace shown in Figure 23-33. *Save* the drawing as **CRNBRACE**. *Plot* at **1=1** scale on an A size sheet.

Figure 23-33

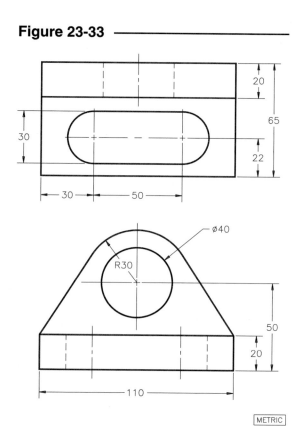

METRIC

3. Draw the Support Bracket (Fig. 23-34) in isometric. The drawing can be *Plotted* at **1=1** scale on an A size sheet. *Save* the drawing and assign the name **SBRACKET**.

Figure 23-34

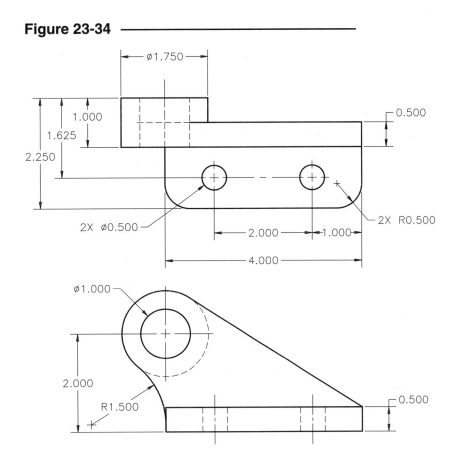

Oblique Drawing

For exercises 4 and 5, create oblique drawings as instructed. To begin, use an appropriate template drawing and draw a *Pline* border and *Insert* the **TBLOCK**.

4. Make an oblique cabinet projection of the Bearing shown in Figure 23-35. Construct all dimensions on the receding axis 1/2 of the actual length. Select the optimum angle for the receding axis to be able to view the 15 x 15 slot. *Plot* at **1=1** scale on an A size sheet. *Save* the drawing as **BEARING**.

Figure 23-35

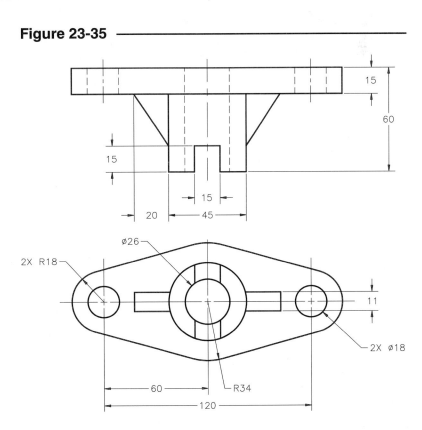

5. Construct a cavalier oblique drawing of the Pulley showing the circular view true size and shape. The illustration in Figure 23-36 gives only the side view. All vertical dimensions in the figure are diameters. *Save* the drawing as **PULLEY** and make a *Plot* on an A size sheet at **1=1**.

Figure 23-36

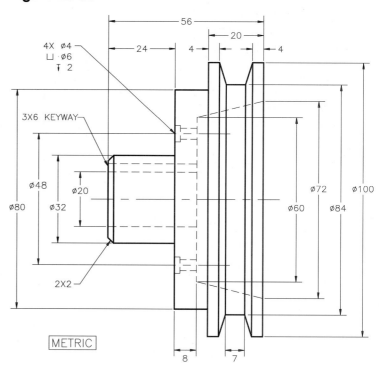

24

SECTION VIEWS

Chapter Objectives

After completing this chapter you should:

1. be able to use the *Bhatch* command to select associative hatch patterns;

2. be able to specify a *Scale* and *Angle* for hatch lines;

3. know how to define a boundary for hatching using the *Pick Points* and *Select Objects* methods;

4. be able to Preview the hatch and make necessary adjustments, then *Apply* the hatch pattern;

5. be able to use *Hatch* to create non-associative hatch lines and discard the boundary;

6. know how to use *Hatchedit* to modify parameters of existing hatch patterns in the drawing;

7. be able to edit hatched areas using *Grips*, selection options, and *Draworder*.

CONCEPTS

A section view is a view of the interior of an object after it has been imaginarily cut open to reveal the object's inner details. A section view is only one of two or more views of a multiview drawing describing the object. For example, a multiview drawing of a machine part may contain three views, one of which is a section view.

Hatch lines (also known as section lines) are drawn in the section view to indicate the solid material that has been cut through. Each combination of section lines is called a hatch pattern, and each pattern is used to represent a specific material. In full and half section views, hidden lines are omitted since the inside of the object is visible.

ANSI (American National Standards Institute) and ISO (International Organization of Standardization) have published standard configurations for section lines, and AutoCAD Release 14 supports those standards. However, ANSI no longer specifies section line pattern standards.

A cutting plane line is drawn in an adjacent view to the section view to indicate the plane that imaginarily cuts through the object. Arrows on each end of the cutting plane line indicate the line of sight for the section view. ANSI dictates that a thick dashed or phantom line be used as the standard cutting plane line.

This chapter discusses the AutoCAD methods used to draw hatch lines for section views and related cutting plane lines. The *Bhatch* (boundary hatch) command allows you to select an enclosed area and select the hatch pattern and the parameters for the appearance of the hatch pattern; then AutoCAD automatically draws the hatch (section) lines. Existing hatch lines in the drawing can be modified using *Hatchedit*. Cutting plane lines are created in AutoCAD by using a dashed linetype. The line itself should be created with the *Pline* command to achieve line width.

DEFINING HATCH PATTERNS AND HATCH BOUNDARIES

A hatch pattern is composed of many lines that have a particular linetype, spacing, and angle. Many standard hatch patterns are provided by AutoCAD for your selection. Rather than having to draw each section line individually, you are required only to specify the area to be hatched and AutoCAD fills the designated area with the selected hatch pattern. An AutoCAD hatch pattern is inserted as one object. For example, you can *Erase* the inserted hatch pattern by selecting only one line in the pattern, and the entire pattern in the area is *Erased*.

Figure 24-1

In a typical section view (Fig. 24-1), the hatch pattern completely fills the area representing the material that has been cut through. With the *Bhatch* command you can define the boundary of an area to be hatched simply by pointing inside of an enclosed area.

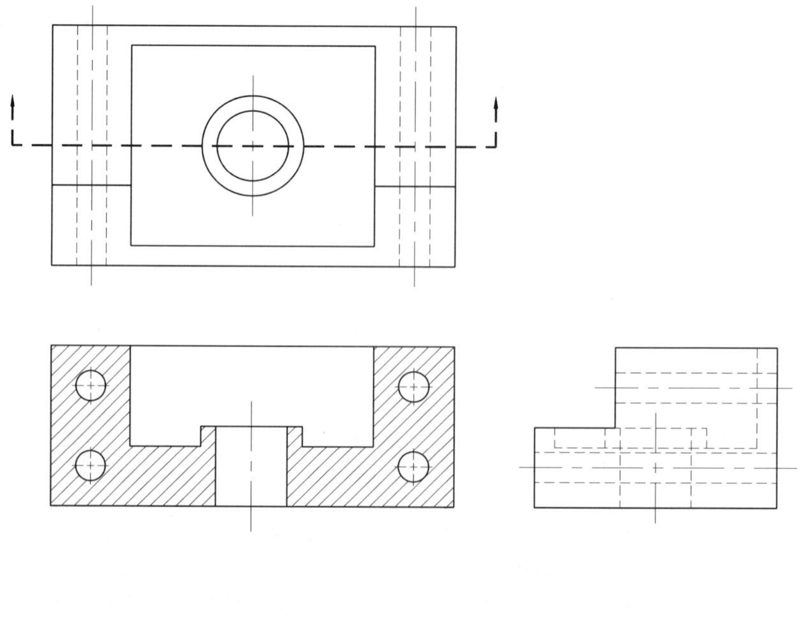

Both the *Hatch* and the *Bhatch* commands fill a specified area with a selected hatch pattern. *Hatch* requires that you select <u>each object</u> defining the boundary, whereas the *Bhatch* command finds the boundary automatically when you point inside it. Additionally, *Hatch* operates in command line format, whereas *Bhatch* operates in dialog box fashion. For these reasons, *Bhatch* is superior to *Hatch* and is recommended in most cases for drawing section views.

Hatch patterns created with *Bhatch* are <u>associative</u>. Associative hatch patterns are associated to the boundary geometry such that when the shape of the boundary changes (by *Stretch*, *Scale*, *Rotate*, *Move*, *Ddmodify*, *Grips*, etc.), the hatch pattern automatically reforms itself to conform to the new shape (Fig. 24-2). For example, if a design change required a larger diameter for a hole, *Ddmodify* could be used to change the diameter of the hole, and the surrounding section lines would automatically adapt to the new diameter.

Figure 24-2

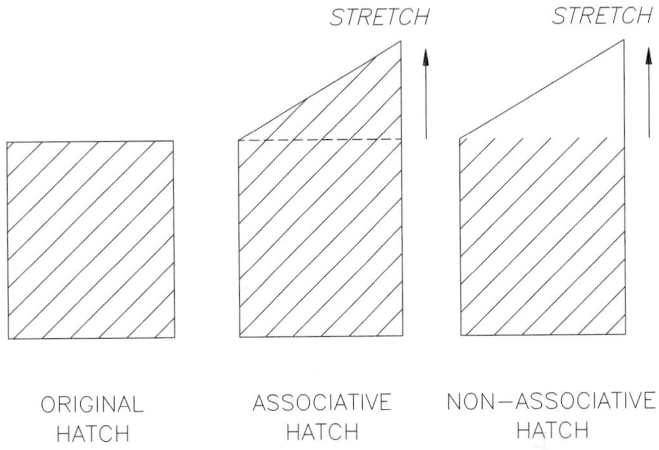

Once the hatch patterns have been drawn, any feature of the existing hatch pattern (created with *Bhatch*) can be changed retroactively using *Hatchedit*. The *Hatchedit* dialog box gives access to the same options that were used to create the hatch (in the *Boundary Hatch* dialog box). Changing the scale, angle, or pattern of any existing section view in the drawing is a simple process.

Steps for Creating a Section View Using the *Bhatch* Command

1. Create the view that contains the area to be hatched using typical draw commands such as *Line*, *Arc*, *Circle*, or *Pline*. If you intend to have text or dimensions inside the area to be hatched, add them before hatching.

2. Invoke the *Bhatch* command. The *Boundary Hatch* dialog box appears (Fig. 24-3).

3. Specify the *Pattern Type* to use. Select the desired pattern by clicking on the image tile until the desired pattern appears, select from the *Pattern* drop-down list or select the *Pattern* tile to allow you to select from the *Hatch pattern palette* image tiles.

4. Specify the *Scale* and *Angle* in the dialog box.

5. Define the area to be hatched by PICKing an internal point (AutoCAD automatically traces the boundary) or by individually selecting the objects.

6. *Preview* the hatch to make sure everything is as expected. Adjust hatching parameters as necessary and *Preview* again.

7. *Apply* the hatch. The hatch pattern is automatically drawn and becomes an associated object in the drawing.

8. If other areas are to be hatched, additional internal points or objects can be selected to define the new area for hatching. The parameters previously used are again applied to the new hatch area by default. You may want to *Inherit Properties* from a previously applied hatch.

9. For mechanical drawings, draw a cutting plane line in a view adjacent to the section view. The *Pline* command with a *Dashed* or *Phantom* linetype is used. Arrows at the ends of the cutting plane line indicate the line of sight for the section view.

10. If any aspect of the hatch lines needs to be edited at a later time, *Hatchedit* can be used to change those properties. If the hatch boundary is changed by *Stretch, Rotate, Scale, Move, Ddmodify*, etc., the hatched area will conform to the new boundary.

BHATCH

	COMMAND (TYPE)	ALIAS (TYPE)	Short-cut	Screen (side) Menu	Tablet Menu
Pull-down Menu					
Draw *Hatch*	*BHATCH*	*BH or H*	...	DRAW 2 *Bhatch*	P,9

Bhatch allows you to create hatch lines for a section view (or for other purposes) by simply PICKing inside a closed boundary. A closed boundary refers to an area completely enclosed by objects. *Bhatch* locates the closed boundary automatically by creating a temporary *Pline* that follows the outline of the hatch area, fills the area with hatch lines, and then deletes the boundary (default option) after hatching is completed. *Bhatch* ignores all objects or parts of objects that are not part of the boundary.

Any method of invoking *Bhatch* yields the *Boundary Hatch* dialog box (Fig. 24-3). Typically, the first step in the *Boundary Hatch* dialog box is the selection of a hatch pattern.

Figure 24-3

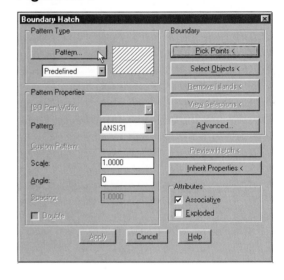

Boundary Hatch Dialog Box Options

Pattern Type
This option allows you to specify the type of the hatch pattern: *Predefined, User-defined*, or *Custom*. Use *Predefined* for standard hatch pattern styles that AutoCAD provides (in the ACAD.PAT file).

Figure 24-4

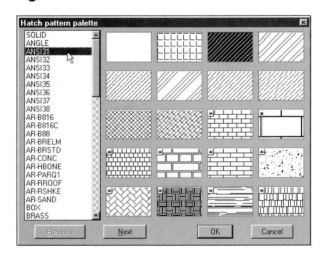

Predefined
There are three ways to select from *Predefined* patterns: (1) select the *Pattern...* tile to produce the *Hatch pattern palette* dialog box displaying hatch pattern names and image tiles (Fig. 24-4), (2) select the *Pattern:* drop-down list to PICK the pattern name, or (3) select the single image tile (in the *Boundary Hatch* dialog box) to sequence through the possible choices.

The *Hatch pattern palette* dialog box (Fig. 24-4) allows you to select a predefined pattern by its image tile or by its name. Use the *Next* and *Previous* buttons to cycle through the four screens.

User-defined

To define a simple hatch pattern "on the fly," select the *User-defined* tile. This causes the *Pattern* and *Scale* options to be disabled and the *Angle, Spacing,* and *Double* options to be enabled. Creating a *User-defined* pattern is easy. Specify the *Angle* of the lines, the *Spacing* between lines, and optionally create *Double* (perpendicular) lines. All *User-defined* patterns have continuous lines.

Custom

Custom patterns are previously created user-defined patterns stored in other than the ACAD.PAT file. Custom patterns can contain continuous, dashed and dotted line combinations. See the AutoCAD Customization Guide for information on creating and saving custom hatch patterns.

Figure 24-5 displays each of the AutoCAD hatch patterns defined in the ACAD.PAT file. Note that the patterns are not shown to scale.

Figure 24-5

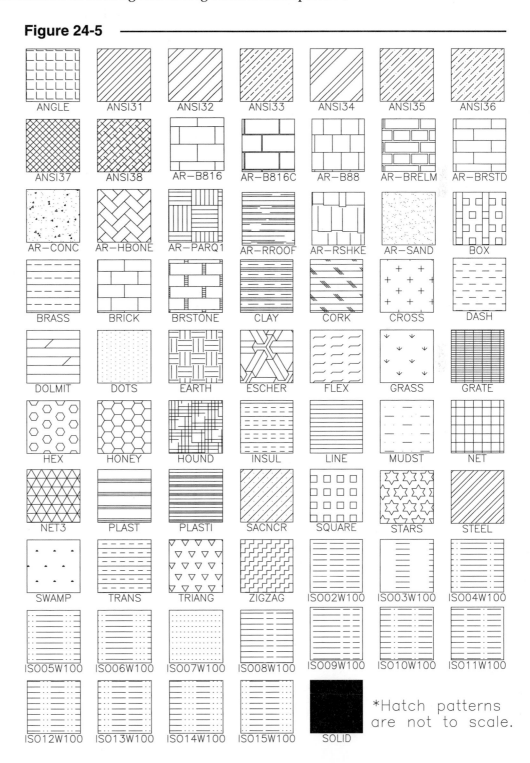

Since hatch patterns have their own linetype, the *Continuous* linetype or a layer with the *Continuous* linetype <u>should be current</u> when hatching. After selecting a pattern, specify the desired *Scale* and *Angle* of the pattern or ISO pen width.

Pattern Properties

This cluster (Fig. 24-3) is used to specify the type of hatch pattern to be drawn and the parameters that govern its appearance.

ISO Pen Width

You must select an ISO hatch pattern for this tile to be enabled. Selecting an *ISO Pen Width* from the drop-down list automatically sets the scale and enters the value in the *Scale* edit box. See *Scale*.

Pattern...

Selecting the *Pattern* drop-down list displays the name of each predefined pattern. Making a selection sets that as the current pattern and causes the pattern to display in the window above. Using the up and down arrow keys causes the name and display (in the image tile) to sequence.

Scale

The value entered in this edit box is a scale factor that is applied to the existing selected pattern. Normally, this scale factor should be changed proportionally with changes in the drawing *Limits*. Like many other scale factors (*LTSCALE, DIMSCALE*), AutoCAD defaults are set to a value of 1, which is appropriate for the default *Limits* of 12 x 9. If you have calculated the drawing scale factor, it can be applied here (see Chapter 13). In the example, *Limits* are set to 22 x 17, so a *Scale* of 1.8 is used. The *Scale* value is stored in the *HPSCALE* system variable.

ISO hatch patterns are intended for use with metric drawings; therefore, the scale (spacing between hatch lines) is much greater than for inch drawings. Because *Limits* values for metric sheet sizes are greater than for inch-based drawings (25.4 times greater than comparable inch drawings), the ISO hatch pattern scales are automatically compensated. If you want to use an ISO pattern with inch-based drawings, calculate a hatch pattern scale based on the drawing scale factor and multiply by .039 (1/25.4).

Angle

The *Angle* specification determines the angle (slant) of the hatch lines. The default angle of 0 represents whatever angle is displayed in the pattern's image tile. Any value entered deflects the existing pattern by the specified value (in degrees). The value entered in this box is held in the *HPANG* system variable.

Spacing

This option is enabled only if *User-defined* pattern is specified. Enter a value for the distance between lines.

Double

Only for a *User-defined* pattern, check this box to have a second set of lines drawn at 90 degrees to the original set.

Boundary—Selecting the Hatch Area

Once the hatch pattern and options have been selected, you must indicate to AutoCAD what area(s) should be hatched. Either the *Pick Points* method, *Select Objects* method, or a combination of both can be used to accomplish this.

Pick Points <

This tile should be selected if you want AutoCAD to automatically determine the boundaries for hatching. You only need to select a point <u>inside</u> the area you want to hatch. The point selected must be inside a <u>completely closed shape</u>. When the *Pick Points* tile is selected, AutoCAD gives the following prompts:

Select internal point: **PICK**
Selecting everything...
Selecting everything visible...
Analyzing the selected data...
Analyzing the internal islands...
Select internal point: **PICK** another area or **Enter**

When an internal point is PICKed, AutoCAD traces and highlights the boundary (Fig. 24-6). The interior area is then analyzed for islands to be included in the hatch boundary. Multiple boundaries can be designated by selecting multiple internal points. Type *U* to undo the last one, if necessary.

Figure 24-6 ────────────

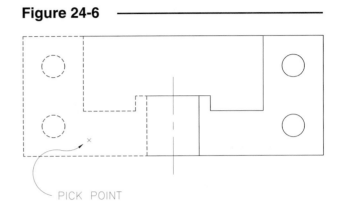

PICK POINT

The location of the point selected is usually not critical. However, the point must be PICKed inside the expected boundary. If there are any gaps in the area, a complete boundary cannot be formed and a boundary error message appears (Fig. 24-7).

Figure 24-7 ────────────

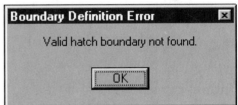

Select Objects<
Alternately, you can designate the boundary with the *Select Objects* method. Using the *Select Objects* method, <u>you</u> specify the boundary objects rather than let AutoCAD locate a boundary. With the *Select Objects* method, no temporary *Pline* boundary is created as with the *Pick Points* method. Therefore, the selected objects must form a closed shape with no gaps or overlaps. If gaps exist (Fig. 24-8) or if the objects extend past the desired hatch area (Fig. 24-9), AutoCAD cannot interpret the intended hatch area correctly, and problems will occur.

Figure 24-8 ────────────

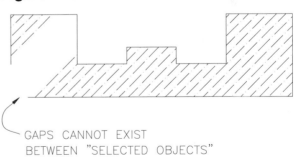

GAPS CANNOT EXIST
BETWEEN "SELECTED OBJECTS"

The *Select Objects* option can be used after the *Pick Points* method to select specific objects for *Bhatch* to consider before drawing the hatch pattern lines. For example, if you had created text objects or dimensions within the boundary found by the *Pick Points* method, you must then use the *Select Objects* option to select the text and dimensions. Using this procedure, the hatch lines are automatically "trimmed" around the text and dimensions (see Fig. 24-18). If you do not use *Select Objects* to indicate these items, *Bhatch* does not recognize the text and dimensions and draws the hatch lines right over them.

Figure 24-9 ────────────

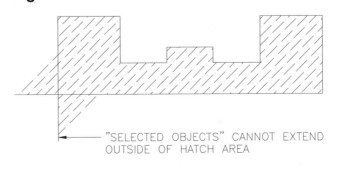

"SELECTED OBJECTS" CANNOT EXTEND
OUTSIDE OF HATCH AREA

Remove Islands <
PICKing this tile allows you to select specific islands to remove from those AutoCAD has found within the outer boundary. If hatch lines have been drawn, be careful to *Zoom* in close enough to select the desired boundary and not the hatch pattern.

View Selections <
Clicking the *View Selections* tile causes AutoCAD to highlight all selected boundaries. This can be used as a check to ensure the desired areas are selected.

Advanced...
This tile invokes the *Advanced Options* dialog box (see *Advanced Options* Dialog Box).

Preview Hatch <
You should always use the *Preview Hatch* option after specifying the hatch parameters and selecting boundaries, but before you *Apply* the hatch. This option allows you to temporarily look at the hatch pattern in your drawing with the current settings applied and allows you to adjust the settings, if necessary, before using *Apply*. After viewing the drawing, pressing the *Continue* tile redisplays the *Boundary Hatch* dialog box, allowing you to make adjustments.

Inherit Properties <
This option allows you to select a hatch pattern from one existing in the drawing. The dialog box disappears and the "Select hatch objects:" prompt appears. PICKing a previously drawn hatch pattern resets all the parameters in the *Boundary Hatch* dialog box for subsequent application to another hatch area.

Associative
This checkbox toggles (on or off) the associative property for newly applied hatch patterns. Associative hatch patterns automatically update by conforming to the new boundary shape when the boundary is changed (see Fig. 24-2). A non-associative hatch pattern is static even when the boundary changes.

Exploded
Bhatch normally draws a hatch pattern as one object, similar to a *Block*. Checking this box causes the hatch patterns to be pre-exploded when inserted. (This is similar to *Inserting* a *Block* with an asterisk [*] prefixing the *Block* name.) Exploded hatch patterns allow you to edit the hatch lines individually. Exploded hatch patterns are <u>non-associative</u>.

Apply
This is typically the <u>last step</u> to the hatching process with *Bhatch*. Selecting *Apply* causes AutoCAD to create the hatch using all the parameters specified in the *Boundary Hatch* dialog box and return to the Command: prompt for further drawing and editing.

Advanced Options Dialog Box

Figure 24-10

Define Boundary Set
By default, AutoCAD examines <u>all objects on the screen</u> when determining the boundaries by the *Pick Points* method. (The *From Everything on Screen* box is checked by default when you begin the *Bhatch* command.) For complex drawings, examining all objects can take some time. In that case, you may want to specify a smaller boundary set for AutoCAD to consider. Clicking the *Make New Boundary Set* tile clears the dialog boxes and permits you to select objects or select a window to define the new set.

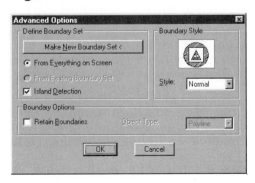

Style

This section allows you to specify how the hatch pattern is drawn when the area inside the defined boundary contains text or closed areas (islands). Select the text from the list or PICK the icons displayed in the window. These options are applicable <u>only</u> when interior objects (islands) have been included in the selection set (considered for hatching). Otherwise, if only the outer shape is included in the selection set, the results are identical to the *Ignore* option.

Normal

This should be used for most applications of *Bhatch*. Text or closed shapes within the outer border are considered in such cases. Hatching will begin at the outer boundary and move inward alternating between applying and not applying the pattern as interior shapes or text are encountered (Fig. 24-11).

Outer

This option causes AutoCAD to hatch only the outer closed shape. Hatching is turned off for all interior closed shapes (Fig. 24-11).

Figure 24-11

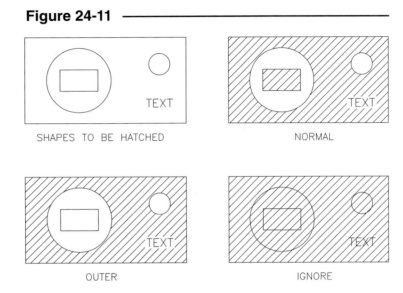

SHAPES TO BE HATCHED

NORMAL

OUTER

IGNORE

Ignore

Ignore draws the hatch pattern from the outer boundary inward ignoring any interior shapes. The resulting hatch pattern is drawn through the interior shapes (Fig. 24-11).

Island Detection

Removing the check in this box causes AutoCAD to trace the outer boundary only and ignore any internal closed areas (islands) when the hatch pattern is applied.

Retain Boundaries

When AutoCAD uses the *Pick Points* method to locate a boundary for hatching, a temporary *Pline* is created for hatching, then discarded after the hatching process. Checking this box forces AutoCAD to <u>keep</u> the boundary. When the box is checked, *Bhatch* creates two objects—the hatch pattern and the boundary object. (You can specify whether you want to create a *Pline* or a *Region* boundary in the *Object Type* drop-down list.) Using this option and erasing the hatch pattern accomplishes the same results as using the *Boundary* command.

Figure 24-12

After using *Bhatch*, the *Pline* or *Region* can be used for other purposes. To test this function, complete a *Bhatch* with the *Retain Boundaries* box checked; then use *Move* (make sure you select the boundary) to reveal the new boundary object (Fig. 24-12).

Object Type

Bhatch creates *Polyline* or *Region* boundaries. This option is enabled only when the *Retain Boundaries* box is checked (see Fig. 24-10).

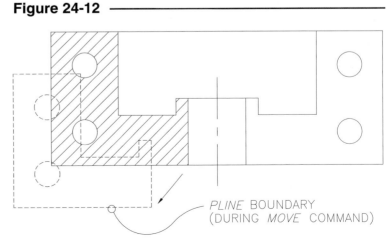

PLINE BOUNDARY
(DURING *MOVE* COMMAND)

HATCH

Pull-down Menu	COMMAND (TYPE)	ALIAS (TYPE)	Short-cut	Screen (side) Menu	Tablet Menu
...	*HATCH*	*-H*	...	...	...

Hatch must be typed at the keyboard since it is not available from the menus. *Hatch* operates in a command line format with many of the options available in the *Bhatch* dialog box. However, *Hatch* creates a <u>non-associative</u> hatch pattern and <u>does not use the *Pick points*</u> method ("select internal point:") to create a boundary. Generally, *Bhatch* would be used instead of *Hatch* except for special cases.

You can use *Hatch* to find a boundary from existing objects. The objects must form a complete closed shape, just as with *Bhatch*. The shape can be composed of <u>one</u> closed object such as a *Pline, Polygon, Spline,* or *Circle* or composed of <u>several</u> objects forming a closed area such as *Lines* and *Arcs*. For example, the shape in Figure 24-13 can be hatched correctly with *Hatch,* whether it is composed of one object (*Pline, Region,* etc.) or several objects (*Lines, Arc,* etc.).

Figure 24-13 ────────────

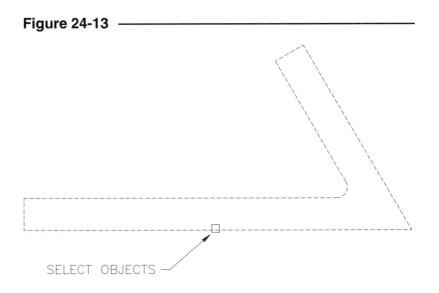

SELECT OBJECTS

If you select several objects to define the boundary, the objects must comprise <u>only the boundary shape</u> and not extend past the desired boundary. If objects extend past the desired boundary (Fig. 24-14) or if there are gaps, the resulting hatch will be <u>incorrect</u> as shown in Figures 24-8 and 24-9.

Figure 24-14 ────────────

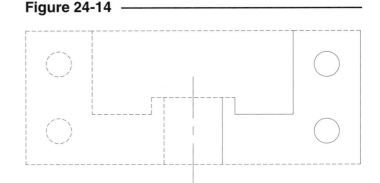

```
Command: hatch
Enter pattern name or [?/Solid/User defined]<ANSI31>: (style or pattern name)
Scale for pattern <1.0000>: (value) or Enter
Angle for pattern <0>: (value) or Enter
Select hatch boundaries or RETURN for direct hatch option
Select objects: PICK
Select objects: Enter
Command:
```

Direct Hatch

With the *Direct Hatch* method of *Hatch*, you specify points to define the boundary, not objects. The significance of this method is that you can create a hatch pattern without using existing objects as the boundary. In addition, you can select whether you want to retain or discard the boundary after the pattern is applied (Fig. 24-15).

Figure 24-15

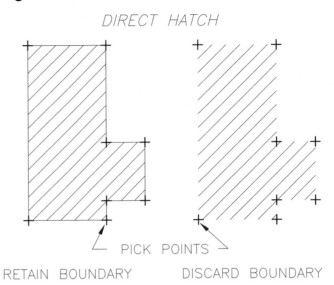

DIRECT HATCH

PICK POINTS

RETAIN BOUNDARY   DISCARD BOUNDARY

```
Command: hatch
Enter pattern name or [?/Solid/User defined] <ANSI31>: (pattern name, style) or Enter
Scale for pattern <1.0000>: (value) or Enter
Angle for pattern <0>: (value) or Enter
Select hatch boundaries or RETURN for direct hatch option
Select objects: Enter
Retain polyline? <N> Y or Enter
From point: PICK or (coordinates)
Arc/Close/Length/Undo/<Next point>: PICK or (coordinates)
Arc/Close/Length/Undo/<Next point>: PICK or (coordinates)
Arc/Close/Length/Undo/<Next point>: PICK or (coordinates)
Arc/Close/Length/Undo/<Next point>: c
From point or RETURN to apply hatch: Enter
Command:
```

For special cases when you do not want a hatch area boundary to appear in the drawing (Fig. 24-16), you can use the *Direct Hatch* method to specify a hatch area, then discard the boundary *Pline*.

Figure 24-16

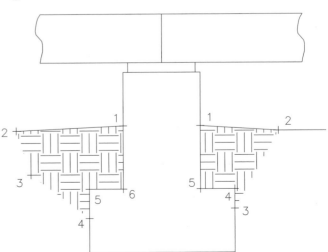

EDITING HATCH PATTERNS AND BOUNDARIES

HATCHEDIT

	COMMAND (TYPE)	ALIAS (TYPE)	Short-cut	Screen (side) Menu	Tablet Menu
Pull-down Menu					
Modify *Object >* *Hatchedit*	*HATCHEDIT*	*HE*	...	*MODIFY1* *Hatchedt*	*Y,16*

Hatchedit allows you to modify an <u>existing associative</u> hatch pattern in the drawing. This feature of AutoCAD makes the hatching process more flexible because you can hatch several areas to quickly create a "rough" drawing, then retroactively fine-tune the hatching parameters when the drawing nears completion with *Hatchedit*.

Invoking the *Hatchedit* command by any method produces the *Hatchedit* dialog box (Fig. 24-17). The dialog box provides options for changing the *Pattern, Scale, Angle,* and *Style* properties of the existing hatch. You can also change the hatch to non-associative and *Exploded*.

Apparent in Figure 24-17, the *Hatchedit* dialog box is essentially the same as the *Boundary Hatch* dialog box with some of the options disabled. For an explanation of the options that are available with *Hatchedit*, see *Boundary Hatch* earlier in this chapter.

Figure 24-17 ─────────────

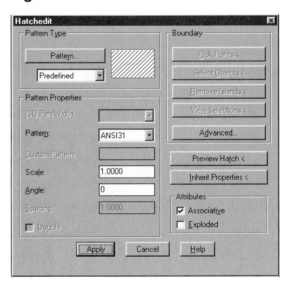

Exploded or Associative Hatches?

Exploded hatches allow you to edit individual hatch lines. For example, you may want to *Trim* an area out of a hatch pattern to place a construction note, etc. To explode a hatch pattern, use either the *Hatchedit* dialog box or the *Explode* command.

Although exploded hatches give you the ability to edit the hatch lines individually, the exploded hatch loses its associativity and therefore loses its ability to be edited with *Hatchedit* or to update when the related boundary is changed.

It is generally preferred to retain hatch associativity. The associative feature usually gives you more flexibility in editing hatch patterns. If dimensions or text objects are <u>created before you use *Bhatch*</u> and you use the *Select Objects* button to indicate those objects when you specify the boundary, the hatch pattern is automatically drawn correctly around those objects as if you had *Trimmed* the hatch pattern around the objects (Fig. 24-18).

Figure 24-18 ─────────────

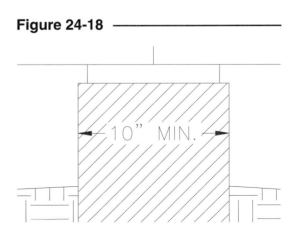

Using Grips with Hatch Patterns

Grips can be used effectively to edit hatch pattern boundaries of associative hatches. Do this by selecting the boundary with the pickbox or automatic window/crossing window (as you would normally activate an object's grips). Associative hatches <u>retain the association</u> with the boundary after the boundary has been changed using grips (see Fig. 24-2, earlier this chapter).

Beware, when selecting the hatch boundary, ensure you <u>select only the boundary and not the hatch object</u>. If you edit the hatch object with grips, the associative feature is lost. This can happen if you select the <u>hatch object</u> grip (Fig. 24-19). In this case, AutoCAD issues the following prompt:

Figure 24-19

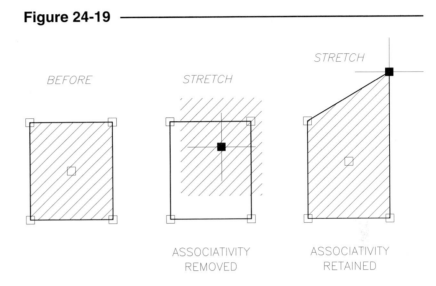

```
Command: (select grip on hatch object)
**STRETCH**
<Stretch to point>/Base point/Copy/Undo/eXit: PICK
Hatch boundary associativity removed.
Command:
```

If you activate grips on <u>both</u> the hatch object and the boundary and make a boundary grip hot, the boundary and related hatch retain associativity (Fig. 24-19).

Object Selection Features of Hatch Patterns

When you select a hatch object for editing (PICK only one line of an entire hatch pattern), you actually select the entire hatch object. By default, when you PICK any element of the hatch pattern, the hatch object can be *Moved, Copied*, grip edited, etc. However, when you edit the hatch object <u>but not the boundary</u> associativity is removed (see Using Grips with Hatches).

In Release 14, you can easily control whether just the hatch object or the <u>hatch object and boundary together</u> are selected when you PICK the hatch object. Do this by selecting *Associative Hatch* in the *Object Selection Settings* dialog box (Fig. 24-20). Type *Ddselect* or choose *Selection…* from the *Tools* pull-down menu to produce the dialog box. Your choice for the *Associative Hatch* option sets the *PICKSTYLE* variable (see Appendix A).

Figure 24-20

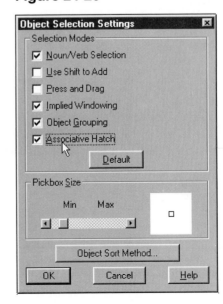

With *Associative Hatch* selected, PICKing the hatch object results in highlighting both the hatch object and boundary; therefore, you always edit the hatch object and the boundary together (Fig. 24-21). When *Associative Hatch* is not selected, PICKing the hatch object results in only the hatch object selection; therefore, you run the risk of editing the hatch object independent of the boundary and losing associativity.

Figure 24-21

OBJECT SELECTION WITH ASSOCIATIVE HATCH CHECKED

OBJECT SELECTION WITH ASSOCIATIVE HATCH NOT CHECKED

Solid *Hatch* Fills

In Release 14, a *Solid* hatch pattern is available. This feature is welcomed by professionals who create hatch objects such as walls for architectural drawings and require the solid filled areas. There are many applications for this feature.

You can use the *Fill* command (or *FILLMODE* system variable) to control the visibility of the solid filled areas. *Fill* also controls the display of solid filled TrueType fonts (see Chapter 18). If you want to display solid filled *Hatch* objects as solid, set *Fill* to *On* (Fig. 24-22). Setting *Fill* to *Off* causes AutoCAD <u>not</u> to display the solid filled areas. *Regen* must be used after *Fill* to display the new visibility state. This solid fill control can be helpful for saving toner or ink during test prints or speeding up plots when much solid fill is used in a drawing.

Figure 24-22

FILL ON

FILL OFF

DRAWORDER

	COMMAND (TYPE)	ALIAS (TYPE)	Short-cut	Screen (side) Menu	Tablet Menu
Pull-down Menu					
Tools *Display Order >*	*DRAWORDER*	*DR*	...	*TOOLS 1* *Drawordr*	*T,9*

New possibilities exist in Release 14 such as using solid hatch patterns and raster images in combination with filled text, other hatch patterns and solid images, etc. With these solid areas and images, some control of which objects are in "front" and "back or "above" and "under" must be provided. This control is provided by the *Draworder* command.

For example, if a company logo were created using filled *Dtext*, a *Sand* hatch pattern, and a *Solid* hatched circle, the object that was created last would appear "in front" (Fig. 24-23, top logo). The *Draworder* command is used to control which objects appear in front and back or above and under. This capability is necessary for creating prints and plots in black and white and in color.

The *Draworder* command is simple to use. Simply select objects and indicate *Front*, *Back*, *Above*, or *Under*.

> Command: **draworder**
> Select objects: **PICK**
> Select objects: **Enter**
> Above object/Under object/Front/<Back>: (**option**)
> Regenerating drawing.
> Command:

In Figure 24-23, three (of four) possibilities are shown for the three objects—*Dtext*, *Solid* hatch and boundary, and *Sand* hatch and boundary.

Figure 24-23

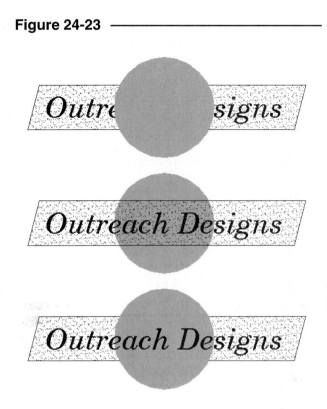

DRAWING CUTTING PLANE LINES

Most section views (full, half, and offset sections) require a cutting plane line to indicate the plane on which the object is cut. The cutting plane line is drawn in a view <u>adjacent</u> to the section view because the line indicates the plane of the cut from its edge view. (In the section view, the cutting plane is per-pendicular to the line of sight, therefore, not visible as a line.)

Standards provide two optional line types for cutting plane lines. In AutoCAD, the two linetypes are *Dashed* and *Phantom*, as shown in Figure 24-24. Arrows at the ends of the cutting plane line indicate the <u>line-of-sight</u> for the section view.

Figure 24-24

Cutting plane lines should be drawn or plotted in a <u>heavy</u> line weight. This is accomplished in AutoCAD by using the *Pline* command to draw the cutting plane line having width assigned using the *Width* option. For plotting full size, a width of .02 for small plots (8.5 x 11, or 17 x 22) to .03 for large plots is recommended. Keep in mind that appearance of the *Pline Width* is affected by the drawing plot scale; and, therefore, *Width* should be increased or decreased by that proportion. For example, if you changed the *Limits*, multiply .02 times the drawing scale factor. (See Chapter 13 for information on drawing scale factor.)

For the example section view, a cutting plane line is created in the top view to indicate the plane on which the object is cut and the line of sight for the section view (Fig. 24-25). (The cutting plane could be drawn in the side view, but the top would be much clearer.) First, a *New* layer named CUT is created and the *Dashed linetype* is assigned to the layer. The layer is *Set* as the *Current* layer. Next, construct a *Pline* with the desired *Width* to represent the cutting plane line.

Figure 24-25

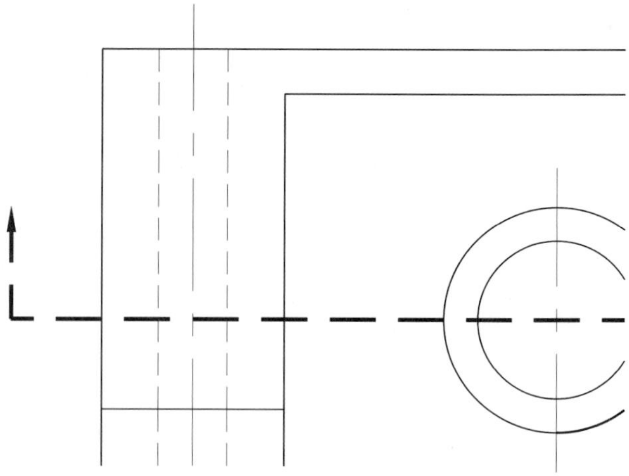

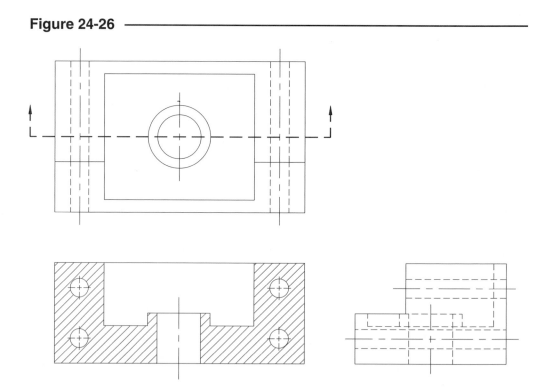

The resulting cutting plane line appears as shown in Figure 24-26 but without the arrow heads. The horizontal center line for the hole (top view) was *Erased* before creating the cutting plane line.

The last step is to add arrowheads to the ends of the cutting plane line. Arrowheads can be drawn by three methods: (1) use the *Solid* command to create a three-sided solid area; (2) use a tapered *Pline* with beginning width of 0 and ending width of .08, for example; or (3) use the *Leader* command to create the arrowhead (see Chapter 26, Dimensioning, for details on *Leader*). The arrowhead you create can be *Scaled*, *Rotated*, and *Copied* if needed to achieve the desired size, orientation, and locations.

The resulting multiview drawing with section view is complete and ready for dimensioning, drawing a border, and inserting a title block (Fig. 24-26).

Figure 24-26

CHAPTER EXERCISES

For the following exercises, create the section views as instructed. Use an appropriate template drawing for each unless instructed otherwise. Include a border and title block for each drawing.

1. ***Open*** the **SADDLE** drawing that you created in Chapter 22. Convert the front view to a full section. ***Save*** the drawing as **SADL-SEC**. ***Plot*** the drawing at **1=1** scale.

2. Make a multiview drawing of the Bearing shown in Figure 24-27. Convert the front view to a full section view. Add the necessary cutting plane line in the top view. ***Save*** the drawing as **BEAR-SEC**. Make a ***Plot*** at full size.

Figure 24-27

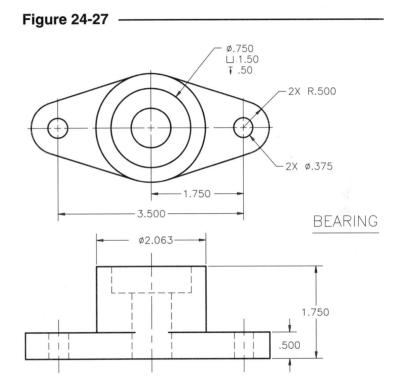

BEARING

3. Create a multiview drawing of the Clip shown in Figure 24-28. Include a side view as a full section view. You can use the **CLIP** drawing you created in Chapter 22 and convert the side view to a section view. Add the necessary cutting plane line in the front view. ***Plot*** the finished drawing at **1=1** scale and ***SaveAs*** **CLIP-SEC**.

Figure 24-28

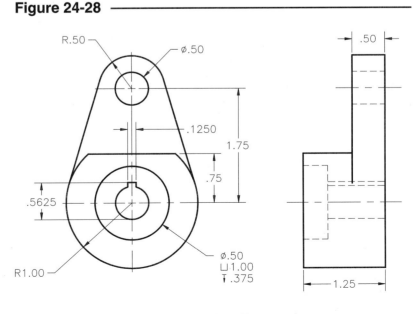

CLIP

4. Create a multiview drawing, including two full sections of the Stop Block as shown in Figure 24-29. Section B–B' should replace the side view shown in the figure. *Save* the drawing as **SPBK-SEC**. *Plot* the drawing at **1=2** scale.

Figure 24-29

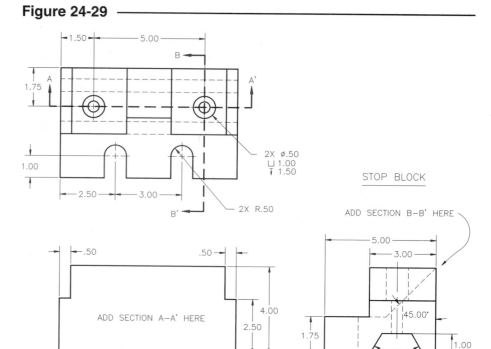

STOP BLOCK

ADD SECTION A–A' HERE

ADD SECTION B–B' HERE

5. Make a multiview drawing, including a half section of the Pulley. All vertical dimensions are diameters. Two views (including the half section) are sufficient to describe the part. Add the necessary cutting plane line. *Save* the drawing as **PUL-SEC** and make a *Plot* at **1:1** scale.

Figure 24-30

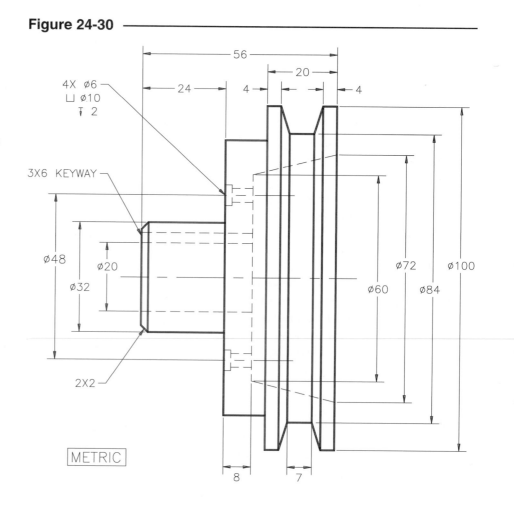

METRIC

25

AUXILIARY VIEWS

Chapter Objectives

After completing this chapter you should:

1. be able to use the *Rotate* option of *Snap* to change the angle of the *SNAP*, *GRID*, and *ORTHO*;

2. know how to use the *Offset* command to create parallel line copies;

3. be able to use *Xline* and *Ray* to create construction lines for auxiliary views.

CONCEPTS

AutoCAD provides no features explicitly for the creation of auxiliary views in a 2D drawing. No new commands are discussed in this chapter. However, three particular features that have been discussed earlier can assist you in construction of auxiliary views. Those features are the *SNAP* rotation, the *Offset* command, and the *Xline* and *Ray* commands.

An auxiliary view is a supplementary view among a series of multiviews. The auxiliary view is drawn in addition to the typical views that are mutually perpendicular (top, front, side). An auxiliary view is one that is normal (the line-of-sight is perpendicular) to an inclined surface of the object. Therefore, the auxiliary view is constructed by projecting in a 90 degree direction from the edge view of an inclined surface in order to show the true size and shape of the inclined surface. The edge view of the inclined

surface could be at any angle (depending on the object), so lines are typically drawn parallel and perpendicular relative to that edge view. Hence, the *SNAP* rotation feature, the *Offset* command, and the *Xline* and *Ray* commands can provide assistance in this task.

Figure 25-1 ───────────

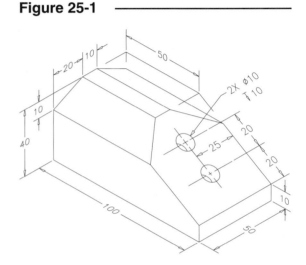

An example mechanical part used for the application of these AutoCAD features related to auxiliary view construction is shown in Figure 25-1. As you can see, there is an inclined surface that contains two drilled holes. To describe this object adequately, an auxiliary view should be created to show the true size and shape of the inclined surface.

This chapter explains the construction of a partial auxiliary view for the example object in Figure 25-1.

CONSTRUCTING AN AUXILIARY VIEW

Setting Up the Principal Views

Figure 25-2 ───────────────────

To begin this drawing, the typical steps are followed for drawing setup (Chapter 13). Because the dimensions are in millimeters, *Limits* should be set accordingly. For example, to provide enough space to draw the views full size and to plot full size on an A sheet, *Limits* of 279 x 216 are specified.

In preparation for the auxiliary view, the principal views are "blocked in," as shown in Figure 25-2. The purpose of this step is to ensure that the desired views fit and are optimally spaced within the allocated

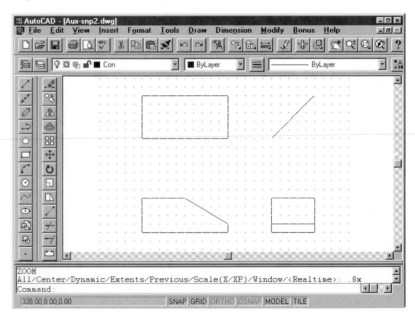

Limits. If there is too little or too much room, adjustments can be made to the *Limits*. Notice that space has been allotted between the views for a partial auxiliary view to be projected from the front view.

Before additional construction on the principal views is undertaken, initial steps in the construction of the partial auxiliary view should be performed. The projection of the auxiliary view requires drawing lines perpendicular and parallel to the inclined surface. One or more of the three alternatives (explained next) can be used.

Using *Snap Rotate*

One possibility to construct an auxiliary view is to use the *Snap* command with the *Rotate* option. This action permits you to rotate the *SNAP* to any angle about a specified base point. The *GRID* automatically follows the *SNAP*. Turning *ORTHO ON* forces *Lines* to be drawn orthogonally with respect to the rotated *SNAP* and *GRID*.

Figure 25-3 displays the *SNAP* and *GRID* after rotation. The command syntax is given below.

In this and the following figures, the cursor size is changed from the default 5% of screen size to 100% of screen size to help illustrate the orientation of the *SNAP, GRID,* and cursor when *SNAP* is *Rotated*. You can change the cursor size in the *Pointer* tab of the *Preferences* dialog box.

Figure 25-3

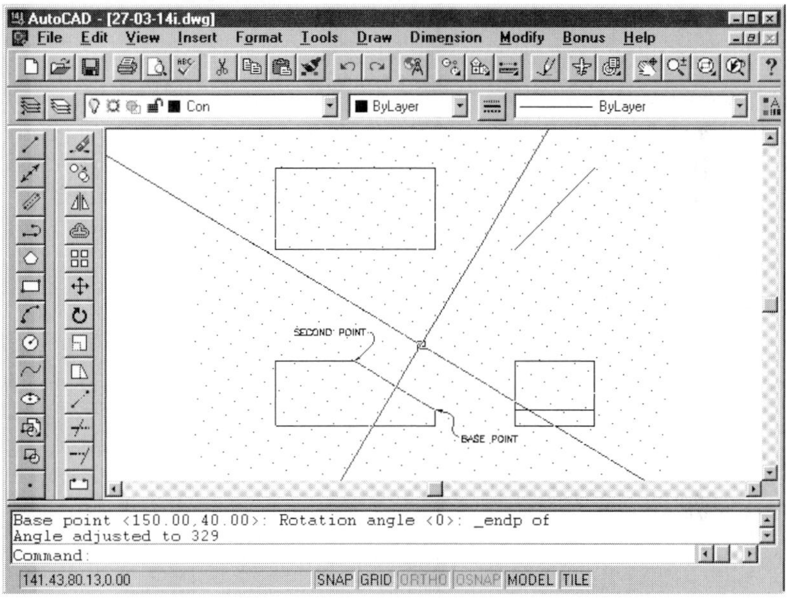

```
Command: snap
Snap spacing or ON/OFF/Aspect/Rotate/Style <current>: r
Base point <0.00,0.00>: PICK or (coordinates) (Starts a rubberband line.)
Rotation angle <0>: PICK or (value) (PICK to specify the second point to define the angle.
See Fig. 25-3.)
Command:
```

PICK (or specify coordinates for) the endpoint of the *Line* representing the inclined surface as the "Base point." At the "Rotation angle:" prompt, a value can be entered or another point (the other end of the inclined *Line*) can be PICKed. Use *OSNAP* when PICKing the *Endpoints*. If you want to enter a value but don't know what angle to rotate to, use *List* to display the angle of the inclined *Line*. The *GRID* and *SNAP* should align with the inclined plane as shown in Figure 25-3.

After rotating the *SNAP* and *GRID*, the partial auxiliary view can be "blocked in," as displayed in Figure 25-4. Begin by projecting *Lines* up from and perpendicular to the inclined surface. (Make sure *ORTHO* is *ON*.) Next, two *Lines* representing the depth of the view should be constructed parallel to the inclined surface and perpendicular to the previous two projection lines. The depth dimension of the object in the auxiliary view is equal to the depth dimension in the top or right view. *Trim* as necessary.

Locate the centers of the holes in the auxiliary view and construct two *Circles*. It is generally preferred to construct circular shapes in the view

Figure 25-4

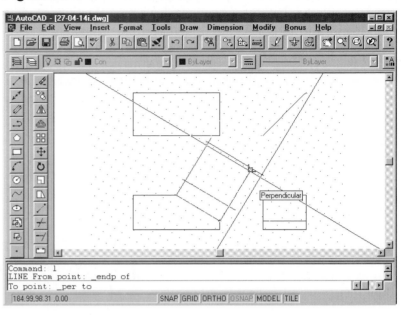

in which they appear as circles, then project to the other views. That is particularly true for this type of auxiliary since the other views contain ellipses. The centers can be located by projection from the front view or by *Offsetting Lines* from the view outline.

Next, project lines from the *Circles* and their centers back to the inclined surface. Use of a hidden line layer can be helpful here. While the *SNAP* and *GRID* are rotated, construct the *Lines* representing the bottom of the holes in the front view. (Alternately, *Offset* could be used to copy the inclined edge down to the hole bottoms; then *Trim* the unwanted portions of the *Lines*.)

Figure 25-5

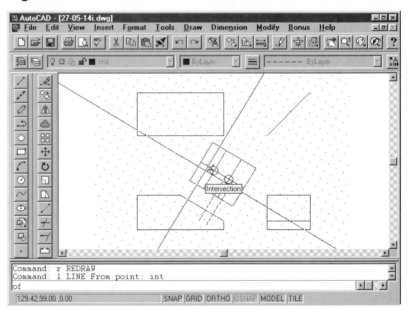

Rotating *SNAP* Back to the Original Position

Before details can be added to the other views, the *SNAP* and *GRID* should be rotated back to the original position. It is very important to rotate back using the <u>same basepoint</u>. Fortunately, AutoCAD remembers the original basepoint so you can accept the default for the prompt. Next, enter a value of **0** when rotating back to the original position. (When using the *Snap Rotate* option, the value entered for the angle of rotation is absolute, not relative to the current position. For example, if the *Snap* was rotated to 45 degrees, rotate back to 0 degrees, not -45.)

> Command: **snap**
> Snap spacing or ON/OFF/Aspect/Rotate/Style <2.00>: **r**
> Base point <150.00,40.00>: **Enter** (AutoCAD remembers the previous basepoint)
> Rotation angle <329>: **0**
> Command:

Construction of multiview drawings with auxiliaries typically involves repeated rotation of the *SNAP* and *GRID* to the angle of the inclined surface and back again as needed.

With the *SNAP* and *GRID* in the original position, details can be added to the other views as shown in Figure 25-6. Since the two circles appear as ellipses in the top and right side views, project lines from the circles' centers and limiting elements on the inclined surface. Locate centers for the two *Ellipses* to be drawn in the top and right side views.

Figure 25-6

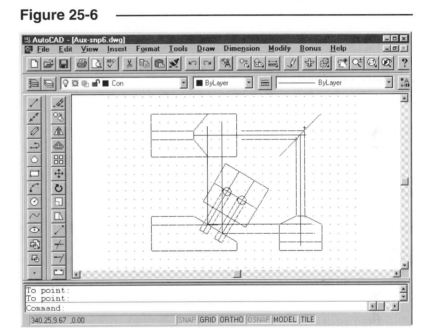

Use the *Ellipse* command to construct the ellipses in the top and right side views. Using the *Center* option of *Ellipse*, specify the center by PICKing with the *Intersection OSNAP* mode. *OSNAP* to the appropriate construction line *Intersection* for the first axis endpoint. For the second axis endpoint (as shown), use the actual circle diameter, since that dimension is not foreshortened.

Figure 25-7

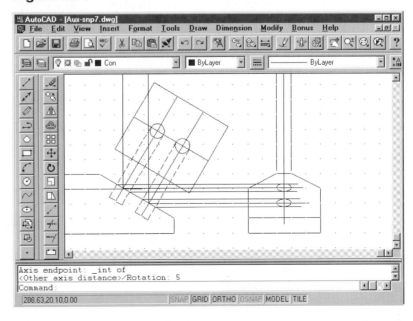

Figure 25-8

The remaining steps for completing the drawing involve finalizing the perimeter shape of the partial auxiliary view and *Copying* the *Ellipses* to the bottom of the hole positions. The *SNAP* and *GRID* should be rotated back to project the new edges found in the front view (Fig. 25-8).

At this point, the multiview drawing with auxiliary view is ready for centerlines, dimensioning, and construction or insertion of a border and title block.

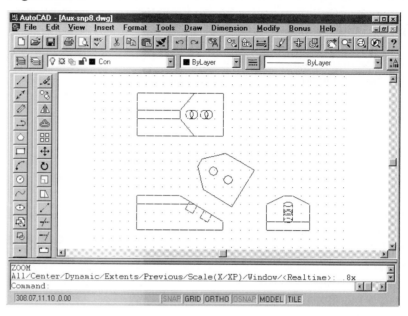

Using the *Offset* Command

Another possibility, and an alternative to the *SNAP* rotation, is to use *Offset* to make parallel *Lines*. This command can be particularly useful for construction of the "blocked in" partial auxiliary view because it is not necessary to rotate the *SNAP* and *GRID*.

OFFSET

Pull-down Menu	COMMAND (TYPE)	ALIAS (TYPE)	Short-cut	Screen (side) Menu	Tablet Menu
Modify *Offset*	*OFFSET*	*O*	...	*MODIFY1* *Offset*	*V,17*

Figure 25-9

Invoke the *Offset* command and specify a distance. The first distance is arbitrary. Specify an appropriate value between the front view inclined plane and the nearest edge of the auxiliary view (20 for the example). *Offset* the new *Line* at a distance of 50 (for the example) or PICK two points (equal to the depth of the view).

Note that the *Offset* lines have lengths equal to the original and therefore require no additional editing (Fig. 25-9).

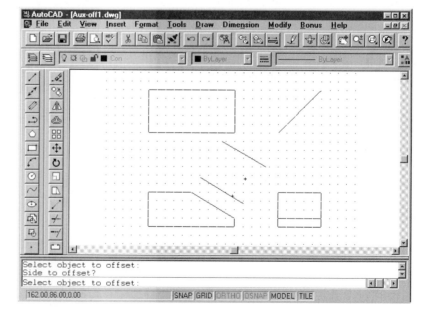

Next, two *Lines* would be drawn between *Endpoints* of the existing offset lines to complete the rectangle. *Offset* could be used again to construct additional lines to facilitate the construction of the two circles in the partial auxiliary view (Fig 25-10).

From this point forward, the construction process would be similar to the example given previously (Figs. 25-5 through 25-8). Even though *Offset* does not require that the *SNAP* be *Rotated*, the complete construction of the auxiliary view could be simplified by using the rotated *SNAP* and *GRID* in conjunction with *Offset*.

Figure 25-10

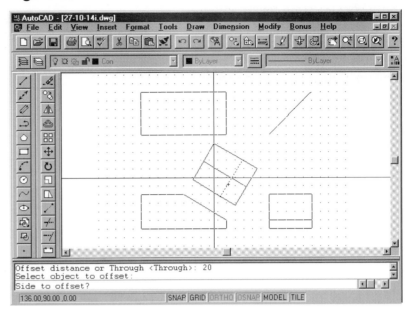

Using the *Xline* and *Ray* Commands

As a third alternative for construction of auxiliary views, the *Xline* and *Ray* commands could be used to create construction lines.

XLINE

Pull-down Menu	COMMAND (TYPE)	ALIAS (TYPE)	Short-cut	Screen (side) Menu	Tablet Menu
Draw *Construction Line*	*XLINE*	*XL*	...	*DRAW 1* *Xline*	*L,10*

RAY

Pull-down Menu	COMMAND (TYPE)	ALIAS (TYPE)	Short-cut	Screen (side) Menu	Tablet Menu
Draw *Ray*	*RAY*	...	...	*DRAW 1* *Ray*	*K,10*

The *Xline* command offers several options shown below:

```
Command: Xline
Hor/Ver/Ang/Bisect/Offset/<From point:>
```

The *Ang* option can be used to create a construction line at a specified angle. In this case, the angle specified would be that of the inclined plane or perpendicular to the inclined plane. The *Offset* option works well for drawing construction lines parallel to the inclined plane, especially in the case where the angle of the plane is not known.

Figure 25-11 illustrates the use of *Xline Offset* to create construction lines for the partial auxiliary view. The *Offset* option operates similarly to the *Offset* command described previously. Remember that an *Xline* extends to infinity but can be *Trimmed*, in which case it is converted to *Ray* (*Trim* once) or to a *Line* (*Trim* twice). See Chapter 15 for more information on the *Xline* command.

Figure 25-11

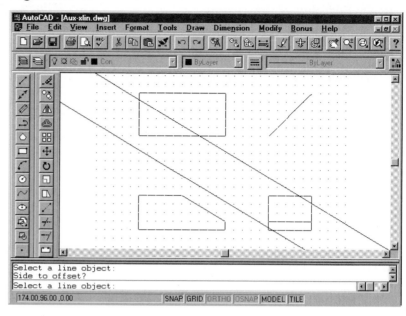

The *Ray* command also creates construction lines; however, the *Ray* has one anchored point and the other end extends to infinity.

```
Command: Ray
From point: PICK or (coordinates)
To point: PICK or (coordinates)
```

Rays are helpful for auxiliary view construction when you want to create projection lines perpendicular to the inclined plane. In Figure 25-12, two *Rays* are constructed from the *Endpoints* of the inclined plane and *Perpendicular* to the existing *Xlines*. Using *Xlines* and *Rays* in conjunction is an <u>excellent method</u> for "blocking in" the view.

There are two strategies for creating drawings using *Xlines* and *Rays*. First, these construction lines can be created on a separate layer and set up as a framework for the object lines. The object lines would then be drawn "on top of" the construction lines using *Osnaps*, but would be drawn on the object layer. The con-

Figure 25-12

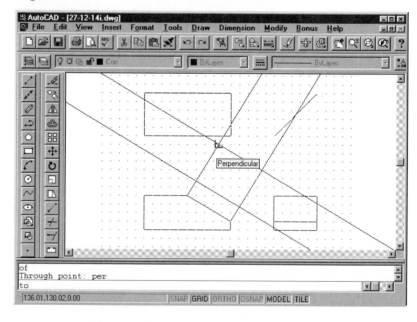

struction layer would be *Frozen* for plotting. The other strategy is to create the construction lines on the object layer. Through a series of *Trims* and other modifications, the *Xlines* and *Rays* are transformed to the finished object lines.

Now that you are aware of several methods for constructing auxiliary views, use any one method or a combination of methods for your drawings. No matter which methods are used, the final lines that are needed for the finished auxiliary view should be the same. It is up to you to use the methods that are the most appropriate for the particular application or are the easiest and quickest for you personally.

Constructing Full Auxiliary Views

The construction of a full auxiliary view begins with the partial view. After initial construction of the partial view, begin the construction of the full auxiliary by projecting the other edges and features of the object (other than the inclined plane) to the existing auxiliary view.

The procedure for constructing full auxiliary views in AutoCAD is essentially the same as that for partial auxiliary views. Use of the *Offset, Xline,* and *Ray* commands and *SNAP* and *GRID* rotation should be used as illustrated for the partial auxiliary view example. Because a full auxiliary view is projected at the same angle as a partial, the same rotation angle and basepoint would be used for the *SNAP*.

CHAPTER EXERCISES

For the following exercises, create the multiview drawing, including the partial or full auxiliary view as indicated. Use the appropriate template drawing based on the given dimensions and indicated plot scale.

1. Make a multiview drawing with a partial auxiliary view of the example used in this chapter. Refer to Figure 25-1 for dimensions. *Save* the drawing as **CH25EX1**. *Plot* on an A size sheet at **1=1** scale.

2. Recreate the views given in Figure 25-13 and add a partial auxiliary view. *Save* the drawing as **CH25EX2** and *Plot* on an A size sheet at **2=1** scale.

Figure 25-13

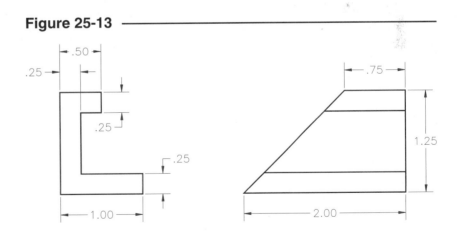

3. Recreate the views shown in Figure 25-14 and add a partial auxiliary view. *Save* the drawing as **CH25EX3**. Make a *Plot* on an A size sheet at **1=1** scale.

Figure 25-14

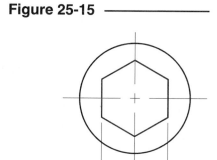

4. Make a multiview drawing of the given views in Figure 25-15. Add a full auxiliary view. *Save* the drawing as **CH25EX4**. *Plot* the drawing at an appropriate scale on an A size sheet.

Figure 25-15

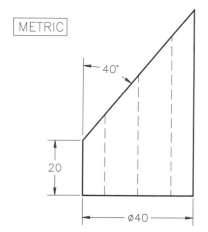

5. Draw the front, top, and a partial auxiliary view of the holder. Make a *Plot* full size. Save the drawing as **HOLDER**.

Figure 25-16

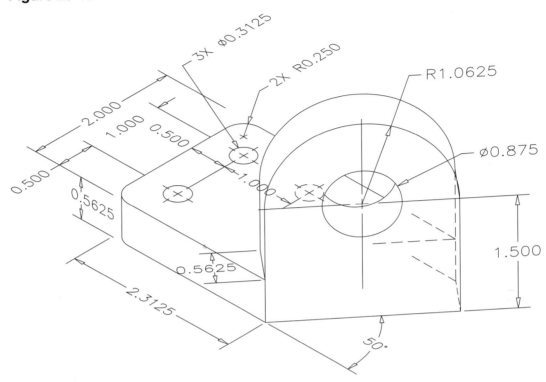

6. Draw three principal views and a full auxiliary view of the V-block shown in Figure 25-17. *Save* the drawing as **VBLOCK**. *Plot* to an accepted scale.

Figure 25-17

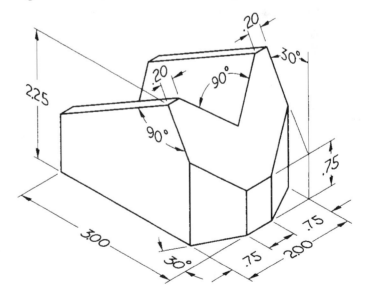

7. Draw two principal views and a partial auxiliary of the angle brace. *Save* as **ANGLBRAC**. Make a plot to an accepted scale on an A or B size sheet.

Figure 25-18

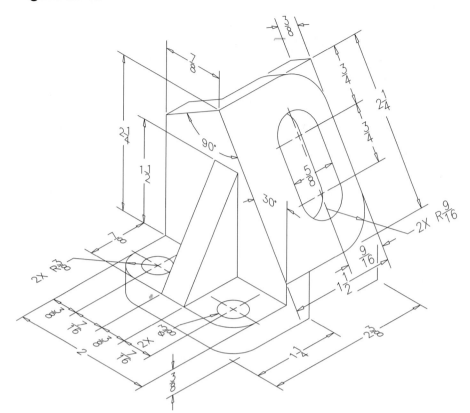

26

DIMENSIONING

Chapter Objectives

After completing this chapter you should:

1. be able to create linear dimensions with *Dimlinear*;

2. be able to append *Dimcontinue* and *Dimbaseline* dimensions to existing dimensions;

3. be able to create *Angular, Diameter,* and *Radius* dimensions;

4. know how to affix notes to drawings with *Leaders*;

5. know the possible methods for editing associative dimensions and dimensioning text;

6. know how to dimension isometric drawings.

CONCEPTS

As you know, drawings created with CAD systems should be constructed with the same dimensions and units as the real-world objects they represent. The importance of this practice is evident when you begin applying dimensions to the drawing geometry in AutoCAD. The features of the object that you specify for dimensioning are automatically measured, and those values are used for the dimensioning text. If the geometry has been drawn accurately, the dimensions will be created correctly.

The main components of a dimension are:

Figure 26-1

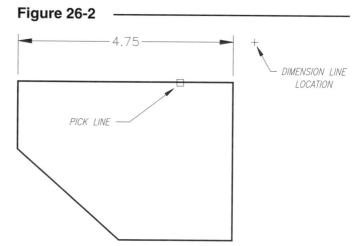

1. Dimension line
2. Extension lines
3. Dimension text (usually a numeric value)
4. Arrowheads or tick marks

AutoCAD dimensioning is <u>semi-automatic</u>. When you invoke a command to create a linear dimension, AutoCAD only requires that you PICK an object or specify the extension line origins (where you want the extension lines to begin) and PICK the location of the dimension line (distance from the object). AutoCAD then measures the feature and draws the extension lines, dimension line, arrowheads, and dimension text.

For linear dimensioning commands, there are <u>two ways</u> to specify placement for a dimension in AutoCAD: you can PICK the <u>object</u> to be dimensioned or you can PICK the two <u>extension line origins</u>. The simplest method is to select the object because it requires only one PICK (Fig. 26-2):

Figure 26-2

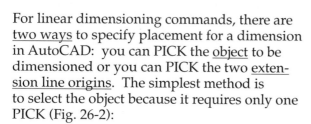

> Command: *dimlinear*
> First extension line origin or press ENTER to select: **Enter**
> Select object to dimension: **PICK**

The other method is to PICK the extension line origins (Fig. 26-3). *Osnaps* should be used to PICK the object (endpoints in this case) so that the dimension is <u>associated</u> with the object.

Figure 26-3

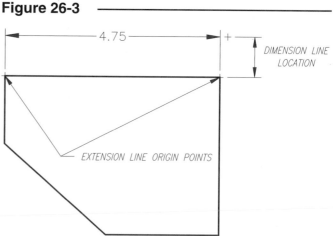

> Command: *dimlinear*
> First extension line origin or press ENTER to select: **PICK**
> Second extension line origin: **PICK**

Once the dimension is attached to the object, you specify how far you want the dimension to be placed from the object (called the "dimension line location").

Dimensioning in AutoCAD is <u>associative</u> (by default). Because the extension line origins are "associated" with the geometry, the dimension text automatically updates if the geometry is *Stretched*, *Rotated*, *Scaled*, or likewise edited using grips.

Because dimensioning is semi-automatic, <u>dimensioning variables</u> are used to control the way dimensions are created. Dimensioning variables can be used to control features such as text or arrow size, direction of the leader arrow for radial or diametrical dimensions, format of the text, and many other possible options. Groups of variable settings can be named and saved as <u>Dimension Styles</u>. Dimensioning variables and dimension styles are discussed in Chapter 27.

Dimensioning commands can be invoked by the typical methods. The *Dimension* pull-down menu contains the dimension creation and editing commands (Fig. 26-4). A *Dimension* toolbar can also be activated by using the *Toolbars* dialog box (Fig. 26-5).

Figure 26-4 ──────────────────────────

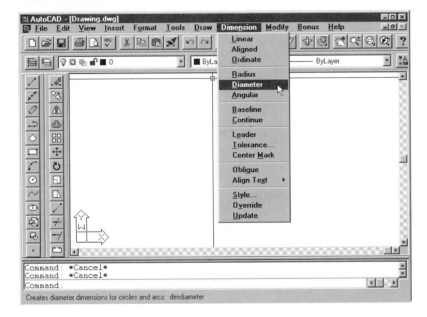

Figure 26-5 ──────────────────────────

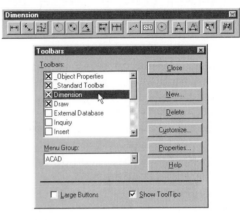

DIMENSION DRAWING COMMANDS

DIMLINEAR

Pull-down Menu	COMMAND (TYPE)	ALIAS (TYPE)	Short-cut	Screen (side) Menu	Tablet Menu
Dimension *Linear*	*DIMLINEAR*	*DIMLIN* *or DLI*	...	*DIMNSION* *Linear*	*W,5*

Dimlinear creates a <u>horizontal, vertical, or rotated</u> dimension. If the object selected is a horizontal line (or the extension line origins are horizontally oriented), the resulting dimension is a horizontal dimension. This situation is displayed in the previous illustrations (Figs. 26-2 and 26-3), or if the selected object or extension line origins are vertically oriented, the resulting dimension is vertical (Fig. 26-6).

Figure 26-6 ──────────────────────────

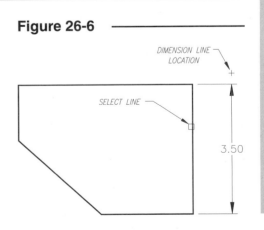

When you dimension an inclined object (or if the selected extension line origins are diagonally oriented), a vertical <u>or</u> horizontal dimension can be made, depending on where you drag the dimension line in relation to the object. If the dimension line location is more to the side, a vertical dimension is created (Fig. 26-7), or if you drag farther up or down, a horizontal dimension results (Fig 26-8).

If you select the extension line origins, it is very important to PICK the <u>object's endpoints</u> if the dimensions are to be truly associative (associated with the geometry). *OSNAP* should be used to find the object's *Endpoint, Intersection,* etc., unless that part of the geometry is located at a *SNAP* point.

Figure 26-7

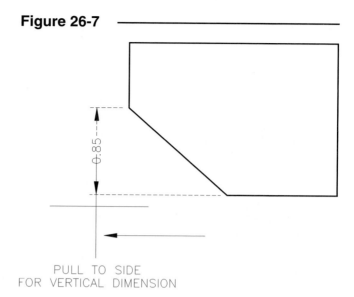

PULL TO SIDE
FOR VERTICAL DIMENSION

Command: **dimlinear**
First extension line origin or press ENTER to select: **endp of PICK**
Second extension line origin: **endp of PICK**
Dimension line location (Mtext/Text/Angle/Horizontal/Vertical/Rotated): **PICK** (where you want the dimension line placed)
Command:

When you pick the location for the dimension line, AutoCAD automatically measures the object and inserts the correct numerical value. The other options are explained next.

Figure 26-8

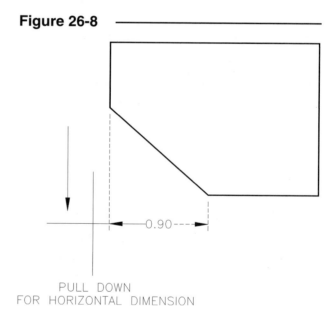

PULL DOWN
FOR HORIZONTAL DIMENSION

Rotated
If you want the dimension line to be drawn at an angle instead of vertical or horizontal, use this option. Selecting an inclined line, as in the previous two illustrations, would normally create a horizontal or vertical dimension. The *Rotated* option allows you to enter an <u>angular value</u> for the dimension line to be drawn. For example, selecting the diagonal line and specifying the appropriate angle would create the dimension shown in Figure 26-9. This object, however, could be more easily dimensioned with the *Dimaligned* command (see *Dimaligned*).

Figure 26-9

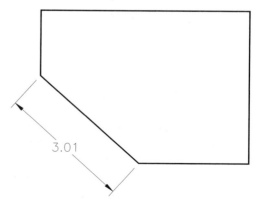

A *Rotated* dimension should be used when the geometry has "steps" or any time the desired dimension line angle is different than the dimensioned feature (when you need extension lines of different lengths). Figure 26-10 illustrates the result of using a *Rotated* dimension to give the correct dimension line angle and extension line origins for the given object. In this case, the extension line origins were explicitly PICKed. The feature of *Rotated* that makes it unique is that you specify the <u>angle</u> that the dimension line will be drawn.

Figure 26-10 ──────────

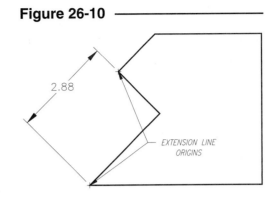

Text
Using the *Text* option allows you to enter any value or characters in place of the AutoCAD-measured text. The measured value is given as a reference at the command prompt.

> Command: **DIMLINEAR**
> First extension line origin or press ENTER to select:
> **Enter**
> Select object to dimension: **PICK**
> Dimension line location (Mtext/Text/Angle/Horizontal/Vertical/Rotated): **T**
> Dimension text <0.25>: (**text** or **Value**)

Figure 26-11 ──────────

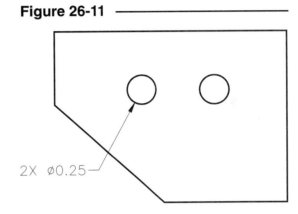

Entering a value or text at the prompt (above) causes AutoCAD to display that value or text instead of the AutoCAD-measured value. If you want to keep the AutoCAD-measured value but add a prefix or suffix, use the < and > (less-than, greater-than) symbols to represent the actual (AutoCAD) value:

> Dimension text <0.25>: **2X <>**
> Dimension line location (Mtext/Text/Angle/Horizontal/Vertical/Rotated): **PICK**
> Dimension text = 0.25
> Command:

The "2X <>" response produces the dimension text shown in Figure 26-11.

NOTE: Changing the AutoCAD-measured value should be discouraged. If the geometry is drawn accurately, the dimensional value is correct. If you specify other dimension text, the text value is <u>not</u> updated in the event of *Stretching, Rotating,* or otherwise editing the associative dimension.

Mtext
This option allows you to change the existing or insert additional text to the AutoCAD-supplied numerical value. The text is entered by the *Multiline Text Editor*. The less-than and greater-than symbols (< >) represent the AutoCAD-supplied dimensional value. Place text or numbers inside the symbols if you

Figure 26-12 ──────────

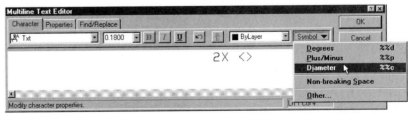

want to override the correct measurement (not advised) or place text outside the symbols if you want to add annotation to the numerical value. For example, entering "2X " before the symbols (Fig. 26-12) creates a dimensional value as shown in Figure 26-11.

Keep in mind—the power of using the *Multiline Text Editor* is the availability of all the text creation and editing features in the dialog box. For example, you can specify fonts, text height, bold, italic, underline, stacked text or fractions, and layer from the *Character* tab. In addition, you can use the *Symbol* drop-down box to insert a degree, plus/minus, or diameter symbol with the text value. The *Properties* tab gives access to text style, justification, and rotation angle options.

Angle

This creates text drawn at the angle you specify. Use this for special cases when the text must be drawn to a specific angle other than horizontal. (It is also possible to make the text automatically align with the angle of the dimension line using the *Dimension Styles* dialog box. See Chapter 27.)

Horizontal

Use the *Horizontal* option when you want to force a horizontal dimension for an inclined line and the desired placement of the dimension line would otherwise cause a vertical dimension.

Vertical

This option forces a *Vertical* dimension for any case.

DIMALIGNED

Pull-down Menu	COMMAND (TYPE)	ALIAS (TYPE)	Short-cut	Screen (side) Menu	Tablet Menu
Dimension Aligned	DIMALIGNED	DIMALI or DAL	...	DIMNSION Aligned	W,4

An *Aligned* dimension is aligned with (at the same angle as) the selected object or the extension line origins. For example, when *Aligned* is used to dimension the angled object shown in Figure 26-13, the resulting dimension aligns with the *Line*. This holds true for either option—PICKing the object or the extension line origins. If a *Circle* is PICKed, the dimension line is aligned with the selected point on the *Circle* and its center. The command syntax for the *Dimaligned* command accepting the defaults is:

Figure 26-13

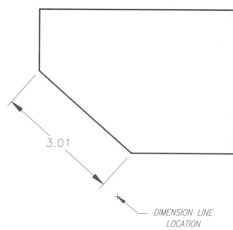

```
Command: dimaligned
First extension line origin or press ENTER to select: PICK
Second extension line origin: PICK
Dimension line location (Mtext/Text/Angle): PICK
Command:
```

The three options (*Mtext/Text/Angle*) operate similar to those for *Dimlinear*.

Mtext

The *Mtext* option calls the *Multiline Text Editor*. You can alter the AutoCAD-supplied text value or modify other visible features of the dimension text such as fonts, text height, bold, italic, underline, stacked text or fractions, layer, symbols, text style, justification, and rotation angle (see *Dimlinear, Mtext*).

Text

You can change the AutoCAD-supplied numerical value or add other annotation to the value in command line format (see *Dimlinear, Text*).

Angle

Enter a value for the angle that the text will be drawn.

The typical application for *Dimaligned* is for dimensioning an angled but <u>straight</u> feature of an object, as shown in Figure 26-13. *Dimaligned* should not be used to dimension an object feature that contains "steps," as shown in Figure 26-10. *Dimaligned* always draws <u>extension lines of equal length</u>.

DIMBASELINE

Pull-down Menu	COMMAND (TYPE)	ALIAS (TYPE)	Short-cut	Screen (side) Menu	Tablet Menu
Dimension Baseline	DIMBASELINE	DIMBASE or DBA	…	DIMNSION Baseline	W,2

Dimbaseline allows you to create a dimension that uses an extension line origin from a previously created dimension. Successive *Dimbaseline* dimensions can be used to create the style of dimensioning shown in Figure 26-14.

A baseline dimension must be connected to an existing dimension. If *Dimbaseline* is invoked immediately after another dimensioning command, you are required only to specify the second extension line origin since AutoCAD knows to use the <u>previous</u> dimension's <u>first</u> extension line origin:

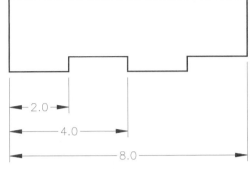

Figure 26-14

```
Command: Dimbaseline
Specify a second extension line origin or (Undo/<Select>): PICK
Dimension text = nn.nnnn
Specify a second extension line origin or (Undo/<Select>): Enter
Command:
```

The <u>previous dimension's first extension line</u> is used also for the baseline dimension (Fig. 26-15). Therefore, you specify only the second extension line origin. Note that you are not required to specify the dimension line location. AutoCAD spaces the new dimension line automatically, based on the setting of the dimension line increment variable (Chapter 27).

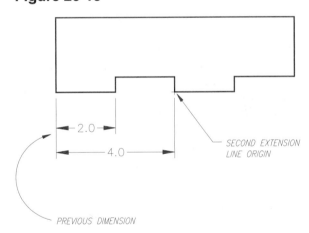

Figure 26-15

If you wish to create a *Dimbaseline* dimension using a dimension other than the one just created, use the "Select" option (Fig. 26-16):

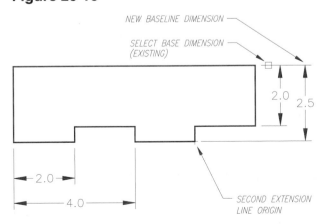

Figure 26-16

```
Command: Dimbaseline
Specify a second extension line origin or
(Undo/<Select>): S
Select base dimension: PICK (existing exten-
sion line)
Specify a second extension line origin or
(Undo/<Select>): PICK
Dimension text = nn.nnnn
Specify a second extension line origin or
(Undo/<Select>): Enter
Command:
```

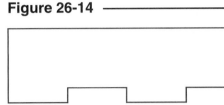

The extension line selected as the base dimension becomes the first extension line for the new *Dimbaseline* dimension.

The *Undo* option can be used to undo the last baseline dimension created in the current command sequence.

Dimbaseline can be used with rotated, aligned, angular, and ordinate dimensions.

DIMCONTINUE

Pull-down Menu	COMMAND (TYPE)	ALIAS (TYPE)	Short-cut	Screen (side) Menu	Tablet Menu
Dimension Continue	*DIMCONTINUE*	*DIMCONT or DCO*	...	*DIMNSION Continue*	*W,1*

Dimcontinue dimensions continue in a line from a previously created dimension. *Dimcontinue* dimension lines are attached to, and drawn the same distance from, the object as an existing dimension.

Dimcontinue is similar to *Dimbaseline* except that an existing dimension's <u>second</u> extension line is used to begin the new dimension. In other words, the new dimension is connected to the <u>second</u> extension line, rather than to the <u>first</u>, as with a *Dimbaseline* dimension (Fig. 26-17).

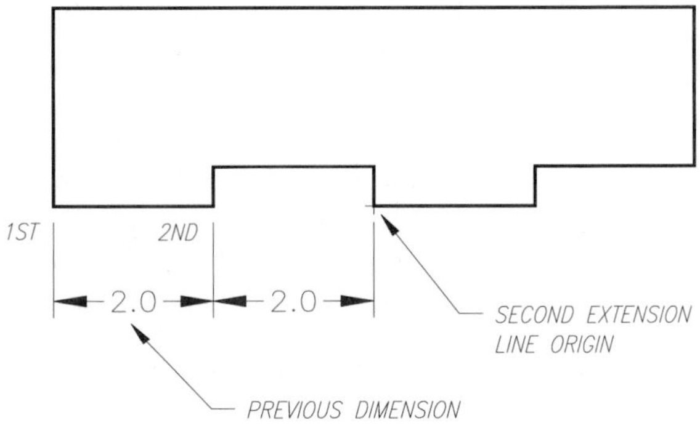

The command syntax is as follows:

Command: **Dimcontinue**
Specify a second extension line origin or
(Undo/<Select>): **PICK**
Dimension text = *nn.nnnn*
Specify a second extension line origin or (Undo/<Select>): **Enter**
Command:

Assuming a dimension was just drawn, *Dimcontinue* could be used to place the next dimension, as shown in Figure 26-17.

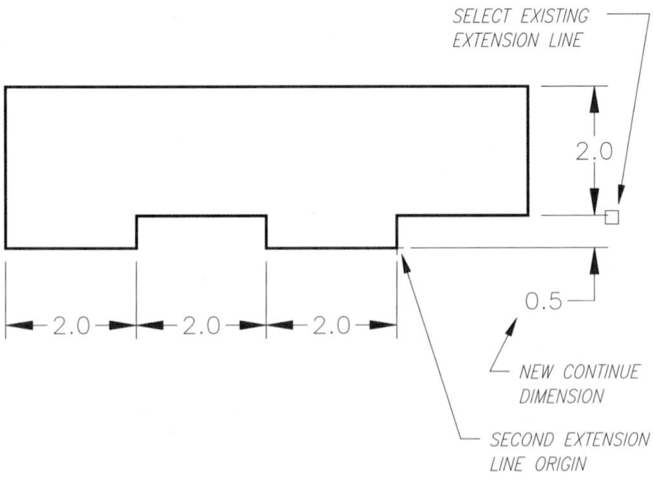

If you want to create continued dimension and attach it to an extension line <u>other</u> than the previous dimension's second extension line, you can use the *Select* option to pick an extension line of any other dimension. Then, use *Dimcontinue* to create a continued dimension from the selected extension line (Fig. 26-18).

The *Undo* option can be used to undo the last continued dimension created in the current command sequence. *Dimcontinue* can be used with rotated, aligned, angular, and ordinate dimensions.

Figure 26-17 —————————

1ST · 2ND · 2.0 · 2.0 · SECOND EXTENSION LINE ORIGIN · PREVIOUS DIMENSION

Figure 26-18 —————————

SELECT EXISTING EXTENSION LINE · 2.0 · 2.0 · 2.0 · 2.0 · 0.5 · NEW CONTINUE DIMENSION · SECOND EXTENSION LINE ORIGIN

DIMDIAMETER

	COMMAND (TYPE)	ALIAS (TYPE)	Short-cut	Screen (side) Menu	Tablet Menu
Pull-down Menu					
Dimension *Diameter*	DIMDIAMETER	DIMDIA or DDI	...	*DIMNSION* *Diameter*	X,4

The *Dimdiameter* command creates a diametrical dimension by selecting any *Circle*. Diametrical dimensions should be used for full 360 degree *Circles* and can be used for *Arcs* of more than 180 degrees.

```
Command: Dimdiameter
Select arc or circle: PICK
Dimension text = nn.nnnn
Dimension line location (Mtext/Text/Angle): PICK
Command:
```

You can PICK the circle at any location. AutoCAD allows you to adjust the position of the dimension line to any angle or length (Fig. 26-19). Dimension lines for diametrical or radial dimensions should be drawn to a regular angle, such as 30, 45, or 60 degrees, never vertical or horizontal.

Figure 26-19

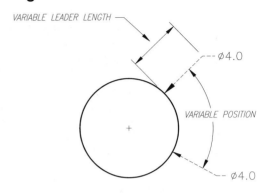

A typical diametrical dimension appears as the example in Figure 26-19. According to ANSI standards, a diameter dimension line and arrow should point inward (toward the center) for holes and small circles where the dimension line and text do not fit within the circle. Use the default variable settings for *Dimdiameter* dimensions such as this.

For dimensioning large circles, ANSI standards suggest an alternate method for diameter dimensions where sufficient room exists for text and arrows inside the circle (Fig. 26-20). To create this style of dimension in Release 14, the variables that control these features should be set to *Arrows Only Fit* and *User Defined* in the *Format* Dimension Style dialog box (see Chapter 27).

Figure 26-20

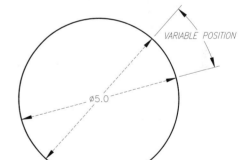

Notice that the *Diameter* command creates center marks at the *Circle's* center. Center marks can also be drawn by the *Dimcenter* command (discussed later). AutoCAD uses the center and the point selected on the *Circle* to maintain its associativity.

Mtext/Text

The *Mtext* or *Text* options can be used to modify or add annotation to the default value. The *Mtext* option summons the *Multiline Text Editor* and the *Text* option uses command line format. Both options operate similar to the other dimensioning commands (see *Dimlinear*, *Mtext*, and *Text*).

Figure 26-21

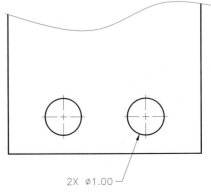

Notice that with diameter dimensions AutoCAD automatically creates the Ø (phi) symbol before the dimensional value. This is the latest ANSI standard for representing diameters. If you prefer to use a prefix before or suffix after the dimension value, it can be accomplished with the *Mtext* or *Text* option. Remember that the < > symbols represent the AutoCAD-measured value so the additional text should be inserted on either side of the symbols. Inserting a prefix by this method does not override the Ø (phi) symbol (Fig. 26-21).

A prefix or suffix can alternately be added to the measured value by using the *Dimension Styles* dialog box and entering text or values in the *Prefix* or *Suffix* edit boxes. Using the dialog box method, however, <u>overrides</u> the Ø symbol (see Chapter 27).

Angle
With this option, you can specify an angle (other than the default) for the text to be drawn by entering a value.

DIMRADIUS

Pull-down Menu	COMMAND (TYPE)	ALIAS (TYPE)	Short-cut	Screen (side) Menu	Tablet Menu
Dimension Radius	*DIMRADIUS*	*DIMRAD or DRA*	...	*DIMNSION Radius*	X,5

Dimradius is used to create a dimension for an arc of anything less than half of a circle. ANSI standards dictate that a *Radius* dimension line should point outward (from the arc's center), unless there is insufficient room, in which case the line can be drawn on the outside pointing inward, as with a leader. The text can be located inside an arc (if sufficient room exists) or is forced outside of small *Arcs* on a leader.

```
Command: Dimradius
Select arc or circle: PICK
Dimension text = nn.nnnn
Dimension line location (Mtext/Text/Angle): PICK
Command:
```

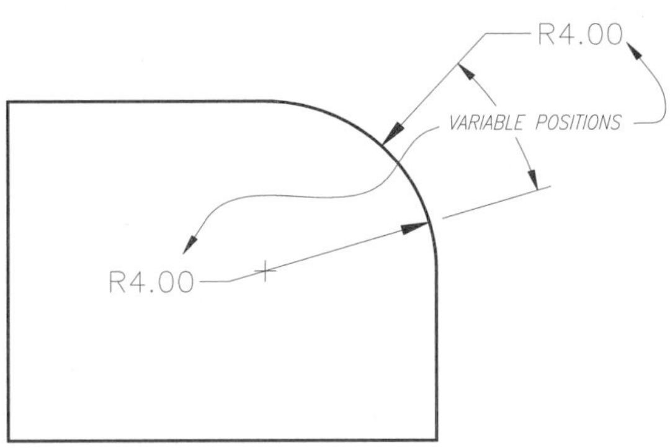

Assuming the defaults, a *Dimradius* dimension can appear on either side of an arc, as shown in Figure 26-22. Placement of the dimension line is variable. Dimension lines for arcs and circles should be positioned at a regular angle such as 30, 45, or 60 degrees, never vertical or horizontal.

When the radius dimension is dragged outside of the arc, a center mark is automatically created (Fig. 26-22). When the radius dimension is dragged inside the arc, no center mark is created. (Center marks can be created using the *Center* command discussed next.)

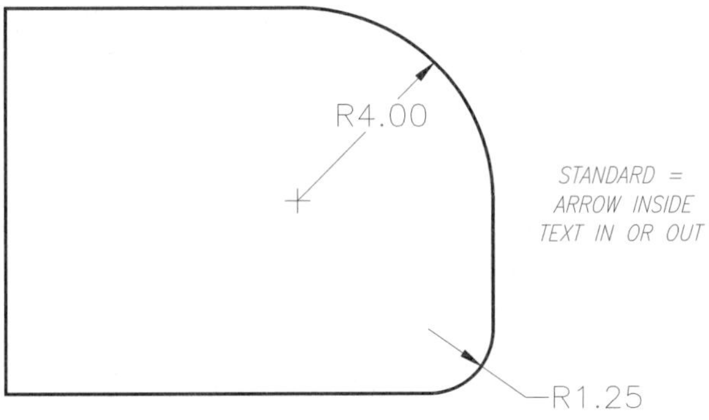

According to ANSI standards, the dimension line and arrow should point <u>outward</u> from the center for radial dimensions (Fig. 26-23). The text can be placed inside or outside of the *Arc*, depending on how much room exists. To create a *Dimradius* dimension to comply with this standard, dimension variables must be changed from the defaults. Using the *Dimension Styles* dialog box, then the *Format* dialog box, check *User Defined* and set *Fit* to *Arrows Only* to achieve *Dimradius* dimensions as shown in Figure 26-23 (see Chapter 27 for more information on these variables).

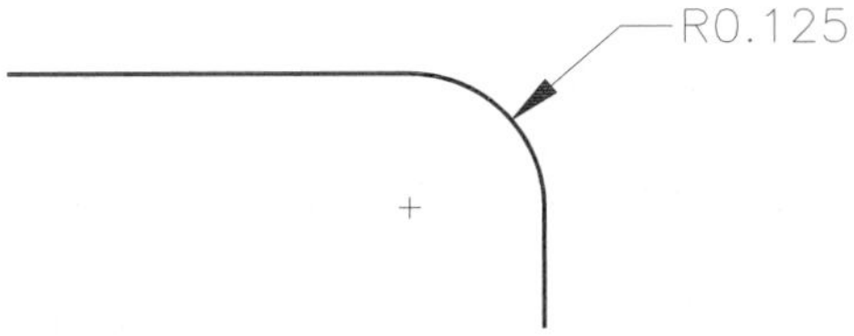

For very small radii, such as that shown in Figure 26-24, there is insufficient room for the text and arrow to fit inside the *Arc*. In this case, AutoCAD automatically forces the text outside with the leader pointing inward toward the center. <u>No changes</u> have to be made to the default settings for this to occur.

Figure 26-22

R4.00
VARIABLE POSITIONS
R4.00

Figure 26-23

R4.00
STANDARD = ARROW INSIDE TEXT IN OR OUT
R1.25

Figure 26-24

R0.125

Mtext/Text

These options can be used to modify or add annotation to the default value. AutoCAD automatically inserts the letter "R" before the numerical text whenever a *Dimradius* dimension is created. This is the correct notation for radius dimensions. The *Mtext* option calls the *Multiline Text Editor*, and the *Text* option uses the command line format for entering text. Remember that AutoCAD uses the < > symbols to represent the AutoCAD-supplied value. Entering text inside the < > symbols overrides the measured value. Entering text before the symbols adds a prefix <u>without overriding the "R" designation</u>. Alternately, text can be added by using the *Prefix* and *Suffix* options of the *Dimension Styles* dialog box series; however, a *Prefix* entered in the edit box <u>replaces</u> the letter "R." (See Chapter 27.)

Angle

With this option, you can specify an angle (other than the default) for the <u>text</u> to be drawn by entering a value.

DIMCENTER

Pull-down Menu	COMMAND (TYPE)	ALIAS (TYPE)	Short-cut	Screen (side) Menu	Tablet Menu
Dimension Center Mark	DIMCENTER	DCE	...	DIMNSION Center	X,2

The *Dimcenter* command draws a center mark on any selected *Arc* or *Circle*. As shown earlier, the *Dimdiameter* command and the *Dimradius* command sometimes create the center marks automatically.

The command requires you only to select the desired *Circle* or *Arc* to acquire the center marks:

    Command: *dimcenter*
    Select arc or circle: **PICK**
    Command:

Figure 26-25 ———————————

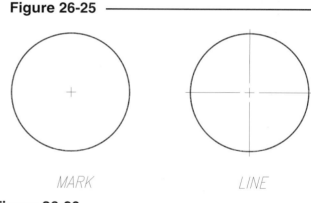

No matter if the center mark is created by the *Center* command or by the *Diameter* or *Radius* commands, the center mark can be either a small cross or complete center lines extending past the *Circle* or *Arc* (Fig 26-25). The type of center mark drawn is controlled by the *Mark* or *Line* setting in the *Geometry* dialog box from the *Dimension Style* series. It is suggested that a new *Dimension Style* be created with these settings just for drawing center marks. (See Chapter 27.)

Figure 26-26 ———————————

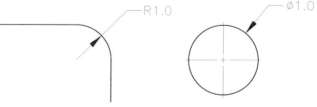

When dimensioning, short center marks should be used for *Arcs* of less than 180 degrees, and full center lines should be drawn for *Circles* and for *Arcs* of 180 degrees or more (Fig. 26-26).

NOTE: Since the center marks created with the *Dimcenter* command are <u>not</u> associative, they may be *Trimmed*, *Erased*, or otherwise edited, as shown in Figure 26-27. The center marks created with the *Dimradius* or *Dimdiameter* commands <u>are</u> associative and cannot be edited.

Figure 26-27 ———————————

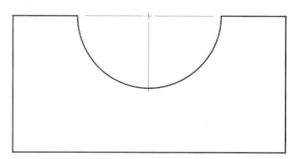

The lines comprising the center marks created with *Dimcenter* can be *Erased* or otherwise edited. In the case of a 180 degree *Arc*, two center mark lines can be shortened using *Break*, and one line can be *Erased* to achieve center lines as shown in Figure 26-27.

DIMANGULAR

Pull-down Menu	COMMAND (TYPE)	ALIAS (TYPE)	Short-cut	Screen (side) Menu	Tablet Menu
Dimension Angular	*DIMANGULAR*	*DIMANG or DAN*	...	*DIMNSION Angular*	X,3

The *Dimangular* command provides many possible methods of creating an angular dimension.

A typical angular dimension is created between two *Lines* that form an angle (of other than 90 degrees). The dimension line for an angular dimension is radiused with its center at the vertex of the angle (Figure 26-28). A *Dimangular* dimension automatically adds the degree symbol (°) to the dimension text. The dimension text format is controlled by the current settings for *Units* in the *Dimension Style* dialog box.

Figure 26-28

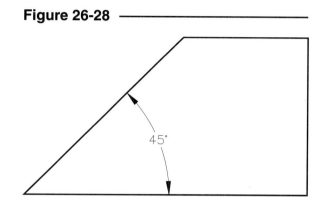

AutoCAD automates the process of creating this type of dimension by offering options within the command syntax. The default options create a dimension, as shown here:

```
Command: Dimangular
Select arc, circle, line, or press ENTER: PICK
Second line: PICK
Dimension arc line location (Mtext/Text/Angle): PICK the desired location of the dimension line
Dimension text = nn.nnnn
Command:
```

Dimangular dimensioning offers some very useful and easy-to-use options for placing the desired dimension line and text location.

At the "Dimension arc line location (Mtext/Text/Angle):" prompt, you can move the cursor around the vertex to dynamically display possible placements available for the dimension. The dimension can be placed in any of four positions as well as any distance from the vertex (Fig. 26-29). Extension lines are automatically created as needed.

The *Dimangular* command offers other options, including dimensioning angles for *Arcs*, *Circles*, or allowing selection of any three points.

Figure 26-29

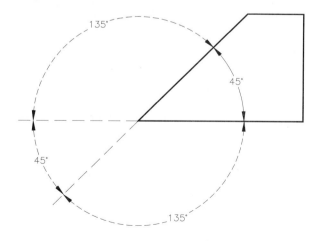

If you select an *Arc* in response to the "Select arc, circle, line, or press ENTER:" prompt, AutoCAD uses the *Arc's* center as the vertex and the *Arc's* endpoints to generate the extension lines. You can select either angle of the *Arc* to dimension (Fig. 26-30).

Figure 26-30 ────────

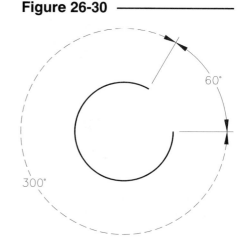

If you select a *Circle*, AutoCAD uses the PICK point as the first extension line origin. The second extension line origin does not have to be on the *Circle*, as shown in Figure 26-31.

Figure 26-31 ────────

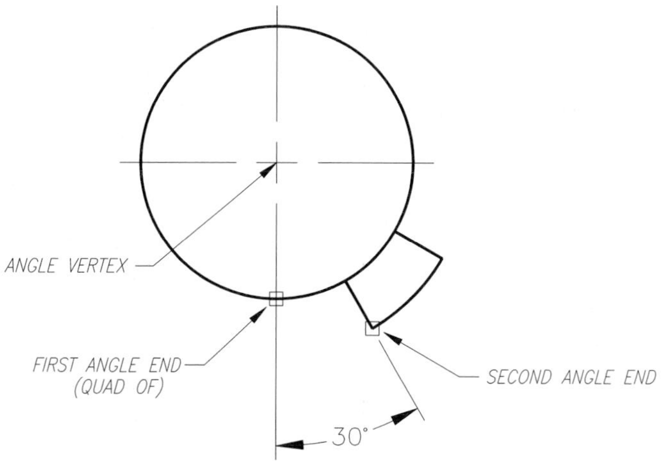

If you press **Enter** in response to the "Select arc, circle, line, or press ENTER:" prompt, AutoCAD responds with the following:

    Angle vertex: **PICK**
    First angle endpoint: **PICK**
    Second angle endpoint: **PICK**

This option allows you to apply an *Angular* dimension to a variety of shapes.

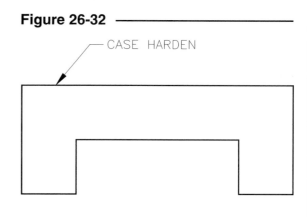

LEADER

	COMMAND (TYPE)	ALIAS (TYPE)	Short-cut	Screen (side) Menu	Tablet Menu
Pull-down Menu					
Dimension Leader	*LEADER*	*LEAD or LE*	...	*DIMNSION Leader*	*R,7*

The *Leader* command allows you to create an associative leader similar to that created with the *Diameter* command. The *Leader* command is intended to give dimensional notes such as the manufacturing or construction specifications shown here:

    Command: **leader**
    From point: **PICK**
    To point: **PICK**
    To point (Format/Annotation/Undo)<Annotation>:
    CASE HARDEN
    <Mtext>: **Enter**
    Command:

Figure 26-32 ────────

CASE HARDEN

At the "From point:" prompt, select the desired location for the arrow. You should use *SNAP* or an *OSNAP* option (such as *Nearest*, in this case) to ensure the arrow touches the desired object.

Figure 26-33

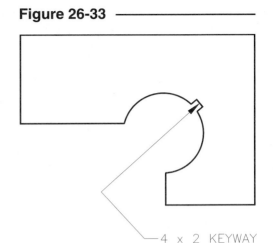

A short horizontal line segment called the "hook line" is automatically added to the last line segment drawn if the leader line is 15 degrees or more from horizontal. Note that command syntax for the *Leader* in Figure 26-32 indicates only one line segment was PICKed. A *Leader* can have as many segments as you desire (Fig. 26-33).

If you do not enter text at the "Annotation:" prompt, another series of options are available:

Command: **leader**
From point: **PICK**
To point: **PICK**
To point (Format/Annotation/Undo)<Annotation>: **Enter** or **(option)** or **(text)**
Annotation (or press ENTER for options): **Enter** or **(text)**
Tolerance/Copy/Block/None/<Mtext>: **Enter** or **(option)**
Command:

Format
This option produces another list of choices:

Spline/STraight/Arrow/None/<Exit>:

Spline/STraight
You can draw either a *Spline* or straight version of the leader line with these options. The resulting *Spline* leader line has the characteristics of a normal *Spline*. An example of a *Splined* leader is shown in Figure 26-34.

Figure 26-34

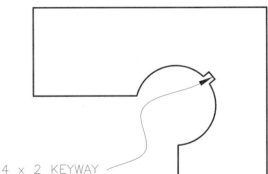
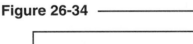

Arrow/None
This option draws the leader line with or without an arrowhead at the start point.

Annotation
This option prompts for text to insert at the end of the *Leader* line.

Mtext
The *Multiline Text Editor* appears with this option. Text can be entered into paragraph form using the *Mtext* text boundary mechanism (see *Mtext*, Chapter 18).

Tolerance
This option produces a feature control frame using the *Geometric Tolerances* dialog boxes (see *Tolerance*).

Copy
You can copy existing *Text, Dtext,* or *Mtext* objects from the drawing to be placed at the end of the *Leader* line. The copied object is associated with the *Leader* line.

Block
An existing *Block* of your selection can be placed at the end of the *Leader* line. The same prompts as the *Insert* command are used.

None
Using this option draws the *Leader* line with no annotation.

Undo
This option undoes the last vertex point of the *Leader* line.

Since a *Leader* is associative, it is affected by the current dimension style settings. You can control the *Leader's* arrowhead type, scale, color, etc., with the related dimensioning variables (see Chapter 27).

DIMORDINATE

Pull-down Menu	COMMAND (TYPE)	ALIAS (TYPE)	Short-cut	Screen (side) Menu	Tablet Menu
Dimension Ordinate	DIMORDINATE	DIMORD or DOR	...	DIMNSION Ordinate	W,3

Ordinate dimensioning is a specialized method of dimensioning used in the manufacturing of flat components such as those in the sheet metal industry. Because the thickness (depth) of the parts is uniform, only the width and height dimensions are specified as Xdatum and Ydatum dimensions.

Dimordinate dimensions give an Xdatum or a Ydatum distance between object "features" and a reference point on the geometry treated as the origin, usually the lower-left corner of the part. This method of dimensioning is relatively simple to create and easy to understand. Each dimension is composed only of one leader line and the aligned numerical value.

Figure 26-35 ───────────

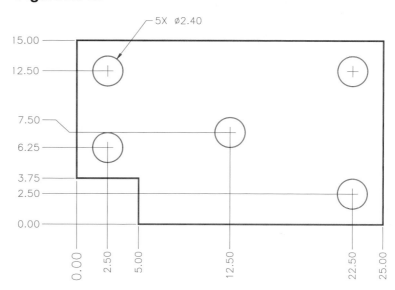

For more information on ordinate dimensioning, see the AutoCAD *User's Guide*, the AutoCAD *Command Reference*, or *AutoCAD 14 Instructor*.

TOLERANCE

Pull-down Menu	COMMAND (TYPE)	ALIAS (TYPE)	Short-cut	Screen (side) Menu	Tablet Menu
Dimension Tolerance...	TOLERANCE	TOL	...	DIMNSION Toleranc	X,1

Geometric dimensioning and tolerancing (GDT) has become an essential component of detail drawings of manufactured parts. Standard symbols are used to define the geometric aspects of the parts and how the parts function in relation to other parts of an assembly. The *Tolerance* command in AutoCAD produces a series of dialog boxes for you to create the symbols, values, and feature control frames needed to dimension a drawing with GDT. Invoking the *Tolerance* command produces the *Symbol* dialog box (Fig. 26-36) and the *Geometric Tolerance* dialog box. These dialog boxes can also be accessed by the *Leader*

Figure 26-36 ───────────

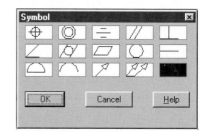

command. Unlike general dimensioning in AutoCAD, geometric dimensioning is <u>not associative</u>; however, you can use *Ddedit* to edit the components of an existing feature control frame.

Some common examples of geometric dimensioning may be a control of the flatness of a surface, the parallelism of one surface relative to another, or the positioning of a group of holes to the outside surfaces of the part. Most of the conditions are controls that cannot be stated with the other AutoCAD dimensioning commands.

Geometric dimensioning and tolerancing is beyond the purpose of this book; however, the AutoCAD command is presented here to recognize this capability. For more information on *Tolerance* and GDT, see the AutoCAD *User's Guide*, the AutoCAD *Command Reference*, or *AutoCAD 14 Instructor*.

DIM

In AutoCAD releases previous to Release 13, dimensioning commands had to be entered at the "Dim:" prompt and could not be entered at the "Command:" prompt. The "Dim:" prompt puts AutoCAD in the dimensioning mode so earlier release dimensioning commands can be used. This convention can also be used in Release 14. The *DIM* command must be typed in Release 14 to produce the "Dim:" prompt. Enter *E* or *Exit* or press Escape to exit the dimensioning mode and return to the "Command:" prompt. The syntax is as follows:

> Command: **dim**
> Dim:

Release 12 and previous dimensioning command names do not have the "dim" prefix; therefore, using the "Dim:" prompt is required for them. Almost all of the Release 12 commands have the same name as the Release 14 version but without the "dim" prefix. For example, *Dimdiameter* is "*Diameter*" in Release 12. In Release 14, you can enter these older dimensioning command names (without the "dim" prefix) at the "Dim:" prompt; for example,

> Command: **dim**
> Dim: **diameter**

With the exception of a few dimensioning commands and options, the Release 12 dimensioning commands are the same as the Release 14 versions without the "dim" prefix. However, several other older commands can be used in Release 14 at the "Dim:" prompt.

UNDO or *U*
Erases the last dimension objects or dimension variable setting.

UPDATE
Performs the same action as *Dimstyle, Apply*.

STYLE
Changes the current <u>text</u> style.

HORIZONTAL
Performs the same action as *Dimlinear, Horizontal*.

VERTICAL

Performs the same action as *Dimlinear, Vertical*.

ROTATED

Performs the same action as *Dimlinear, Rotate*.

Another use for the "Dim:" prompt is for the *Viewport* option of *DIMLFAC*, which is not available by any other method.

EDITING DIMENSIONS

Associative Dimensions

AutoCAD creates associative dimensions by default. Associative dimensions contain "definition points" that define points in the dimensions such as extension line origins, placement of dimension line, the selected points of *Circles* and *Arcs*, and the centers of those *Circles* and *Arcs*. If the geometry is modified by certain editing commands (more specifically, if the definition points are changed), the dimension components, including the numerical value, automatically update. When you create the first associative dimension, AutoCAD automatically creates a new layer called DEFPOINTS that contains all the definition points. The layer should be kept in a *Frozen* state. If you alter the points on this layer, existing dimensions lose their associativity and you eliminate the possibility of automatic editing.

All of the dimensioning commands are associative by default, and the resulting dimensions would be affected by the editing commands listed below. The *DIMASO* variable can be changed to make new dimensions <u>unassociative</u> if you desire (see Chapter 27). The commands that affect associative dimensions are *Extend, Mirror, Rotate, Scale, Stretch, Trim* (linear dimensions only), *Array* (if rotated in a *Polar* array), and grip editing options.

These commands cause the changed dimension to adjust to the changed angle or length. For example, if you use *Stretch* to change some geometry with associative dimensions, the numerical values and extension lines automatically update as the related geometry changes (Fig. 26-37). Remember to use a <u>Crossing Window</u> for selection with *Stretch* and to include the extension line <u>origin</u> (definition point) in the selection set.

Figure 26-37

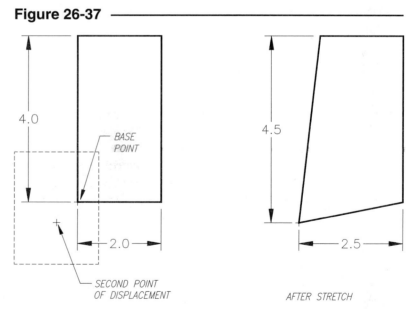

Another feature of associative dimensions is the ability to use *Trim* and *Extend*. For example, you can *Trim* an object and at the same time *Trim* its associated dimension (Fig. 26-38). In this case select the geometry to *Trim* as you would normally, then <u>also select the dimension to *Trim*</u> as shown.

Figure 26-38

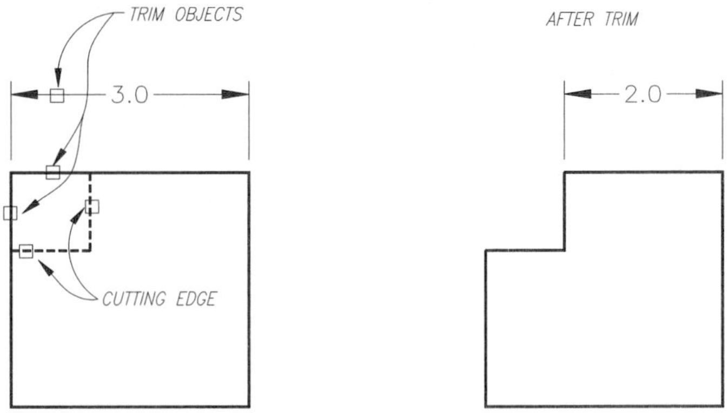

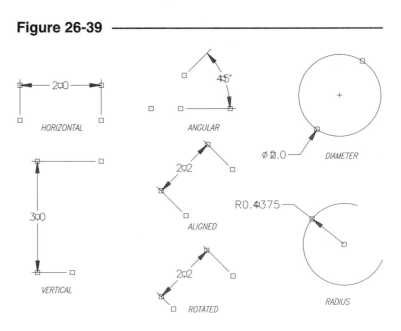

Grip Editing Dimensions

Grips can be used effectively for editing dimensions. Any of the grip options (STRETCH, MOVE, ROTATE, SCALE, and MIRROR) is applicable. Depending on the type of dimension (linear, radial, angular, etc.), grips appear at several locations on the dimension when you activate the grips by selecting the dimension at the Command: prompt. Associative dimensions offer the most powerful editing possibilities, although non-associative dimension components can also be edited with grips. There are many ways in which grips can be used to alter the measured value and configuration of dimensions.

Figure 26-39 shows the grips for each type of associative dimension. Linear dimensions (horizontal, vertical, and rotated) and aligned and angular dimensions have grips at each extension line origin, the dimension line position, and a grip at the text. A diameter dimension has grips defining two points on the diameter as well as one defining the leader length. The radius dimension has center and radius grips as well as a leader grip.

With dimension grips, a wide variety of editing options are possible. Any of the grips can be PICKed to make them **hot** grips. All grip options are valid methods for editing dimensions.

Figure 26-39

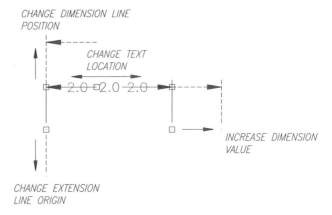

For example (Fig. 26-40), a horizontal dimension value can be increased by stretching an extension line origin grip in a horizontal direction. A vertical direction movement changes the length of the extension line. The dimension line placement is changed by stretching its grips. The dimension text can be stretched to any position by manipulating its grip.

An angular dimension can be increased by stretching the extension line origin grip. The numerical value automatically updates.

Figure 26-40

Stretching a rotated dimension's extension line origin allows changing the length of the dimension as well as the length of the extension line. An aligned dimension's extension line origin grip also allows you to change the aligned <u>angle</u> of the dimension.

Figure 26-41

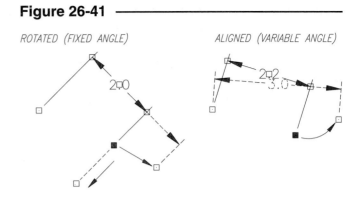

Rotating the <u>center</u> grip of a radius dimension (with the ROTATE option) allows you to reposition the location of the dimension around the *Arc*. Note that the text remains in its original horizontal orientation.

Figure 26-42

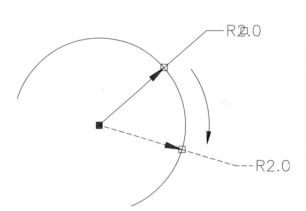

You can move the dimension text independent of the dimension line with grips for dimensions that have a *DIMFIT* variable setting of 5. Use an existing dimension, and change the *DIMFIT* setting to 5 (or use the *Dimension Style* dialog box, then the *Format* dialog box, and set *Fit* to *No leader*). The dimension text can be moved to any location without losing its associativity. See *Fit: (DIMFIT)*, Chapter 27.

Many other possibilities exist for editing dimensions using grips. Experiment on your own to discover some of the possibilities that are not shown here.

Exploding Associative Dimensions

Associative dimensions are treated as one object. If *Erase* is used with an associative dimension, the entire dimension (dimension line, extension lines, arrows, and text) is selected and *Erased*.

Explode can be used to break an associative dimension into its component parts. The individual components can then be edited. For example, after *Exploding*, an extension line can be *Erased* or text can be *Moved*.

There are two main drawbacks to *Exploding* associative dimensions. First, the associative property is lost. Editing commands or *Grips* cannot be used to change the entire dimension but affect only the component objects. More important, Dimension Styles cannot be used with unassociative dimensions. Second, the dimension is *Exploded* onto Layer 0 and loses its *color* and *linetype* properties. This can include additional work to reestablish the desired layer, *color*, and *linetype*.

Dimension Editing Commands

Several commands are provided to facilitate easy editing of existing dimensions in a drawing. Most of these commands are intended to allow variations in the appearance of dimension <u>text</u>. These editing commands operate <u>only</u> with <u>associative</u> dimensions.

DIMTEDIT

Pull-down Menu	COMMAND (TYPE)	ALIAS (TYPE)	Short-cut	Screen (side) Menu	Tablet Menu
Dimension Align Text >	DIMTEDIT	DIMTED	...	*DIMNSION Dimtedit*	Y,2

Dimtedit (text edit) allows you to change the position or orientation of the text for a single associative dimension. To move the position of text, this command syntax is used:

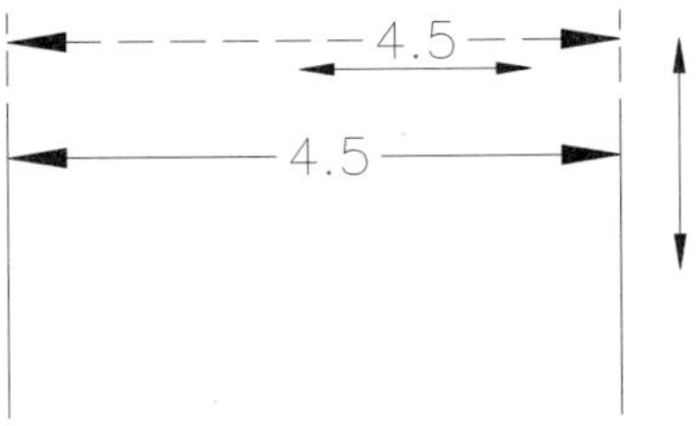

    Command: **Dimtedit**
    Select dimension: **PICK**
    Enter text location (Left/Right/Home/Angle): **PICK**

At the "Enter text location" prompt, drag the text to the desired location. The selected text and dimension line can be changed to any position, while the text and dimension line retain their associativity.

Figure 26-43 ───────

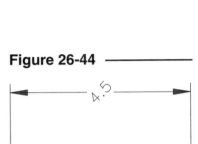

Angle
The *Angle* option works with any *Horizontal, Vertical, Aligned, Rotated, Radius,* or *Diameter* dimensions. You are prompted for the new <u>text</u> angle (Fig. 26-44):

    Command: **dimtedit**
    Select dimension: **PICK**
    Enter text location (Left/Right/Home/Angle): **a**
    Enter text angle: **45**
    Command:

Figure 26-44 ───────

Home
The text can be restored to its original (default) rotation angle with the *Home* option. The text retains its right/left position.

Right/Left
The *Right* and *Left* options automatically justify the dimension text at the extreme right and left ends, respectively, of the dimension line. The arrow and a short section of the dimension line, however, remain between the text and closest extension line (Fig. 26-45).

Center
The *Center* option brings the text to the center of the dimension line and sets the rotation angle to 0 (if previously assigned; Fig. 26-45).

Figure 26-45 ───────

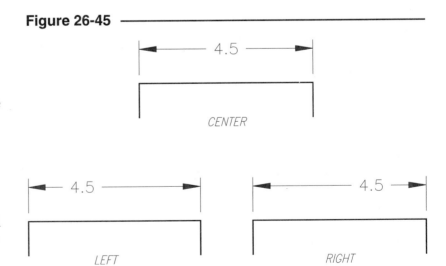

DIMEDIT

	Pull-down Menu	COMMAND (TYPE)	ALIAS (TYPE)	Short-cut	Screen (side) Menu	Tablet Menu
	Dimension *Oblique*	*DIMEDIT*	*DIMED* or *DED*	...	*DIMNSION* *Dimedit*	*Y,1*

The *Dimedit* command allows you to change the angle of the extension lines to an obliquing angle and provides several ways to edit dimension text. Two of the text editing options (*Home, Rotate*) duplicate those of the *Dimtedit* command. Another feature (*New*) allows you to change the text value and annotation.

```
Command: dimedit
Dimension Edit (Home/New/Rotate/Oblique) <Home>:
```

Home
This option moves the dimension text back to its original angular orientation angle without changing the left/right position. *Home* is a duplicate of the *Dimtedit* option of the same name.

New
You can change the text value and annotation of existing text. The same mechanism appears when you create the dimension and use the *Mtext* option—that is, the *Multiline Text Editor* appears. Remember that AutoCAD draws the dimension value in place of the < > characters. Add a prefix before these characters or a suffix after them. Entering text to replace the < > characters (erase them) causes your new text to appear instead of the AutoCAD-supplied value.

Placing text or numbers <u>inside</u> the symbols overrides the correct measurement and creates static, non-associative text. The original AutoCAD-measured value <u>can be restored</u>, however, using the *New* option. Simply invoke *Dimedit* and *New* but do not enter any value in the *Multiline Text Editor*. After selecting the desired dimension, the AutoCAD-measured value is restored.

Rotate
AutoCAD prompts for an angle to rotate the text. Enter an absolute angle (relative to angle 0). This option is identical to the *Angle* option of *Dimtedit*.

Oblique
This option is unique to *Dimedit*. Entering an angle at the prompt affects the extension lines. Enter an absolute angle:

```
Enter obliquing angle (press ENTER for none):
```

Normally, the extension lines are perpendicular to the dimension lines. In some cases, it is desirable to set an obliquing angle for the extension lines, such as when dimensions are crowded and hard to read (Fig. 26-46) or for dimensioning isometric drawings (see Dimensioning Isometric Drawings).

Figure 26-46 ————————————

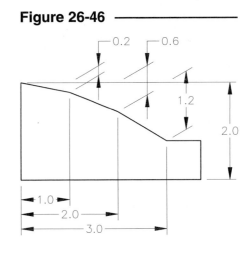

DDMODIFY

	Pull-down Menu	COMMAND (TYPE)	ALIAS (TYPE)	Short-cut	Screen (side) Menu	Tablet Menu
	Modify Properties...	DDMODIFY	MO	...	*MODIFY1 Modify*	Y,14

The *Ddmodify* command (discussed in Chapter 16) can be used effectively for a wide range of dimension editing purposes. Selecting a dimension at the "Select objects:" prompt produces the *Modify Dimension* dialog box (Fig. 26-47). This dialog box provides access to the *Mtext*, *Geometry*, *Format*, and *Annotation* dialog boxes.

Use the *Contents:* edit box to edit the dimension text. Overwriting or deleting the < > symbols results in creating new, non-associative text (not recommended). Add a prefix before or suffix after the < > symbols without affecting the AutoCAD-measured value or associativity feature.

Figure 26-47

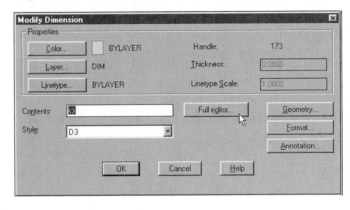

You can also edit text and all aspects of the text (height, font, style, bold, italic, etc.) by selecting the *Full editor...* button. This action produces the *Multiline Text Editor*.

If you want to change other aspects of the dimension object, such as the appearance of the dimension lines, extension lines, or how the text value is displayed, select the *Geometry*, *Format*, or *Annotation* tiles. This action produces a series of dialog boxes that change <u>dimension variables</u>. Keep in mind that when you use this interface for changing dimension variables, the change creates a <u>dimension style override</u> and applies the change to the selected dimension object. For information on dimension variables, dimension styles, and dimension style overrides, read Chapter 27, Dimension Styles and Dimension Variables.

Customizing Dimensioning Text

As discussed earlier, you can specify dimensioning text other than what AutoCAD measures and supplies as the default text, using the *Text* option of the individual dimension creation commands. In addition, the text can be modified at a later time using the *Ddmodify* command or the *New* option of *Dimedit*, both of which provide access to the *Multiline Text Editor*.

The less-than and greater-than symbols (< >) represent the AutoCAD-supplied dimensional value. Placing text or numbers <u>inside</u> the symbols overrides the correct measurement (not advised) and creates static text. Text placed inside the < > symbols is not associative and is <u>not</u> updated in the event of *Stretching*, *Rotating*, or otherwise editing the associative dimension. The original text can only be retrieved using the *New* option of *Dimedit*.

Text placed outside of the < > symbols, however, acts only as a prefix/suffix to the measured value. In the case of a diameter or radius dimension, AutoCAD automatically inserts a diameter symbol (Ø) or radius designator (R) before the dimension value. Inserting a prefix with the *Multiline Text Editor* <u>does not override</u> the AutoCAD symbols; however, entering a *Prefix* or *Suffix* using the *Dimension Styles* dialog box <u>overrides</u> the symbols. This feature is important in correct ANSI standard dimensioning for entering the number of times a feature occurs. For example, one of two holes having the same diameter would be dimensioned "2X Ø1.00" (see Fig. 26-21).

R14

You can also enter special characters by using the Unicode values or "%%" symbols. The following codes can be entered <u>outside</u> the < > symbols for *Dimedit New*, *Mtext*, or the *Text* option of dimensioning commands:

Enter:	Enter:	Result	Description
\U+2205	%%c	ø	diameter (metric)
\U+00b0	%%d	°	degrees
	%%o	――――	overscored text
	%%u	____	underscored text
\U+00b1	%%p	±	plus or minus
	%%*nnn*	varies	ASCII text character number
\U+*nnnn*		varies	Unicode text hexadecimal value

DIMENSIONING ISOMETRIC DRAWINGS

Dimensioning isometric drawings in AutoCAD is accomplished using *Dimaligned* dimensions and then adjusting the angle of the extension lines with the *Oblique* option of *Dimedit*. The technique follows two basic steps:

1. Use *Dimaligned* or the *Vertical* option of *Dimlinear* to place a dimension along one edge of the isometric face. Isometric dimensions should be drawn on the isometric axes lines (vertical or at a 30° rotation from horizontal; Fig. 26-48).

Figure 26-48 ―――――――――――――――――

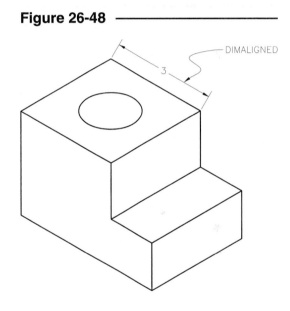

2. Use the *Oblique* dimensioning option of *Dimedit* and select the dimension just created. When prompted to "Enter obliquing angle," enter the desired value (**30** in this case) or PICK two points designating the desired angle. The extension lines should change to the designated angle. In AutoCAD, the possible obliquing angles for isometric dimensions are **30**, **150**, **210**, or **330**.

Figure 26-49 ―――――――――――――――――

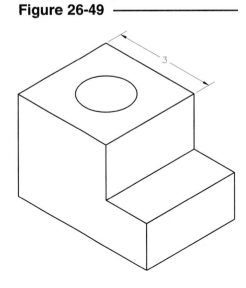

Place isometric dimensions so that they align with the face of the particular feature. Dimensioning the object in the previous illustrations would continue as follows.

Create a *Dimlinear, vertical* dimension along a vertical edge.

Figure 26-50

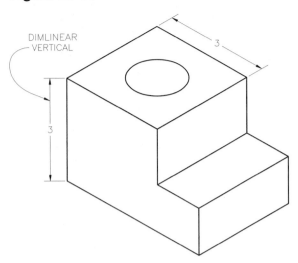

Use *Dimedit, Oblique* to force the dimension to an isometric axis orientation. Enter a value of **150** or PICK two points in response to the "Enter obliquing angle:" prompt.

Figure 26-51

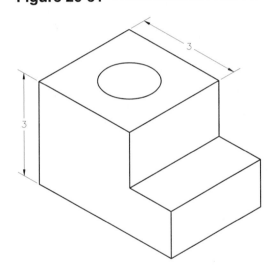

The dimensioning text can be treated two ways. (1) For quick and simple isometric dimensioning, <u>unidirectional</u> dimensioning (all values read from the bottom) is preferred since there is no automatic method for drawing the numerical values in an isometric plane (Fig. 26-51). (2) As a better alternative, text *Styles* could be created with the correct obliquing angles (30 and -30 degrees). The text must also be rotated to the correct angle using the *Rotate* option of the dimension commands or by using *Dimedit Rotate* (Fig. 26-52). Optionally, *Dimension Styles* could be created with the correct variables set for text style and rotation angle for dimensioning on each isoplane.

Figure 26-52

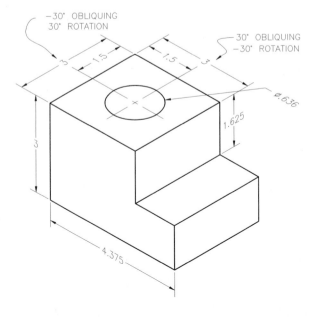

CHAPTER EXERCISES

Only four exercises are offered in this chapter to give you a start with the dimensioning commands. Many other dimensioning exercises are given at the end of Chapter 27, Dimension Styles and Dimension Variables. Since most dimensioning practices in AutoCAD require the use of dimensioning variables and dimension styles, that information should be discussed before you can begin dimensioning effectively.

The units that AutoCAD uses for the dimensioning values are based on the *Units* and *Precision* settings in the *Annotation* dialog box accessed through the *Dimension Styles* dialog box. You can dimension these drawings using the default settings, or if you are adventurous, you can change the *Units* and *Precision* settings for each drawing to match the dimensions shown in the figures. (Type *DDIM*, select *Annotation...*, then *Units*.)

Other dimensioning features may appear different than those in the figures because of the variables set in the AutoCAD default STANDARD dimension style. For example, the default settings for diameter and radius dimensions may draw the text and dimension lines differently than you desire. After reading Chapter 27, those features in your exercises can be changed retroactively by changing the dimension style.

1. ***Open*** the **PLATES** drawing that you created in Chapter 10 Exercises. ***Erase*** the plate on the right. Use ***Move*** to move the remaining two plates apart, allowing 5 units between. Create a ***New*** layer called **DIM** and make it ***Current***. Create a ***Text Style*** using ***Romans*** font. Dimension the two plates, as shown in Figure 26-53. ***Save*** the drawing as **PLATES-D**.

Figure 26-53 ——

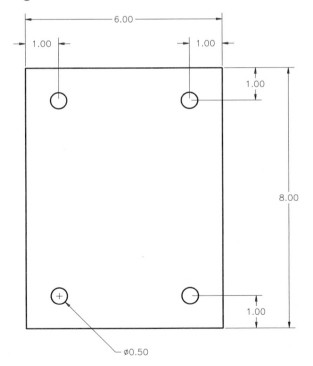

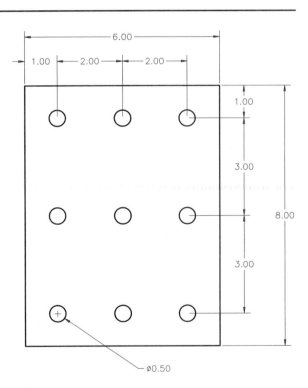

2. *Open* the **PEDIT1** drawing. Create a *New* layer called **DIM** and make it *Current*. Create a *Text Style* using *Romans* font. Dimension the part as shown in Figure 26-54. *Save* the drawing as **PEDIT1-DIM**. Draw a *Pline* border of .02 *width* and *Insert* **TBLOCK**. *Plot* on an A size sheet using *Scale to Fit*.

Figure 26-54 ───────────

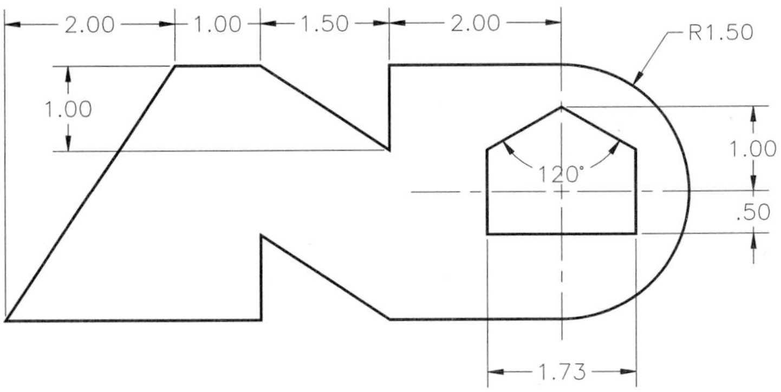

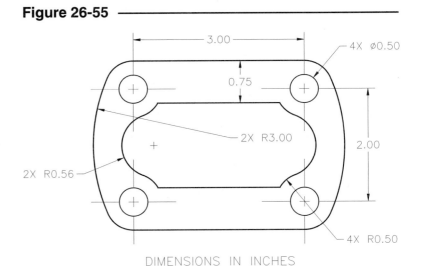

3. *Open* the **GASKETA** drawing that you created in Chapter 9 Exercises. Create a *New* layer called **DIM** and make it *Current*. Create a *Text Style* using *Romans* font. Dimension the part as shown in Figure 26-55. *Save* the drawing as **GASKETD**.

    Draw a *Pline* border with .02 *width* and *Insert* **TBLOCK** with an **8/11** scale factor. *Plot* the drawing *Scaled to Fit* on an A size sheet.

Figure 26-55 ───────────

4. *Open* the **BARGUIDE** multiview drawing that you created in Chapter 22 Exercises. Create the dimensions on the **DIM** layer. Keep in mind that you have more possibilities for placement of dimensions than are shown in Figure 26-56.

Figure 26-56 ───────────

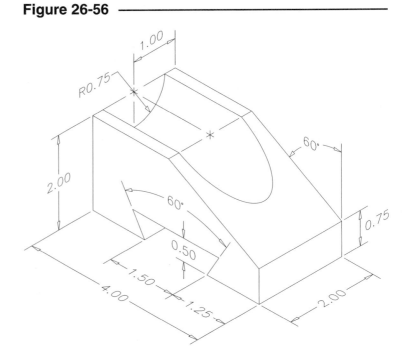

27

DIMENSION STYLES AND DIMENSION VARIABLES

Chapter Objectives

After completing this chapter you should:

1. be able to *Save* and restore dimension styles with the *Dimension Styles* dialog box or in command line format with the *Dimstyle* command;

2. know how to create dimension style families and specify variables for each child;

3. know how to create and apply dimension style overrides by either command line or dialog box method;

4. be able to control dimension variable settings using the *Geometry, Format,* and *Annotation* dialog boxes;

5. be able to modify dimensions using *Update, Dimoverride,* and *Matchprop;*

6. know the guidelines for dimensioning.

CONCEPTS

Dimension Variables

Since a large part of AutoCAD's dimensioning capabilities are automatic, some method must be provided for you to control the way dimensions are drawn. A set of 59 <u>dimension variables</u> allows you to affect the way dimensions are drawn by controlling sizes, distances, appearance of extension and dimension lines, and dimension text formats.

An example of a dimension variable is *DIMSCALE* (if typed) or *Overall Scaling* (if selected from a dialog box, Fig. 27-1). This variable controls the overall size of the dimension features (such as text, arrowheads, and gaps). Changing the value from 1.00 to 1.50, for example, makes all of the size-related features of the drawn dimension 1.5 times as large as the default size of 1.00 (Fig. 27-2). Other examples of features controlled by dimension variables are arrowhead type, orientation of the text, text style, units and precision, suppression of extension lines, fit of text and arrows (inside or outside of extension lines), and direction of leaders for radii and diameters (pointing in or out). The dimension variable changes that you make affect the <u>appearance</u> of the dimensions that you create.

Figure 27-1 ───────

There are two basic ways to control dimensioning variables:

1. Use the *Dimension Styles* dialog box series (Fig. 27-3).

2. Type the dimension variable name in command line format.

Figure 27-2 ──────────────────────────

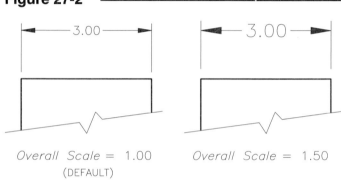

Overall Scale = 1.00
(DEFAULT)

Overall Scale = 1.50

The dialog boxes employ "user-friendly" terminology and selection, while the command line format uses the formal dimension variable names.

Changes to dimension variables are usually made <u>before</u> you create the affected dimensions. Dimension variable changes are <u>not automatically retroactive</u>, in contrast to *LTSCALE*, for example, which can be continually modified to adjust the spacing of existing non-continuous lines. Changes to dimensioning variables affect only <u>newly created</u> dimensions unless those changes are *Saved* to an existing *Dimension Style* that was in effect when previous dimensions were created. Generally, dimensioning variables should be set <u>before</u> creating the desired dimensions although it is possible to modify dimensions and *Dimension Styles* retroactively.

Dimension Styles

All associative dimensions are part of a <u>dimension style</u>. The default (and the only supplied) dimension style is named STANDARD (Fig. 27-3). Logically, this dimension style has all of the default dimension variable settings for creating a dimension with the typical size and appearance. Similar to layers, you can create, name, and specify settings for any number of dimension styles. Each dimension style contains the dimension variable settings that you select. <u>A dimension style is a group of dimension variable settings that has been saved under a name you assign.</u> When you select a dimension style from the *Current:* list, AutoCAD remembers and resets that <u>particular combination of dimension variable settings</u>.

Figure 27-3 ──────

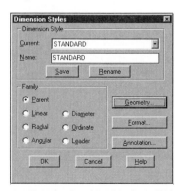

R13

Imagine dimensioning a complex drawing <u>without</u> having dimension styles (Fig. 27-4). In order to create a dimension using limit dimensioning (as shown in the diameter dimension), for example, you would change the desired variable settings, then "draw" the dimension. To draw another dimension without extension lines (as shown in the interior slot), you would have to reset the previous variables and make changes to other variables in order to place the special dimensions as you prefer. This same process would be repeated each time you want to create a new type of dimension. If you needed to add another dimension with the limits, you would have to reset the same variables as before.

Figure 27-4

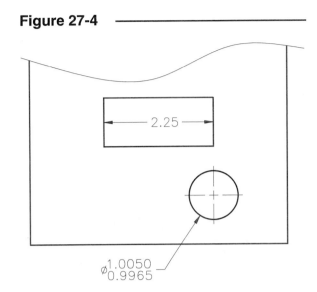

To simplify this process, you can create and save a dimension style for each particular "style" of dimension. Each time you want to draw a particular style of dimension, select the style name from the list and begin drawing. A dimension style could contain all the default settings plus only one or two dimension variable changes or a large number of variable changes.

NOTE: When you make dimension variable changes that you want to keep, make sure you save them to an existing or new style name (by PICKing the *Save* tile).

Another advantage of using dimension styles is that <u>existing</u> dimensions in the drawing can be globally modified by making a change and saving it to the dimension style(s). Knowing that all dimensions are part of a dimension style, making a change in the variable settings using the *Geometry..., Format..., or Annotation...*dialog boxes and *Saving* that change results in an <u>automatic update</u> for all existing dimensions drawn with that style.

Dimension Style Families

In the previous chapter, you learned about the various <u>classifications</u> of dimensions, such as linear, angular, radial, diameter, ordinate, and leader. You may want the appearance of each of these classifications of dimensions to vary, for example, all radius dimensions to appear with the dimension line pointing out and all diameter dimensions to appear with the dimension line inside pointing in. It is logical to assume that a new dimension style would have to be created for each variation. However, a new dimension style name is not necessary for each classification of dimension. AutoCAD provides <u>dimension style families</u> for the purpose of providing variations within each named dimension style.

The dimension style <u>names</u> that you create are assigned to a dimension style <u>family</u>. Each dimension style family has a <u>parent</u> and can have six <u>children</u>. The children are <u>linear, angular, diameter, radial, ordinate, and leader</u>. If the *Parent* button is selected in the *Dimension Styles* dialog box (Fig. 27-3) when a variable change is made, the variable setting generally affects the entire family. If any one of the buttons for the children (*Linear, Angular, Radial*, etc.) is selected when saving a variable, the variable affects only that child (classification of dimension) (Fig. 27-6). When a dimension is drawn, AutoCAD knows what classification of dimension is created and applies the selected variables to that child.

AutoCAD refers to these children as $0, $2, $3, $4, $6, and $7 as suffixes appended to the parent name (see following table). AutoCAD automatically applies the appropriate suffix code to the dimension style name when a child is created. Although the codes do not appear in the *Dimension Styles* dialog box, you can display the code by using the *List* command and selecting an existing child dimension.

Child Type	Suffix Code
linear	$0
angular	$2
diameter	$3
radial	$4
ordinate	$6
leader	$7

For example, you may want to set the DETAILS dimension style to draw the arrows and dimension lines inside the arc but drag the dimension text outside of the arc for *Radial* dimensions (as shown in Fig. 27-5). First, click on the *Radial* button. Second, make the necessary variable changes in the *Format* dialog box (explained later, see *Radius* and *Diameter* Variable Settings). Third, save the changes to the dimension style, assigning a name (DETAILS) and selecting the *Save* button in the *Dimension Styles* dialog box (as shown in Fig. 27-6). When a *Radius* dimension is drawn, the dimension line, text, and arrow should appear as shown in Figure 27-5. Using *List* to display information about the dimension reports the dimension style name as DETAILS$4.

Figure 27-5 ————————

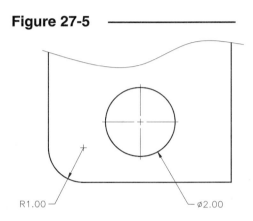

R1.00 —   — ⌀2.00

In summary, a dimension style family is simply a set of dimension variables related by name. Variations within the family are allowed, in that each child (classification of dimension) can possess its own subset of variables. Therefore, each child inherits all the variables of the parent in addition to any others that may be assigned individually.

NOTE: Once a child has been created (special variables have been set for one type of dimension), then variables are changed and saved for the parent; the changes are automatically updated for the existing dimensions (of that style) in the drawing but are <u>not</u> updated for existing or new children (of that style). For example, let's assume you set variables and saved them to a dimension style family (*Parent*), then set other variables and saved them for a child of that style (*Radial*, for example). Later you changed the *Overall Scale* and *Saved* it for the *Parent*. All the dimensions of that style in the drawing automatically update to the new setting except the radial dimensions. To minimize this problem: (1) assure you make all familywide settings before creating children, and (2) when new settings are made to the style (*Parent*) after drawing dimensions, make the same changes and *Save* them for each child or use *Dimoverride* or *Update* to apply new variable settings to existing child dimensions in the drawing. Alternately, you could create a separate dimension style for radial dimensions, etc. instead of creating a radial child. (See Modifying Existing Dimensions, later in this chapter.)

Dimension Style Overrides

When you make the variable change and <u>do not *Save*</u> it to the current dimension style, AutoCAD remembers the setting nevertheless. The change is appended to the <u>current</u> style as a <u>dimension style override</u>. This is particularly helpful if you need a special feature on one or two dimensions but do not want to update the dimension style (and all existing dimensions of the style) with the new change.

A dimension style override is created by <u>changing a variable setting but not saving it</u> to the current dimension style. You can create an override by using the *Dimension Styles* dialog box series to set a variable and exiting without using *Save* or by changing a variable setting by command line format. (If you type the name of a dimension variable and change a value at the command prompt "on the fly," the current dimension style is used as a base style.) When an override is made, a new style name is created with a "+" placed as a prefix and the new style is set current. Each new dimension created with the new modified style applies the overridden variables. If you list one of these dimensions (using *List*), the new style name and the overridden variables are displayed. Dimension style overrides <u>can be applied only to the *Parent* style</u> and cannot be created for an individual child.

DIMENSION STYLES

DDIM

Pull-down Menu	COMMAND (TYPE)	ALIAS (TYPE)	Short-cut	Screen (side) Menu	Tablet Menu
Dimension Style...	*DDIM*	*D*	*...*	*DIMNSION Ddim*	*Y,5*

The *Ddim* command produces the *Dimension Styles* dialog box (Fig. 27-6). This is the primary interface for creating new dimension styles and making existing dimension styles current. This dialog box also provides access to the three boxes that enable you to change dimension variables. The radio buttons are used to select a classification of dimension (child) for applying variable changes. (Please read the previous sections of this chapter for an introduction to these concepts.) The available options are explained next.

Figure 27-6 ——————

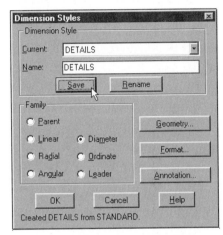

Current:
Select the desired <u>existing</u> style from the drop-down list to set it as the current style. The dimension style appearing in the box is the current style that is used for drawing subsequent dimensions. Additionally, the current style is the template used for any operations within the *Dimension Style* dialog box series, such as (1) creating a new style, (2) making modifications to the current style by changing dimension variables, then using *Save*, or (3) renaming the style. A name appearing with a "+" (plus) symbol before the name, "+STANDARD," for example, means that a variable has been changed but not yet *Saved* to the dimension style (see Dimension Style Overrides).

Name:
Enter a name in the box to be used for creating a new style or renaming an existing style. The name appearing in the *Current:* box is the template for the new style or renamed style.

Save
Select this tile after creating a new style or making the desired variable changes to the current style. Changing a variable setting and exiting the dialog box without saving creates a dimension style override.

Rename
Select this option to replace the old style name (appearing in the *Current:* box) with the new name (appearing in the *Name:* box). Enter the desired new name in the *Name:* box, then select *Rename*.

Geometry...
This tile produces the *Geometry* dialog box (explained in detail in the following section). The *Geometry* box enables you to make dimension variable changes that affect the appearance of <u>arrowheads</u>, <u>dimension lines</u>, and <u>extension lines</u>.

Format...
This tile produces the *Format* dialog box (explained in detail in the following section). You can make variable changes related to <u>location and orientation of dimension text</u> in the *Format* box.

Annotation...
The *Annotation* dialog box is the interface for dimension variables controlling the <u>content of the dimension text</u>.

Family
Use this section to specify the classification of dimension that you want to affect using the variable controls in the *Geometry, Format,* and *Annotation* boxes. Select the desired family member <u>first</u>; then make the desired adjustments. The family members are:

Parent
Select this button to make familywide dimension variable changes.

Linear, Angular, Radial, Diameter, Ordinate, and Leader
Select the desired classification (child) for subsequent dimension variable changes.

Creating a New Dimension Style Family Using the *Dimension Styles* Dialog Box

1. Select the *Parent* button in the *Dimension Styles* dialog box (see Fig. 27-6).

2. Select the desired dimension style name from the *Current:* list to use as a template. Whatever style appears in this edit box is used as a template. The new style will contain all of the variable settings of the template plus any changes you make and *Save.* Initially, the STANDARD style is used as a template.

3. Type over or modify the name in the *Name:* edit box to assign the new name.

4. Change the desired dimension variables from the *Geometry..., Format...,* or *Annotation...* dialog boxes (discussed in detail later). Familywide variables such as *Overall Scale* are set at this stage. Select the *OK* tile in each of these boxes to keep your changes.

5. After returning to the *Dimension Styles* dialog box, PICK *Save* to assign the variable changes to the new dimension style. The new style becomes current. Then select the *OK* tile. *Cancel* cancels the creation of the new dimension style.

NOTE: Do not change the STANDARD style or you lose your ability to easily restore all of the default settings. Generally, make dimension styles with new names using STANDARD as a <u>template</u>.

Setting Dimension Variables for a Child Using the *Dimension Styles* Dialog Box

1. Select the desired <u>existing</u> dimension style family name from the *Current:* list (see Fig. 27-6).

2. Select the desired classification of dimension (child) such as *Linear, Angular,* or *Radial.*

3. Change the desired dimension variables from the *Geometry..., Format...,* or *Annotation...* dialog boxes (discussed in detail later). Select the *OK* tile in each of these boxes to keep your changes.

4. After returning to the *Dimension Styles* dialog box, PICK *Save* to assign the variable changes to the dimension style family. Then select the *OK* tile.

Restoring a Dimension Style Using the *Dimension Styles* Dialog Box

1. Select the desired style name from the *Current:* list. The style should then appear in the *Name:* edit box. Select the *OK* tile.

2. The selected dimension style becomes the current style and remains so until another is made current. Any dimensions created while the style is current contain the features dictated by the style's variable settings.

Modifying a Dimension Style and Updating Existing Dimensions

1. Select the desired style name from the *Current:* list in the *Dimension Styles* dialog box.

2. Change the desired dimension variables from the *Geometry...*, *Format...*, or *Annotation...* dialog boxes. Select the *OK* tile in each of these boxes to keep your changes. When you select the *Save* tile in the main dialog box, the current dimension style is updated.

3. As you exit the dialog box, you will notice some activity in the Drawing Editor. Most dimensions originally created with that style are <u>automatically</u> updated with the new variables.

4. Dimensions that were created as children (special variables set and saved for that type of dimension) do not automatically update when a variable change is made and *Saved* for the style (*Parent*). To update existing children, either (1) select the child (*Radial, Diameter,* etc.), make the same variable changes and select *Save,* or (2) use *Dimoverride* or *Update* to apply new variable settings to existing child dimensions in the drawing.

(See Modifying Existing Dimensions later in this chapter.)

DIMSTYLE

Pull-down Menu	COMMAND (TYPE)	ALIAS (TYPE)	Short-cut	Screen (side) Menu	Tablet Menu
Dimension *Update*	*DIMSTYLE*	*DIMSTY* *or DST*	...	*DIMNSION* *Dimstyle*	*Y,3*

The *Dimstyle* command is the command line format equivalent to the *Dimension Style* dialog box. Operations that are performed by the dialog box can also be accomplished with the *Dimstyle* command. The *Apply* option offers one other important feature that is <u>not</u> available in the dialog box:

```
Command: dimstyle
dimension style: STANDARD
Dimension Style Edit (Save/Restore/STatus/Variables/Apply/?) <Restore>:
```

Save
Use this option to <u>create a new</u> dimension style. The new style assumes all of the current dimension variable settings comprised of those from the current style plus any other variables changed by command line format (overrides). The new dimension style becomes the current style.

```
Dimension Style Edit (Save/Restore/STatus/Variables/Apply/?) <Restore>: s
?/Name for new dimension style:
```

Restore
This option prompts for an existing dimension style name to be restored. The restored style becomes the current style. You can use the "press ENTER to select:" option to PICK a dimension object that references the style you want to restore.

STatus

Status gives the current settings for all dimension variables. The displayed list comprises the settings of the current dimension style and any overrides. All 59 dimension variables are listed:

Dimension Style Edit (Save/Restore/STatus/Variables/Apply/?) <Restore>: **st**

DIMALT Off	Alternate units selected
DIMALTD 2	Alternate unit decimal places
DIMALTF 25.4000	Alternate unit scale factor
DIMALTTD 2	Alternate tolerance decimal places
etc.	

Variables

You can use this option to list the dimension variable settings for any existing dimension style. You cannot modify the settings.

Dimension Style Edit (Save/Restore/STatus/Variables/Apply/?) <Restore>: **v**
?/Enter dimension style name or press ENTER to select dimension:

Enter the name of any style or "press ENTER to select:" to PICK a dimension object that references the desired style. A list of all variables and settings appears.

~stylename

This variation can be entered in response to the *Variables* option. Entering a dimension style name preceded by the tilde symbol (~) displays the differences between the current style and the (~) named style.

For example, assume that you wanted to display the differences between STANDARD and the current dimension style (LIMITS-1, for example). Use *Variables* with the ~ symbol. (The ~ symbol is used as a wildcard to mean "all but.")

Command: **dimstyle**
dimension style: LIMITS-1
Dimension Style Edit (Save/Restore/STatus/Variables/Apply/?) <Restore>: **v**
?/Enter dimension style name or press ENTER to select dimension: **~standard**
Differences between STANDARD and current settings:

	STANDARD	Current Setting
DIMLIM	Off	On
DIMTFAC	1.0000	0.7000
DIMTM	0.0000	0.1000
DIMTP	0.0000	0.0500

?/Enter dimension style name or press ENTER to select dimension: *Cancel*
Command:

This option is very useful for keeping track of your dimension styles and their variable settings.

Apply

Use *Apply* to update dimension objects that you PICK with the current variable settings. The current variable settings can contain those of the current dimension style plus any overrides. The selected object loses its reference to its original style and is "adopted" by the applied dimension style family.

Dimension Style Edit (Save/Restore/STatus/Variables/Apply/?) <Restore>: **a**
Select objects:

This is a useful tool for <u>changing an existing dimension object from one style to another</u>. Simply make the desired dimension style current; then use *Apply* and select the dimension objects to change to that style.

Using the *Apply* option of *Dimstyle* is identical to using the *Update* command. The *Update* command can be invoked by using the *Dimension Update* button, using *Update* from the *Dimension* pull-down menu, or typing *Dim*: (Enter), then *Update*.

DIMENSION VARIABLES

Now that you understand how to create and use dimension styles and dimension style families, let's explore the dimension variables. <u>Two methods</u> can be used to set dimension variables: the *Dimension Styles* dialog box series and command line format. First, we will examine the *Geometry, Format,* and *Annotation* dialog boxes (accessible through the *Dimension Styles* dialog box). These dialog boxes contain an interface for setting the dimension variables. Dimension variables can alternately be set by typing the formal name of the variable and changing the desired value (discussed after this section on dialog boxes). However, the dialog boxes offer a more understandable terminology, buttons, check-boxes, lists, and several image tiles that automatically change the variables when you click on the image.

Changing Dimension Variables Using the Dialog Boxes

In this section, the three main dialog boxes for changing variables are explained. The headings for the paragraphs below include the <u>option</u> title in the dialog box and the related <u>dimension variable name</u> in parentheses.

The following figures usually display the AutoCAD default setting for the variable and an example of changing the setting to another value. These AutoCAD default settings (in the STANDARD dimension style) are for the ACAD.DWT template drawing, which is used for *Start from Scratch, English Default Settings*. If you use the *Start from Scratch, Metric Default Settings; Quick Setup Wizard, Advanced Setup Wizard;* or other template drawings, dimension variables may be automatically reset or other dimension styles may exist. See Using *Setup Wizards* and Template Drawings later in this chapter.

Geometry...

Choosing the *Geometry...* button in the *Dimension Styles* dialog box (Fig. 27-6) opens the *Geometry* dialog box (Fig. 27-7). Use this box to change variables that control the <u>appearance of arrowheads, dimension lines, and extension lines</u>. The *Geometry* dialog box contains five specific areas, which are explained in the following sections:

Figure 27-7 ────────

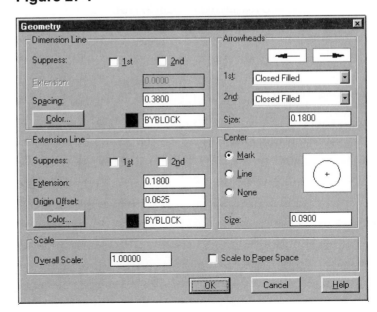

Overall Scale (DIMSCALE)

The *Overall Scale* value globally affects the scale of <u>all size-related features</u> of dimension objects, such as arrowheads, text height, extension line gaps (from the object), extensions (past dimension lines), etc. All other size-related values (variables) appearing in the dialog box series are <u>multiplied by the</u> *Overall Scale*. Therefore, to keep all features proportional, <u>change this one setting rather than each of the others individually</u>.

Although this area is located at the bottom of the box, it is probably the most important option. Because the *Overall Scale* should be set as a familywide variable, enter this value as the <u>first step</u> in creating a dimension style. Select *Parent* family member before making this setting.

Figure 27-8

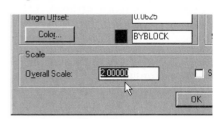

Changes in this variable should be based on the *Limits* and plot scale. You can use the drawing scale factor to determine this value. (See Drawing Scale Factor, Chapter 13.) The *Overall Scale* value is stored in the *DIMSCALE* variable.

Scale to Paper Space
Checking this box forces dimensions to appear in the same size for all viewports (Paper Space viewports created with *Mview*). Toggling this on sets *DIMSCALE* to 0.

Dimension Line
Suppress 1st, 2nd **(DIMSD1, DIMSD2)**
This area allows you to suppress (not draw) the *1st* or *2nd* dimension line or both (Fig. 27-9). The <u>first</u> dimension line would be on the "First extension line origin" side or nearest the end of object PICKed in response to "Select object to dimension." These toggles change the *DIMSD1* and *DIMSD2* dimension variables.

Figure 27-9

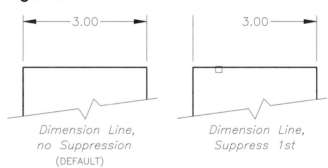

Dimension Line, no Suppression (DEFAULT) *Dimension Line, Suppress 1st*

Extension *(DIMDLE)*
The *Extension* edit box is disabled unless the *Oblique* or *Architectural Tick* arrowhead type is selected (in the *Arrowheads* section). The *Extension* value controls the length of dimension line that extends past the dimension line for *Oblique* or *Architectural Tick* "arrows" (Fig. 27-10).

Figure 27-10

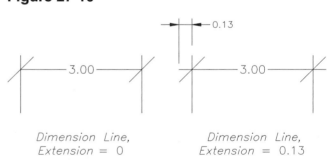

Dimension Line, Extension = 0 *Dimension Line, Extension = 0.13*

Spacing *(DIMDLI)*
The *Spacing* edit box reflects the value that AutoCAD uses in baseline dimensioning to "stack" the dimension line above or below the previous one (Fig. 27-11). This value is held in the *DIMDLI* variable.

Figure 27-11

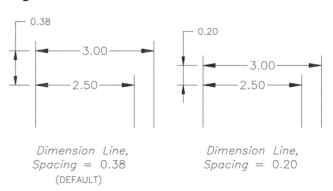

Dimension Line, Spacing = 0.38 (DEFAULT) *Dimension Line, Spacing = 0.20*

Color (DIMCLRD)

The *Color* button allows you to choose the color for the dimension line (Fig. 27-12). Assigning a specific color to the dimension lines, extension lines, and dimension text enables you to print or plot these features with different line widths or colors (pen assignments are made in the *Print/Plot Configuration* dialog box). This feature corresponds to the *DIMCLRD* variable.

Figure 27-12

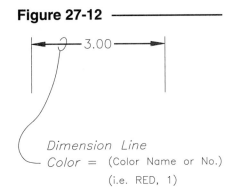

Extension Line

Suppress 1st, 2nd (DIMSE1, DIMSE2)

This area is similar to the *Dimension Line* area and can be easily confused with it, so be careful! This area allows you to suppress the *1st* or *2nd* extension line or both (Fig. 27-13). These options correspond to the *DIMSE1* and *DIMSE2* dimension variables.

Figure 27-13

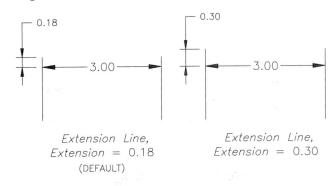

Extension (DIMEXE)

The *Extension* edit box reflects the value that AutoCAD uses to set the distance for the extension line to extend beyond the dimension line (Fig. 27-14). This value is held in the *DIMEXE* variable. Generally, this value does not require changing since it is automatically multiplied by the *Overall Scale*.

Figure 27-14

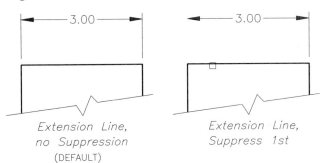

Origin Offset (DIMEXO)

The *Origin Offset* value specifies the distance between the origin points and the extension lines (Fig. 27-15). This offset distance allows you to PICK the object corners, yet the extension lines maintain the required gap from the object. This value rarely requires input since it is affected by *Overall Scale*. The value is stored in the *DIMEXO* variable.

Figure 27-15

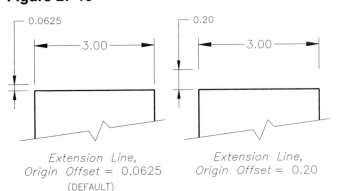

Color (DIMCLRE)

The *Color* button allows you to choose the color for the extension lines (Fig. 27-16). This feature corresponds to the *DIMCLRE* variable.

Figure 27-16

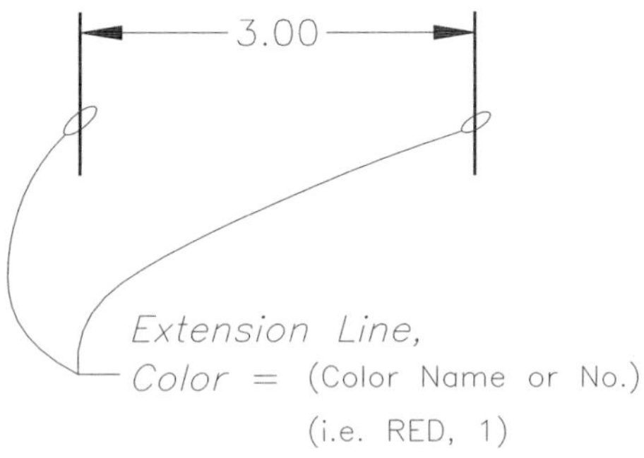

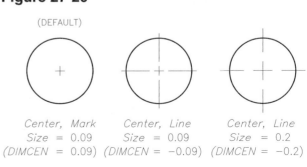

Extension Line,
Color = (Color Name or No.)
(i.e. RED, 1)

Arrowheads
1st, 2nd (DIMBLK, DIMBLK1, DIMBLK2)

This area contains two drop-down lists of various arrowhead types, including dots and ticks. Each list corresponds to the *1st* or *2nd* arrowhead created in the drawn dimension. The image tiles display each arrowhead type selected. Click in the first image tile to change both arrowheads, or click in each to change them individually (Fig. 27-17). The variables affected are *DIMBLK, DIMBLK1, DIMBLK2,* and *DIMSAH* (see Appendix A for details).

Figure 27-17

Size (DIMASZ)

The size of the arrow can be specified in the *Size* edit box (Fig. 27-18). The *DIMASZ* variable holds the value. Remember that *Size* is multiplied by *Overall Scale.*

Figure 27-18

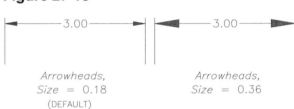

Arrowheads,
Size = 0.18
(DEFAULT)

Arrowheads,
Size = 0.36

Center
Mark, Line, None (DIMCEN)

The radio buttons determine how center marks are drawn when the dimension commands *Dimcenter, Dimdiameter,* or *Dimradius* are used (Fig. 27-19). The image tile displays the *Mark, Line,* or *None* radio button feature specified. This area actually controls the value of <u>one</u> dimension variable, *DIMCEN* (by using a 0, positive, or negative value; see Fig. 27-20).

Figure 27-19

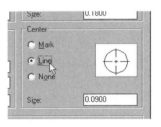

Size (DIMCEN)

The *Size* edit box controls the size of the short dashes and extensions past the arc or circle. The value is stored in the *DIMCEN* variable (Fig. 27-20). Only positive values can be (and need to be) entered in the <u>dialog box</u>.

Figure 27-20

(DEFAULT)

Center, Mark
Size = 0.09
(DIMCEN = 0.09)

Center, Line
Size = 0.09
(DIMCEN = −0.09)

Center, Line
Size = 0.2
(DIMCEN = −0.2)

Format...

The *Format* dialog box (Fig. 27-21) provides control of the <u>location and orientation of the text with respect to the dimension line</u>.

Figure 27-21

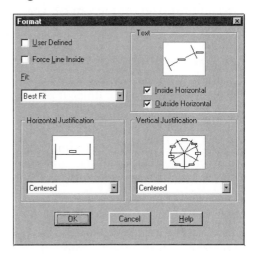

User Defined (DIMUPT)

Checking the *User Defined* box enables you to move the dimension text value lengthwise along the dimension line when you select the location for the dimension line. The default position (not checked) always keeps the text in the center of the dimension line. In Release 14, *Diameter* and *Radius* dimension text can automatically be dragged assuming the default variable settings. If you want to draw *Radius* dimensions with the arrow and dimension line inside the arc (the correct standard for most applications), you must enable *User Defined* to allow dragging the text outside of the arc, as on a leader (see *Radius* and *Diameter* Variable Settings). The value for this switch is stored in the *DIMUPT* variable.

Figure 27-22

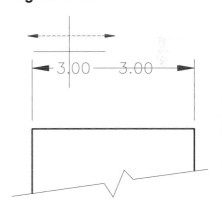

User Defined CHECKED

Force Line Inside (DIMTOFL)

The *Force Line Inside* toggle sets the value of *DIMTOFL* on or off. When this variable is on, a dimension line is drawn between the extension lines when the arrows are drawn outside the extension lines (Fig. 27-23).

Figure 27-23

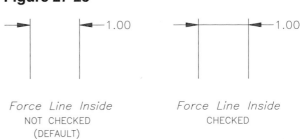

Force Line Inside
NOT CHECKED
(DEFAULT)

Force Line Inside
CHECKED

Fit (DIMFIT)

For *Linear, Aligned, Angular, Baseline, and Continue* dimensions, the *Fit* drop-down list determines how text and arrows are treated <u>only if there is insufficient room</u> between extension lines for text, arrows, and extension lines. In most cases where space permits all components to fit inside, the *Fit* setting has no effect on the placement. If there is absolutely not enough room for text or arrows, all options behave similarly—text and arrows are placed outside. The six *Fit* options set the value of the dimension variable *DIMFIT* to 0 through 5.

For *Diameter* and *Radius* dimensions, the text and arrows are normally dragged outside of the arc or circle, but the *Fit* option can be used to force dimension line and arrows inside (see *Radius* and *Diameter* Variable Settings).

The *Text and Arrows* option <u>keeps the text and arrows together</u> always. If space does not permit both features to fit between the extension lines, it places the text and arrows outside the extension lines (Fig. 27-24). (*DIMFIT* = 0.)

The *Text Only* option keeps the <u>text between the extension lines</u> and the arrows on the outside. If the text absolutely cannot fit, it is placed outside the extension lines (Fig. 27-24). (*DIMFIT* = 1.)

The *Arrows Only* option places the text on the outside and the <u>arrows on the inside</u> unless the arrows cannot fit, in which case they are placed on the outside as well (Fig. 27-24). (*DIMFIT* = 2.)

The *Best Fit* option behaves similarly to *Text Only*, except that if the text cannot fit, the text is placed outside the extension lines and the arrows are placed inside. However, if the arrows cannot fit either, both text and arrows are placed outside the extension lines. This is the default setting for the STANDARD style. (*DIMFIT* = 3.)

Figure 27-24

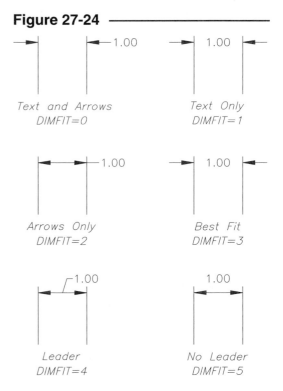

The *Leader* option enables the creation of leaders. If text cannot fit between the extension lines, then leaders to the text are created automatically (Fig. 27-24). (*DIMFIT* = 4.)

A setting of *No Leader* (*DIMFIT*=5) causes the dimension text to appear above the dimension line, similar to the *Leader* option (*DIMFIT*=4) when there is insufficient room for the dimensioning components (Fig. 27-24). A setting of 5 (or any other *DIMFIT* setting) has <u>no effect</u> on the <u>appearance</u> of the dimension when there is enough room for the dimensioning components.

There is an <u>important benefit</u> to creating a dimension with a *Fit* setting of *No Leader*; that is, the text can be edited with *Dimtedit* or with grips as if it were "detached" from the dimension line. In Figure 27-25, a dimension was created with *Fit* setting of *No Leader*. Although there is no apparent difference between this dimension and one created with another *Fit* setting (when sufficient room exists), the dimension text can be edited independent of the other dimension components. Using **grips** or the **Dimtedit** command (but <u>not</u> *Dimedit*), the text component can be moved independently to any location and the dimension retains its associativity. Figure 27-25 illustrates the use of grips to edit the dimension text.

Figure 27-25

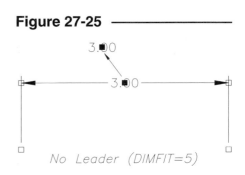

Radius and Diameter Variable Settings
For creating mechanical drawing dimensions according to ANSI standards, the default settings in AutoCAD Release 14 are correct for creating *Diameter* dimensions but not for *Radius* dimensions. The following variable settings are recommended for creating *Radius* and *Diameter* dimensions for mechanical applications.

For *Diameter* dimensions, the default settings produce ANSI-compliant dimensions that suit most applications—that is, text and arrows are on the outside of the circle or arc pointing inward toward the center. For situations where large circles are dimensioned, *Fit* (*DIMFIT*) and *User Defined* (*DIMUPT*) can be changed to force the dimension inside the circle (Fig. 27-26).

Figure 27-26

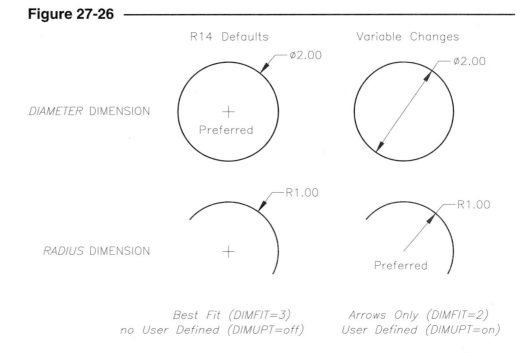

For *Radius* dimensions the default settings produce incorrect dimensioning practices. Normally (when space permits) you want the dimension line and arrow to be inside the arc, while the text can be inside or outside. To produce ANSI-compliant *Radius* dimensions, set *Fit* (*DIMFIT*) and *User Defined* (*DIMUPT*) as shown in Figure 27-26. Radius dimensions can be outside the arc in cases where there is insufficient room inside.

Text–Inside/Outside Horizontal (DIMTIH, DIMTOH)

The *Text* section in the upper-right of the dialog box sets the orientation of the dimension text to horizontal or aligned (with the dimension line). The setting can be selected from the drop-down list or by clicking on the image tile (Fig. 27-27). The results of the changes are obvious by the configuration in the image tile. Your choice sets the value for variables *DIMTIH* and *DIMTOH*.

Figure 27-27 —

Horizontal Justification (DIMJUST)

The *Horizontal Justification* section determines the horizontal location of the text with respect to the dimension line. Change the setting with the drop-down list or by clicking the image tile. The choice sets the value for *DIMJUST*.

Figure 27-28 —

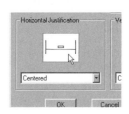

Vertical Justification (DIMTAD, DIMTIH)

The *Vertical Justification* section operates similarly to the *Horizontal Justification* feature. Changes can be made through selection from the drop-down list or by clicking on the image tile. The possible orientation options are displayed in the image tile. The selections modify the values of the *DIMTAD* (four possible settings) and *DIMTIH* variables together; therefore, eight different images are possible. Experiment with these settings and examine the changes to the image tile.

Figure 27-29 —

Annotation...

The *Annotation* dialog box (Fig. 27-30) controls the <u>format of the annotation (text) of a dimension</u>. You can vary the AutoCAD-supplied numerical value that is drawn in the dimension in several ways. Features such as the text style and height, prefix and/or suffix, alternate units (for inch and metric notation), and several variations of tolerance and limit dimensions are possible. These features are displayed in the dialog box in four separate areas.

Figure 27-30

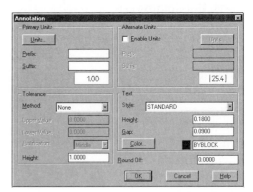

NOTE: The image tile in the *Primary Units* section of the *Annotation* dialog box (just below the *Suffix*: edit box) <u>does not</u> display the units and precision selected. It displays only the options selected from the *Tolerance* section below and should therefore be located in the *Tolerance* section.

Figure 27-31

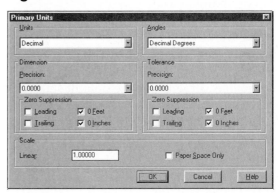

Primary Units (DIMUNIT)

The *Primary Units* section (top left) controls the display of units, precision, and prefix and/or suffix. Selecting the *Units* tile summons the *Primary Units* dialog (Fig. 27-31).

You can select the *Unit* type from the drop-down list (Fig. 27-32). These are the same unit types available with the *Units* and *Ddunits* command (*Decimal, Scientific, Engineering, Architectural,* and *Fractional*) with the addition of *Architectural (stacked)* and *Fractional (stacked)* and *Windows Desktop*. The *Windows Desktop* option displays AutoCAD units based on the settings made for units display in Windows 95/NT Control Panel (settings for decimal separator and number grouping symbols). Remember that your selection affects the <u>units drawn in dimension objects,</u> <u>not the global drawing units.</u>

Figure 27-32

Examples of the resulting dimension objects for the two fractional and two architectural unit choices are shown in Figure 27-33. The choice for *Primary Units* is stored in the *DIMUNIT* variable. This drop-down list is disabled for an *Angular* family member.

Figure 27-33

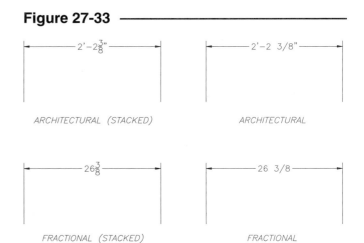

ARCHITECTURAL (STACKED)   ARCHITECTURAL

FRACTIONAL (STACKED)   FRACTIONAL

Dimension, Precision (DIMDEC)

The *Precision* drop-down list in the *Dimension* section (Fig. 27-34) specifies the number of places for decimal dimensions or denominator for fractional dimensions. This setting does not alter the drawing units precision. This value is stored in the *DIMDEC* variable.

Figure 27-34

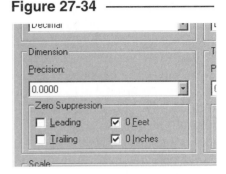

Dimension, Zero Suppression (DIMZIN)

The *Zero Suppression* section controls how zeros are drawn in a dimension when they occur. A check in one of these boxes means that zeros are <u>not drawn for that case</u>. This sets the value for *DIMZIN*.

Angles (DIMAUNIT)

The *Angles* section (Fig. 27-31) sets the unit type for angular dimensions from a drop-down list, including *Decimal degrees, Deg/Min/Sec, Grads, Radians,* and *Surveyor* (units). This list is enabled only for *Parent* and *Angular* family members. The variable used for the angular units is *DIMAUNIT*.

DIMADEC (Decimal Places for Angular Units)

DIMADEC is the only new dimensioning variable for Release 14. This variable sets the number of places of precision (decimal places) for angular dimension text. *DIMADEC* can be set only for *Parent* and *Angular* family members.

To make the desired setting for angular dimensions, set the *DIMADEC* variable for the entire family (although only *Angular* dimensions are affected) by typing *DIMADEC* <u>at the command prompt.</u> For example, to display 2 places of precision for angular dimensions, change *DIMADEC* from -1 to 2. Then use *Dimstyle, Save* (in command line format) to save the new setting to the current dimension style. A positive value sets the precision to that number of places for the entire family, although only angular dimensions are affected. A setting of -1 (default) dictates that the angular units are controlled by the *DIMDEC* variable (*Primary Units Precision*) setting.

There is no method for changing this variable for the *Parent* using the *Dimension Style* dialog boxes. You can, however, select the *Angular* member in the *Family* list (in the *Dimension Style* dialog box) and make the desired change in the *Precision* drop-down list. This action creates an *Angular* child and sets the *DIMADEC* variable for that child. Creating an *Angular* child to set this variable adds unnecessary complexity since a *Parent* setting affects only angular dimensions; therefore, it is recommended that you use the command line format to set the variable for the entire dimension style.

Tolerance (DIMTDEC, DIMTZIN)

Use the *Tolerance* cluster area of the *Primary Units* dialog box (Fig. 27-35) to set values when drawing tolerance dimensions (see *Tolerance*). The *Precision* and *Zero Suppression* operate identically to the *Dimension* section (see *Dimension, Precision* and *Dimension, Zero Suppression*). The affected variables are *DIMTDEC* and *DIMTZIN*.

Figure 27-35

Scale, Linear (DIMLFAC)

The value appearing in the *Linear Scale* section of the *Primary Units* dialog box is a <u>multiplier</u> for the AutoCAD-measured numerical value. The default is 1. Entering a 2 would cause AutoCAD to draw a value two times the actual measured value (Fig. 27-36). This feature is used when a drawing is created in some scale other than the actual size, such as in an enlarged detail created in the same drawing as the full view. For example, if you wanted to make a detail (enlarged view) of a portion of a part, create the detail two times the actual size, enter .5 in the *Linear Scale* edit box, dimension the detail, and AutoCAD will display the actual part measurements rather than the actual measured value of the enlarged AutoCAD objects. This value is stored in the *DIMLFAC* variable.

Figure 27-36

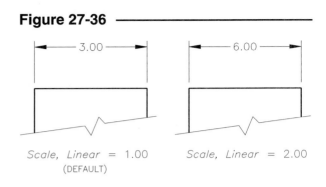

Scale, Linear = 1.00
(DEFAULT)

Scale, Linear = 2.00

Alternate Units (DIMALTx)

You can specify that AutoCAD draws <u>alternate units</u> (a <u>second set</u> of numerical values) when placing a dimension (Fig. 27-37). In the *Annotation* dialog box (Fig. 27-30), the *Enable Units* toggle can be checked to enable the *Units...* button of the *Alternate Units* cluster. The *Alternate Units* dialog box features are nearly the same as *Primary Units* with the exception of the *Linear Scale*. The value appearing in this field is a multiplier for the AutoCAD-measured value. Note that *Linear Scale* in the *Aternate Units* dialog box is set to 25.4, the inch-to-milimeter conversion factor (Fig. 27-38). In many industries, both inch units and metric units are required for all dimensions.

Figure 27-37

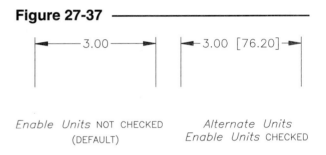

Enable Units NOT CHECKED
(DEFAULT)

Alternate Units
Enable Units CHECKED

Figure 27-38

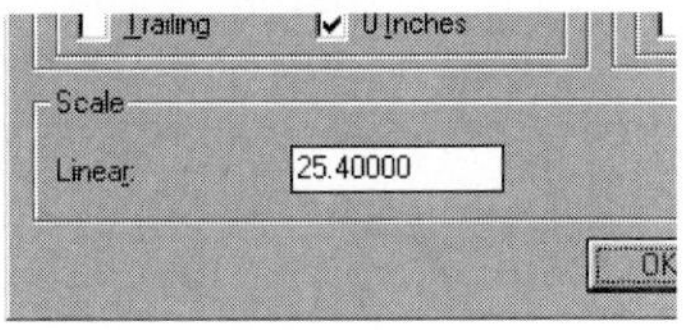

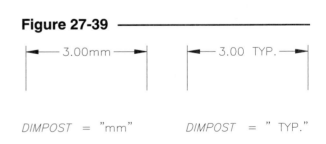

Because the alternate unit options parallel the primary unit options, a parallel set of variables is used to store the related values. The variable names are *DIMALTU* for alternate units, *DIMALTD* for alternate units decimal places, *DIMALTF* for alternate linear scale, *DIMALTZ* for alternate units zero suppression, *DIMALTTZ* for alternate units tolerance zero suppression, and *DIMALTTD* for alternate units tolerance decimal places. *DIMALT* stores the on/off toggle for enabling alternate units.

Prefix/Suffix (DIMPOST, DIMAPOST)

In both the *Primary Units* and *Alternate Units* sections of the *Annotation* dialog box (Fig. 27-30), the *Prefix* and *Suffix* edit box holds any text that you want to add to the AutoCAD-supplied dimensional value. For example, entering the string "mm" or a "TYP." in the *Suffix* edit box would produce text as shown in Figure 27-39. (In such a case, don't forget the space between the numerical value and the suffix.) The string is stored in the variable *DIMPOST* and in *DIMAPOST* for alternate units prefix or suffix.

Figure 27-39

3.00mm

3.00 TYP.

DIMPOST = "mm"

DIMPOST = " TYP."

If you use the *Prefix* box to enter letters or values, any AutoCAD-supplied symbols (for radius and diameter dimensions) are overridden (not drawn). For example, if you want to specify that a specific hole appears twice, you should indicate by designating a "2X" before the diameter dimension. However, doing so by this method overrides the phi (Ø) symbol that AutoCAD inserts before the value. Instead, use the *Mtext/Text* options within the dimensioning command or use *Dimedit* to add a prefix to an existing dimension having an AutoCAD-supplied symbol (see Fig. 26-21 and related text, and *Dimedit*).

Tolerance
Method **(DIMTOL, DIMLIM, DIMGAP)**

Figure 27-40

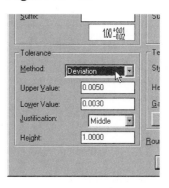

The *Tolerance Method* displays a drop-down list with five types: *None*, *Symmetrical*, *Deviation*, *Limits*, and *Basic*. Each format is displayed in the image tile <u>above</u> this section (actually in the *Primary Units* section) (Fig. 27-40). You can click in the image tile above to cycle through the formats. The four possibilities (other than *None*) are illustrated in Figure 27-41. The *Symmetrical* and *Deviation* methods create plus/minus dimensions and turn on the *DIMTOL* variable. The *Limits* method creates limit dimensions and turns the *DIMLIM* variable on. The *Basic* method creates a basic dimension by drawing a box around the dimensional value, which is accomplished by changing the *DIMGAP* to a negative value.

Upper Value/Lower Value **(DIMTP, DIMTM)**

Figure 27-41

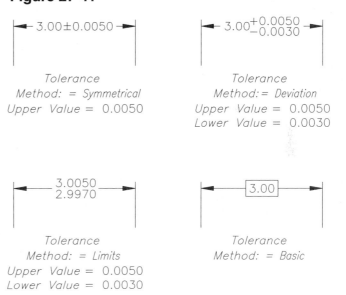

An *Upper Value* and *Lower Value* can be entered in the edit boxes for the *Deviation* and *Limits* method types. In these cases, the *Upper Value* (*DIMTP*—DIM Tolerance Plus) is added to the measured dimension and the *Lower Value* (*DIMTP*—DIM Tolerance Minus) is subtracted (Fig. 27-41). An *Upper Value* only is needed for *Symmetrical* and is applied as both the plus and minus value.

Height **(DIMTFAC)**

Figure 27-42

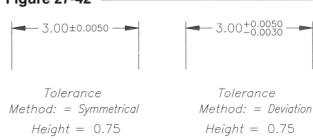

The height of the <u>tolerance text</u> is controlled with the *Height* edit box value. The entered value is a <u>proportion</u> of the primary dimension value height. For example, a value of .75 would draw the tolerance text at 75% of the primary text height (Fig. 27-42). The *Tolerance Height* value is <u>not reflected in the image tile above</u>. The setting affects the *Symmetrical*, *Deviation*, and *Limits* tolerance methods. The value is stored in the *DIMTFAC* (Text FACtor) variable.

Justification (DIMTOLJ)

Justification places the tolerance dimension values with either a top, middle, or bottom alignment with the primary dimension value. The result is displayed in the image tile. The *Justification* feature is enabled only for *Symmetrical* and *Deviation* dimension methods (Fig. 27-43).

Figure 27-43

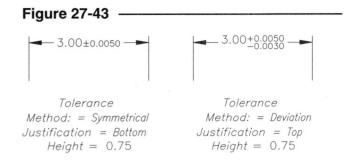

Text

Style (DIMTXSTY)

The *Text* cluster controls the text *Style*, *Height*, and *Color* of the dimension text. The text styles are chosen from a drop-down list of existing styles. The advantage is different dimension styles can have different text styles. The text style used for dimensions remains constant (as defined by the dimension style) and does not change when other text styles in the drawing are made current (as defined by *Style*, *Dtext*, *Text*, or *Mtext*). The *DIMTXSTY* variable holds the text style name for the dimension style.

Figure 27-44

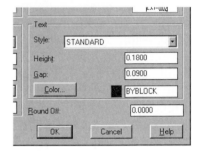

Height (DIMTXT)

This value specifies the primary text height; however, this value is multiplied by the *Overall Scale (DIMSCALE)* to determine the actual drawn text height. Change *Height* if you want to increase or decrease only the text height (Fig. 27-45). The value is stored in *DIMTXT*. Change instead the *Overall Scale* to change all size-related features (arrows, gaps, text, etc.) proportionally.

Figure 27-45

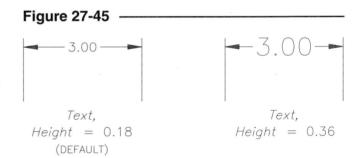

Gap (DIMGAP)

The *Gap* edit box sets the distance between the dimension text and its dimension line (Fig. 27-46). The "gap" is actually an invisible box around the text. Changing the *DIMGAP* variable to a negative value makes the box visible and is used for creating a *Basic* dimension (see *Tolerance, Method*).

Figure 27-46

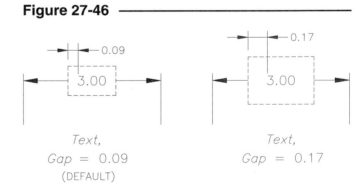

Color (DIMCLRT)

Select this edit box to activate the standard *Select Color* dialog box. The color choice is assigned to the dimension text only (Fig. 27-47). This is useful for printing or plotting the dimension text in a darker color or wider pen.

Figure 27-47

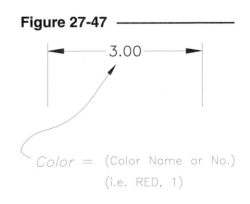

Round Off (DIMRND)

The *Round Off* edit box rounds the primary dimension value to the decimal place (or fraction) specified (Fig. 27-48). Normally, AutoCAD values are kept to 14 significant places but are rounded to the place dictated by the dimension *Primary Units, Precision* (DIMDEC). Use this feature to round up or down appropriately to the specified decimal or fraction.

Figure 27-48

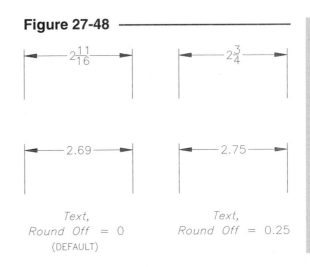

Changing Dimension Variables Using the Command Line Format

(VARIABLE NAME)	Pull-down Menu	COMMAND – (TYPE)	ALIAS (TYPE)	Short-cut	Screen (side) Menu	Tablet Menu
	Tools *Inquiry >* *Set Variable*	*(VARIABLE NAME)*	...	...	*TOOLS 1* *Setvar*	*U,10*

As an alternative to setting dimension variables through the *Geometry, Format,* and *Annotation* dialog boxes, you can type the dimensioning variable name at the Command: prompt. There is a noticeable difference in the two methods—the dialog boxes use different nomenclature than the formal dimensioning variable names used in command line format; that is, the dialog boxes use descriptive terms that, if selected, make the appropriate change to the dimensioning variable. The formal dimensioning variable names accessed by command line format, however, all begin with the letters *DIM*, are limited to eight characters, and are accessible only by typing.

Another important but subtle difference in the two methods is the act of saving dimension variable settings to a dimension style. Remember that all drawn dimensions are part of a dimension style, whether it is STANDARD or some user-created style. The act of saving variable changes made through the *Dimension Styles* dialog boxes is accomplished by selecting the *Save* tile upon exiting. When you change dimension variables by the command line format, the changes become overrides until you use the *Save* option of the *Dimstyle* command or *Dimension Styles* dialog box. When a variable change is made, it becomes an override that is applied to the current dimension style and affects only the newly drawn dimensions. Variable changes must be *Saved* to become a permanent part of the style and to retroactively affect all dimensions created with that style.

In order to access and change a variable's setting by name, simply type the variable name at the Command: prompt. For dimension variables that require distances, you can enter the distance (in any format accepted by the current *Units* settings) or you can designate by (PICKing) two points:

For example, to change the value of the *DIMSCALE* to **.5**, this command syntax is used:

```
Command: dimscale
New value for dimscale <1.0000>: .5
Command:
```

See Appendix A for the complete list of dimension variables, descriptions, and default settings.

Using *Setup Wizards* and **Template Drawings**

The previous dimension variable table gives default settings for the ACAD.DWT template drawing. This drawing is basically the same that is traditionally used when you begin a drawing session in previous releases of AutoCAD. In Release 14, the ACAD.DWT drawing is the template or basis used when you begin AutoCAD with the *Startup* dialog box toggled off, choose *Start from Scratch, English Default Settings* in the *Create New Drawing* dialog box, select the ACAD.DWT as a template drawing, or choose all the defaults in the *Setup Wizards*. In other cases, one or more dimension variables are automatically changed. In some template drawings, pre-made dimension styles are available.

Start from Scratch, Metric Default Settings or **ACADISO.DWT**

When you choose *Start from Scratch, Metric Default Settings,* AutoCAD uses the ACADISO.DWT template drawing. The *DIMSCALE* retains a setting of 1, but many other dimension variables are changed to accommodate metric sizes (English default setting x 25.4). The following settings are made to the following dimension variables (that differ from the ACAD.DWT setting):

DIMASZ	2.5000
DIMCEN	2.5000
DIMDLI	3.7500
DIMEXE	1.2500
DIMEXO	0.6250
DIMGAP	0.6250
DIMTXT	2.5000

Setup Wizards

When you use a *Setup Wizard,* many system variables are changed according to the values you enter for *Limits* (*Step 2: Units* in *Quick Setup* or *Step 5: Units* in *Advanced Setup*). DIMSCALE is the only dimensioning variable changed by the *Setup Wizards*. The affected system variables are changed from the default settings based on the following formula:

$$N=D(Y/9)$$

where N is the new system variable setting,
   D is the system variable default value,
   Y is the Area (*Limits*) Y value that you enter, and
   9 is the default *Area* (*Limits*) Y value.

For example, entering *Limits* values of 24 x 18 would set *DIMSCALE* to 2, or using the formula above, 2=1(18/9). Beware, settings made by the *Setup Wizards* are correct for printing or plotting only on an A size sheet and are incorrect for any other sheet size. See Chapters 6, 13, and 14 for discussion on Drawing Scale Factor and Plotting to Scale.

Template Drawings

Template drawings may or may not include dimensioning variable changes or dimension styles, depending on the selected template. In general, the ANSI templates (Ansi_a through Ansi_e) use the same default settings (no changes) for all dimension variables, including *DIMSCALE*, as the ACAD.DWT template. The *Din, ISO,* and *JIS* templates include one or more *Dimension Styles* with appropriate variable settings. See Table of AutoCAD-Supplied Template Drawing Settings, Chapter 13, for a more complete description.

R14

MODIFYING EXISTING DIMENSIONS

Even with the best planning, a full understanding of dimension variables, and the correct use of *Dimension Styles*, it is probable that changes will have to be made to existing dimensions in the drawing due to design changes, new plot scales, or new industry/company standards. There are several ways that changes can be made to existing dimensions while retaining associativity and membership to a dimension style. The possible methods are discussed in this section.

Modifying a Dimension Style

Existing dimensions in a drawing can be modified by making one or more variable changes to the dimension style family or child, then *Saving* those changes to the dimension style. This process can be accomplished using either the *Dimension Style* dialog box series or command line format. When you use the *Save* option (in the *Dimension Style* dialog box or in the *Dimstyle* command), the existing dimensions in the drawing <u>automatically update</u> to display the new variable settings. For the individual steps for this procedure, see Modifying a Dimension Style and Updating Existing Dimensions in the Dimension Styles section earlier in this chapter.

This method is generally preferred in cases where all dimensions in a style need updating. One drawback of this method is that children created previously (special settings for *Radial, Diameter,* etc.) in the dimension style being modified <u>do not automatically update.</u> The solution is to make the same variable changes for the child (select *Radial, Diameter,* etc. in the *Dimension Styles* dialog box, then make variable changes and *Save*) or use *Dimoverride* or *Update* (discussed later).

Creating Dimension Style Overrides and Using *Update*

You can modify existing dimensions in a drawing without making permanent changes to the dimension style by creating a dimension style override, then using *Update* to apply the new setting to an existing dimension. This method is preferred if you wish <u>to modify one or two dimensions</u> without modifying all dimensions referencing (created in) the style.

To do this, use either the command line format or *Dimension Styles* dialog box series to set the new variable. In the *Dimension Styles* dialog box, make the desired variable settings in the *Geometry, Format,* or *Annotation* dialog boxes, but <u>do not use *Save*—use *OK*</u> instead. This creates an override to the current style. In command line format, simply enter the formal dimension variable name and make the change to create an override to the current style. Next, use *Update* to apply the current style settings plus the override settings to existing dimensions that you PICK. The overrides remain in effect for the current style unless the variables are reset to the original values. You can <u>clear overrides</u> for a dimension style by selecting the original style name (without the "+" prefix) in the *Dimension Style* dialog box to make it current, then answer *No* to "save changes to current style?"

UPDATE

Pull-down Menu	COMMAND (TYPE)	ALIAS (TYPE)	Short-cut	Screen (side) Menu	Tablet Menu
Dimension *Update*	*DIM* *UPDATE*	*DIM* *UP*	...	*DIMNSION* *Update*	...

Update can be used to update existing dimensions in the drawing to the current settings. The current settings are determined by the current *Dimension Style* and any dimension variable overrides that are in effect (see previous explanation, Creating Dimension Style Overrides and Using *Update*). This is an excellent method of modifying one or more existing dimensions without making permanent changes to the dimension style that the selected dimensions reference. *Update* has the same effect as using *Dimstyle, Apply.*

Update is actually a Release 12 command. In Release 12, dimensioning commands could only be entered at the Dim: prompt. For example, to create a linear dimension, you had to type *Dim* and press Enter, then type *Linear*. In Release 13, all dimensioning commands were upgraded to top-level commands with the *Dim-* prefix added, so you can type *Dimlinear* at the command prompt, for example. *Update*, although very useful, was never upgraded. In Release 14 the command is given prominence by making available an *Update* button and an *Update* option in the *Dimension* pull-down menu. However, if you prefer to type, you must first type *Dim*, press Enter, and then enter *Update*. (See *DIM* in Chapter 26.) The command syntax is as follows:

```
Command: Dim
Dim: Update
Select objects: PICK (select a dimension object to update)
Select objects: Enter
Dim: press Esc or type Exit
Command:
```

For example, if you wanted to change the *DIMSCALE* of several existing dimensions to a value of 2, change the variable by typing *DIMSCALE* or change *Overall Scale* in the *Geometry* dialog box (do not *Save*). Next use *Update* and select the desired dimension. That dimension is updated to the new setting. The command syntax is the following:

```
Command: Dimscale
New value for DIMSCALE <1.0000>: 2
Command: Dim
Dim: Update
Select objects: PICK  (select a dimension object to update)
Select objects: PICK  (select a dimension object to update)
Select objects: Enter
Dim: Exit
Command:
```

Beware, after *Updating* the selected dimensions, the overrides for that style remain in effect. You should then reset the variable to its original value unless you want to keep the override for creating other new dimensions. You can clear overrides for a dimension style by selecting the original style name (without the "+" prefix) in the *Dimension Style* dialog box to make it current, then answer *No* to "save changes to current style?"

DIMOVERRIDE

Pull-down Menu	COMMAND (TYPE)	ALIAS (TYPE)	Short-cut	Screen (side) Menu	Tablet Menu
Dimension Override	DIMOVERRIDE	DIMOVER or DOV	...	...	Y,4

Dimoverride grants you a great deal of control to edit existing dimensions. The abilities enabled by *Dimoverride* are unique in that no other dimensioning commands or options allow you to retroactively modify dimensions in such a way.

Dimoverride enables you to make variable changes to dimensions that exist in your drawing without changing the dimension style that the dimension references (was created under). For example, using *Dimoverride*, you can make a variable change and select existing dimension objects to apply the change. The existing dimension does not lose its reference to the parent dimension style nor is the dimension style changed in any way. In effect, you can override the dimension styles for selected dimension objects. There are two steps: set the desired variable and select dimension objects to alter.

```
Command: dimoverride
Dimension variable to override (or Clear to remove overrides): (variable name)
Current value <current  value> New value: (value)
Dimension variable to override:  (variable name) or Enter
Select objects: PICK
Select objects: PICK or Enter
Command:
```

The *Dimoverride* feature differs from creating overrides in two ways: overrides (variable changes that are made but not *Saved*) (1) are appended to the *Parent* dimension style and (2) are applied to newly created dimensions only and are not retroactive. In contrast, *Dimoverride* affects only the selected existing dimensions and does not change the parent dimension style.

Dimoverride is useful as a "backdoor" approach to dimensioning. Once dimensions have been created, you may want to make a few modifications, but you do not want the changes to affect dimension styles (resulting in an update to all existing dimensions that reference the dimension styles). *Dimoverride* offers that capability. You can even make one variable change to affect all dimensions globally without having to change all the dimension styles. For example, you may be required to make a test plot of the drawing in a different scale than originally intended, necessitating a new global *Dimscale*. Use *Dimoverride* to make the change for the plot:

```
Command: dimoverride
Dimension variable to override (or Clear to remove overrides): dimscale
Current value <1.0000> New value: .5
Dimension variable to override: Enter
Select objects: (window entire drawing) Other corner: 128 found
Select objects: Enter
Command:
```

This action results in having all the existing dimensions reflect the new *DIMSCALE*. No other dimension variables or any dimension styles are affected. Only the selected objects contain the overrides. *Dimoverride* does not append changes (overrides) to the original dimension styles, so no action must be taken to clear the overrides from the styles. After making the plot, *Dimoverride* can be used with the *Clear* option to change the dimensions (by object selection) back to their original appearance. The *Clear* option is used to clear overrides from dimension objects, not from dimension styles.

Clear

The *Clear* option removes the overrides from the selected dimension objects. It does not remove overrides from the current dimension style:

```
Command: dimoverride
Dimension variable to override (or Clear to remove overrides): c
Select objects: PICK
Select objects: Enter
Command:
```

The dimension then displays the variable settings as specified by the dimension style it references without any overrides (as if the dimension were originally created without the overrides). Using *Clear* does not remove the overrides that are appended to the dimension style so that if another dimension is drawn, the overrides apply.

Dimoverride is a very powerful and useful tool because it provides capabilities that are not available by any other method. You should experiment with it now so you can use it later to simplify otherwise difficult dimension editing situations.

Other Methods for Modifying Existing Dimensions

MATCHPROP

Matchprop can be used to "convert" an existing dimension to the style (including overrides) of another dimension in the drawing. For example, if you have two linear dimensions, one has *Oblique* arrows and *Romans* text font (*Dimension Style* = "Oblique") and one is a typical linear dimension (*Dimension Style* = "Standard") as in Fig. 27-49, before. You can convert the typical dimension to the "Oblique" style by using *Matchprop*, selecting the "Oblique" dimension as the "source object" (to match), then selecting the typical dimension as the "destination object" (to convert). The typical dimension then references the "Oblique" dimension style and changes appearance accordingly (Fig. 27-49, after). Note that *Matchprop* does not alter the dimension text value.

Figure 27-49

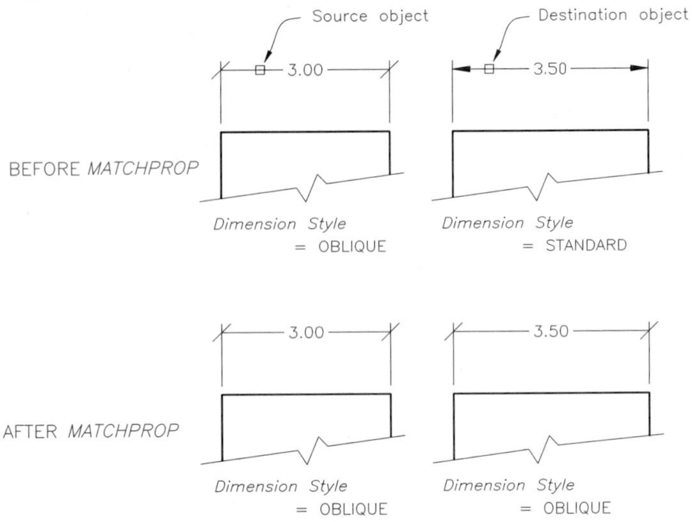

Using the *Settings* option of the *Matchprop* command, you can display the *Property Settings* dialog box (Fig. 27-50). The *Dimension* box under *Special Properties* must be checked to "convert" existing dimensions as illustrated above.

Using *Matchprop* is a fast and easy method for modifying dimensions from one style to another. However, this method is applicable only if you have existing dimensions in the drawing with the desired appearance that you want others to match.

Figure 27-50

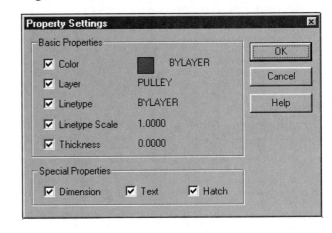

DDMODIFY

Remember that *Ddmodify* can be used to edit existing dimensions. Using *Ddmodify* and selecting one dimension invokes the *Modify Dimension* dialog box. Here you can modify any aspect of the dimension, including text, and access is given to the *Geometry, Format,* and *Annotation* dialog boxes. Any changes to dimension variables through *Ddmodify* result in <u>overrides to the dimension object only</u> but do not affect the dimension style. *Ddmodify* has essentially the same result as *Dimoverride*, except only one dimension object at a time can be modified. *Ddmodify* can also be used to modify the dimension text value. *Ddmodify* with dimensions is discussed in Chapter 26.

DIMTEDIT, DIMEDIT

Dimension text of existing dimensions can be modified using either *Dimtedit* or *Dimedit*. These commands allow you to change the text position with respect to the dimension line and the angle of the text. *Dimtedit* can be used with the *New* option to restate the original AutoCAD-measured text value if needed. See Chapter 26 for a full explanation of these commands.

Grips

Grips can be used to effectively alter dimension length, position, text location and more. Keep in mind the possibilities of moving the dimension text with grips for dimensions with *DIMFIT* of 5. See Chapter 26 for a full discussion of the possibilities of Grip editing dimensions.

GUIDELINES FOR DIMENSIONING IN AutoCAD

Listed in this section are some guidelines to use for dimensioning a drawing using dimensioning variables and dimension styles. Although there are other strategies for dimensioning, two strategies are offered here as a framework so you can develop an organized approach to dimensioning.

In almost every case, dimensioning is one of the last steps in creating a drawing, since the geometry must exist in order to dimension it. You may need to review the steps for drawing setup, including the concept of drawing scale factor (Chapter 13).

Strategy 1. Dimensioning a Single Drawing

This method assumes that the fundamental steps have been taken to set up the drawing and create the geometry. Assume this has been accomplished:

Drawing setup is completed: *Units, Limits, Snap, Grid, Ltscale, Layers,* border, and title block.
The drawing geometry (objects comprising the subject of the drawing) has been created.

Now you are ready to dimension the drawing subject (of the multiview drawing, pictorial drawing, floor plan, or whatever type of drawing).

1. Create a *Layer* (named DIM, or similar) for dimensioning if one has not already been created. Set *Continuous* linetype and appropriate color. Make it the *current* layer.

2. Set the *Overall Scale (DIMSCALE)* based on drawing *Limits* and expected plotting size.

   For plotted dimension text of 3/16":

   Multiply *Overall Scale (DIMSCALE)* times the drawing scale factor. The default *Overall Scale* is set to 1, which creates dimensioning text of approximately 3/16" (default *Text Height:* or *DIMTXT* =.18) when plotted full size. All other size-related dimensioning variables' defaults are set appropriately.

   For plotted dimension text of 1/8":

   Multiply *Overall Scale* times the drawing scale factor, <u>times .7</u>. Since the *Overall Scale* times the scale factor produces dimensioning text of .18, then .18 x .7 = .126 or approximately 1/8". (See Optional Method for Fixed Dimension Text Height.) (*Overall Scale* [*DIMSCALE*] is one variable that usually remains constant throughout the drawing and therefore should generally have the same value for every dimension style created.)

3. Make the other dimension variable changes you expect you will need to create <u>most dimensions in the drawing</u> with the appearance you desire. When you have the basic dimension variables set as you want, *Save* this new dimension style family (select *Parent*) to the TEMPLATE (or other descriptive name) style. (Remember you must select *Save* and exit the dialog box to save the new dimension style.) If you need to make special settings for types of dimensions (*Linear, Diameter,* etc.), create "children" at this stage and *Save* the changes to the style. This style is the fundamental style to use for creating most dimensions and should be used as a template for creating other dimension styles. If you need to reset or list the original (default) settings, the STANDARD style can be restored.

4. Create all the relatively simple dimensions first. These are dimensions that are easy and fast and require <u>no other dimension variable changes</u>. Begin with linear dimensions; then progress to the other types of dimensions.

5. Create the special dimensions next. These are dimensions that require variable changes. Create appropriate dimension styles by changing the necessary variables, then *Save* each set of variables (relating to a particular style of dimension) to an appropriate dimension style name. Specify dimension variables for the classification of dimension (children) in each dimension style when appropriate. The dimension styles can be created "on the fly" or as a group before dimensioning. Use TEMPLATE as your base dimension style when appropriate.

    For the few dimensions that require variable settings unique to that style, you can create dimension variable overrides. Change the desired variable(s), but do not *Save* to the style so the other dimensions in the style are not affected.

6. When all of the dimensions are in place, make the final adjustments. Several methods can be used:

    a. If modifications need to be made <u>familywide,</u> change the appropriate variables and *Save* the changes to the dimension style. This action automatically updates existing dimensions that reference that style.

    b. To modify the appearance of selected dimensions, you can create dimension style overrides, then use *Update* or *Dimstyle, Apply* to update the selected dimensions to the new settings. Keep in mind that the dimension style overrides are still in effect and are applied to new dimensions. Clear the overrides by setting the original style current.

    c. Alternately, use *Ddmodify* to change variable settings for selected dimensions. These changes are made as overrides to the selected object only and do not affect the dimension style. This action has the same result as using *Dimoverride* but for only one dimension.

    d. To change one dimension to adopt the appearance of another, use *Matchprop*. Select the dimension to match first, then the dimension(s) to convert. *Dimstyle, Apply* can be used for this same purpose.

    e. Use *Dimoverride* to make changes to selected dimensions or to all dimensions <u>globally</u> by windowing the entire drawing. *Dimoverride* has the advantage of allowing you to assign the variables to change and selecting the objects to change all in one command. What is more, the changes are applied as overrides only to the selected dimensions but <u>do not alter the original dimension styles</u> in any way.

    f. If you want to change only the dimension text value, use *Ddmodify* or *Dimedit, New. Dimedit, New* can also be used to reapply the AutoCAD-measured value if text was previously changed. The location of the text can be changed with *Dimtedit* or Grips.

    g. Grips can be used effectively to change the location of the dimension text or to move the dimension line closer or further from the object. Other adjustments are possible, such as rotating a *Radius* dimension text around the arc.

Strategy 2. Creating Dimension Styles as Part of Template Drawings

1. Begin a *New* drawing or *Open* an existing *Template*. Assign a descriptive name.

2. Create a DIM *Layer* for dimensioning with *continuous* linetype and appropriate color (if one has not already been created).

3. Set the *Overall Scale* accounting for the drawing *Limits* and expected plotting size. Use the guidelines given in Strategy 1, step 2. Make any other dimension variable changes needed for general dimensioning or required for industry or company standards.

4. Next, *Save* a dimension style named "TEMPLATE." This should be used as a template when you create most new dimension styles. The *Overall Scale* is already set appropriately for new dimension styles in the drawing.

5.   Create the appropriate dimension styles for expected drawing geometry. Use TEMPLATE dimension style as a base style when appropriate.

6.   *Save* and *Exit* the newly created template drawing.

7.   Use this template in the future for creating new drawings. Restore the desired dimension styles to create the appropriate dimensions.

Using a template drawing with prepared dimension styles is a preferred alternative to repeatedly creating the same dimension styles for each new drawing.

Optional Method for Fixed Dimension Text Height in Template Drawings

To summarize Strategy 1, step 2., the default *Overall Scale* (*DIMSCALE* =1) times the default *Text Height* (*DIMTXT* =.18) produces dimensioning text of approximately 3/16" when plotted to 1=1. To create 1/8" text, multiply *Overall Scale* times .7 (.18 x .7 = .126). As an alternative to this method, try the following.

For 1/8" dimensions, for example, multiply the initial values of the size-related variables by .7; namely

Text Height	(*DIMTXT*)	.18 x .7 = .126
Arrow Size	(*DIMASZ*)	.18 x .7 = .126
Extension Line Extension	(*DIMEXE*)	.18 x .7 = .126
Dimension Line Spacing	(*DIMDLI*)	.38 x .7 = .266
Text Gap	(*DIMGAP*)	.09 x .7 = .063

Save these settings in your template drawing(s). When you are ready to dimension, simply multiply *Overall Scale* (1) times the drawing scale factor.

CHAPTER EXERCISES

For each of the following exercises, use the existing drawings, as instructed. Create dimensions on the DIM (or other appropriate) layer. Follow the Guidelines for Dimensioning given in the chapter, including setting an appropriate *Overall Scale* based on the drawing scale factor. Use dimension variables and create and use dimension styles when needed.

Figure 27-51 ———————————————

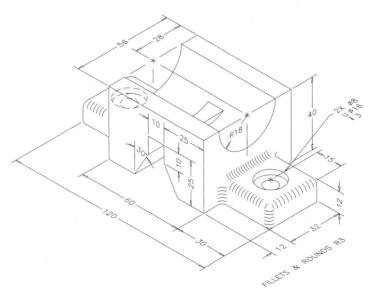

1.   **Dimensioning a Multiview**

Open the **SADDLE** drawing that you created in Chapter 22 Exercises. Set the appropriate dimensional **Units** and **Precision**. Add the dimensions as shown. Because the illustration in Figure 27-51 is in isometric, placement of the dimensions can be improved for your multiview. Use optimum placement for the dimensions. *Save* the drawing as **SADDL-DM** and make a *Plot* to scale.

2.   **Architectural Dimensioning**

Open the **OFFICE** drawing that you completed in Chapter 20. Dimension the floor plan as shown in Figure 20-18. Add *Text* to name the rooms. *Save* the drawing as **OFF-DIM** and make a *Plot* to an accepted scale and sheet size based on your plotter capabilities.

3.   **Dimensioning an Auxiliary**

Open the **ANGLBRAC** drawing that you created in Chapter 25. Dimension as shown in Figure 27-52, but <u>convert the dimensions to</u> *Decimal* with *Precision* of *.000*. Use the Guidelines for Dimensioning. Dimension the slot width as a *Limit* dimension—*.6248/.6255*. *Save* the drawing as **ANGL-DIM** and *Plot* to an accepted scale.

Figure 27-52 ————————————————

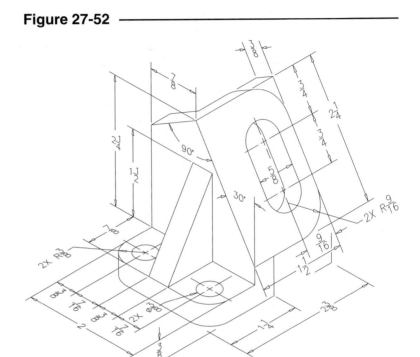

4.   **Isometric Dimensioning**

Dimension the support bracket that you created as an isometric drawing named **SBRACKET** in the Chapter 23 Exercises. All of the dimensions shown in Figure 27-53 should appear on your drawing in the optimum placement. *Save* the drawing as **SBRCK-DM** and *Plot* to **1=1** on and A size sheet.

Figure 27-53 ————————————————

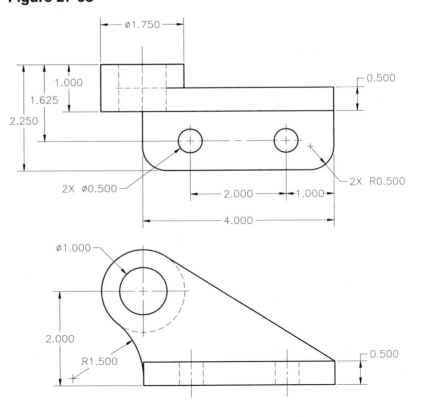

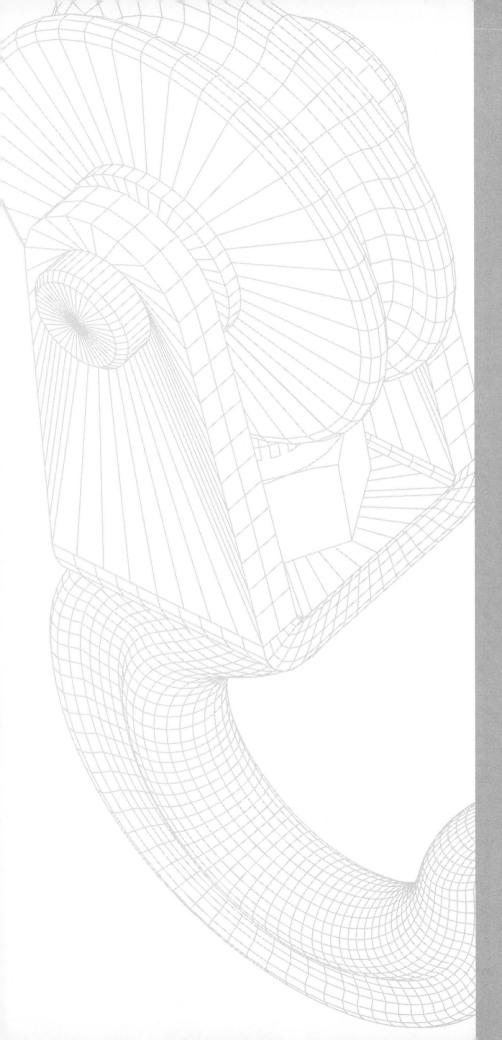

28

3D MODELING BASICS

Chapter Objectives

After completing this chapter you should:

1. know the characteristics of wireframe, surface, and solid models;

2. know the six formats for 3D coordinate entry;

3. understand the orientation of the World Coordinate System (WCS);

4. be able to use the right-hand rule for orientation of the X, Y, and Z axes and for determining positive and negative rotation about an axis;

5. be able to control the appearance and positioning of the Coordinate System icon with the *Ucsicon* command.

CONCEPTS

Three basic types of 3D (three-dimensional) models created by CAD systems are used to represent actual objects. They are:

1. Wireframe models
2. Surface models
3. Solid models

These three types of 3D models range from a simple description to a very complete description of an actual object. The different types of models require different construction techniques, although many concepts of 3D modeling are the same for creating any type of model on any type of CAD system.

Wireframe Models

"Wireframe" is a good descriptor of this type of modeling. A wireframe model of a cube is like a model constructed of 12 coat-hanger wires. Each wire represents an <u>edge</u> of the actual object. The <u>surfaces</u> of the object are <u>not</u> defined; only the boundaries of surfaces are represented by edges. No wires exist where edges do not exist. The model is see-through since it has no surfaces to obscure the back edges. A wireframe model has complete dimensional information but contains no volume. Examples of wireframe models are shown in Figures 28-1 and 28-2.

Wireframe models are relatively easy and quick to construct; however, they are not very useful for visualization purposes because of their "transparency." For example, does Figure 28-1 display the cube as if you are looking towards a top-front edge or looking toward a bottom-back edge? Wireframe models tend to have an optical illusion effect, allowing you to visualize the object from two opposite directions unless another visual clue such as perspective is given.

With AutoCAD a wireframe model is constructed by creating 2D objects in 3D space. The *Line, Circle, Arc,* and other 2D *Draw* and *Edit* commands are used to create the "wires," but 3D coordinates must be specified. The cube in Figure 28-1 was created with 12 *Line* segments. AutoCAD provides all the necessary tools to easily construct, edit, and view wireframe models.

Figure 28-1 ———

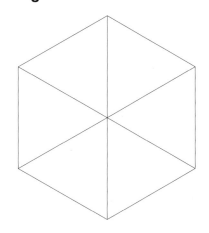

Figure 28-2 ———

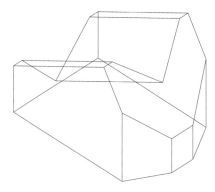

A wireframe model offers many advantages over a 2D engineering drawing. Wireframe models are useful in industry for providing computerized replicas of actual objects. A wireframe model is dimensionally complete and accurate for all three dimensions. Visualization of a wireframe is generally better than a 2D drawing because the model can be viewed from any position or a perspective can be easily attained. The 3D database can be used to test and analyze the object <u>three dimensionally</u>. The sheet metal industry, for example, uses wireframe models to calculate flat patterns complete with bending allowances. A wireframe model can also be used as a foundation for construction of a surface model. Because a wireframe describes edges but not surfaces, wireframe modeling is appropriate to describe objects with planar or single-curved surfaces, but not compound curved surfaces.

Surface Models

Surface models provide a better description of an object than a wireframe, principally because the surfaces as well as the edges are defined. A surface model of a cube is like a cardboard box—all the surfaces and edges are defined, but there is nothing inside. Therefore, a surface model has volume but no mass. A surface model provides an excellent visual representation of an actual 3D object because the front surfaces obscure the back surfaces and edges from view. Figure 28-3 shows a surface model of a cube, and Figure 28-4 displays a surface model of a somewhat more complex shape. Notice that a surface model leaves no question as to which side of the object you are viewing.

Figure 28-3

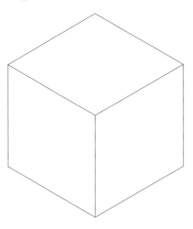

Surface models require a relatively tedious construction process. Each surface must be constructed individually. Each surface must be created in, or moved to, the correct orientation with respect to the other surfaces of the object. In AutoCAD, a surface is constructed by defining its edges. Often, wireframe models are used as a framework to build and attach surfaces. The complexity of the construction process of a surface is related to the number and shapes of its edges. AutoCAD Release 14 is not a complete surface modeler. The tools provided allow construction of simple planar and single curved surfaces, but there are few capabilities for construction of double-curved or other complex surfaces. No NURBS (Non-Uniform Rational B-Splines) surfacing capabilities exist in Release 14, although these capabilities are available in other Autodesk products such as AutoSurf™ and Mechanical Desktop™.

Figure 28-4

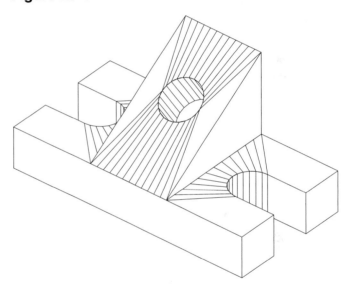

Most CAD systems, including AutoCAD, can display surface and solid models in both wireframe and "hidden" representation. Figure 28-4 shows the object "hidden" (with the *Hide* command). Figure 28-5 displays the same object in "wireframe" (the default) representation. Surface modeling systems use wireframe representation during the construction and editing process to speed computing time. The user activates the "hidden" representation after completing the model to enhance visibility. A reasonable amount of computing time is required to calculate the surface model "visibility." That is, the process required to determine which surfaces would obscure other surfaces for every position of the model would require noticeable computing time, so wireframe display is used during the model construction process.

Figure 28-5

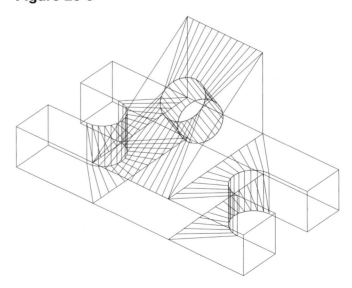

Solid Models

Solid modeling is the most complete and descriptive type of 3D modeling. A solid model is a complete computerized replica of the actual object. A solid model contains the complete surface and edge definition, as well as description of the interior features of the object. If a solid model is cut in half (sectioned), the interior features become visible. Since a solid model is "solid," it can be assigned material characteristics and is considered to have mass. Because solid models have volume and mass, most solid modeling systems include capabilities to automatically calculate volumetric and mass properties.

Solid model construction techniques are generally much simpler (and much more fun) than those of surface models. AutoCAD's solid modeler, called ACIS, is a <u>hybrid</u> modeler. That is, ACIS is a combination CSG (Constructive Solid Geometry) and B-Rep (Boundary Representation) modeler. CSG is characterized by its simple and straightforward construction techniques of combining primitive shapes (boxes, cylinders, wedges, etc.) utilizing Boolean operations (*Union*, *Subtract*, and *Intersect*, etc.). Boundary Representation modeling defines a model in terms of its edges and surfaces (boundaries) and determines the solid model based on which side of the surfaces the model lies. The user interface and construction techniques used in ACIS (primitive shapes combined by Boolean operations) are CSG-based, whereas the B-Rep capabilities are invoked automatically to display models in mesh representation and are transparent to the user. Figure 28-6 displays a solid model constructed of simple primitive shapes combined by Boolean operations.

Figure 28-6 ————————————

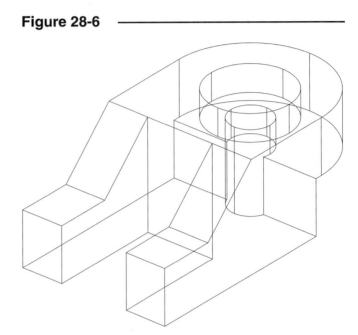

The CSG modeling techniques offer you the advantage of complete and relatively simple editing. CSG construction typically begins by specifying dimensions for simple primitive shapes such as boxes or cylinders, then combines the primitives using Boolean operations to create a "composite" solid. Other primitives and/or composite solids can be combined by the same process. Several repetitions of this process can be continued until the desired solid model is finally achieved. CSG construction techniques are discussed in detail in Chapter 31.

Solid models, like surface models, are capable of either wireframe or "hidden" display. Generally, wireframe display is used during construction (Fig. 28-6), and hidden representation is used to display the finished model (Fig. 28-7).

Figure 28-7 ————————————

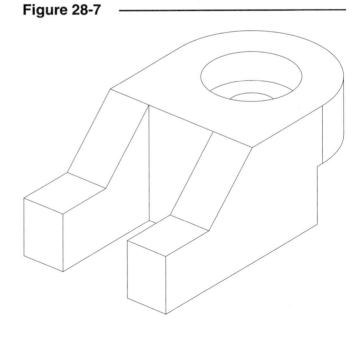

3D COORDINATE ENTRY

Figure 28-8

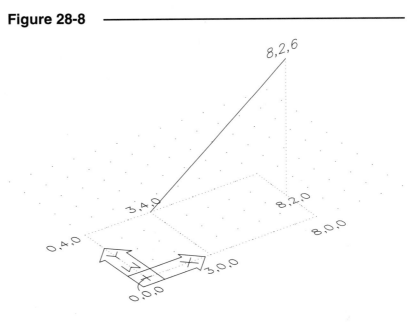

When creating a model in three-dimensional drawing space, the concept of the X and Y coordinate system, which is used for two-dimensional drawing, must be expanded to include the third dimension, Z, which is measured from the origin in a direction perpendicular to the plane defined by X and Y. Remember that two-dimensional CAD systems use X and Y coordinate values to define and store the location of drawing elements such as *Lines* and *Circles*. Likewise, a three-dimensional CAD system keeps a database of X, Y, and Z coordinate values to define locations and sizes of two- and three-dimensional elements. For example, a *Line* is a two-dimensional object, yet the location of its endpoints in three-dimensional space must be specified and stored in the database using X, Y, and Z coordinates (Fig. 28-8). The X, Y, and Z coordinates are always defined in that order, delineated by commas. The AutoCAD Coordinate Display (*Coords*), however, displays only X and Y values, even though a three-dimensional model can be displayed in the Drawing Editor.

The icon that appears in the lower-left corner of the AutoCAD Drawing Editor is the Coordinate System icon (sometimes called the UCS icon) (Fig. 28-9). The Coordinate System icon displays the directions for the X and the Y axes and the orientation of the XY plane and can be made to locate itself at the origin, 0,0. The X coordinate values increase going to the right along the X axis, and the Y values increase going upward along the Y axis.

Figure 28-9

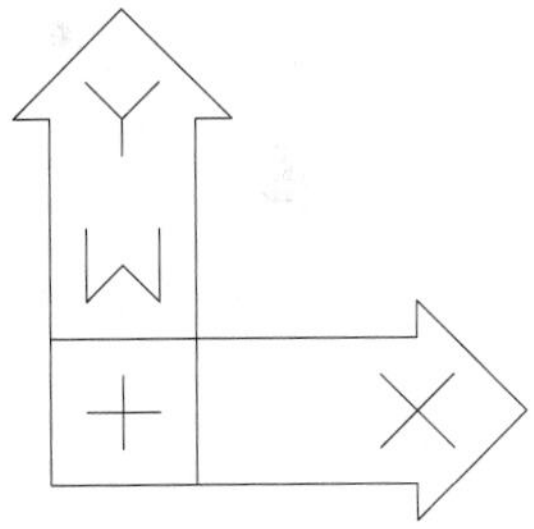

The Z axis is not indicated by the Coordinate System icon, but is assumed to be in a direction <u>perpendicular</u> to the XY plane. In other words, in the default orientation (when you begin a new drawing), the XY plane is <u>parallel</u> with the screen and the Z axis is <u>perpendicular</u> to, and out of, the screen (Fig. 28-10).

Figure 28-10

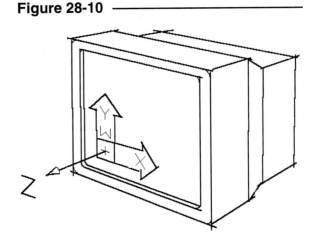

3D Coordinate Entry Formats

Because construction in three dimensions requires the definition of X, Y, and Z values, the methods of coordinate entry used for 2D construction must be expanded to include the Z value. The five methods of command entry used for 2D construction are valid for 3D coordinates with the addition of a Z value specification. Relative polar coordinate specification (@dist<angle) is expanded to form two other coordinate entry methods available explicitly for 3D coordinate entry. The six methods of coordinate entry for 3D construction follow:

1. **Interactive coordinates**   **PICK**   Use the cursor to select points on the screen. *OSNAP* or point filters must be used to select a point in 3D space; otherwise, points selected are <u>on the XY plane</u>.

2. **Absolute coordinates**   **X,Y,Z**   Enter explicit X, Y, and Z values relative to point 0,0.

3. **Relative rectangular coordinates**   **@X,Y,Z**   Enter explicit X, Y, and Z values relative to the last point.

4. **Cylindrical coordinates (relative)**   **@dist<angle,Z**   Enter a distance value, an angle in the XY plane value, and a Z value, all relative to the last point.

5. **Spherical coordinates (relative)**   **@dist<angle<angle**   Enter a distance value, an angle <u>in</u> the XY plane value, and an angle <u>from</u> the XY plane value, all relative to the last point.

6. **Direct distance entry**   **dist,direction**   Enter (type) a value, and move the cursor in the desired direction. To draw in 3D effectively, <u>ORTHO</u> must be <u>On</u>.

Cylindrical and spherical coordinates can be given <u>without</u> the @ symbol, in which case the location specified is relative to point 0,0,0 (the origin). This method is useful if you are creating geometry centered around the origin. Otherwise, the @ symbol is used to establish points in space relative to the last point.

Examples of each of the six 3D coordinate entry methods are illustrated in the following section. In the illustrations, the orientation of the observer has been changed from the default plan view in order to enable the visibility of the three dimensions.

Interactive Coordinate Specification
Figure 28-11 illustrates using the <u>interactive</u> method to PICK a location in 3D space. *OSNAP* <u>must</u> be used in order to PICK in 3D space. Any point PICKed with the input device without *OSNAP* will result in a location <u>on the XY plane</u>. In this example, the *Endpoint OSNAP* mode is used to establish the "To point:" of a second *Line* by snapping to the end of an existing vertical *Line* at 8,2,6:

```
Command: Line
From point: 3,4,0
To point: Endpoint of PICK
```

Figure 28-11

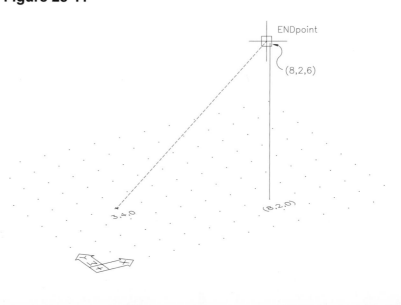

Absolute Coordinates

Figure 28-12 illustrates the <u>absolute</u> coordinate entry to draw the *Line*. The endpoints of the *Line* are given as explicit X,Y,Z coordinates:

```
Command: Line
From point: 3,4,0
To point: 8,2,6
Command:
```

Figure 28-12 ───────────────

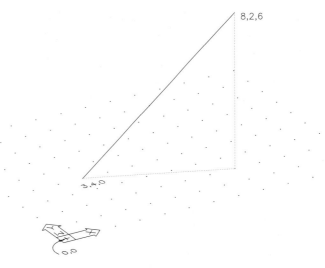

Relative Rectangular Coordinates

Relative rectangular coordinate entry is displayed in Figure 28-13. The "From point:" of the *Line* is given in absolute coordinates, and the "To point:" end of the *Line* is given as X,Y,Z values <u>relative</u> to the last point:

```
Command: Line
From point: 3,4,0
To point: @5,-2,6
Command:
```

Figure 28-13 ───────────────

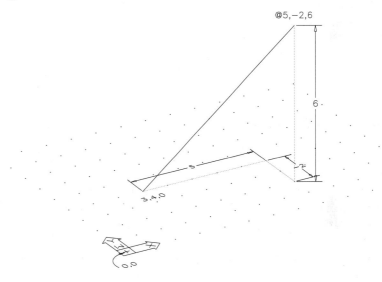

Cylindrical Coordinates (Relative)

Cylindrical and spherical coordinates are an extension of polar coordinates with a provision for the third dimension. Relative cylindrical coordinates give the distance <u>in</u> the XY plane, angle <u>in</u> the XY plane, and Z dimension and can be relative to the last point by prefixing the @ symbol. The *Line* in Figure 28-14 is drawn with absolute and relative cylindrical coordinates. (The *Line* established is approximately the same *Line* as in the previous figures.)

```
Command: Line
From point: 3,4,0
To point: @5<-22,6
Command:
```

Figure 28-14 ───────────────

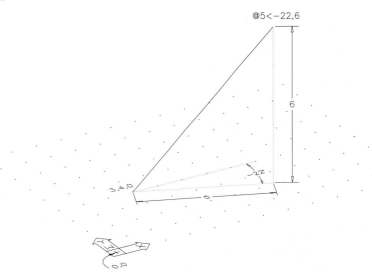

Spherical Coordinates (Relative)
Spherical coordinates are also an extension of polar coordinates with a provision for specifying the third dimension in angular format. Spherical coordinates specify a distance, an angle <u>in</u> the XY plane, and an angle <u>from</u> the XY plane and can be relative to the last point by prefixing the @ symbol. The distance specified is a <u>3D distance</u>, not a distance in the XY plane. Figure 28-15 illustrates the creation of approximately the same line as in the previous figures using absolute and relative spherical coordinates.

Figure 28-15

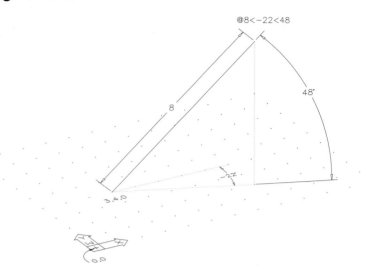

```
Command: Line
From point: 3,4,0
To point: @8<-22<48
Command:
```

Direct Distance Entry
Direct distance entry operates as it does in 2D, except <u>ORTHO must be On</u> in order to draw effectively by this method in 3D space. When *ORTHO* is *On*, objects drawn in 3D space are limited to a <u>plane parallel with the current XY plane.</u> For example, to draw a square in 3D space with sides of 3 units (Fig. 28-16), begin by using *Line* and specifying the "From point:" at the *Endpoint* of the existing diagonal *Line* shown (point 8,2,6). To draw the first 3 unit segment, turn on *ORTHO*, move the cursor in the desired direction, and enter "3." The *ORTHO* feature "locks" the *Line* to a plane parallel to the current XY plane. Next, move the cursor in the desired (90 degree) direction and enter "3," and so on. The command syntax follows:

Figure 28-16

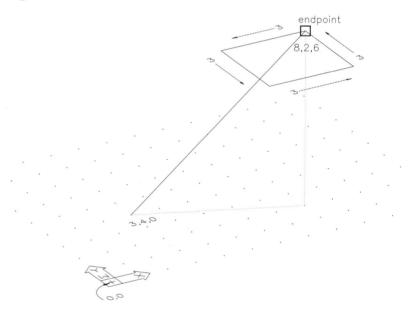

```
Command: LINE
From point: end of
To point: <Ortho on>  3
To point: 3
To point: 3
To point: 3
To point: Enter
Command:
```

COORDINATE SYSTEMS

In AutoCAD two kinds of coordinate systems can exist, the <u>World Coordinate System</u> (WCS) and one or more <u>User Coordinate Systems</u> (UCS). The World Coordinate System <u>always</u> exists in any drawing and cannot be deleted. The user can also create and save multiple User Coordinate Systems to make construction of a particular 3D geometry easier. <u>Only one coordinate system can be active</u> at any one time, either the WCS or one of the user-created UCSs.

The World Coordinate System (WCS) and WCS Icon

The World Coordinate System (WCS) is the default coordinate system in AutoCAD for defining the position of drawing objects in 2D or 3D space. The WCS is always available and cannot be erased or removed but is deactivated temporarily when utilizing another coordinate system created by the user (UCS). The icon that appears (by default) at the lower-left corner of the Drawing Editor (Fig. 28-9) indicates the orientation of the WCS. The coordinate system icon indicates only X and Y directions, so Z is assumed to be perpendicular to the XY plane. The icon, whose appearance (*ON, OFF*) is controlled by the *Ucsicon* command, appears for the WCS and for any UCS. However, the letter "W" appears in the icon slightly above the origin <u>only</u> when the <u>WCS</u> is active.

Figure 28-17 ————————————————

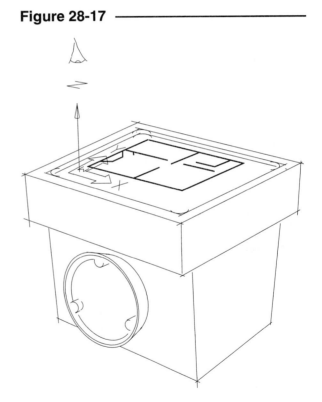

The orientation of the WCS with respect to Earth may be different among CAD systems. In AutoCAD the WCS has an architectural orientation such that the XY plane is a horizontal plane with respect to Earth, making Z the height dimension. A 2D drawing (X and Y coordinates only) is thought of as being viewed from above, sometimes called a <u>plan</u> view. Therefore, in a 3D AutoCAD drawing, X is the width dimension, Y is the depth dimension, and Z is height. This default orientation is like viewing a floor <u>plan</u>—from above (Fig. 28-17).

Some CAD systems that have a mechanical engineering orientation align their World Coordinate Systems such that the XY plane is a vertical plane intended for drawing a front view. In other words, some mechanical engineering CAD systems define X as the width dimension, Y as height, and Z as depth.

User Coordinate Systems (UCS) and Icons

There are no User Coordinate Systems that exist as part of the AutoCAD default template drawing (ACAD.DWT) as it comes "out of the box." UCSs are created to suit the 3D model when and where they are needed.

Creating geometry is relatively simple when having to deal only with X and Y coordinates, such as in creating a 2D drawing, or when creating simple 3D geometry with uniform Z dimensions. However,

3D models containing complex shapes on planes not parallel with the XY plane are good candidates for UCSs (Fig. 28-18).

A User Coordinate System is thought of as a construction plane created to simplify creation of geometry on a specific plane or surface of the object. The user creates the UCS, aligning its XY plane with a surface of the object, such as along an inclined plane, with the UCS origin typically at a corner or center of the surface.

Figure 28-18 —————————————

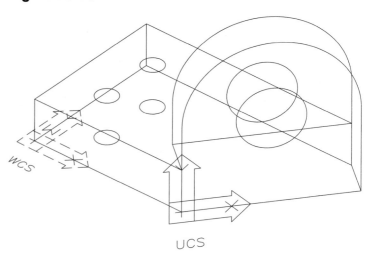

The user can then create geometry aligned with that plane by defining only X and Y coordinate values of the current UCS (Fig. 28-19). The *SNAP* and *GRID* automatically align with the current coordinate system, providing *SNAP* points and enhancing visualization of the construction plane. Practically speaking, it is easier in some cases to specify only X and Y coordinates with respect to a specific plane on the object rather than calculating X, Y, and Z values with respect to the World Coordinate System.

Figure 28-19 —————————————

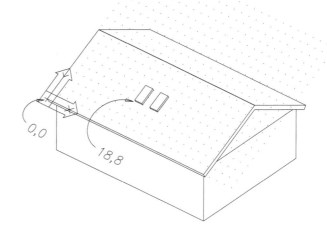

User Coordinate Systems can be created by any of several options of the *UCS* command. Once a UCS has been created, it becomes the current coordinate system. Only one coordinate system can be active; therefore, it is suggested that UCSs be saved (by using the *Save* option of *UCS*) for possible future geometry creation or editing.

When a UCS is created, the icon at the lower-left corner of the screen can be made to automatically align itself with the UCS along with *SNAP* and *GRID*. The letter "W," however, appears on the icon only when the WCS is active (when the WCS is the current coordinate system). When creating 3D geometry, it is recommended that the *ORigin* option of the *Ucsicon* command be used to place the icon always at the origin of the current UCS, rather than in the lower-left corner of the screen. Since the origin of the UCS is typically specified as a corner or center of the construction plane, aligning the Coordinate System icon with the current origin aids your visualization of the UCS orientation. The Coordinate Display (*Coords*) in the Status line always displays the X, Y and Z values of the current coordinate system, whether it is WCS or UCS.

THE RIGHT-HAND RULE

AutoCAD complies with the right-hand rule for defining the orientation of the X, Y, and Z axes. The right-hand rule states that if your right hand is held partially open, the thumb, first, and middle fingers define positive X, Y, and Z directions, respectively, and positive rotation about any axis is like screwing in a light bulb.

More precisely, if the thumb and first two fingers are held out to be mutually perpendicular, the thumb points in the positive X direction, the first finger points in the positive Y direction, and the middle finger points in the positive Z direction (Fig. 28-20).

Figure 28-20

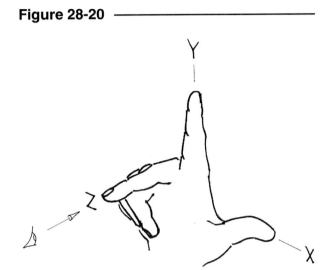

In this position, looking toward your hand from the tip of your middle finger is like the default AutoCAD viewing orientation—positive X is to the right, positive Y is up, and positive Z is toward you.

Positive rotation about any axis is <u>counterclockwise looking down the axis toward the origin</u>. For example, when you view your right hand, positive rotation about the X axis is as if you look down your thumb toward the hand (origin) and twist your hand counterclockwise (Fig. 28-21).

Figure 28-21

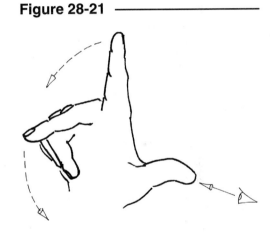

Figure 28-22 shows the Coordinate System icon in a +90 degree rotation about the X axis (like the previous figure). This orientation is typical for setting up a <u>front</u> view UCS for drawing on a plane parallel with the front surface of an object.

Figure 28-22

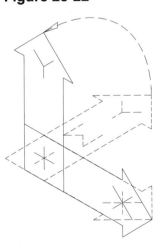

Positive rotation about the Y axis would be as if looking down your first finger toward the hand (origin) and twisting counterclockwise (Fig. 28-23).

Figure 28-23 ————————

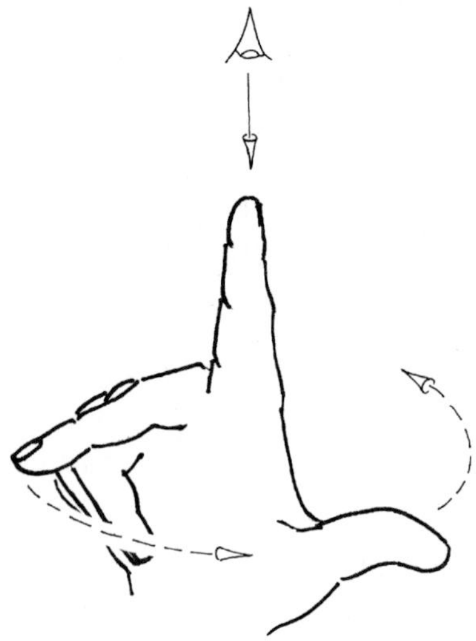

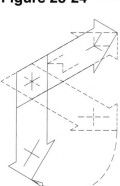

Figure 28-24 illustrates the Coordinate System icon with a +90 degree rotation about the Y axis (like the previous figure). To avoid confusion in relating this figure to Figure 28-23, consider that the icon is oriented on the XY plane as a horizontal plane in its original (highlighted) position, whereas Figure 28-23 shows the hand in an upright position. (Compare Figures 28-17 and 28-10.)

Figure 28-24 ——

Positive rotation about the Z axis would be as if looking toward your hand from the end of your middle finger and twisting counterclockwise (Fig. 28-25).

Figure 28-25 ————————

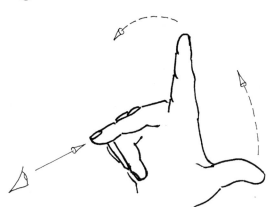

Figure 28-26 shows the Coordinate System icon with a +90 degree rotation about the Z axis. Again notice the orientation of the icon is horizontal, whereas the hand (Fig. 28-25) is upright.

Figure 28-26 ————————

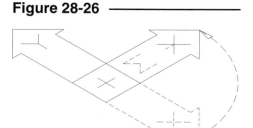

UCS ICON CONTROL

UCSICON

Pull-down Menu	COMMAND (TYPE)	ALIAS (TYPE)	Short-cut	Screen (side) Menu	Tablet Menu
View *Display >* *UCS Icon >*	*UCSICON*	...	...	VIEW 2 UCSicon	L,2

The *Ucsicon* command controls the appearance and positioning of the Coordinate System icon. In order to aid your visualization of the current UCS or the WCS, it is <u>highly</u> recommended that the Coordinate System icon be turned *ON* and positioned at the *ORigin*.

```
Command: ucsicon
ON/OFF/All/Noorigin/ORigin<ON>: on
Command:
```

This option causes the Coordinate System icon to appear (Fig. 28-27).

Figure 28-27 ————————————

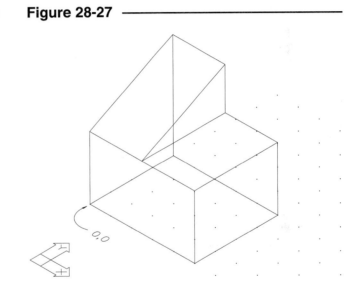

The *Ucsicon* command must be invoked again to use the *ORigin* option:

```
Command: ucsicon
ON/OFF/All/Noorigin/ORigin<ON>: or
Command:
```

This setting causes the icon to move to the origin of the current coordinate system (Fig. 28-28).

Figure 28-28 ————————————

The icon does not always appear at the origin after using some viewing commands like *Vpoint*. This is because *Vpoint* causes a *Zoom Extents*, forcing the geometry against the border of the graphics screen (or viewport) area, which prevents the icon from aligning with the origin. In order to cause the icon to appear at the origin, try using *Zoom* with a *.9X* magnification factor. This action usually brings the geometry slightly in from the border and allows the icon to align itself with the origin.

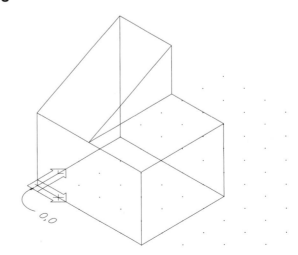

Options of the *Ucsicon* command are:

ON   Turns the Coordinate System icon on.

OFF   Turns the Coordinate System icon off.

All   Causes the *Ucsicon* settings to be effective for all viewports.

Noorigin Causes the icon to appear always in the lower-left corner of the screen, not at the origin.

ORigin   Forces the placement and orientation of the icon to align with the origin of the current coordinate system.

Now that you know the basics of 3D modeling, you will develop your skills in viewing and displaying 3D models (Chapter 29, 3D Viewing and Display). It is imperative that you are able to view objects from different viewpoints in 3D space before you learn to construct them.

CHAPTER EXERCISES

1. What are the three types of 3D models?

2. What characterizes each of the three types of 3D models?

3. What are the five formats for 3D coordinate specification?

4. Examine the 3D geometry shown in Figure 28-29. Specify the designated coordinates in the specified formats below.

    A. Give the coordinate of corner D in absolute format.

    B. Give the coordinate of corner F in absolute format.

    C. Give the coordinate of corner H in absolute format.

    D. Give the coordinate of corner J in absolute format.

    E. What are the coordinates of the line that define edge F-I?

    F. What are the coordinates of the line that define edge J-I?

    G. Assume point E is the "last point." Give the coordinates of point I in relative rectangular format.

    H. Point I is now the last point. Give the coordinates of point J in relative rectangular format.

    I. Point J is now the last point. Give the coordinates of point A in relative rectangular format.

    J. What are the coordinates of corner G in cylindrical format (from the origin)?

    K. If E is the "last point," what are the coordinates of corner C in relative cylindrical format?

Figure 28-29

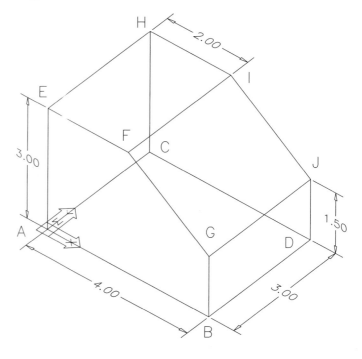

29

3D VIEWING AND DISPLAY

Chapter Objectives

After completing this chapter you should:

1. be able to recognize a 3D model in wireframe, perspective, *Hide*, *Shade*, and *Render* display representations;

2. be able to use *3D Viewpoint* to quickly attain the desired view;

3. be able to use *Vpoint* to view a 3D model from various viewpoints;

4. be able to use the *Vector*, *Tripod*, and *Rotate* options of *Vpoint*;

5. be able to create several configurations of tiled viewports with the *Vports* command;

6. know how to suppress "hidden lines" of a solid or surface model with the *Hide* command;

7. know how to *Shade* a solid model using four *SHADEDGE* options.

AutoCAD'S 3D VIEWING AND DISPLAY CAPABILITIES

Viewing Commands

These commands allow you to change the direction from which you view a 3D model or otherwise affect the view of the 3D model:

Vpoint   *Vpoint* allows you to change your viewpoint of a 3D model. The object remains stationary while the viewpoint of the observer changes. Three options are provided: *Vector, Tripod,* and *Rotate.* Many useful *3D Viewpoint* options are available in Release 14.

Plan   The *Plan* command automatically gives the observer a plan (top) view of the object. The plan view can be with respect to the WCS (World Coordinate System) or an existing UCS (User Coordinate System).

Dview   This command allows you to dynamically (interactively) rotate the viewpoint of the observer about 3D objects. *Dview* also allows generation of perspective views.

Zoom   Although *Zoom* operates in 3D just as in 2D, the *Zoom Previous* option restores the previous 3D viewpoint.

View   The *View* command can be used with 3D viewing to *Save* and *Restore* 3D viewpoints.

Vports   The *Vports* command creates tiled viewports. *Vports* does not actually change the viewpoint of the model but divides the screen into several viewports and allows you to display several viewpoints or sizes of the model on one screen.

Display Commands

In most CAD systems, <u>surface and solid models are shown in a wireframe display</u> during the construction and editing process. A "hidden" display may hinder your ability to select needed edges of the object during the construction process and would require additional computing time. Once the model is completed and the desired viewpoint has been attained, these commands can change the appearance of a 3D surface or solid model from the default wireframe representation by displaying the surfaces:

Hide   This command removes normally "hidden" edges and surfaces from a solid or surface model, making it appear as an opaque rather than transparent object.

Shade   The *Shade* command fills the surfaces with the object's color and calculates light reflection by applying gradient shading (variable gray values) to the surfaces.

Render   *Render* allows you to create and place lights in 3D space, adjust the light intensity, and assign materials (color and reflective qualities) to the surfaces. This is the most sophisticated of the visualization capabilities offered in AutoCAD.

When using the viewing commands *Vpoint, Dview,* and *Plan,* it is important to imagine that the <u>observer moves about the object</u> rather than imagining that the object rotates. The object and the coordinate system (WCS and icon) always remain <u>stationary</u> and always keep the same orientation with respect to Earth. Since the observer moves and not the geometry, the objects' coordinate values retain their integrity, whereas if the geometry rotated within the coordinate system, all coordinate values of the objects would change as the object rotated. The *Vpoint, Dview,* and *Plan* commands change only the viewpoint of the <u>observer</u>.

Figure 29-1 ────────────────────

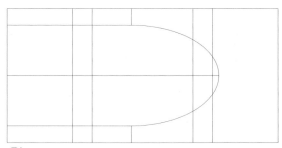

Plan

Figure 29-2 ────────────

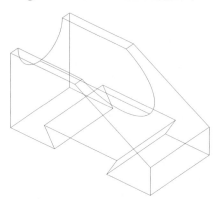

Vpoint (wireframe representation)

Figure 29-3 ────────────

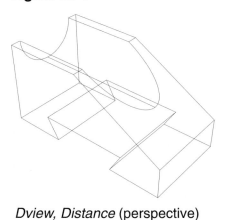

Dview, Distance (perspective)

Figure 29-4 ────────────

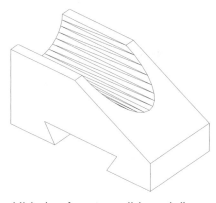

Hide (surface or solid model)

Figure 29-5 ────────────

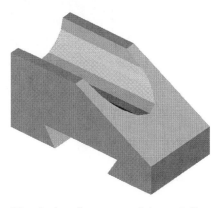

Shade (surface or solid model)

Figure 29-6 ────────────

Render (surface or solid model)

As the 3D viewing commands are discussed, it would be helpful if you opened a 3D drawing to practice using the viewing commands. The FORKLIFT drawing is an excellent 3D model to practice using the 3D viewing commands. The FORKLIFT.DWG is not copied to your computer during installation but is located on the AutoCAD Release 14 CD-ROM. It is located in the Acad/Bonus/Sample/History directory along with many other interesting and useful sample drawings. When you open the drawing, remember to set the *Ucsicon* to the *Origin*.

R14

3D VIEWING COMMANDS

3D VIEWPOINT

Pull-down Menu	COMMAND (TYPE)	ALIAS (TYPE)	Short-cut	Screen (side) Menu	Tablet Menu
View *3D Viewpoint >*	...	...	...	*VIEW 1* *Vpoint*	*O,3–Q,5*

AutoCAD Release 13 introduced *3D Viewpoint*, which really speeds up the process of attaining common views of 3D objects. All of these options actually use the *Vpoint* command and automatically enter in coordinate values. The other options of *Vpoint* can be used for other particular viewing angles (see *Vpoint* next). These options can be selected from the *View* pull-down menu (Fig. 29-7).

Figure 29-7

The *3D Viewpoint* options are accessible from the Standard toolbar (Fig. 29-8). If you prefer a separate toolbar, one can be invoked from the toolbar list and made to float or dock.

Each of the *3D Viewpoint* options is described next. The FORKLIFT drawing is displayed for several viewpoints (the *Hide* command was used in these figures for clarity). It is important to remember that for each view, imagine you are viewing the object from the indicated position in space. The object does not rotate.

Figure 29-8

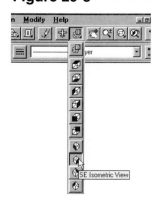

Top

The object is viewed from the top (Fig. 29-9). Selecting this option shows the XY plane from above (the default orientation when you begin a *New* drawing). Notice the position of the WCS icon. This orientation should be used periodically during construction of a 3D model to check for proper alignment of parts.

Figure 29-9

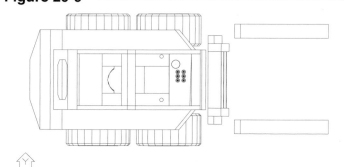

Bottom

This option displays the object as if you are looking up at it from the bottom. AutoCAD uses the *Vpoint* command and automatically enters in coordinates to show this view. The coordinate system icon appears backwards when viewed from the bottom.

Left

This is looking at the object from the left side. Again the *Vpoint* command is used by AutoCAD to give this view.

Right

Imagine looking at the object from the right side (Fig. 29-10). This view is used often as one of the primary views for mechanical part drawing. The coordinate system icon does not appear in this view, but a "broken pencil" is displayed instead (meaning that it is not a good idea to draw on the XY plane from this view).

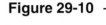
Figure 29-10

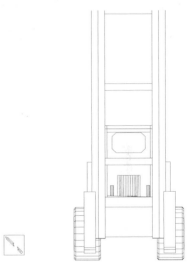

Front

Selecting this option displays the object from the front view. This is a common view that can be used often during the construction process. The front view is usually the "profile" view for mechanical parts (Fig. 29-11).

Figure 29-11

Back

Back displays the object as if the observer is behind the object. Remember that the object does not rotate; the observer moves.

SW Isometric

Isometric views are used more than the "orthographic" views (front, top, right, etc.) for constructing a 3D model. Isometric viewpoints give more information about the model because you can see three dimensions instead of only two dimensions (as in the previously discussed views).

SE Isometric

 The southeast isometric is generally the first choice for displaying 3D geometry. If the object is constructed with its base on the XY plane so that X is length, Y is depth, and Z equals height, this orientation shows the front, top, and right sides of the object. Try to use this viewpoint as your principal mode of viewing during construction. Note the orientation of the WCS icon (Fig. 29-12).

NE Isometric

 The northeast isometric shows the right side, top, and back (if the object is oriented in the manner described earlier).

NW Isometric

 This viewpoint allows the observer to look at the left side, top, and back of the 3D object (Fig. 29-13).

NOTE: *3D Viewpoint* always displays the object (or orients the observer) with respect to the World Coordinate System. For example, the *Top* option always shows the plan view of the WCS XY plane. Even if another coordinate system (UCS) is active, AutoCAD temporarily switches back to the WCS to attain the selected viewpoint.

Figure 29-12 ──────────────

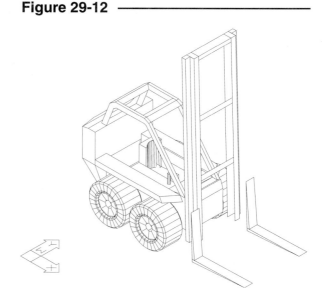

Figure 29-13 ──────────────

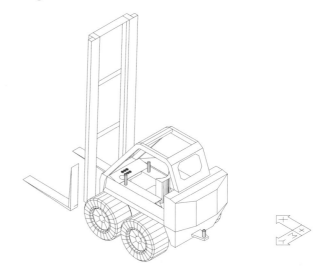

VPOINT

Pull-down Menu	COMMAND (TYPE)	ALIAS (TYPE)	Short-cut	Screen (side) Menu	Tablet Menu
View 3D Viewpoint >	VPOINT	-VP	...	VIEW 1 Vpoint	N,4

The *3D Viewpoint* options actually use the *Vpoint* command to generate the typical views of a 3D object. In some cases, it is necessary to display a 3D object from a direction other than the typical *Vpoints* attained through the presets. For example, an object with 45 degree angled planes or with regular proportions would present a visually confusing description if a perfect isometric view were given (some lines in the view would overlap and angles would align). Instead, a viewpoint slightly off of a pure isometric would show the angled surfaces and regular proportions more clearly. In cases such as these, one of the three options of the *Vpoint* command should be used: *Rotate*, *Vector*, or *Tripod* (or *Axes*).

The *Vpoint* command displays a parallel projection rather than a perspective projection. In a parallel projection, all visual rays from the observer to the object are parallel as if the observer were (theoretically) an infinite distance from the object. The resulting display shows all parallel edges of the object as parallel in the display. This projection differs from a perspective projection, where parts of the object that are farther from the observer appear smaller and parallel edges converge to a point.

Rotate

The name of this option is somewhat misleading because the object is not rotated. The *Rotate* option prompts for two angles in the WCS (by default) which specify a vector indicating the direction of viewing. The two angles are (1) the angle in the XY plane and (2) the angle from the XY plane. The observer is positioned along the vector looking toward the origin.

Command: **Vpoint**
Rotate <View point><0,0,1>: **R**
Enter angle in XY plane from X axis <270>: **315**
Enter angle from XY plane <90>: **35**
Command:

The first angle is the angle in the XY plane at which the observer is positioned looking toward the origin. This angle is just like specifying an angle in 2D. The second angle is the angle that the observer is positioned up or down from the XY plane. The two angles are given with respect to the WCS (Fig. 29-14).

Figure 29-14 ————————————

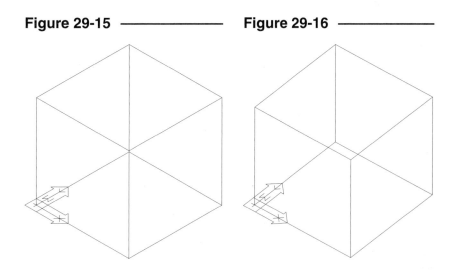

Angles of **315** and **35** specified in response to the *Rotate* option display an almost perfect isometric viewing angle. An isometric drawing often displays some of the top, front, and right sides of the object. For some regularly proportioned objects, perfect isometric viewing angles can cause visualization difficulties, while a slightly different angle can display the object more clearly. Figure 29-15 and Figure 29-16 display a cube from an almost perfect isometric viewing angle (**315** and **35**) and from a slightly different viewing angle (**310** and **40**), respectively.

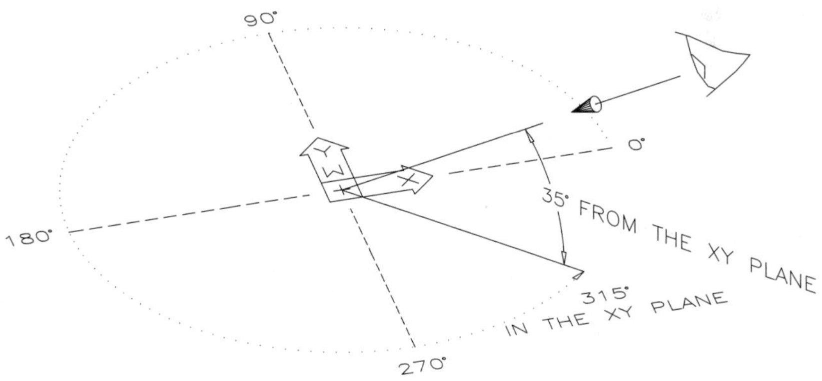

Figure 29-15 ——————— **Figure 29-16** ———————

The *Vpoint Rotate* option can be used to display 3D objects from a front, top, or side view. *Rotate* angles for common views are:

Top	270, 90
Front	270, 0
Right side	0,0
Southeast isometric	315, 35.27
Southwest isometric	225, 35.27

Vector

Another option of the *Vpoint* command is to enter X,Y,Z coordinate values. The coordinate values indicate the position in 3D space at which the observer is located. The coordinate values do not specify an absolute position but rather specify a <u>vector</u> passing through the coordinate position and the origin. In other words, the observer is located at <u>any point along the vector looking toward the origin.</u>

Because the *Vpoint* command generates a parallel projection and since parallel projection does not consider a distance (which is considered only in perspective projection), the magnitude of the coordinate <u>values</u> is of no importance, only the relationship among the values. Values of 1,-1,1 would generate the same display as 2,-2,2.

The command syntax for specifying a perfect isometric viewing angle is as follows:

```
Command: Vpoint
Rotate <View point><0,0,1>: 1,-1,1
Command:
```

Coordinates of 1,-1,1 generate a display from a perfect isometric viewing angle.

Figure 29-17 illustrates positioning the observer in space, using coordinates of 1,-1,1. Using *Vpoint* coordinates of 1,-1,1 generates an isometric display similar to *Rotate* angles of 315, 35, or *SE Isometric*.

Figure 29-17

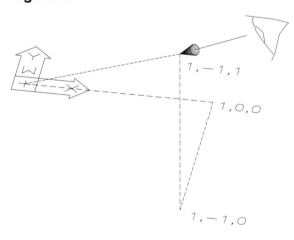

Other typical views of an object can be easily achieved by entering coordinates at the *Vpoint* command. Consider the following coordinate values and the resulting displays:

Coordinates	Display
0,0,1	Top
0,-1,0	Front
1,0,0	Right side
1,-1,1	Southeast isometric
-1,-1,1	Southwest isometric

Tripod or Axes

This option displays a three-pole axes system which rotates dynamically on the screen as the cursor is moved within a "globe." The axes represent the X, Y, and Z axes of the WCS. When you PICK the desired viewing position, the geometry is displayed in that orientation. Because the viewing direction is specified by PICKing a point, it is difficult to specify an exact viewpoint.

The *Tripod* option does not appear on the command line as one of the possible *Vpoint* methods. This option must be invoked by pressing Enter at the "Rotate<Viewpoint> <(coordinates)>:" prompt or by selecting *Axes* or *Tripod* from one of the menus. When the *Tripod* method is invoked, the current drawing temporarily disappears and a three-pole axes system appears at the center of the screen. The axes are dynamically rotated by moving the cursor (a small cross) in a small "globe" at the upper-right of the screen (Fig. 29-18).

Figure 29-18

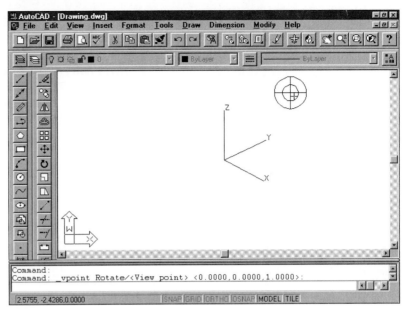

The three-pole axes indicate the orientation of the X, Y, and Z axes for the new *Vpoint*. The center of the "globe" represents the North Pole, so moving the cursor to that location generates a plan, or top, view. The small circle of the globe represents the Equator, so moving the cursor to any location on the Equator generates an elevation view (front, side, back, etc.). The outside circle represents the South Pole, so locating the cursor there shows a bottom view. When you PICK, the axes disappear and the current drawing is displayed from the new viewpoint. The command format for using the *Tripod* method is as follows:

Command: **Vpoint**
Rotate <Viewpoint><(current coordinates)>: **Enter** (axes appear)
PICK (Select desired cursor location.)
Regenerating drawing. (Current drawing appears showing the new viewpoint.)

Using the *Tripod* method of *Vpoint* is quick and easy. Because no exact *Vpoint* can be given, it is difficult to achieve the exact *Vpoint* twice. Therefore, if you are working with a complex drawing that requires a <u>specific</u> viewpoint of a 3D model, use another option of *Vpoint* or save the desired *Vpoint* as a named *View*.

Using *Zoom .9X* with *Vpoint*

In the FORKLIFT drawing (Figs. 29-9, 29-12, and 29-13), the Coordinate System icon was previously turned *ON* and forced to appear at the *ORigin*. This action was accomplished by the *Ucsicon* command (see *Ucsicon*, Chapter 28). The icon does not <u>always</u> appear at the origin after using the *Vpoint* command. This is because *Vpoint* causes a *Zoom Extents*, forcing the geometry near the border of the graphics screen area. This sometimes prevents the icon from aligning with the origin since the origin can also be against the border of the graphics screen area. In order to cause the icon to appear at the origin, try using *Zoom* with a *.9X* magnification factor. This action usually brings the geometry slightly away from the border and allows the icon to "pop" into place at the origin.

DDVPOINT

Pull-down Menu	COMMAND (TYPE)	ALIAS (TYPE)	Short-cut	Screen (side) Menu	Tablet Menu
View *3D Viewpoint >* *Select...*	*DDVPOINT*	*VP*	...	*VIEW 1* *Ddvpoint*	*N,5*

The *Ddvpoint* command produces the *Viewpoint Presets* dialog box (Fig. 29-19). This tool is an interface for attaining viewpoints that could otherwise be attained with the *3D Viewpoint* or the *Vpoint* command.

Ddvpoint serves the same function as the *Rotate* option of *Vpoint*. You can specify angles *From: X Axis* and *From: XY Plane*. Angular values can be entered in the edit boxes, or you can PICK anywhere in the image tiles to specify the angles. PICKing in the enclosed boxes results in a regular angle (e.g., 45, 90, or 10, 30) while PICKing near the sundial-like "hands" results in irregular angles (see pointer in Fig. 29-19). The *Set to Plan View* tile produces a plan view.

Figure 29-19

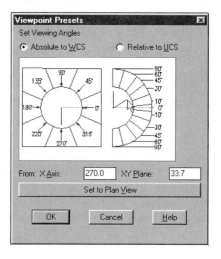

The *3D Viewpoint* options are newer and faster for producing standard views and are suggested for most cases. *Ddvpoint*, however, offers one option not available by other means. The *Relative to UCS* radio button calculates the viewing angles with respect to the current UCS rather than the WCS. Normally, the viewing angles should be absolute to WCS, but certain situations may require this alternative. Viewing angles relative to the current UCS can produce some surprising viewpoints if you are not completely secure with the model and observer orientation in 3D space.

NOTE: When you use this tool, ensure that the *Absolute to WCS* radio button is checked unless you are sure the specified angles should be applied relative to the current UCS.

PLAN

Pull-down Menu	COMMAND (TYPE)	ALIAS (TYPE)	Short-cut	Screen (side) Menu	Tablet Menu
View *3D Viewpoint >* *Plan View >*	*PLAN*	...	...	*VIEW 1* *Plan*	*N,3*

This command is useful to quickly display a plan view of any UCS:

    Command: **plan**
    <Current UCS>/Ucs/World: (**letter**) or **Enter**

Responding by pressing Enter causes AutoCAD to display the plan (top) view of the current UCS. Typing *W* causes the display to show the plan view of the World Coordinate System. The *World* option does <u>not</u> cause the WCS to become the active coordinate system but only displays its plan view. Invoking the *UCS* option displays the following prompt:

    ?/Name of UCS:

The *?* displays a list of existing UCSs. Entering the name of an existing UCS displays a plan view of that UCS.

ZOOM

The *Zoom* command can be used effectively with 3D models just as with 2D drawings. However, two options of *Zoom* are particularly applicable to 3D work: *Previous* and *Scale*.

Previous

This option of *Zoom* restores the previous display. When you are using *3D Viewpoint, Vpoint, Dview,* and *Plan* for viewing a 3D model, *Zoom Previous* will display the previous viewpoint.

Scale

When a *Vpoint* is specified, AutoCAD automatically causes a *Zoom Extents* that leaves the 3D model nearly against the borders of the screen. If you are using the *Ucsicon* with the *ORigin* option, invoking a *Zoom .9X* or *.8X* usually allows room for the icon to "pop" back to the origin position.

VIEW

The *View* command can also be used effectively for 3D modeling. When a desirable viewpoint is achieved by the *Vpoint* or *Dview* commands, the viewpoint can be saved with the *Save* option of *View*. *Restoring* the *View* is often easier than using *Vpoint* or *Dview* again.

VPORTS

Pull-down Menu	COMMAND (TYPE)	ALIAS (TYPE)	Short-cut	Screen (side) Menu	Tablet Menu
View *Tiled Viewports >*	*VPORTS*	...	...	*VIEW 1* *Vports*	*M,3*

The *Vports* command allows <u>tiled</u> viewports to be created on the screen. *Vports* allows the screen to be divided into several areas. Tiled viewports are different than paper space viewports. (See Chapter 19, Tiled and Paper Space Viewports.)

Vports (tiled viewports) are available only when the *TILEMODE* variable is set to **1**. Tiled *Vports* affect <u>only the screen display</u>. The viewport configuration <u>cannot be plotted</u>. If the *Plot* command is used, only the <u>current</u> viewport display is plotted. Figure 29-20 displays the AutoCAD drawing editor after the *Vports* command was used to divide the screen into tiled viewports.

After you select the desired layout, the previous display appears in <u>each</u> of the viewports. For example, if a top view of the 3D model is displayed when you use the *Vports* command or the related dialog box, the resulting display in each of the viewports would be the top view (see Fig. 29-20).

Figure 29-20

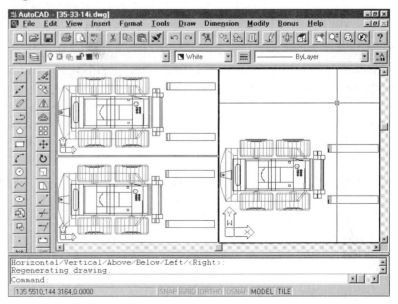

It is up to you then to use viewing commands (*Vpoint*, *Dview*, *Plan*, *Zoom*, etc.) to specify what viewpoints or areas of the model you want to see in each viewport. There is no automatic viewpoint configuration option with *Vports*.

A popular arrangement of viewpoints for construction and editing of 3D models is a combination of principal views (top, front, side, etc.) and an isometric or other pictorial type of viewpoint (Fig. 29-21). You can use *Zoom* with magnification factors to size the model in each viewport and use *PAn* to "align" the views. There is, however, <u>no automatic method</u> of alignment of views to achieve a true orthogonal projection. You cannot draw or project from one viewport to another.

A viewport is made active by PICKing a point in it. Any <u>display</u> commands (*Zoom*, *Vpoint*, *Redraw*, etc.) and drawing aids (*SNAP*, *GRID*, *ORTHO*) used affect only the <u>current viewport</u>. <u>Draw and edit</u> commands that affect the model are potentially apparent in *all* <u>viewports</u> (for every display of the affected part of the model). *Redrawall* and *Regenall* can be used to redraw and regenerate all viewports.

Figure 29-21

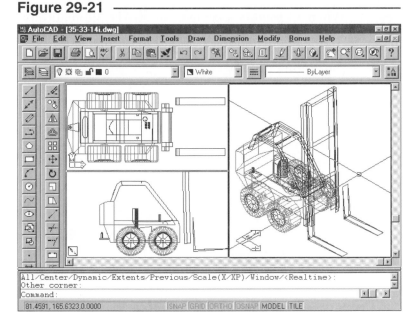

The *Vports* command can be used most effectively when constructing and editing 3D geometry, whereas paper space viewports are generally used when the model is complete and ready to prepare a plot. *Vports* is typically used to display a different *Vpoint* of the 3D model in each viewport, as in Figure 29-21. In this way, you can create 3D geometry and view the construction from different *Vpoint*s in order to enhance your visibility in all three dimensions.

3D DISPLAY COMMANDS

Commands that are used for changing the appearance of a surface or solid model in AutoCAD are *Hide*, *Shade*, and *Render*. By default, surface and solid models are shown in wireframe representation in order to speed computing time during construction. As you know, wireframe representation can somewhat hinder your visualization of a model because it presents the model as transparent. To display surfaces and solid models as opaque and to remove the normally obscured edges, *Hide*, *Shade*, or *Render* can be used.

Wireframe models are not affected by these commands since they do not contain surfaces. Wireframe models can only be displayed in wireframe representation.

HIDE

Pull-down Menu	COMMAND (TYPE)	ALIAS (TYPE)	Short-cut	Screen (side) Menu	Tablet Menu
View *Hide*	*HIDE*	*HI*	...	VIEW 2 *Hide*	*M,2*

The *Hide* command suppresses the hidden lines (lines that would normally be obscured from view by opaque surfaces) for the current display or viewport:

Command: *hide*

There are no options for the command. The current display may go blank for a period of time depending on the complexity of the model(s); then the new display with hidden lines removed temporarily appears.

The hidden line display is maintained only for the current display. Once a regeneration occurs, the model is displayed in wireframe representation again. You cannot *Plot* the display generated by *Hide*. Instead, use the *Hide Lines* option of the *Print/Plot Configuration* dialog box (or *Hideplot* option of *Mview* for paper space viewports).

Figure 29-22 displays the FORKLIFT model in the default wireframe representation, and Figure 29-23 illustrates the use of *Hide* with the same drawing.

Figure 29-22 ————————————————————— **Figure 29-23** ————————————————————

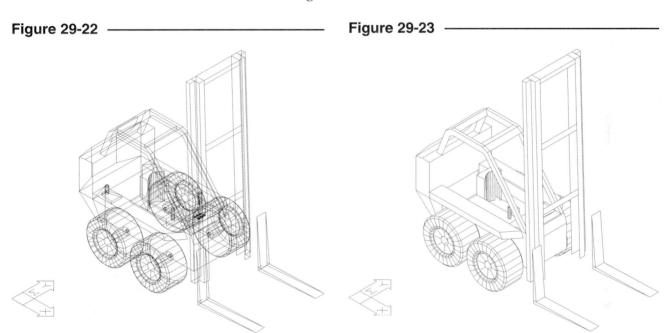

SHADE

Pull-down Menu	COMMAND (TYPE)	ALIAS (TYPE)	Short-cut	Screen (side) Menu	Tablet Menu
View Shade >	SHADE	SHA	...	VIEW 2 Shade	N,2

The *Shade* command allows you to shade a surface or solid model in a full-screen display or in the current viewport. A shaded image is present only for the duration of the current display (until a *Regen*) and cannot be plotted. The image can be saved using the *Mslide* command. You can think of the shaded display as being projected in front of the current drawing. You cannot draw or edit the drawing while a shaded image is displayed. If you want to view or edit the original drawing again, use *Regen*.

Shade automatically positions one light in space and sets a default intensity. The light is located at the camera position. Several options allow you to fill the model surfaces and/or edges with the object color. Two system variables control how the shade appears: *SHADEDGE* and *SHADEDIF*. *SHADEDGE* controls how the surfaces and edges are shaded. The only lighting control is a percentage of diffuse and ambient light adjusted by *SHADEDIF*. Any changes made to the variables are displayed on the next use of the *Shade* command.

If you select *Shade* from the *View* pull-down menu, the *SHADEDGE* options appear on a cascading menu (Fig. 29-24) and prevent you from having to change the variable setting by other means. By contrast, if you <u>type</u> the *Shade* command, the *SHADEDGE* variable must be previously set. The four possible settings are given next (with the integer values used when typing).

Figure 29-24

256 Color, (0)
The object surfaces are shaded in gradient values of the object color (Fig. 29-25). The object's visible edges are not highlighted. This option requires 256 color display capabilities of the monitor and video card. (Keep in mind that these figures have been converted to gray scale.)

Figure 29-25

256 Color Edge Highlight, (1)
This option shades the surfaces in gradient values of the object color and highlights the visible edges in the background color (Fig. 29-26). This option also requires a 256 color display.

Figure 29-26

16 Color Hidden Line, (2)

This option simulates hidden line removal. The resulting
display is similar to using the *Hide* command. All surfaces are painted
in the background color, and the visible edges display the object color
(Fig. 29-27). This display works well for monitors with only 16 colors
or for monochrome displays.

Figure 29-27 ————

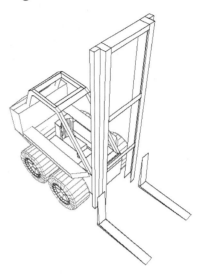

16 Color Filled, (3)

The surfaces are filled solid with the object's color <u>without</u> gradient
shading. The visible edges are drawn in the background color (Fig. 29-
28). This display also works well for display capabilities of 16 colors and
for monochrome displays.

Figure 29-28 ————

SHADEDIF

When you use 256 color shading, *SHADEDIF* can be used to control the contrast of those values. Two
kinds of lights are used with *Shade:* ambient and diffuse. The *SHADEDIF* variable controls the amount
of diffuse (reflected) lighting versus the ambient (overall) light. The default setting of 70 specifies 70%
diffuse and 30% ambient light. The <u>higher the *SHADEDIF*</u> setting, the greater the reflected light and the
<u>greater the contrast</u> of surfaces on the object. The lower the *SHADEDIF* setting, the lower the contrast
and the more all surfaces are lighted equally.

AutoCAD has sophisticated capabilities for generating a photo-realistic surface or solid object. With
these capabilities, you can place lights in 3D space, assign light intensity and color, attach materials, and
assign backgrounds in the model. *Render* is the command that causes the current rendering parameters
to be applied and generates the rendered display. Although much time and effort can be spent adjusting
parameters, *Render* can be used anytime to generate a display using the default parameters.

CHAPTER EXERCISES

1. In this exercise, you use an AutoCAD sample drawing to practice with *3D Viewpoint*. **Open** the **CAMPUS** drawing from the Sample directory. (Assuming AutoCAD is installed using the default location, find the sample drawings in C:/Program Files/AutoCAD R14/Sample.)

    A. Use any method to generate a *Top* view. Examine the view and notice the orientation of the coordinate system icon.

    B. Generate a *SE Isometric* viewpoint to orient yourself. Examine the icon and find north (Y axis). Consider how you (the observer) are positioned as if viewing <u>from</u> the southeast.

    C. Produce a *Front* view. This is a view looking north. Notice the new icon. Remember that you cannot see the XY plane, so it is not a good idea to draw on that plane from this viewpoint.

    D. Generate a *Right* view. Produce a *Top* view again.

    E. Next, view the CAMPUS from a *SE Isometric* viewpoint. Your view should appear like that in Figure 29-29. Notice the position of the WCS icon.

    Figure 29-29 ——————————————————

    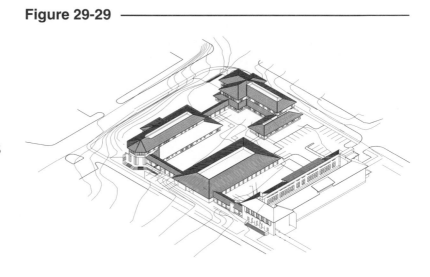

    F. Finally, generate a *NW Isometric* and *NE Isometric*. In each case, examine the WCS icon to orient your viewing direction.

    G. Experiment and practice more with *3D Viewpoint* if you desire. <u>Do not *Save*</u> the drawing.

2. The FORKLIFT sample drawing is an excellent 3D model for practice using the 3D viewing commands because it is a relatively small file (fast to generate) and is easily recognizable from any view (front, back, top, etc.). The FORKLIFT.DWG is not copied to your computer during installation but is located on the AutoCAD Release 14 CD-ROM. It is located in the Acad/Bonus/Sample/History directory along with many other interesting and useful sample drawings.

    A. *Open* the **FORKLIFT** drawing. Notice the *Ucsicon* is *On*. Force the **Ucsicon** to the *Origin* (the cross appears at the intersection of the X and Y arrows only when the icon is at the origin).

    B. Generate a *SE Isometric* viewpoint so you can see the top, front, and side of the forklift. *Zoom* with a *Window* and read the name of the creator of the drawing (Lansing Pugh, Architect). Use *Zoom Previous* to achieve the previous display.

    C. Next, generate a *Top* view, then a *Front* view. Often the front view is the profile view, not necessarily the front of the object. For mechanical applications, the front view should show the most about the object. Find which view (of the *3D Viewpoint* options) displays the actual front of the forklift.

D. Generate a *SW Isometric* view. *Zoom* (out) until the icon orients itself at the origin. Next, generate a *NW Isometric*. Is the icon at the origin?

E. Use *Vpoint* with the *Tripod* option to view the forklift looking from the southeast direction and from slightly above. HINT: PICK the lower-right (southeast) quadrant inside the small circle.

F. Use *Vpoint* with the *Tripod* option again to view the forklift from the southwest and slightly above. View the forklift again from the back side (PICK on the "equator" at the top of the small circle).

G. <u>Do not *Save*</u> the drawing.

3. *Open* the **CAMPUS** drawing again.

Figure 29-30

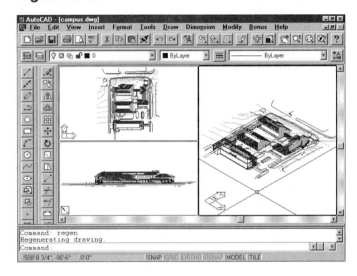

A. Generate a *Top* view. Use *Vports* or the *Tiled Viewports Layout* dialog box to generate 3 viewports with the large vertical viewport on the *right*. The CAMPUS should appear in a top view in all viewports.

B. Use the *Vpoint* command or *3D Viewpoint* to create a specific *Vpoint* for each viewport, as given below and shown in Figure 29-30:

Upper-left	*Top* view
Lower-left	*Front* view
Right	*SE Isometric*

4. *Open* the **FORKLIFT** drawing again.

Figure 29-31

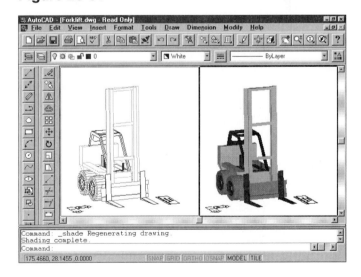

A. Use the *Hide* command to remove hidden edges.

B. *Shade* the forklift with each of the *SHADEDGE* options:

256 Color
256 Color Edge Highlight
16 Color Hidden Line
16 Color Filled

C. Next, use *Vports* or the *Tiled Viewports Layout* dialog box to generate 2 *Vertical* viewports. *Zoom Extents* in each viewport. In the left viewport, use *Hide*, and in the right, *Shade* the forklift with the most realistic option. The drawing should look similar to that in Figure 29-31.

5. In this exercise, you will create a simple solid model that you can use for practicing creation of User Coordinate Systems in Chapter 30. You will also get an introduction to some of the solid modeling construction techniques discussed in Chapter 31.

A. Begin a *New* drawing and name it *SOLID1*. Turn *On* the *Ucsicon* and force it to appear at the *ORigin*.

Figure 29-32 ——————————————

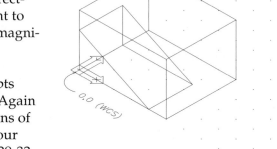

B. Type *Box* to create a solid box. When the prompts appear, use **0,0,0** as the "Corner of box." Next, use the *Length* option and give dimensions for the Length, Width, and Height as **5, 4,** and **3**. A rectangle should appear. Change your viewpoint to *SE Isometric* to view the box. *Zoom* with a magnification factor of *.6X.*

C. Type the *Wedge* command. When the prompts appear, use **0,0,0** as the "Corner of wedge." Again use the *Length* option and give the dimensions of **4, 2.5,** and **2** for *Length, Width,* and *Height.* Your solid model should appear as that in Figure 29-32.

D. Use *Rotate* and select <u>only the wedge</u>. Use **0,0** as the "Base point" and enter **-90** as the "Rotation angle." The model should appear as Figure 29-33.

Figure 29-33 ——————————————————

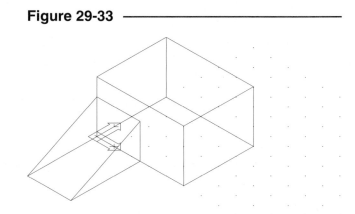

E. Now use *Move* and select the wedge for moving. Use **0,0** as the "Base point" and enter **0,4,3** as the "Second point of displacement."

F. Finally, type *Union*. When prompted to "Select objects:", PICK <u>both</u> the wedge and the box. The finished solid model should look like Figure 29-34. *Save* the drawing.

Figure 29-34 ————————————————

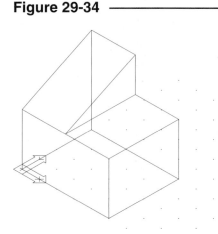

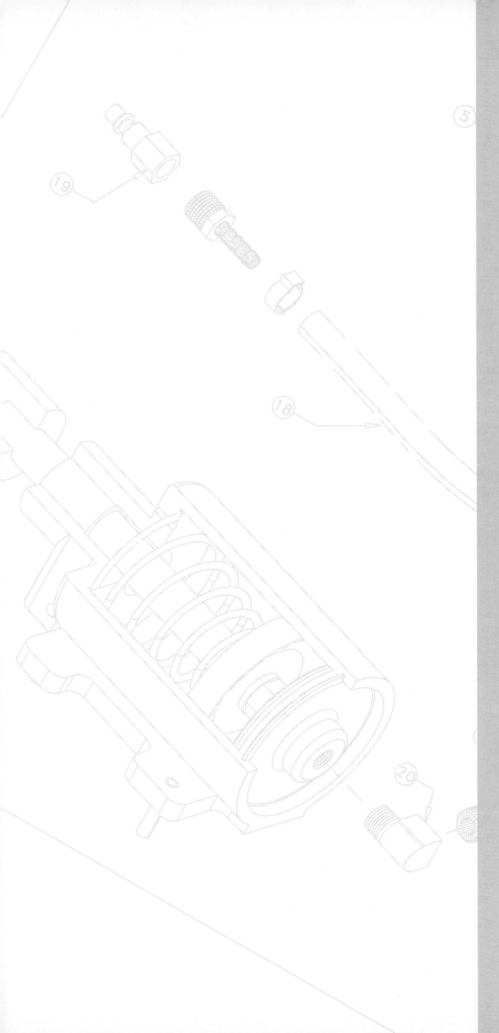

30

USER COORDINATE SYSTEMS

Chapter Objectives

After completing this chapter you should:

1. know how to create, *Save*, *Restore*, and *Delete* User Coordinate Systems;

2. be able to create UCSs by the *Origin*, *Zaxis*, *3point*, *Object*, *View*, *X*, *Y*, or *Z* methods;

3. know how to create UCSs using the *UCS Orientation* dialog box;

4. be able to use the *UCSFOLLOW* variable to automatically display a plan view of new UCSs.

CONCEPTS

No User Coordinate Systems exist as part of the AutoCAD default template drawing (ACAD.DWT) as it comes "out of the box." UCSs are created to simplify construction of the 3D model when and where they are needed.

When you create a multiview or other 2D drawing, creating geometry is relatively simple since you only have to deal with X and Y coordinates. However, when you create 3D models, you usually have to consider the Z coordinates, which makes the construction process more complex. Constructing some geometries in 3D can be very difficult, especially when the objects have shapes on planes not parallel with, or perpendicular to, the XY plane or not aligned with the WCS (World Coordinate System).

Figure 30-1

UCSs are created when needed to simplify the construction process of a 3D object. For example, imagine specifying the coordinates for the centers of the cylindrical shapes in Figure 30-1, using only world coordinates. Instead, if a UCS were created on the face of the object containing the cylinders (the inclined plane), the construction process would be much simpler. To draw on the new UCS, only the X and Y coordinates (of the UCS) would be needed to specify the centers, since anything drawn on that plane has a Z value of 0. The Coordinate Display (*Coords*) in the Status line always displays the X, Y, and Z values of the <u>current</u> coordinate system, whether it is the WCS or a UCS.

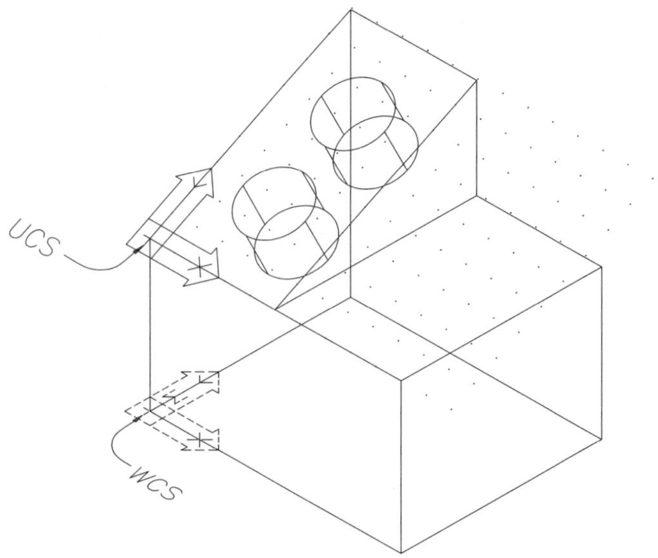

Generally, you would create a UCS aligning its XY plane with a surface of the object, such as along an inclined plane, with the UCS origin typically at a corner or center of the surface. Any coordinates that you specify (in any format) are assumed to be user coordinates (coordinates that lie in or align with the current UCS). Even when you PICK points interactively, they fall on the XY plane of the current UCS. The *SNAP* and *GRID* automatically align with the current coordinate system XY plane, providing *SNAP* points and enhancing your visualization of the construction plane.

UCS COMMANDS AND VARIABLES

User Coordinate Systems are created by using any of several options of the *UCS* command. Once a UCS has been created, it becomes the current coordinate system. <u>Only one coordinate system can be active</u>. If you switch among several UCSs and the WCS, you should save the UCSs when you create them (by using the *Save* option of *UCS*) and restore them at a later time (using the *Restore* option).

When creating 3D models, it is recommended that the <u>*ORigin* option of the *Ucsicon* command be used</u> to place the icon always at the origin of the current UCS, rather than in the lower-left corner of the screen. Since the origin of the UCS is typically specified as a corner or center of the construction plane, aligning the Coordinate System icon with the current origin aids the user's visualization of the UCS orientation.

UCS

Pull-down Menu	COMMAND (TYPE)	ALIAS (TYPE)	Short-cut	Screen (side) Menu	Tablet Menu
Tools *UCS >*	*UCS*	...	...	*TOOLS 2* *UCS*	*W,7*

The *UCS* command allows you to create, save, restore, and delete UCSs. There are many possibilities for creating UCSs to align with any geometry:

Command: **ucs**
Origin/ZAxis/3point/OBject/View/X/Y/Z/Previous/Save/Restore /Delete/?<World>: (**letter**) (Enter capitalized letters of desired option.)

The *UCS* command options are available through the *Tools* pull-down menu.

Keep in mind, <u>UCSs have no association with views or viewpoints.</u> Creating or restoring a UCS <u>does not</u> change the view (unless the *UCS-FOLLOW* variable is *On*), and <u>changing a viewpoint never changes the UCS.</u>

Figure 30-2 ———————————

The *UCS* icon group is available on the Standard toolbar (Fig. 30-3). If you prefer, a separate *UCS* toolbar can be brought on the screen and made to float or dock.

When you create UCSs, AutoCAD prompts for points. These points can be entered as coordinate values at the keyboard, or points on existing geometry can be PICKed. *OSNAP*s should be used to PICK points in 3D space. Understanding the right-hand rule is imperative when creating UCSs.

(If you completed the SOLID1 drawing from Chapter 29, Exercise 5, *Open* the drawing and try the *UCS* options as you read along through this chapter.)

Figure 30-3 ———————————

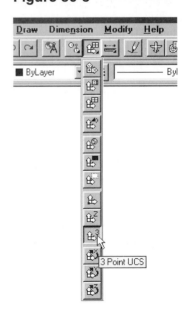

Origin

This option defines a new UCS by specifying a new X,Y,Z location for the origin. The <u>orientation</u> of the UCS (direction of X,Y,Z axes) remains the same as in its previous position; only the location of the origin changes. AutoCAD prompts:

Origin point<0,0,0>:

Coordinates may be specified in any format or PICKed (use *OSNAPs* in 3D space).

Figure 30-4 ─────

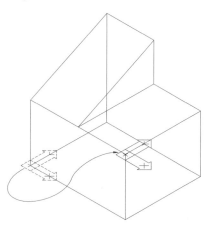

ZAxis

You can define a new UCS by specifying an <u>origin</u> and a direction for the <u>Z axis</u>. Only the two points are needed. The X <u>or</u> Y axis generally remains parallel with the current UCS XY plane, depending on how the Z axis is tilted. AutoCAD prompts:

Origin point <0,0,0>:
Point on positive portion of the Z axis <default>:

Figure 30-5 ─────

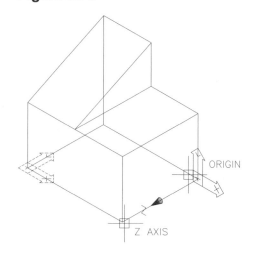

3point

The new UCS is defined by (1) the <u>origin</u>, (2) a point on the <u>X axis</u> (positive direction), and (3) a point on the <u>Y axis</u> (positive direction) or <u>XY plane</u> (positive Y). This is the most universal of all the UCS options (works for most cases). It helps if you have geometry established that can be PICKed with *OSNAP* to establish the points. The prompts are:

Origin point <0,0,0>:
Point on positive portion of the X axis <default>:
Point on positive-Y portion of the UCS XY plane <default>:

Figure 30-6 ─────

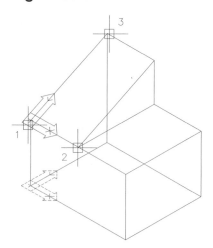

Object

This option creates a new UCS aligned with the selected object. You need to PICK only one point to designate the object. The orientation of the new UCS is based on the type of object selected and the XY plane that was current when the object was created. This option is intended for use primarily with <u>wireframe</u> or <u>surface</u> model objects.

AutoCAD prompts:

> Select object to align UCS:

The following list gives the orientation of the UCS using the *Object* option for each type of object.

Figure 30-7

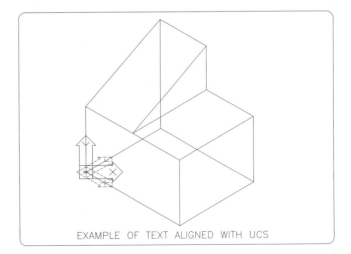

Object	Orientation of New UCS
Line	The end nearest the point PICKed becomes the new UCS origin. The new X axis aligns with the *Line*. The XY plane keeps the same orientation as the previous UCS.
Circle	The center becomes the new UCS origin with the X axis passing through the PICK point.
Arc	The center becomes the new UCS origin. The X axis passes through the endpoint of the *Arc* that is closest to the pick point.
Point	The new UCS origin is at the *Point*. The X axis is derived by an arbitrary but consistent "arbitrary axis algorithm." (See the *AutoCAD Customization Guide*.)
Pline (2D)	The *Pline* start point is the new UCS origin with the X axis extending from the start point to the next vertex.
2D Solid	The first point of the *Solid* determines the new UCS origin. The new X axis lies along the line between the first two points.
Dimension	The new UCS origin is the middle point of the dimension text. The direction of the new X axis is parallel to the X axis of the UCS that was current when the dimension was drawn.
3Dface	The new UCS origin is the first point of the *3Dface*, the X axis aligns with the first two points, and the Y positive side is on that of the first and fourth points. (See Chapter 40, Surface Modeling, *3Dface*.)
Text, Insertion, or *Attribute*	The new UCS origin is the insertion point of the object, while the new X axis is defined by the rotation of the object around its extrusion direction. Thus, the object you PICK to establish a new UCS will have a rotation angle of 0 in the new UCS.

View

This *UCS* option creates a UCS parallel with the screen (perpendicular to the viewing angle). The UCS <u>origin</u> remains unchanged. This option is handy if you wish to use the current viewpoint and include a border, title, or other annotation. There are no options or prompts.

Figure 30-8

EXAMPLE OF TEXT ALIGNED WITH UCS

X   Y   Z

Each of these options rotates the UCS about the indicated axis according to the right-hand rule. The command prompt is:

Figure 30-9

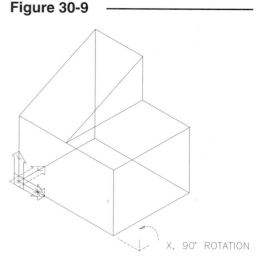

X, 90° ROTATION

Rotation angle about *n* axis:

The angle can be entered by PICKing two points or by entering a value. This option can be repeated or combined with other options to achieve the desired location of the UCS. It is imperative that the right-hand rule is followed when rotating the UCS about an axis (see Chapter 28).

Previous

Use this option to restore the previous UCS. AutoCAD remembers the ten previous UCSs used. *Previous* can be used repeatedly to step back through the UCSs.

World

Using this option makes the WCS (World Coordinate System) the current coordinate system.

Save

Invoking this option prompts for a name for saving the current UCS. Up to 31 characters can be used in the name. Entering a name causes AutoCAD to save the current UCS. The *?* option and the *Dducs* command list all previously saved UCSs.

Restore

Any *Saved* UCS can be restored with this option. The *Restored* UCS becomes the current UCS. The *?* option and the *Dducs* command list the previously saved UCSs.

Delete

You can remove a *Saved* UCS with this option. Entering the name of an existing UCS causes AutoCAD to delete it.

?

This option lists the origin and X, Y, and Z axes for any saved UCS you specify.

It is important to remember that changing to another UCS does not change the display (unless *UCSFOLLOW* is activated). Only one coordinate system (WCS or a UCS) can be current at a time. If you are using viewports, several *Viewpoints* can be displayed, but only one UCS can be current and it appears in all viewports.

DDUCSP

	Pull-down Menu	COMMAND (TYPE)	ALIAS (TYPE)	Short-cut	Screen (side) Menu	Tablet Menu
	Tools UCS > Preset UCS...	DDUCSP	UCP	...	TOOLS 2 Dducsp	W,9

If you prefer to do things a bit more automatically, the *Dducsp* command (UCS presets) lets you set up new UCSs simply by selecting an image tile in the *UCS Orientation* dialog box (Fig. 30-10). These options result in the same functions accomplished by the *UCS* command.

The images displayed in this dialog box are somewhat misleading. The selections <u>do not necessarily reposition the UCS on the face of the object</u> indicated in the images but <u>rotate the coordinate system to the correct orientation</u>.

Figure 30-10 ───────

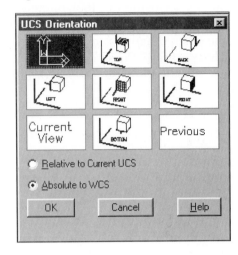

For example, if the current coordinate system were the WCS, selecting the *Right* option would rotate the UCS to the correct orientation <u>keeping the same origin</u>, but would not move the UCS to the rightmost face of the object (Fig. 30-11).

Figure 30-11 ─────────────

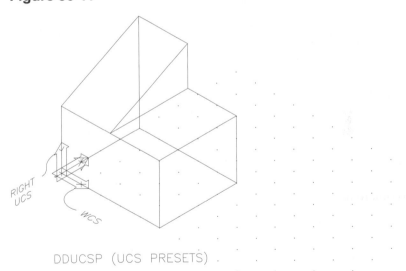

DDUCSP (UCS PRESETS)

NOTE: Always check the settings of the two buttons near the bottom of the dialog box. <u>Generally, UCSs are established *Absolute to the WCS*.</u> Be careful when creating a UCS *Relative to the Current UCS*—you may get some unexpected results.

DDUCS

Pull-down Menu	COMMAND (TYPE)	ALIAS (TYPE)	Short-cut	Screen (side) Menu	Tablet Menu
Tools UCS > Named UCS...	DDUCS	UC	...	TOOLS 2 Dducs	W,8

The *Dducs* command invokes the *UCS Control* dialog box (Fig. 30-12). By using this tool, named UCSs (those that have been previously *Saved*) can be made the current UCS by highlighting the desired name from the list and checking the *Current* tile. This performs the same action as *UCS, Restore*. Named UCSs can also be *Renamed*, *Deleted*, or *Listed* from this dialog box.

Figure 30-12 ─────

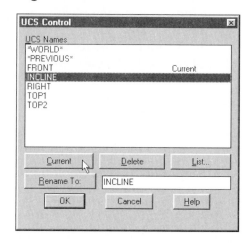

UCSFOLLOW

Pull-down Menu	COMMAND (TYPE)	ALIAS (TYPE)	Short-cut	Screen (side) Menu	Tablet Menu
...	UCSFOLLOW	...	...	...	...

The *UCSFOLLOW* system variable, if set to 1 (*On*), causes the plan view of a UCS to be displayed automatically when a UCS is made current. The default setting for *UCSFOLLOW* is 0 (*Off*). *UCSFOLLOW* can be set separately for each viewport.

For example, consider the case shown in Figure 30-13. Two tiled viewports (*Vports*) are being used, and the *UCS-FOLLOW* variable is turned *On* for the left viewport only. Notice that a UCS created on the inclined surface is current. The left viewport shows the plan view of this UCS automatically, while the right viewport shows the model in the same orientation no matter what coordinate system is current. If the WCS were made active, the right viewport would keep the same viewpoint, while the left would automatically show a plan view when the change was made.

Figure 30-13 ─────

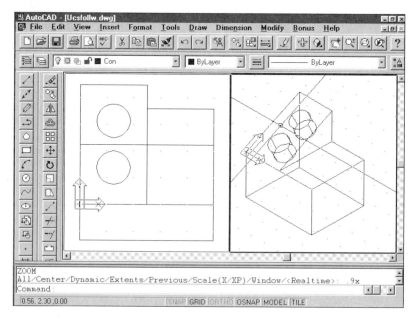

NOTE: When *UCSFOLLOW* is set to 1, the display does not change <u>until</u> a new UCS is created or restored.

UCSFOLLOW is the only variable or command that directly links a change in UCS to an automatic change in viewpoint. Normally, changing to a different current UCS does not change the display, except in this case (when *UCSFOLLOW* is set to 1). On the other hand, making a change in viewpoint <u>never</u> causes a change in UCS.

3D Basics and Solid Modeling

Now that you understand the basics of 3D modeling, you are ready to progress to constructing and editing solid models. Almost all the topics discussed in Chapters 28, 29, and 30 are applicable to <u>all</u> <u>types</u> of 3D modeling.

CHAPTER EXERCISES

1. Using the model in Figure 30-14, assume a UCS was created by rotating about the X axis 90 degrees from the existing orientation (World Coordinate System).

    A. What are the absolute coordinates of corner F?

    B. What are the absolute coordinates of corner H?

    C. What are the absolute coordinates of corner J?

Figure 30-14 ——————————

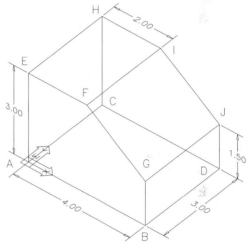

2. Using the model in Figure 30-14, assume a UCS was created by rotating about the X axis 90 degrees from the existing orientation (WCS) and then rotating 90 degrees about the (new) Y.

    A. What are the absolute coordinates of corner F?

    B. What are the absolute coordinates of corner H?

    C. What are the absolute coordinates of corner J?

Figure 30-15 ——————————

3. ***Open*** the **SOLID1** drawing that you created in Chapter 28, Exercise 5. View the object from a ***SE Isometric Vpoint***. Make sure that the ***UCSicon*** is ***On*** and set to the ***ORigin***.

    A. Create a ***UCS*** with a vertical XY plane on the front surface of the model as shown in Figure 30-15. ***Save*** the UCS as **FRONT**.

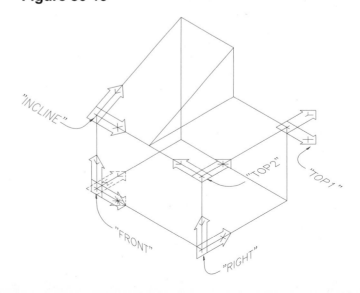

B. Change the coordinate system back to the *World*. Now, create a *UCS* with the XY plane on the top horizontal surface and with the orientation as shown in Figure 30-15 as TOP1. *Save* the UCS as **TOP1**.

C. Change the coordinate system back to the *World*. Next, create the *UCS* shown in the figure as TOP2. *Save* the UCS under the name **TOP2**.

D. Activate the *WCS* again. Create and *Save* the **RIGHT** UCS.

E. Use *SaveAs* and change the drawing name to **SOLIDUCS**.

4. Using the same drawing, make the *WCS* the active coordinate system. In this exercise, you will create a model and screen display like that in Figure 30-13.

A. Create *2 Vertical* tiled viewports. Use *Zoom* in each viewport if needed to display the model at an appropriate size. Make sure the *UCSicon* appears at the origin. Activate the <u>left</u> viewport. Set *UCSFOLLOW On* in that viewport only. (The change in display will not occur until the next change in UCS setting.)

B. Use the right viewport to create the **INCLINE UCS,** as shown in Figure 30-15. *Regenall* to display the new viewpoint.

C. In either viewport create two cylinders as follows: Type *cylinder*; specify **1.25,1** as the *center*, **.5** as the *radius*, and **-1** (negative 1) as the *height*. The new cylinder should appear in both viewports.

D. Use *Copy* to create the second cylinder. Specify the existing center as the "Basepoint, of displacement." To specify the basepoint, you can enter coordinates of **1.25,1** or **PICK** the center with *Osnap*. Make the copy **1.5** units above the original (in a Y direction). Again, you can PICK interactively or use relative coordinates to specify the second point of displacement.

E. Use *SaveAs* and assign the name **SOLID2**.

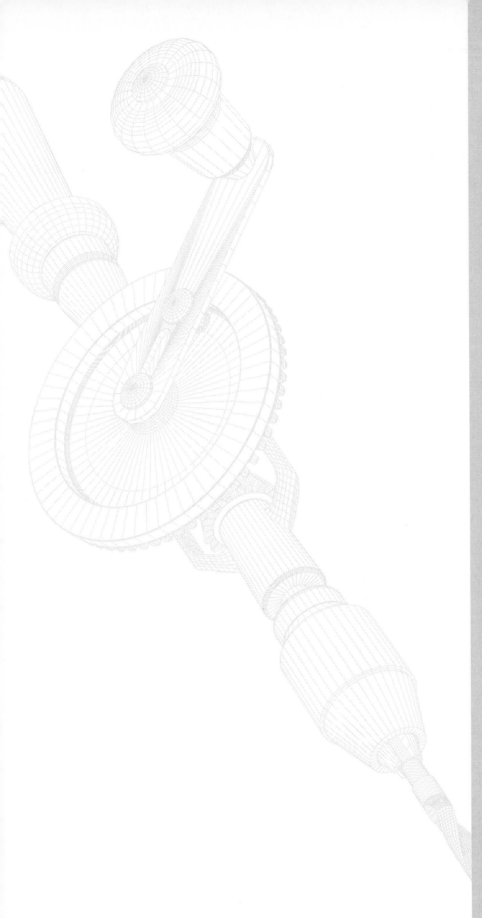

31

SOLID MODELING CONSTRUCTION

Chapter Objectives

After completing this chapter you should:

1. be able to create solid model primitives using the following commands: *Box, Wedge, Cone, Cylinder, Torus,* and *Sphere*;

2. be able to create swept solids from existing 2D shapes using the *Extrude* and *Revolve* commands;

3. be able to *Move* objects in 3D space;

4. know how to assemble a solid to another solid in 3D space using *Align*;

5. be able to rotate a solid about any axis in 3D space with *Rotate3D*;

6. be able to mirror a solid about any axis in 3D space with *Mirror3D*;

7. be able to create rectangular and circular copies with *3Darray*;

8. be able to combine multiple primitives into one composite solid using *Union, Subtract,* and *Intersect;*

9. know how to create beveled edges and rounded corners using *Chamfer* and *Fillet.*

CONCEPTS

The ACIS solid modeler is included in AutoCAD Release 14. With the ACIS modeler, you can create complex 3D parts and assemblies using Boolean operations to combine simple shapes, called *"primitives."* This modeling technique is often referred to as CSG or Constructive Solid Geometry modeling. ACIS also provides a method for analyzing and sectioning the geometry of the models (Chapter 32).

The techniques used with ACIS for construction of many solid models follow three general steps:

1. Construct simple 3D <u>primitive</u> solids, or create 2D shapes and convert them to 3D solids by extruding or revolving.

2. Create the primitives <u>in location</u> relative to the associated primitives or <u>move</u> the primitives into the desired location relative to the associated primitives.

3. Use <u>Boolean operations</u> (such as *Union*, *Subtract*, or *Intersect*) to combine the primitives to form a <u>composite solid</u>.

A solid model is an informationally complete representation of the shape of a physical object. Solid modeling differs from wireframe or surface modeling in two fundamental ways: (1) the information is more complete in a solid model and (2) the method of construction of the model itself is relatively easy to construct and edit. Solid model construction is accomplished by creating regular geometric 3D shapes such as boxes, cones, wedges, cylinders, etc. (primitive solids) and combining them using union, subtract, and intersect (Boolean operations) to form composite solids.

CONSTRUCTIVE SOLID GEOMETRY TECHNIQUES

AutoCAD uses Constructive Solid Geometry (CSG) techniques for construction of solid models. CSG is characterized by solid primitives combined by Boolean operations to form composite solids. The CSG technique is a relatively fast and intuitive way of modeling that imitates the manufacturing process.

Primitives

Solid primitives are the basic building blocks that make up more complex solid models. The ACIS primitives commands are:

BOX	Creates a solid box or cube
CONE	Creates a solid cone with a circular or elliptical base
CYLINDER	Creates a solid cylinder with a circular or elliptical base
EXTRUDE	Creates a solid by extruding (adding a Z dimension to) a closed 2D object (*Pline, Circle, Region*)
REVOLVE	Creates a solid by revolving a shape about an axis
SHPERE	Creates a solid sphere
TORUS	Creates a solid torus
WEDGE	Creates a solid wedge

Primitives can be created by entering the command name or by selecting from the menus or icons.

Boolean Operations

Primitives are combined to create complex solids by using Boolean operations. The ACIS Boolean operators are listed below. An illustration and detailed description are given for each of the commands.

UNION   Unions (joins) selected solids.

SUBTRACT   Subtracts one set of solids from another.

INTERSECT   Creates a solid of intersection (common volume) from the selected solids.

Primitives are created at the desired location or are moved into the desired location before using a Boolean operator. In other words, two or more primitives can occupy the same space (or part of the same space), yet are separate solids. When a Boolean operation is performed, the solids are combined or altered in some way to create one solid. AutoCAD takes care of deleting or adding the necessary geometry and displays the new composite solid complete with the correct configuration and lines of intersection.

Consider the two solids shown in Figure 31-1. When solids are created, they can occupy the same physical space. A Boolean operation is used to combine the solids into a composite solid and it interprets the resulting utilization of space.

Figure 31-1

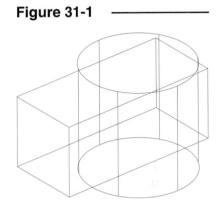

UNION
Union creates a union of the two solids into one composite solid (Fig. 31-2). The lines of intersection between the two shapes are calculated and displayed by AutoCAD. (*Hide* has been used with *DISPSILH*=1 for this figure.)

Figure 31-2

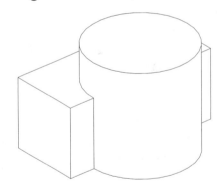

SUBTRACT
Subtract removes one or more solids from another solid. ACIS calculates the resulting composite solid. The term "difference" is sometimes used rather than "subtract." In Figure 31-3, the cylinder has been subtracted from the box.

Figure 31-3

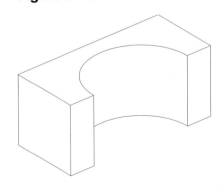

INTERSECT

Figure 31-4

Intersect calculates the intersection between two or more solids. When *Intersect* is used with *Regions* (2D surfaces), it determines the shared <u>area</u>. Used with solids, as in Figure 31-4, *Intersect* creates a solid composed of the shared <u>volume</u> of the cylinder and the box. In other words, the result of *Intersect* is a solid that has only the volume which is part of both (or all) of the selected solids.

SOLID PRIMITIVES COMMANDS

This section explains the commands that allow you to create primitives used for construction of composite solid models. The commands allow you to specify the dimensions and the orientation of the solids. Once primitives are created, they are combined with other solids using Boolean operations to form composite solids.

Figure 31-5

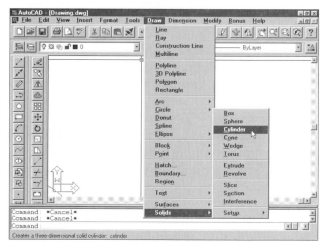

NOTE: If you use a pointing device to PICK points, use *OSNAP* when possible to PICK points in 3D space. If you do not use *OSNAP*, the selected points are located on the current XY construction plane, so the true points may not be obvious. It is recommended that you use *OSNAP* or enter values.

The solid modeling commands are located in groups. The commands for creating solid primitives are located in a group in the *Draw* menus under *Solids* (Fig. 31-5). Commands for moving solids and the Boolean commands are found in the *Modify* menus. You can bring the *Solids* toolbar and the *Modify II* toolbar (which contains the Boolean commands) to the screen by selecting *Toolbars...* from the *View* pull-down menu (Fig. 31-6).

Figure 31-6

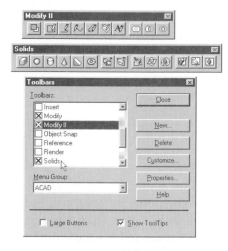

NOTE: When you begin to construct a solid model, <u>create the primitives at location 0,0,0</u> (or some other <u>known point</u>) rather than PICKing points anywhere in space. It is very helpful to know where the primitives are located in 3D space so they can be moved or rotated in the correct orientation with respect to other primitives when they are assembled to composite solids.

BOX

	COMMAND (TYPE)	ALIAS (TYPE)	Short-cut	Screen (side) Menu	Tablet Menu
Pull-down Menu					
Draw *Solids >* *Box*	BOX	...	...	DRAW 2 SOLIDS Box	J,7

Box creates a solid box primitive to your dimensional specifications. You can specify dimensions of the box by PICKing or by entering values. The box can be defined by (1) giving the corners of the base, then height, (2) by locating the center and height, or (3) by giving each of the three dimensions. The base of the box is oriented parallel to the current XY plane:

Command: *box*
Center/<Corner of box> <0,0,0>: **PICK** or (**coordinates**) or (**letter**) or **Enter**

Options for the *Box* command are listed as follows.

Corner of box

Pressing **Enter** begins the corner of a box at 0,0,0 of the current coordinate system. In this case, the box can be moved into its desired location later. As an alternative, a coordinate position can be entered or PICKed as the starting corner of the box. AutoCAD responds with:

Cube/Length/<Other corner>:

The other corner can be PICKed or specified by coordinates. The *Cube* option requires only one dimension to define the cube.

Length

The *Length* option prompts you for the three dimensions of the box in the order of X, Y, and Z. AutoCAD prompts for *Length* (X dimension), *Width* (Y dimension), and *Height* (Z dimension).

Figure 31-7 shows a box created at 0,0,0 with width, depth, and height dimensions of 5, 4, and 3.

Figure 31-7

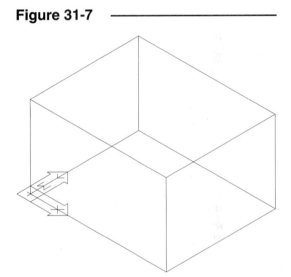

Center

With this option, you first locate the center of the box, then specify the three dimensions of the box. AutoCAD prompts:

Center of box: **PICK** or (**coordinates**)

Specify the center location. AutoCAD responds with:

Cube/Length/<corner of box>:

The resulting solid box is centered about the specified point (0,0,0 for the example; Fig. 31-8).

Figure 31-8

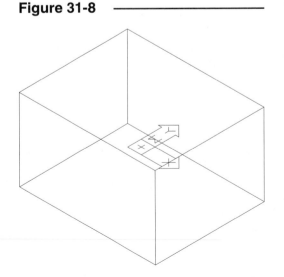

Figure 31-9 illustrates a *Box* created with the *Center* option using *OSNAP* to snap to the *Center* of the top of the cylinder. Note that the center of the box is the <u>volumetric</u> center, not the center of the base.

Figure 31-9

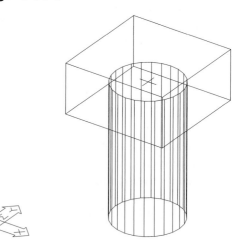

CONE

Pull-down Menu	COMMAND (TYPE)	ALIAS (TYPE)	Short-cut	Screen (side) Menu	Tablet Menu
Draw *Solids >* *Cone*	*CONE*	...	...	DRAW 2 SOLIDS *Cone*	M,7

Cone creates a right circular or elliptical solid cone ("right" means the axis forms a right angle with the base). You can specify the center location, radius (or diameter), and height. By default, the orientation of the cylinder is determined by the current UCS so that the base lies on the XY plane and height is perpendicular (in a Z direction). Alternately, the orientation can be defined by using the *Apex* option.

Center Point

Using the defaults (center at 0,0,0, PICK a radius, enter a value for height), the cone is generated in the orientation shown in Figure 31-10. The cone here may differ in detail (number of contour lines), depending on the current setting of the *ISOLINES* variable. The default prompts are shown as follows:

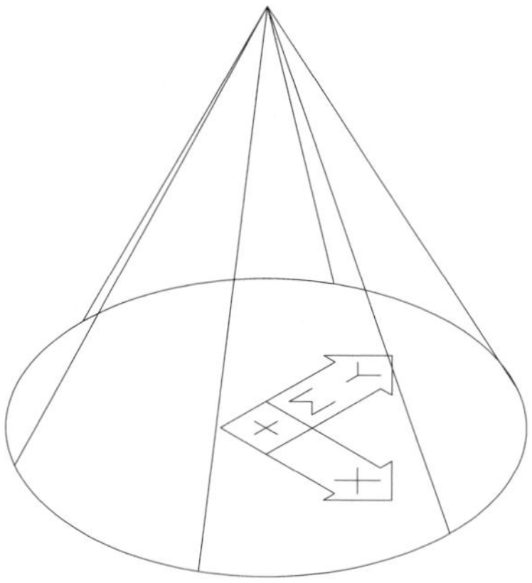

```
Command: cone
Elliptical/<Center point> <0,0,0>: PICK or (coordinates)
Diameter/<Radius>: PICK or (coordinates)
Apex/<Height>: PICK or (value)
Command:
```

Figure 31-10

Invoking the *Apex* option (after *Center point* of the base and *Radius* or *Diameter* have been specified) displays the following prompt:

```
Apex: PICK or (coordinates)
```

Locating a point for the apex defines the height and orientation of the cone (Fig. 31-11). The axis of the cone is aligned with the line between the specified center point and the *Apex* point, and the height is equal to the distance between the two points.

Figure 31-11

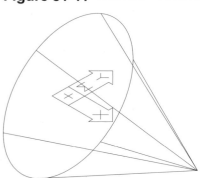

The solid model in Figure 31-12 was created with a *Cone* and a *Cylinder*. The cylinder was created first, then the cone was created using the *Apex* option of *Cone*. The orientation of the *Cone* was generated by PICKing the *Center* of one end of the cylinder for the "Center point" of the base of the cone and the *Center* of the cylinder's other end for the *Apex*. *Union* created the composite model.

Figure 31-12 ————

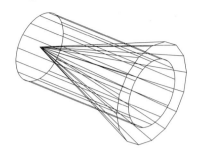
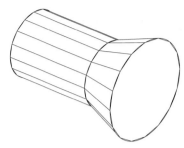
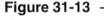

Elliptical

This option draws a cone with an elliptical base (Fig. 31-13). You specify two axis endpoints to define the elliptical base. An elliptical cone can be created using the *Center, Apex,* or *Height* options:

Figure 31-13 ————

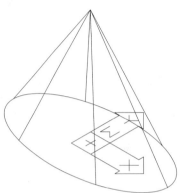

```
Elliptical/<center point> <0,0,0>: e
Center/<Axis endpoint>: c
Center of ellipse <0,0,0>: PICK or (coordinates)
Axis endpoint: PICK or (coordinates)
Other axis distance: PICK or (coordinates)
Apex/<Height>:
```

CYLINDER

Pull-down Menu	COMMAND (TYPE)	ALIAS (TYPE)	Short-cut	Screen (side) Menu	Tablet Menu
Draw *Solids >* *Cylinder*	*CYLINDER*	...	...	*DRAW 2* *SOLIDS* *Cylinder*	*L,7*

Cylinder creates a cylinder with an elliptical or circular base with a center location, diameter, and height you specify. Default orientation of the cylinder is determined by the current UCS, such that the circular plane is coplanar with the XY plane and height is in a Z direction. However, the orientation can be defined otherwise by the *Center of other end* option.

Center Point

The default options create a cylinder in the orientation shown in Figure 31-14:

Figure 31-14 ————

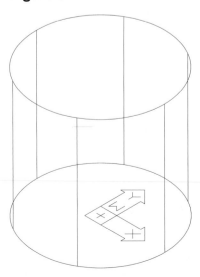

```
Command:  cylinder
Elliptical/<Center point> <0,0,0>: PICK or (coordinates) or
Enter
Diameter/<Radius>: PICK or (coordinates)
Center other end/<Height>: Pick or (coordinates)
```

Elliptical

This option draws a cylinder with an elliptical base (Fig. 31-15). Specify two axis endpoints to define the elliptical base. An elliptical cylinder can be created using the *Center, Center other end,* or *Height* options.

Figure 31-15

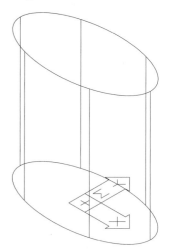

    Elliptical/<center point> <0,0,0>: **e**
    Center/<Axis endpoint>: **c**
    Center of ellipse <0,0,0>: **PICK** or **(coordinates)**
    Axis endpoint: **PICK** or **(coordinates)**
    Other axis distance: **PICK** or **(coordinates)**
    Center other end/<Height>:

The *Center other end* option of *Cylinder* is similar to *Cone Apex* option in that the height and orientation are defined by the *Center point* and the *Center other end*. In Figure 31-16, a hole is created in the *Box* by using *Center other end* option and *OSNAP*-ing to the diagonal lines' *Midpoints*, then *Subtract*ing the *Cylinder* from the *Box*.

Figure 31-16

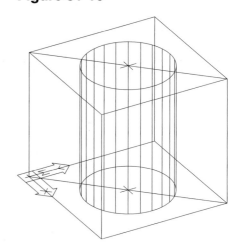

WEDGE

Pull-down Menu	COMMAND (TYPE)	ALIAS (TYPE)	Short-cut	Screen (side) Menu	Tablet Menu
Draw *Solids >* *Wedge*	*WEDGE*	*WE*	...	*DRAW 2* *SOLIDS* *Wedge*	*N,7*

Corner of Wedge

Wedge creates a wedge solid primitive. The base of the wedge is always parallel with the current UCS XY plane, and the <u>slope of the wedge is along the X axis.</u>

Figure 31-17

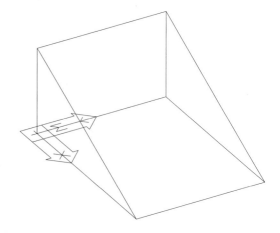

    Command: **wedge**
    Center/<Corner of wedge> <0,0,0>: **PICK** or
    (coordinates)
    Cube/Length/<other corner>: **PICK** or **(coordinates)**

Accepting all the defaults, a *Wedge* can be created as shown in Figure 31-17, with the slope along the X axis.

Invoking the *Length* option prompts you for the *Length, Width,* and *Depth.* More precisely, AutoCAD means width (X dimension), depth (Y dimension), and height (Z dimension).

Center

The point you specify as the center is actually in the center of an imaginary box, half of which is occupied by the wedge. Therefore, the center point is actually at the <u>center of the</u> <u>sloping side of the wedge</u> (Fig. 31-18).

Figure 31-18 ─────────

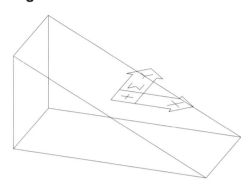

SPHERE

Pull-down Menu	COMMAND (TYPE)	ALIAS (TYPE)	Short-cut	Screen (side) Menu	Tablet Menu
Draw *Solids >* *Sphere*	*SPHERE*	...	...	*DRAW 2* *SOLIDS* *Sphere*	*K,7*

Sphere allows you to create a solid sphere by defining its center point and radius or diameter:

```
Command: sphere
<Center point> <0,0,0>: PICK or (coordinates)
Diameter/<radius>: PICK or (coordinates)
Command:
```

Creating a *Sphere* with the default options would yield a sphere similar to that in Figure 31-19.

Figure 31-19 ─────────

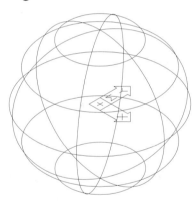

TORUS

Pull-down Menu	COMMAND (TYPE)	ALIAS (TYPE)	Short-cut	Screen (side) Menu	Tablet Menu
Draw *Solids >* *Torus*	*TORUS*	*TOR*	...	*DRAW 2* *SOLIDS* *Tours*	*O,7*

Torus creates a torus (donut shaped) solid primitive using the dimensions you specify. Two dimensions are needed: (1) the radius or diameter of the tube and (2) the radius or diameter from the axis of the torus to the center of the tube. AutoCAD prompts:

```
Command: torus
<Center of torus> <0,0,0>: PICK or (coordinates) or Enter
Diameter/<radius> of torus: PICK or (coordinates)
Diameter/<radius> of tube: PICK or (coordinates)
Command:
```

Figure 31-20 shows a *Torus* created using the default orientation with the axis of the tube aligned with the Z axis of the UCS and the center of the torus at 0,0,0. (*Hide* was used for this display.)

Figure 31-20

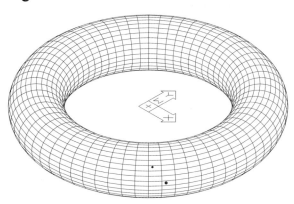

A self-intersecting torus is allowed with the *Torus* command. A self-intersecting torus is created by specifying a torus radius less than the tube radius. Figure 31-21 illustrates a torus with a torus radius of 3 and a tube radius of 4.

Figure 31-21

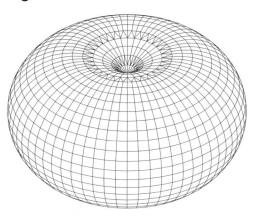

EXTRUDE

Pull-down Menu	COMMAND (TYPE)	ALIAS (TYPE)	Short-cut	Screen (side) Menu	Tablet Menu
Draw *Solids >* *Extrude*	*EXTRUDE*	*EXT*	...	DRAW 2 SOLIDS *Extrude*	P,7

Extrude (like *Revolve*) is a "sweeping operation." Sweeping operations use existing 2D objects to create a solid.

Extrude means to add a Z (height) dimension to an otherwise 2D shape. This command extrudes an existing closed 2D shape such as a *Circle, Polygon, Ellipse, Pline, Spline,* or *Region*. Only <u>closed 2D shapes</u> can be extruded. The closed 2D shape cannot be self-intersecting (crossing over itself).

Two methods determine the direction of extruding: <u>perpendicular</u> to the shape and along a <u>*Path*</u>. With the default method, the selected 2D shape is extruded <u>perpendicular to the plane</u> of the shape <u>regardless</u> of the current UCS orientation. Using the *Path* method, you can extrude the existing closed 2D shape along any existing path determined by a *Line, Arc, Spline,* or *Pline*.

The versatility of this command lies in the fact that <u>any closed shape</u> that can be created by (or converted to) a *Pline, Spline, Region,* etc., no matter how complex, can be transformed into a solid and can be extruded perpendicularly or along a *Path*. *Extrude* can simplify the creation of many solids that may otherwise take much more time and effort using typical primitives and Boolean operations. Create the closed 2D shape first, then invoke *Extrude*:

Command: *extrude*
Select objects: **PICK**
Path/<Height of Extrusion>: (**value**)
Extrusion taper angle <0>: (**value**) or **Enter**
Command:

Figure 31-22 shows a *Pline* before and after using *Extrude*.

Figure 31-22 ——————————————

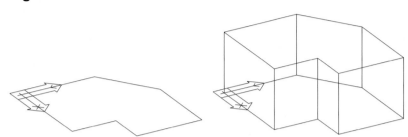

A taper angle can be specified for the extrusion. The resulting solid has sides extruded inward at the specified angle (Fig. 31-23). This is helpful for developing parts for molds that require a slight draft angle to facilitate easy removal of the part from the mold. Entering a negative taper angle results in the sides of the extrusion sloping outward.

Figure 31-23 ——————————————

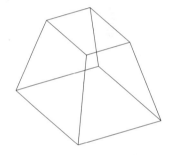

Many complex shapes based on closed 2D *Splines* or *Plines* can be transformed to solids using *Extrude* (Fig. 31-24).

Figure 31-24 ——————————————

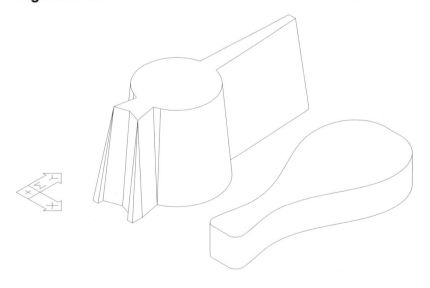

The *Path* option allows you to sweep the 2D shape along an existing line or curve called a "path." The *Path* can be composed of a *Line, Arc, Ellipse, Pline,* or *Spline* (or can be different shapes converted to a *Pline*). This path must lie in a plane. Figure 31-25 shows a closed *Pline* extruded along a *Pline* path and a curved *Spline* path:

Figure 31-25

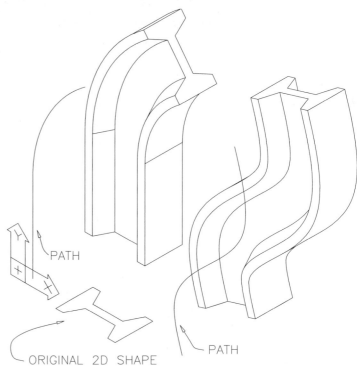

```
Command: extrude
Select objects: PICK
Path/<Height of Extrusion>: path
Select path: PICK
Command:
```

The path cannot lie in the same plane as the 2D shape to be extruded, since the plane of the <u>2D shape is always extruded perpendicular along the path</u>. If one of the endpoints of the path is not located on the plane of the 2D shape, AutoCAD will automatically move the path to the center of the profile temporarily. Notice how the original 2D shape was extruded perpendicularly to the path (Fig. 31-26).

Figure 31-26

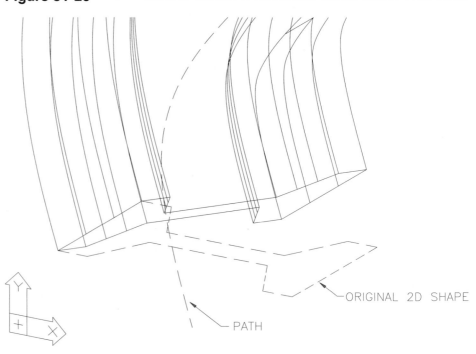

REVOLVE

	Pull-down Menu	COMMAND (TYPE)	ALIAS (TYPE)	Short-cut	Screen (side) Menu	Tablet Menu
	Draw Solids > Revolve	*REVOLVE*	*REV*	...	DRAW 2 SOLIDS *Revolve*	Q,7

Revolve creates a swept solid. *Revolve* creates a solid by revolving a 2D shape about a selected axis. The 2D shape to revolve can be a *Pline, Polygon, Circle, Ellipse, Spline,* or a *Region* object. Only one object at a time can be revolved. *Splines* or *Plines* selected for revolving <u>must be closed</u>. The command syntax for *Revolve* (accepting the defaults) is:

```
Command: revolve
Select objects: PICK
Select objects: Enter (Indicates completion of selection process.)
Axis of revolution - Object/X/Y/<Start point of axis>: PICK
End point of axis: PICK
Angle of revolution <full circle>: (value) or Enter
Command:
```

Figure 31-27 illustrates a possibility for an existing *Pline* shape and the resulting *revolved* shape generated through a full circle.

Because *Revolve* acts on an existing object, the 2D shape intended for revolution should be created in or moved to the desired orientation. There are multiple options for selecting an axis of revolution.

Object
A *Line* or single segment *Pline* can be selected for an axis. The positive axis direction is from the closest endpoint PICKed to the farthest.

X or Y
Uses the positive *X* or *Y* axis of the current UCS as the positive axis direction.

Start point of axis
Defines two points in the drawing to use as an axis (length is irrelevant). Select any two points in 3D space. The two points do <u>not</u> have to be coplanar with the 2D shape.

If the *Object* or *Start point* options are used, the selected object or the two indicated points do <u>not</u> have to be coplanar with the 2D shape. The axis of revolution used is <u>always on the plane of the 2D shape</u> to revolve and is aligned with the direction of the object or selected points.

Figure 31-27 ———————————————————

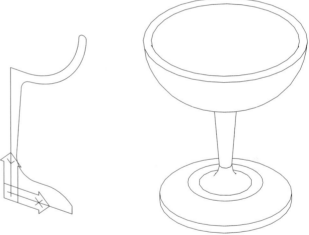

Figure 31-28 demonstrates a possible use of the *Object* option of *Revolve*. In this case, a *Pline* square is revolved about a *Line* object. Note that the *Line* used for the axis is <u>not</u> on the same plane as the *Pline*. *Revolve* uses an axis <u>on the plane</u> of the revolved shape aligned with the direction of the selected axis. In this case, the endpoints of the *Line* are 0,-1,-1 and 0,1,1 so the shape is actually revolved about the Y axis.

Figure 31-28

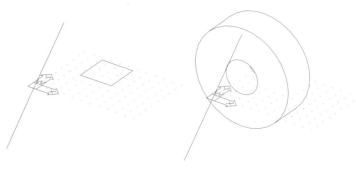

After defining the axis of revolution, *Revolve* requests the number of degrees for the object to be revolved. Any angle can be entered.

Figure 31-29

Another possibility for revolving a 2D shape is shown in Figure 31-29. The shape is generated through 270 degrees.

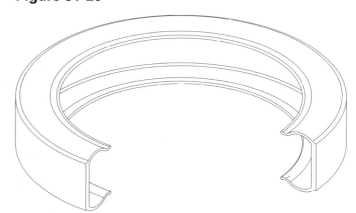

COMMANDS FOR MOVING SOLIDS

When you create the desired primitives, Boolean operations are used to construct the composite solids. However, the primitives must be in the correct position and orientation with respect to each other before Boolean operations can be performed. You can either create the primitives in the desired position during construction (by using UCSs) or move the primitives into position after their creation. Several methods that allow you to move solid primitives are explained in this section.

When constructing 3D geometry, it is critical that the objects are located in space at some <u>known</u> position. Do <u>not</u> create primitives at any convenient place in the drawing; know the position. The location is important when you begin the process of moving primitives to assemble 3D composite solids. Of course, *OSNAPs* can be used, but sometimes it is necessary to use coordinate values in absolute, rectangular, or polar format. A good practice is to <u>create primitives at the final location</u> if possible (use UCSs when needed) or <u>create the primitives at 0,0,0</u>, then <u>move</u> them preceding the Boolean operations.

MOVE

Pull-down Menu	COMMAND (TYPE)	ALIAS (TYPE)	Short-cut	Screen (side) Menu	Tablet Menu
Modify *Move*	*MOVE*	*M*	...	*MODIFY2* *Move*	*V,19*

The *Move* command that you use for moving 2D objects in 2D drawings can also be used to move 3D primitives. Generally, *Move* is used to change the position of an object in one plane (translation), which is typical of 2D drawings. *Move* can also be used to move an ACIS primitive in 3D space <u>if</u> *OSNAPs* or 3D coordinates are used.

Move operates in 3D just as you used it in 2D. Previously, you used *Move* only for repositioning objects in the XY plane, so it was only necessary to PICK or use X and Y coordinates. Using *Move* in 3D space requires entering X, Y, and Z values or using *OSNAPs*.

For example, to create the composite solid used in the figures in Chapter 30, a *Wedge* primitive was *Moved* into position on top of the *Box*. The *Wedge* was created at 0,0,0, then rotated. Figure 31-30 illustrates the movement using absolute coordinates described in the syntax below:

Figure 31-30 ────────────────

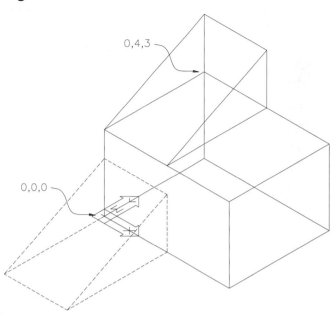

> Command: **move**
> Select objects: **PICK**
> Select objects: **Enter**
> <Base point of displacement>/Multiple: **0,0,0**
> Second point of displacement: **0,4,3**
> Command:

Alternately, you can use *OSNAPs* to select geometry in 3D space. Figure 31-31 illustrates the same *Move* operation using *Endpoint OSNAPs* instead of entering coordinates.

Figure 31-31 ────────────────

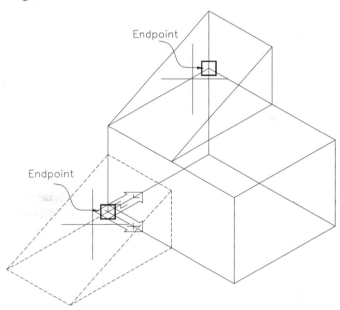

ALIGN

Pull-down Menu	COMMAND (TYPE)	ALIAS (TYPE)	Short-cut	Screen (side) Menu	Tablet Menu
Modify *3D Operations >* *Align*	*ALIGN*	*AL*	...	*MODIFY2* *Align*	*X,14*

Align is an ADS application that is loaded automatically when typing or selecting this command from the menus. *Align* is discussed in Chapter 16 but only in terms of 2D alignment.

Align is, however, a very powerful 3D command because it automatically performs 3D <u>translation and rotation</u> if needed. *Align* is more intuitive and in many situations is easier to use than *Move*. All you have to do is select the points on two 3D objects that you want to align (connect).

Align provides a means of aligning one shape (an object, a group of objects, a block, a region, or a 3D solid) with another shape. The alignment is accomplished by connecting source points (on the shape to be moved) to destination points (on the stationary shape). You can use *OSNAP* modes to select the source and destination points, assuring accurate alignment. Either a 2D or 3D alignment can be accomplished with this command. The command syntax for 3D alignment is as follows:

> Command: **align**
> Select objects: **PICK** (Select object to move, not the stationary object.)
> Select objects: **Enter**
> 1st source point: **PICK** (use *OSNAP*)
> 1st destination point: **PICK** (use *OSNAP*)
> 2nd source point: **PICK** (use *OSNAP*)
> 2nd destination point: **PICK** (use *OSNAP*)
> 3rd source point: **PICK** (use *OSNAP*)
> 3rd destination point: **PICK** (use *OSNAP*)

After the source and destination points have been designated, lines connecting those points temporarily remain until *Align* performs the action (Fig. 31-32).

Align performs a translation (like *Move*) and two rotations (like *Rotate*), each in separate planes to align the points as designated. The motion automatically performed by *Align* is actually done in three steps.

Figure 31-32 ───────────

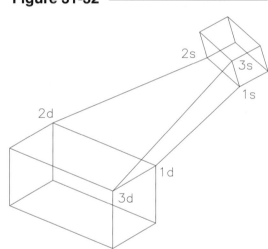

Initially, the first source point is connected to the first destination point (translation). These two points always physically <u>touch</u>.

Next, the vector defined by the first and second source points is aligned with the vector defined by the first and second destination points. The length of the segments between the first and second points on each object is of no consequence because AutoCAD only considers the <u>vector direction</u>. This second motion is a rotation along one axis.

Figure 31-33 ───────────

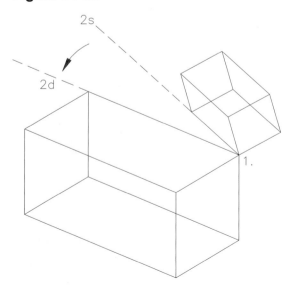

Finally, the third set of points are aligned similarly. This third motion is a rotation along the other axis, completing the alignment.

Figure 31-34

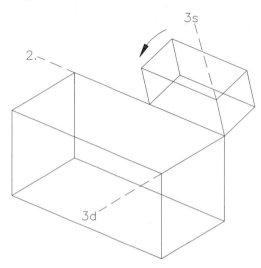

In some cases, such as when cylindrical objects are aligned, only two sets of points have to be specified. For example, if aligning a shaft with a hole (Fig. 31-35), the first set of points (source and destination) specify the attachment of the base of the shaft with the bottom of the hole (use *Center OSNAPs*). The second set of points specify the alignment of the axes of the two cylindrical shapes. A third set of points is <u>not</u> required because the radial alignment between the two objects is not important. When only two sets of points are specified, AutoCAD asks if a 2D or 3D alignment is desired. The command syntax is as follows:

Figure 31-35

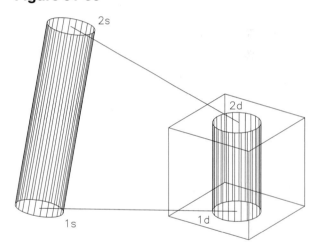

    Command: *align*
    Select objects: **PICK**
    Select objects: **Enter**
    1st source point: cen of **PICK**
    1st destination point: cen of **PICK**
    2nd source point: cen of **PICK**
    2nd destination point: cen of **PICK**
    3rd source point: **Enter**
    <2d> or 3d transformation: **3**
    Command:

ROTATE3D

Pull-down Menu	COMMAND (TYPE)	ALIAS (TYPE)	Short-cut	Screen (side) Menu	Tablet Menu
Modify *3D Operations >* *Rotate 3D*	*ROTATE3D*	...	...	*MODIFY2* *Rotate3D*	*W,22*

Rotate3D is very useful for any type of 3D modeling, particularly with CSG, where primitives must be moved, rotated, or otherwise aligned with other primitives before Boolean operations can be performed.

Rotate3D allows you to rotate a 3D object about any axis in 3D space. Many alternatives are available for defining the desired rotational axis. Following is the command sequence for rotating a 3D object using the default (*2points*) option:

> Command: **rotate3d**
> Select objects: **PICK**
> Select objects: **Enter** (Indicates completion of selection process.)
> Axis by Object/Last/View/Xaxis/Yaxis/Zaxis/<2points>: **PICK** or (**coordinates**) (Select the first point to define the rotational axis.)
> 2nd point on axis: **PICK** or (**coordinates**) (Select the second point to define rotational axis.)
> <Rotation angle>/Reference: **PICK** or (**value**) (Select two points to define the rotation angle or enter a value. If two points are PICKed, the angle between the points in the XY plane of the current UCS determine the angle of rotation.)

The options are explained next.

2points

The power of *Rotate3D* (over *Rotate*) is that any points or objects in 3D space can be used to define the axis for rotation. When using the default (*2points* option), remember you can use *OSNAP* to select points on existing 3D objects. Figure 31-36 illustrates the *2points* option used to select two points with *OSNAP* on the solid object selected for rotating.

Figure 31-36

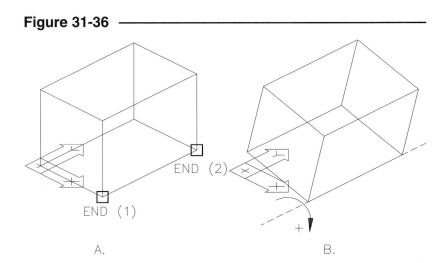

A.

B.

Object

This option allows you to rotate about a selected 2D object. You can select a *Line, Circle, Arc,* or 2D *Pline* segment. The rotational axis is aligned with the selected *Line* or *Pline* segment. Positive rotation is determined by the right-hand rule and the "arbitrary axis algorithm." When selecting *Arc* or *Circle* objects, the rotational axis is perpendicular to the plane of the *Arc* or *Circle* passing through the center. You <u>cannot</u> select the edge of a solid object with this option.

Last

This option allows you to rotate about the axis used for the last rotation.

View

The *View* option allows you to pick a point on the screen and rotates the selected object(s) about an axis perpendicular to the screen and passing through the selected point (Fig. 31-37).

Figure 31-37

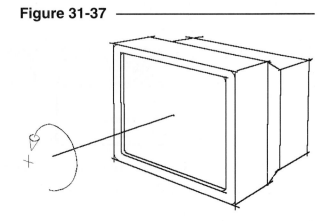

Xaxis

With this option, you can rotate the selected objects about the X axis of the current UCS or any axis parallel to the X axis of the current UCS. You are prompted to pick a point on the X axis. The point selected defines an axis for rotation parallel to the current X axis passing through the selected point. You can use *OSNAP* to select points on existing 3D objects (Fig. 31-38). The current X axis can be used if the point you select is on the X axis.

Figure 31-38

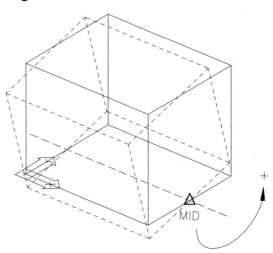

Yaxis

This option allows you to use the Y axis of the current UCS or any axis parallel to the Y axis of the current UCS as the axis of rotation. The point you select defines a rotational axis parallel to the current Y axis passing through the point. *OSNAP* can be used to snap to existing geometry (Fig. 31-39).

Figure 31-39

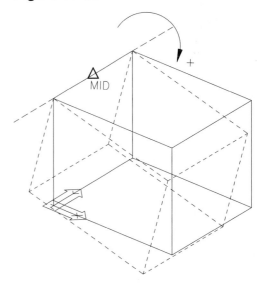

Zaxis

With this option, you can use the Z axis of the current UCS or any axis parallel to the Z axis of the current UCS as the axis of rotation. The point you select defines a rotational axis parallel to the current Z axis passing through the point. Figure 31-40 indicates the use of the *Midpoint OSNAP* to establish a vertical (parallel to Z) rotational axis.

Figure 31-40

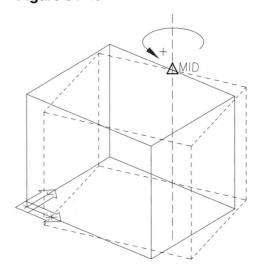

Reference

After you have specified the axis for rotation, you must specify the rotation angle. You are presented with the following prompt:

> <Rotation angle>/Reference: ***r*** (Indicates the *Reference* option.)
> Reference angle: **PICK** or (**value**) (PICK two points; *OSNAP*s can be used. You can enter a value.)
> New angle: **PICK** or (**value**) (Specify the new angle by either method.)

The angle you specify for the reference (relative) is used instead of angle 0 (absolute) for the starting position. You can enter either a value or PICK two points to specify the angle. You then specify a new angle. The *Reference* angle you select is rotated to the absolute angle position you specify as the "New angle."

Figure 31-41 illustrates how *Endpoint OSNAP*s are used to select a *Reference* angle. The "New angle" is specified as **90**. AutoCAD rotates the reference angle to the 90 degree position.

Figure 31-41

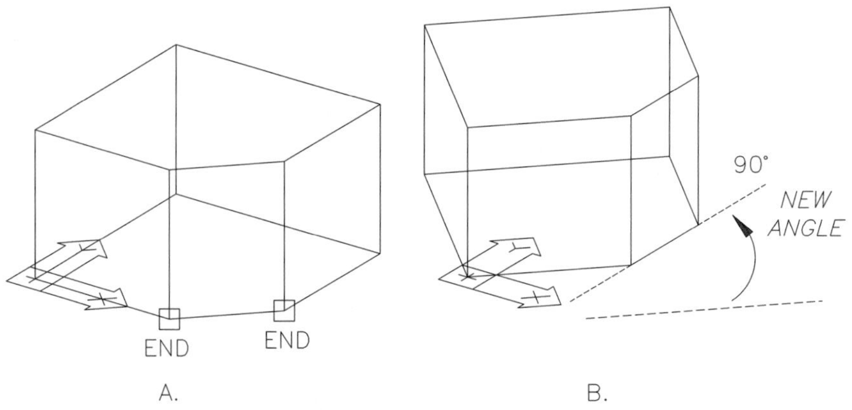

MIRROR3D

Pull-down Menu	COMMAND (TYPE)	ALIAS (TYPE)	Short-cut	Screen (side) Menu	Tablet Menu
Modify *3D Operations >* *Mirror 3D*	MIRROR3D	...	...	MODIFY2 *Mirror3D*	W,21

Mirror3D operates similar to the 2D version of the command *Mirror* in that mirrored replicas of selected objects are created. With *Mirror* (2D) the selected objects are mirrored about an axis. The axis is defined by a vector lying in the XY plane. With *Mirror3D*, selected objects are mirrored about a plane. *Mirror3D* provides multiple options for specifying the plane to mirror about:

> Command: ***mirror3d***
> Select objects: **PICK** (Select one or multiple objects to mirror.)
> Select objects: **Enter** (Indicate completion of the selection process.)
> Plane by Object/Last/Zaxis/View/XY/YZ/ZX/<3points>: **PICK** or (**letters**) (PICK points for the default option or enter a letter(s) to designate option.)

The options are listed and explained next. A phantom icon is shown in the following figures <u>only</u> to aid your visualization of the mirroring plane.

3points

The *3points* option mirrors selected objects about the plane you specify by selecting three points to define the plane. You can PICK points (with or without *OSNAP*) or give coordinates. *Midpoint OSNAP* is used to define the 3 points in Figure 31-42 (A) to achieve the result in (B).

Figure 31-42

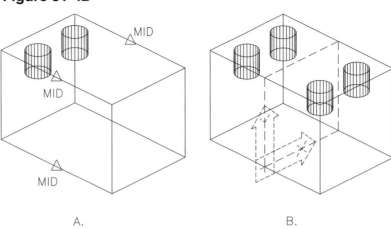

A.   B.

Object

Using this option establishes a mirroring plane with the plane of a 2D object. Selecting an *Arc* or *Circle* automatically mirrors selected objects using the plane in which the *Arc* or *Circle* lies. The plane defined by a *Pline* segment is the XY plane of the *Pline* when the *Pline* segment was created. Using a *Line* object or edge of an ACIS solid is not allowed because neither defines a plane. Figure 31-43 shows a box mirrored about the plane defined by the *Circle* object. Using *Subtract* produces the result shown in (B).

Figure 31-43

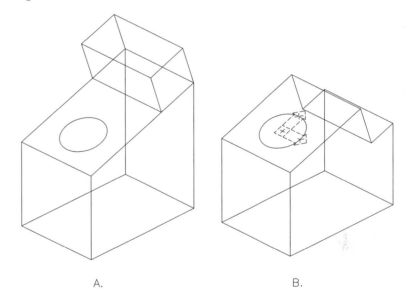

A.   B.

Last

Selecting this option uses the plane that was last used for mirroring.

Zaxis

With this option, the mirror plane is the <u>XY</u> plane perpendicular to a Z vector you specify. The first point you specify on the Z axis establishes the location of the XY plane origin (a point through which the plane passes). The second point establishes the Z axis and the orientation of the XY plane (perpendicular to the Z axis). Figure 31-44 illustrates this concept. Note that this option requires only two PICK points.

Figure 31-44

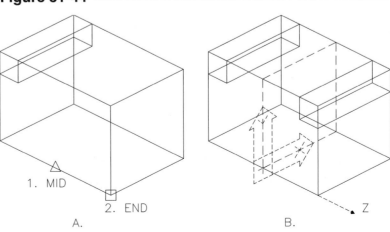

A.   B.

View

The *View* option of *Rotate3D* uses a mirroring plane <u>parallel</u> with the screen and perpendicular to your line of sight based on your current viewpoint. You are required to select a point on the plane. Accepting the default (0,0,0) establishes the mirroring plane passing through the current origin. Any other point can be selected. You must change *Vpoint* to "see" the mirrored objects (Fig. 31-45).

Figure 31-45 ────────────

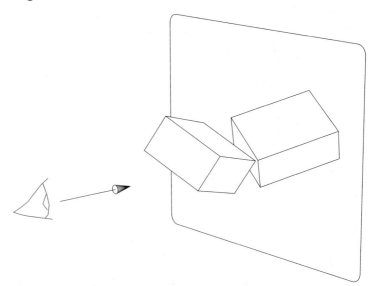

XY

This option situates a mirroring plane parallel with the current XY plane. You can specify a point through which the mirroring plane passes. Figure 31-46 represents a plane established by selecting the *Center* of an existing solid.

Figure 31-46 ────────────

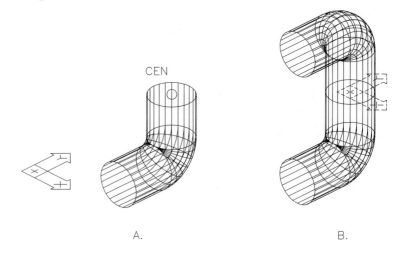

A.   B.

YZ

Using the *YZ* option constructs a plane to mirror about that is parallel with the current YZ plane. Any point can be selected through which the plane will pass (Fig. 31-47).

Figure 31-47 ────────────

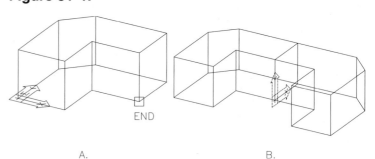

A.   B.

ZX

This option uses a plane parallel with
the current ZX plane for mirroring.
Figure 31-48 shows a point selected on the
Midpoint of an existing edge to mirror two
holes.

Figure 31-48

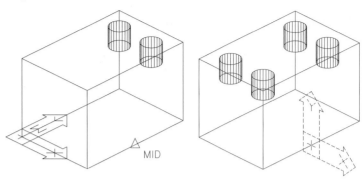

A.   B.

3DARRAY

Pull-down Menu	COMMAND (TYPE)	ALIAS (TYPE)	Short-cut	Screen (side) Menu	Tablet Menu
Modify *3D Operations >* *3D Array*	*3DARRAY*	*3A*	...	*MODIFY2* *3Darray*	*W,20*

Rectangular

With this option of *3Darray*, you create a 3D array specifying three dimensions—the number of and
distance between rows (along the Y axis), the number/distance of columns (along the X axis), and the
number/distance of levels (along the Z axis). Technically, the result is an array in a <u>prism</u> configuration <u>rather than a rectangle</u>.

```
Command: 3darray
Initializing... 3DARRAY loaded.
Select objects: PICK
Select objects: Enter
Rectangular or Polar array (R/P): r
Number of rows (---) <1>: (value)
Number of columns (|||) <1>: (value)
Number of levels (...) <1>: (value)
Distance between rows (---): PICK or (value)
Distance between columns (|||):  PICK or (value)
Distance between levels (...):  PICK or (value)
Command:
```

The selection set can be one or more objects. The entire set is treated as one object for arraying. All
values entered must be positive.

Figures 31-49 and 31-50 illustrate creating a *Rectangular 3Darray* of a cylinder with 3 rows, 4 columns, and 2 levels.

Figure 31-49

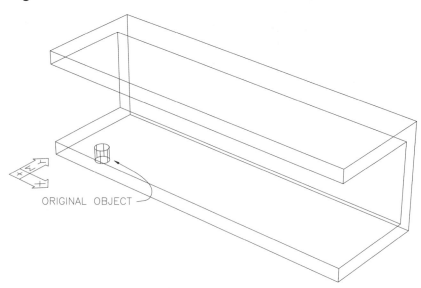

ORIGINAL OBJECT

The cylinders are *Subtracted* from the extrusion to form the finished part.

Figure 31-50

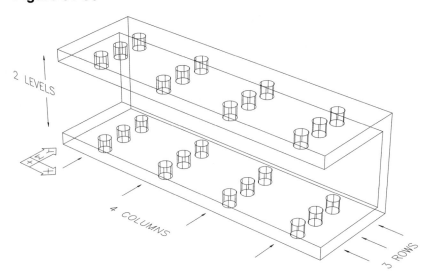

2 LEVELS

4 COLUMNS

3 ROWS

Polar

Similar to a *Polar Array* (2D), this option creates an array of selected objects in a circular fashion. The only difference in the 3D version is that an array is created about an axis of rotation (3D) rather than a point (2D). Specification of an axis of rotation requires two points in 3D space:

```
Command: 3darray
Select objects: PICK
Select objects: Enter
Rectangular or Polar array (R/P): p
Number of items: 8
Angle to fill <360>: Enter or (value)
Rotate objects as they are copied? <Y>: Enter or N
Center point of array: PICK or (coordinates)
Second point on axis of rotation: PICK or (coordinates)
Command:
```

In Figures 31-51 and 31-52, a *3Darray* is created to form a series of holes from a cylinder. The axis of rotation is the center axis of the large cylinder specified by PICKing the *Center* of the top and bottom circles.

Figure 31-51

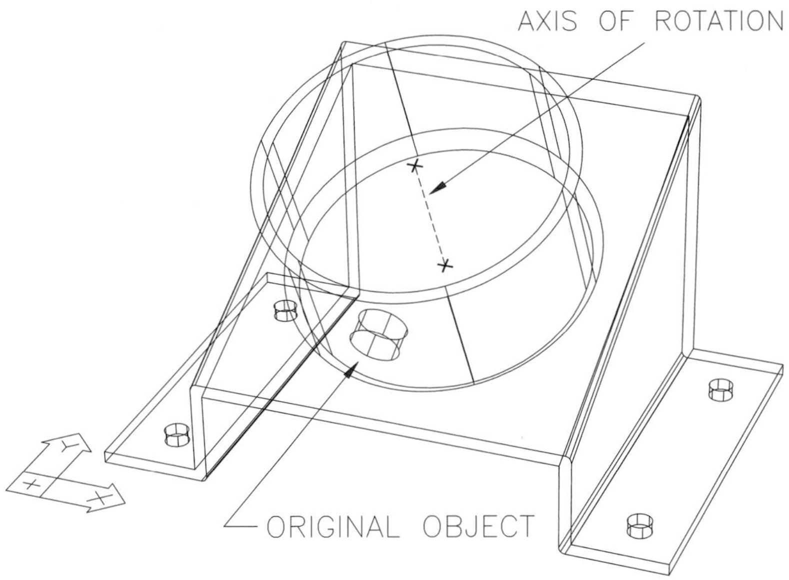

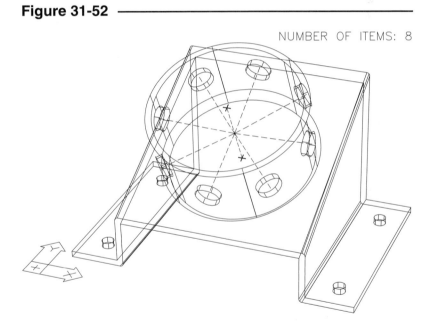

After the eight items are arrayed, the small cylinders are subtracted from the large cylinder to create the holes.

Figure 31-52

BOOLEAN OPERATION COMMANDS

Once the individual 3D primitives have been created and moved into place, you are ready to put together the parts. The primitives can be "assembled" or combined by Boolean operations to create composite solids. The Boolean operations found in AutoCAD are listed in this section: *Union*, *Subtract*, and *Intersect*.

UNION

	COMMAND (TYPE)	ALIAS (TYPE)	Short-cut	Screen (side) Menu	Tablet Menu
Pull-down Menu					
Modify *Boolean >* *Union*	*UNION*	*UNI*	...	*MODIFY2* *Union*	*X,15*

Union joins selected primitives or composite solids to form one composite solid. Usually, the selected solids occupy portions of the same space, yet are separate solids. *Union* creates one solid composed of the total encompassing volume of the selected solids. (You can union solids even if the solids do not overlap.) All lines of intersections (surface boundaries) are calculated and displayed by AutoCAD. Multiple solid objects can be unioned with one *Union* command:

```
Command: solunion
Select objects: PICK (Select two or more solids.)
Select objects: Enter (Indicate completion of the selection process.)
Command:
```

Two solid boxes are combined into one composite solid with *Union* (Fig 31-53). The original two solids (A) occupy the same physical space. The resulting union (B) consists of the total contained volume. The new lines of intersection are automatically calculated and displayed. *Hide* was used to enhance visualization in (B).

Figure 31-53

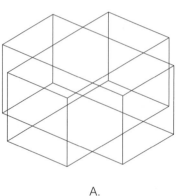
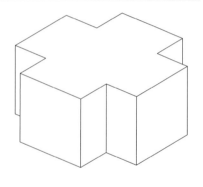

A.   B.

Because the volume occupied by any one of the primitives is included in the resulting composite solid, any redundant volumes are immaterial. The two primitives in Figure 31-54 (A) yield the same enclosed volume as the composite solid (B). (*Hide* has been used with *DISPSILH*=1 for this figure.)

Figure 31-54

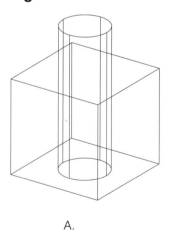
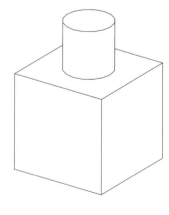

A.   B.

Multiple objects can be selected in response to the *Union* "Select objects:" prompt. It is not necessary, nor is it efficient, to use several successive Boolean operations if one or two can accomplish the same result.

Two primitives that have coincident faces (touching sides) can be joined with *Union*. Several "blocks" can be put together to form a composite solid.

Figure 31-55 illustrates how several primitives having coincident faces (A) can be combined into a composite solid (B). Only one *Union* is required to yield the composite solid.

Figure 31-55

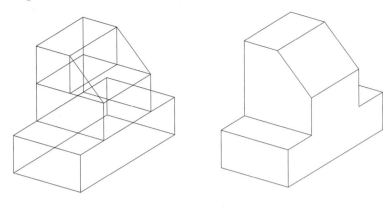

A.   B.

SUBTRACT

Pull-down Menu	COMMAND (TYPE)	ALIAS (TYPE)	Short-cut	Screen (side) Menu	Tablet Menu
Modify Boolean > Subtract	*SUBTRACT*	*SU*	...	MODIFY2 *Subtract*	X,16

Subtract takes the difference of one set of solids from another. *Subtract* operates with *Regions* as well as solids. When using solids, *Subtract* subtracts the volume of one set of solids from another set of solids. Either set can contain only one or several solids. *Subtract* requires that you first select the set of solids that will remain (the "source objects"), then select the set you want to subtract from the first:

    Command: subtract
    Select solids and regions to subtract from...
    Select objects: PICK
    Select objects: Enter
    Select solids and regions to subtract...
    Select objects: PICK
    Select objects: Enter
    Command:

The entire volume of the solid or set of solids that is subtracted is completely removed, leaving the remaining volume of the source set.

To create a box with a hole, a cylinder is located in the same 3D space as the box (see Figure 31-56). *Subtract* is used to subtract the entire volume of the cylinder from the box. Note that the cylinder can have any height, as long as it is at least equal in height to the box.

Because you can select more than one object for the objects "to subtract from" and the objects "to subtract," many possible construction techniques are possible.

Figure 31-56

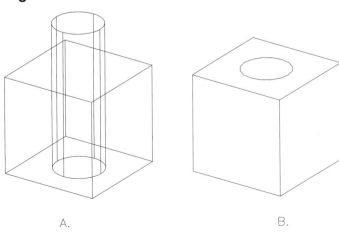

A.   B.

If you select multiple solids in response to the select objects "to subtract from" prompt, they are <u>automatically</u> unioned. This is known as an <u>*n*-way Boolean</u> operation. Using *Subtract* in this manner is very efficient and fast.

Figure 31-57 illustrates an *n*-way Boolean. The two boxes (A) are selected in response to the objects "to subtract from" prompt. The cylinder is selected as the objects "to subtract...". *Subtract* joins the source objects (identical to a *Union*) and subtracts the cylinder. The resulting composite solid is shown in (B). (*Hide* has been used with *DISPSILH*=1 for this figure.)

Figure 31-57 ────────────────

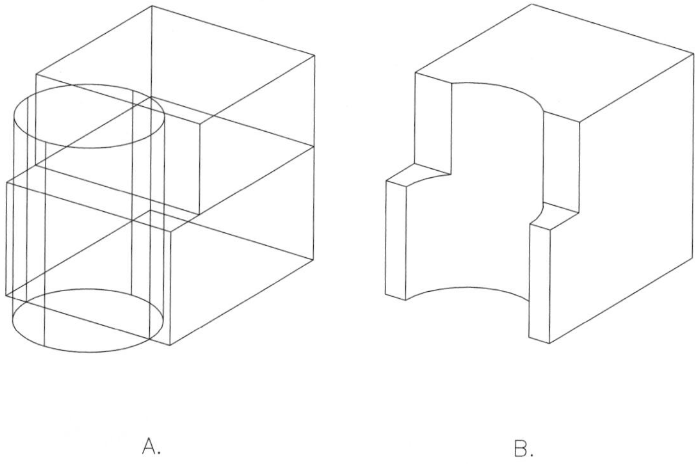

A.   B.

INTERSECT

Pull-down Menu	COMMAND (TYPE)	ALIAS (TYPE)	Short-cut	Screen (side) Menu	Tablet Menu
Modify *Boolean >* *Intersect*	INTERSECT	IN	...	MODIFY2 *Intrsect*	X,17

Intersect creates composite solids by calculating the intersection of two or more solids. The intersection is the common volume <u>shared</u> by the selected objects. Only the 3D space that is <u>part of all</u> of the selected objects is included in the resulting composite solid. *Intersect* requires only that you select the solids from which the intersection is to be calculated.

```
Command: intersect
Select objects: PICK (Select all desired solids.)
Select objects: Enter (Indicates completion of the selection process.)
Command:
```

An example of *Intersect* is shown in Figure 31-58. The cylinder and the box share common 3D space (A). The result of the *Intersect* is a composite solid that represents that common space (B). (*Hide* has been used with *DISPSILH*=1 for this figure.)

Figure 31-58 ────────────────

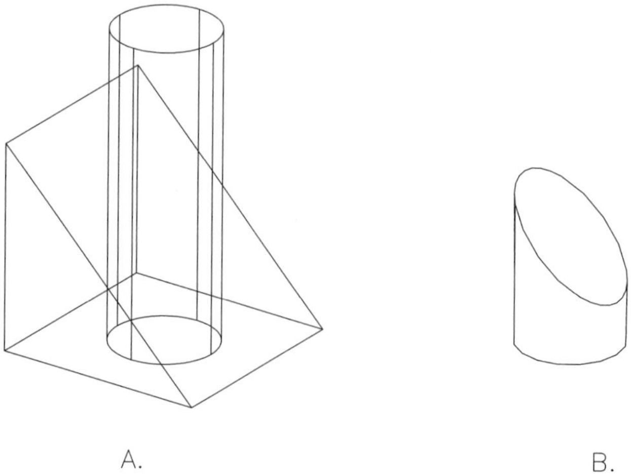

A.   B.

Intersect can be very effective when used in conjunction with *Extrude*. A technique known as <u>reverse drafting</u> can be used to create composite solids that may otherwise require several primitives and several Boolean operations. Consider the composite solid shown in Figure 31-55 A. Using *Union*, the composite shape requires four primitives.

A more efficient technique than unioning several box primitives is to create two *Pline* shapes on vertical planes (Fig. 31-59). Each *Pline* shape represents the outline of the desired shape from its respective view: in this case, the front and side views. The *Pline* shapes are intended to be extruded to occupy the same space. It is apparent from this illustration why this technique is called reverse drafting.

Figure 31-59 ───────────

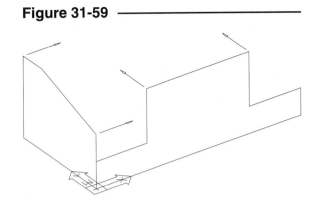

The two *Pline* "views" are extruded with *Extrude* to comprise the total volume of the desired solid (Fig. 31-60 A). Finally, *Intersect* is used to calculate the common volume and create the composite solid (B.).

Figure 31-60 ───────────

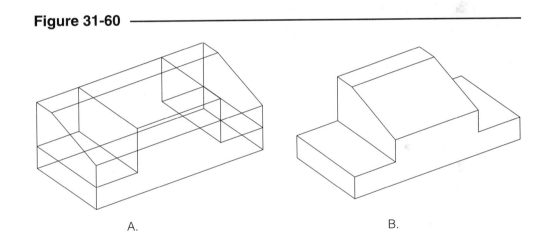

A. B.

CHAMFER

Pull-down Menu	COMMAND (TYPE)	ALIAS (TYPE)	Short-cut	Screen (side) Menu	Tablet Menu
Modify *Chamfer*	*CHAMFER*	*CHA*	...	*MODIFY2* *Chamfer*	*W,18*

Chamfering is a machining operation that bevels a sharp corner. *Chamfer* chamfers selected edges of an AutoCAD solid as well as 2D objects. Technically, *Chamfer* (used with a solid) is a Boolean operation because it creates a wedge primitive and then adds to or subtracts from the selected solid.

When you select a solid, *Chamfer* recognizes the object as a solid and <u>switches to the solid version of prompts and options</u>. Therefore, all of the 2D options are not available for use with a solid, <u>only the "distances" method</u>. When using *Chamfer*, you must both select the "base surface" and indicate which edge(s) on that surface you wish to chamfer:

>   Command: *chamfer*
>   (TRIM mode) Current chamfer Length = 0.5000, Angle = 30.0000
>   Polyline/Distance/Angle/Trim/Method/<Select first line>: **PICK** (Select solid)
>   Select base surface: **PICK** (Select any edge of the desired solid face)
>   Next/<OK>: **N** or **Enter**
>   Enter base surface distance <1.0000>: **Enter** or **(value)**
>   Enter other surface distance <0.5000>: **Enter** or **(value)**
>   Loop/<Select edge>: **PICK** (Select edge to be chamfered)
>   Loop/<Select edge>: **Enter**
>   Command:

When AutoCAD prompts to select the "base surface," only an edge can be selected since the solids are displayed in wireframe. When you select an edge, AutoCAD highlights one of the two surfaces connected to the selected edge. Therefore, you must use the "Next/<OK>:" option to indicate which of the two surfaces you want to chamfer (Fig. 31-61 A). The two distances are applied to the object, as shown in Figure 31-61 B.

Figure 31-61

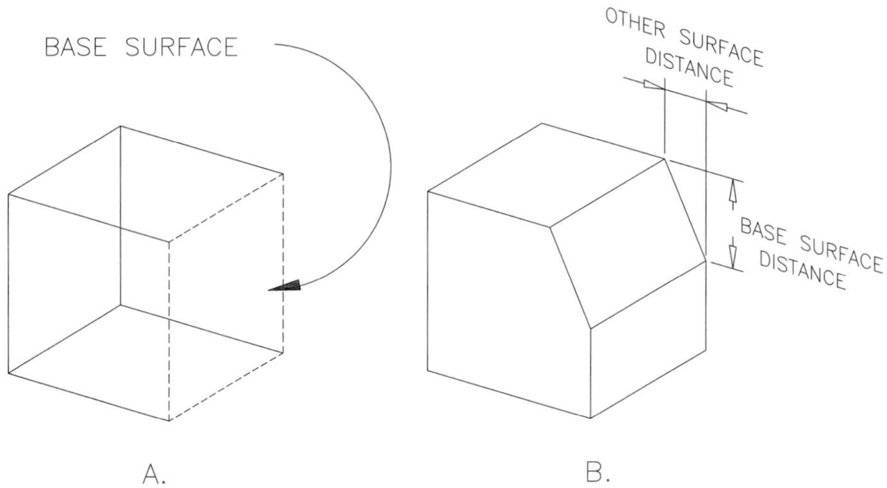

A.

B.

You can chamfer multiple edges of the selected "base surface" simply by PICKing them at the "<Select Edge>:" prompt (Fig. 31-62). If the base surface is adjacent to cylindrical edges, the bevel follows the curved shape.

Figure 31-62

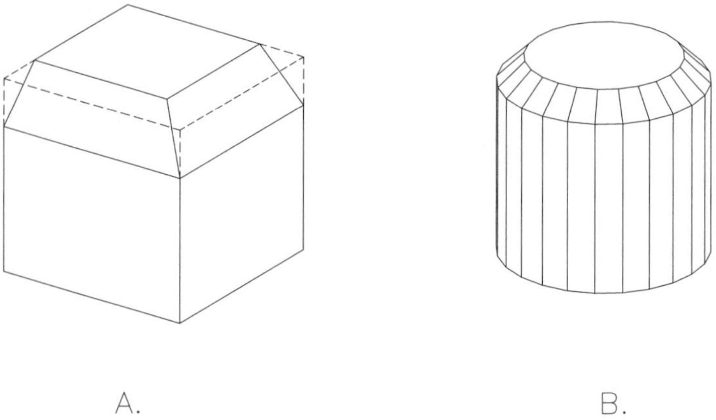

A.

B.

Loop
The *Loop* option chamfers the entire perimeter of the base surface. Simply PICK any edge on the base surface.

Figure 31-63

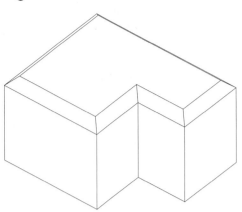

```
Loop/<Select edge>: l
Edge/<Select edge loop>: PICK
Edge/<Select edge loop>: Enter
Command:
```

Edge
The *Edge* option switches back to the "Select edge" method.

FILLET

Pull-down Menu	COMMAND (TYPE)	ALIAS (TYPE)	Short-cut	Screen (side) Menu	Tablet Menu
Modify *Fillet*	*FILLET*	*F*	...	*MODIFY 2* *Fillet*	*W,19*

Fillet creates fillets (concave corners) or rounds (convex corners) on selected solids, just as with 2D objects. Technically, *Fillet* creates a rounded primitive and automatically performs the Boolean needed to add or subtract it from the selected solids.

When using *Fillet* with a solid, the command <u>switches</u> to a special group of prompts and options for 3D filleting, and the <u>2D options become invalid</u>. After selecting the solid, you must specify the desired radius and then select the edges to fillet. When selecting edges to fillet, the edges must be PICKed individually. Figure 31-64 depicts concave and convex fillets created with *Fillet*. The selected edges are highlighted.

Figure 31-64

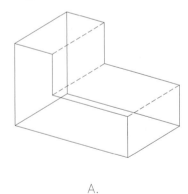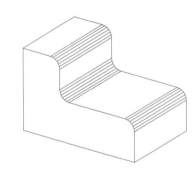

A.

B.

```
Command: fillet
(TRIM mode) Current fillet radius = 0.5000
Polyline/Radius/Trim/<Select first object>: PICK  (Select solid)
Chain/Radius/<Select edge>: Enter radius:  (value)
Chain/Radius/<Select edge>: PICK
Chain/Radius/<Select edge>: PICK
Chain/Radius/<Select edge>: Enter
n edges selected for fillet.
Command:
```

Curved surfaces can be treated with *Fillet*, as shown in Figure 31-65. If you want to fillet intersecting concave or convex edges, *Fillet* handles your request, providing you specify all edges in <u>one</u> use of the command. Figure 31-65 shows the selected edges (highlighted) and the resulting solid. Make sure you select <u>all</u> edges together (in one *Fillet* command).

Figure 31-65

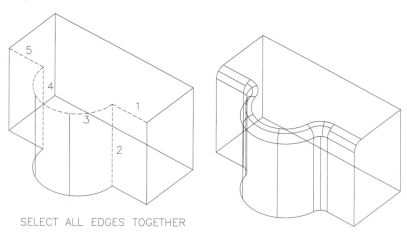

SELECT ALL EDGES TOGETHER

Chain

The *Chain* option allows you to fillet a series of connecting edges. Select the edges to form the chain (Fig. 31-66). If the chain is obvious (only one direct path), you can PICK only the ending edges, and AutoCAD will find the most direct path (series of connected edges):

Figure 31-66

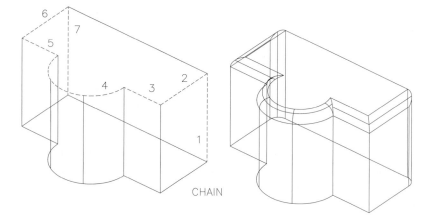

CHAIN

    Chain/Radius/<Select edge>: *c*
    Edge/Radius/<Select edge
    chain>: **PICK**

Edge
This option cycles back to the "<Select edge>:" prompt.

Radius
This method returns to the "Enter radius:" prompt.

DESIGN EFFICIENCY

Now that you know the complete sequence for creating composite solid models, you can work toward improving design efficiency. The typical construction sequence is (1) create primitives, (2) ensure the primitives are in place by using UCSs or any of several move and rotate options, and (3) combine the primitives into a composite solid using Boolean operations. The typical step-by-step, "building-block" strategy, however, may not lead to the most efficient design. In order to minimize computation and construction time, you should <u>minimize the number of Boolean operations</u> and, if possible, the <u>number of primitives</u> you use.

For any composite solid, usually several strategies could be used to construct the geometry. You should plan your designs ahead of time, striving to minimize primitives and Boolean operations.

For example, consider the procedure shown in Figure 31-55. As discussed, it is more efficient to accomplish all unions with one *Union*, rather than each union as a separate step. Even better, create a closed *Pline* shape of the profile; then use *Extrude*. Figure 31-57 is another example of design efficiency based on using an *n*-way Boolean. Multiple solids can be unioned automatically by selecting them at the select objects "to subtract from" prompt of *Subtract*. Also consider the strategy of reverse drafting, as shown in Figure 31-59. Using *Extrude* in concert with *Intersect* can minimize design complexity and time.

In order to create efficient designs and minimize Boolean operations and primitives, keep these strategies in mind:

- Execute as many subtractions, unions, or intersections as possible within one *Subtract*, *Union*, or *Intersect* command.

- Use *n*-way Booleans with *Subtract*. Combine solids (union) automatically by selecting <u>multiple</u> objects "to subtract from," and then select "objects to subtract."

- Make use of *Plines* or regions for complex profile geometry; then *Extrude* the profile shape. This is almost always more efficient for complex curved profile creation than using multiple Boolean operations.

- Make use of reverse drafting by extruding the "view" profiles (*Plines* or *Regions*) with *Extrude*, then finding the common volume with *Intersect*.

CHAPTER EXERCISES

1. What are the typical three steps for creating composite solids?

2. Consider the two solids in Figure 31-67. They are two extruded hexagons that overlap (occupy the same 3D space).

   A. Sketch the resulting composite solid if you performed a *Union* on the two solids.

   B. Sketch the resulting composite solid if you performed an *Intersect* on the two solids.

   C. Sketch the resulting composite solid if you performed a *Subtract* on the two solids.

Figure 31-67 ——————————————

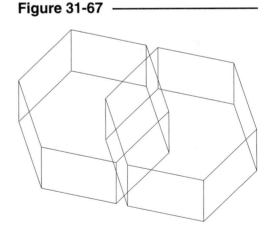

For the following exercises, use a *Template* drawing or begin a *New* drawing. Turn *On* the *Ucsicon* and set it to the *Origin*. Set the *Vpoint* with the *Rotate* option to angles of **310, 30**.

3.   Begin a drawing and assign the name **CH31EX3**.

   Figure 31-68 ─────────────

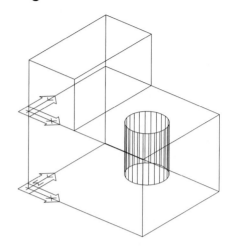

   A. Create a *box* with the lower-left corner at **0,0,0**. The *Lengths* are **5, 4**, and **3**.

   B. Create a second *box* at a new UCS as shown in Figure 31-68 (use the *ORigin* option). The *box* dimensions are **2 x 4 x 2**.

   C. Create a *Cylinder*. Use the same UCS as in the previous step. The *cylinder Center* is at **3.5,2** (of the *UCS*), the *Diameter* is **1.5**, and the *Height* is **-2**.

   D. *Save* the drawing.

   E. Perform a *Union* to combine the two boxes. Next, use *Subtract* to subtract the cylinder to create a hole. The resulting composite solid should look like that in Figure 31-69. *Save* the drawing.

   Figure 31-69 ─────────────

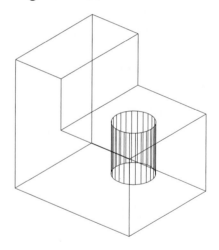

4.   Begin a drawing and assign the name **CH31EX4**.

   Figure 31-70 ─────────────

   A. Create a *Wedge* at point **0,0,0** with the *Lengths* of **5, 4, 3**.

   B. Create a *3point UCS* option with an orientation indicated in Figure 31-70. Create a *Cone* with the *Center* at **2,3** (of the *UCS*) and a *Diameter* of **2** and a *Height* of **-4**.

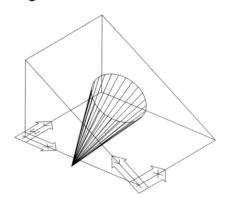

C. *Subtract* the cone from the wedge. The resulting composite solid should resemble Figure 31-71. *Save* the drawing.

Figure 31-71

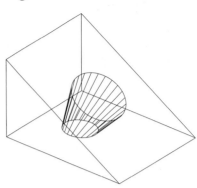

5. Begin a drawing and assign the name **CH31EX5**. Display a *Plan* view.

Figure 31-72

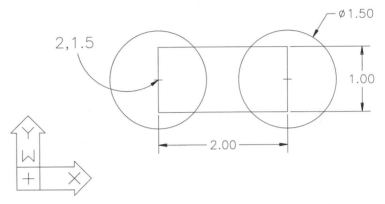

A. Create 2 *Circles* as shown in Figure 31-72, with dimensions and locations as specified. Use *Pline* to construct the rectangular shape. Combine the 3 shapes into a *Region* by using the *Region* and *Union* commands <u>or</u> converting the <u>outside</u> shape into a *Pline* using *Trim* and *Pedit*.

B. Change the display to an isometric-type *Vpoint*. *Extrude* the *Region* or *Pline* with a *Height* of **3** (no *taper angle*).

C. Create a *Box* with the lower-left corner at **0,0**. The *Lengths* of the box are **6, 3, 3**.

D. *Subtract* the extruded shape from the box. Your composite solid should look like that in Figure 31-73. *Save* the drawing.

Figure 31-73

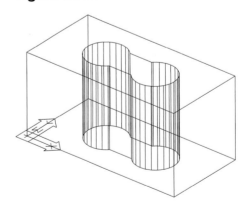

6. Begin a drawing and assign the name **CH31EX6**.
    Display a *Plan* view.

    A. Create a closed *Pline* shape symmetrical
        about the X axis with the locational and
        dimensional specifications given in Fig-
        ure 31-74.

    B. Change to an isometric-type *Vpoint*. Use
        Revolve to generate a complete circular
        shape from the closed *Pline*. Revolve about
        the **Y** axis.

    C. Create a *Torus* with the *Center* at **0,0**. The
        Radius of torus is **3** and the *Radius of tube* is
        .5. The two shapes should intersect.

    D. Use *Hide* to generate a display like Figure 31-75.

    E. Create a *Cylinder* with the *Center* at **0,0,0**, a *Radius* of **3**, and
        a *Height* of **8**.

    F. Use *Rotate3D* to rotate the revolved *Pline* shape **90** degrees
        about the X axis (the *Pline* shape that was previously con-
        verted to a solid—not the torus). Next, move the shape up
        (positive Z) **6** units with *Move*.

    G. Move the torus up **4** units with *Move*.

    H. The solid primitives should appear as those in Figure 31-76. (*Hide*
        has been used for the figure.)

Figure 31-74

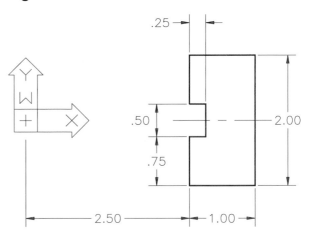

Figure 31-75

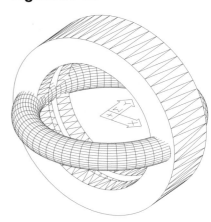

Figure 31-76

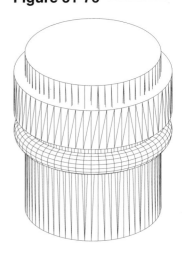

I. Use *Subtract* to subtract both revolved shapes from the cylinder. Use *Hide*. The solid should resemble that in Figure 31-77. (*Hide* has been used with *DISPSILH*=1 for this figure.) *Save* the drawing.

Figure 31-77 —————

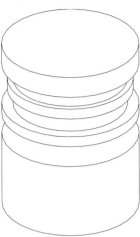

7. Begin a *New* drawing or use a *Template*. Assign the name **FAUCET**.

A. Draw 3 closed *Pline* shapes, as shown in Figure 31-78. Assume symmetry about the longitudinal axis. Use the WCS and create 2 new *UCS*s for the geometry. Use *3point Arcs* for the "front" profile.

Figure 31-78 —————————————————

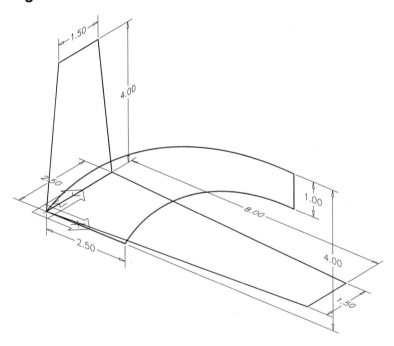

B. *Extrude* each of the 3 profiles into the same space. Make sure you specify the correct positive or negative *Height* value.

Figure 31-79 —————————————————

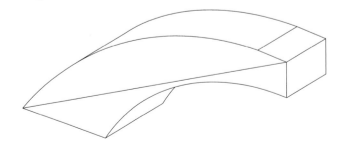

C. Finally, use *Intersect* to create the composite solid of the Faucet. *Save* the drawing.

D. (Optional) Create a nozzle extending down from the small end. Then create a channel for the water to flow (through the inside) and subtract it from the faucet.

8. Construct a solid model of the bar guide in Figure 31-80. Strive for the most efficient design. It is possible to construct this object with one *Extrude* and one *Subtract*. Save the model as **BGUID-SL**.

Figure 31-80

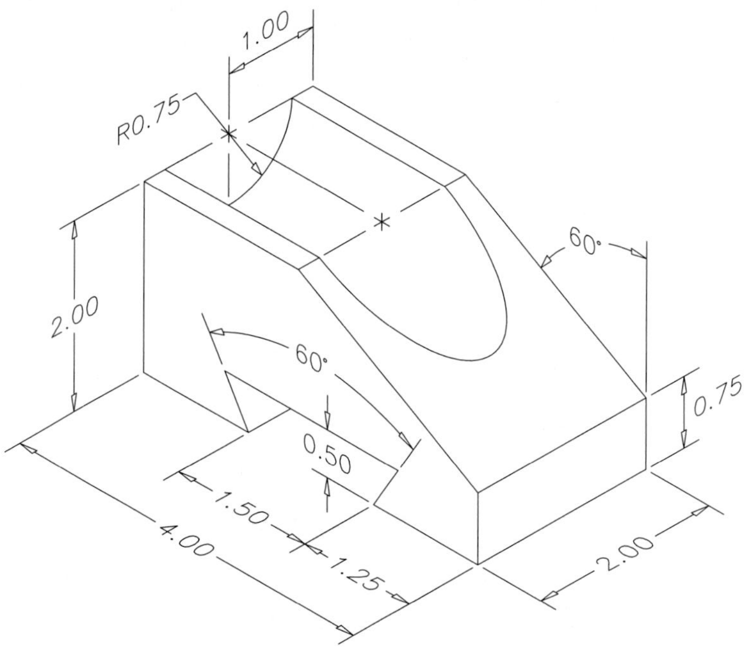

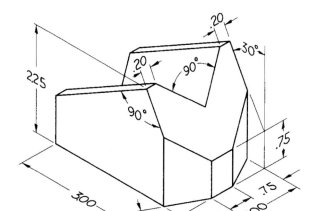

9. Make a solid model of the V-block shown in Figure 31-81. Several strategies could be used for construction of this object. Strive for the most efficient design. Plan your approach by sketching a few possibilities. Save the model as **VBLOK-SL**.

Figure 31-81

10. Construct a composite solid model of the support bracket using efficient techniques. Save the model as **SUPBK-SL**.

Figure 31-82

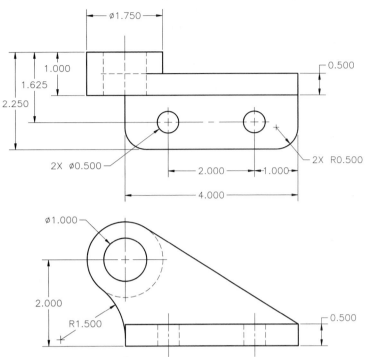

11. Construct the swivel shown in Figure 31-83. The center arm requires *Extruding* the 1.00 x 0.50 rectangular shape along an arc path through 45 degrees. Save the drawing as **SWIVEL**.

Figure 31-83

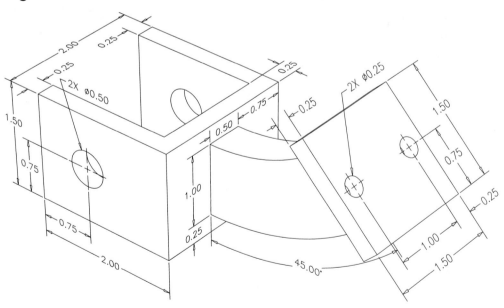

12. Construct a solid model of the angle brace shown in Figure 31-84. Use efficient design techniques. Save the drawing as **AGLBR-SL**.

Figure 31-84

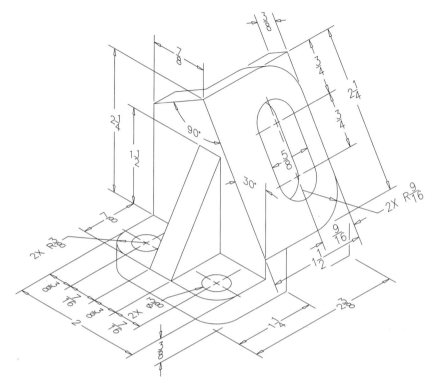

13. Construct a solid model of the saddle shown in Figure 31-85. An efficient design can be utilized by creating *Pline* profiles of the top "view" and the front "view," as shown in Figure 31-86. Use *Extrude* and *Intersect* to produce a composite solid. Additional Boolean operations are required to complete the part. The finished model should look like Figure 31-87 (with *Hide* performed). Save the drawing as **SADL-SL**.

Figure 31-85

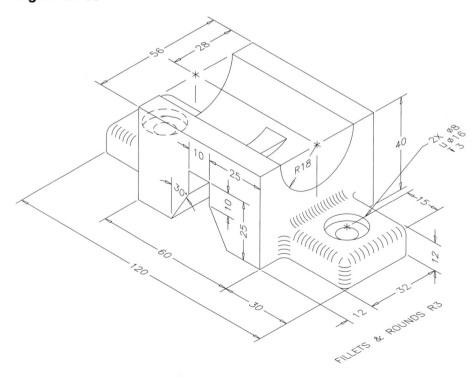

Figure 31-86

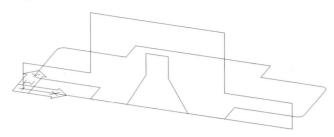

Figure 31-87

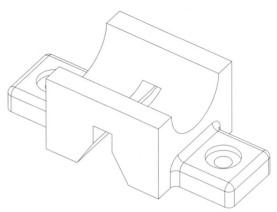

14. Construct a solid model of a bicycle handle bar. Create a centerline (Fig. 31-88) as a *Path* to extrude a *Circle* through. Three mutually perpendicular coordinate systems are required: the **WORLD**, the **SIDE**, and the **FRONT**. The centerline path consists of three separate *Plines*. First, create the 370 length *Pline* with 60 radii arcs on each end on the WCS. Then create the drop portion of the bars using the SIDE UCS. Create <u>three</u> *Circles* using the FRONT UCS, and extrude one along each *Path*.

Figure 31-88

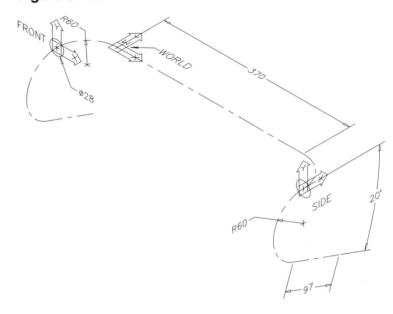

Plot the bar and *Hide Lines* as shown in Figure 31-89. *Save* the drawing as **DROPBAR**.

Figure 31-89

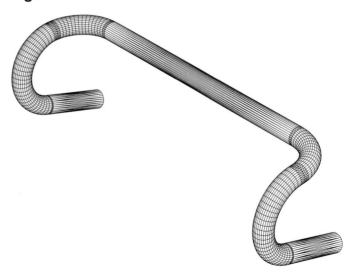

15. Create a solid model of the pulley. All vertical dimensions are diameters. Orientation of primitives is critical in the construction of this model. Try creating the circular shapes on the XY plane (circular axis aligns with Z axis of the WCS). After the construction, use *Rotate3D* to align the circular axis of the composite solid with the Y axis of the WCS. *Save* the drawing as **PULLY-SL**.

Figure 31-90

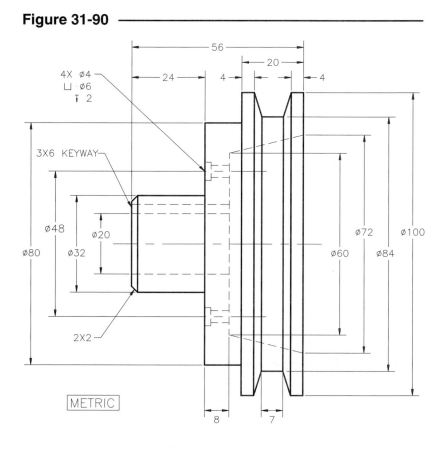

16. Create a composite solid model of the adjustable mount (Fig. 31-91). Use *Move* and other methods to move and align the primitives. Use of efficient design techniques is extremely important with a model of this complexity. Assign the name **ADJMT-SL**.

Figure 31-91

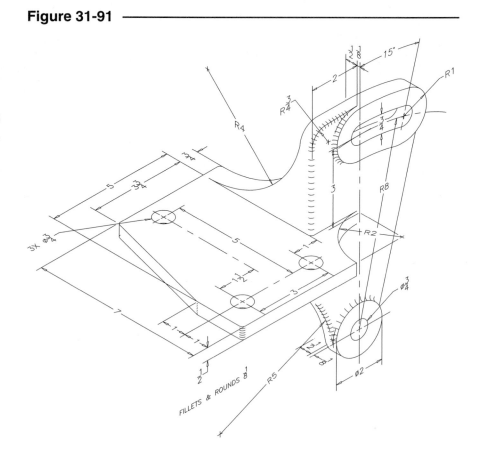

32

ADVANCED SOLIDS FEATURES

Chapter Objectives

After completing this chapter you should be able to:

1. use the *ISOLINES*, *DISPSILH*, and *FACETRES* variables to control the display of tessellation lines, silhouette lines, and mesh density for solid models;

2. calculate mass properties of a solid model using *Massprop*;

3. determine if *Interference* exists between two or more solids and create a solid equal in volume to the interference;

4. create a 2D section view for a solid model using *Section* and *Bhatch*;

5. use *Slice* to cut a solid model at any desired cutting plane and retain one or both halves;

6. convert AME (AutoCAD Release 12) solid models to AutoCAD Release 14 ACIS solid models with *AMECONVERT*.

CONCEPTS

Several topics related to solid modeling capabilities for AutoCAD Release 14 ACIS models are discussed in this chapter. The topics are categorized in the following sections:

> Solid Modeling Display Variables
> Analyzing Solid Models
> Creating Sections from Solids
> Converting Solids

SOLID MODELING DISPLAY VARIABLES

AutoCAD solid models are displayed in wireframe representation by default. Wireframe representation requires less computation time and less complex file structure, so your drawing time can be spent more efficiently. When you use *Hide* or *Shade*, the solid models are automatically meshed before they are displayed with hidden lines removed or as a shaded image. This meshed version of the model is apparent when you use *Hide* on cylindrical or curved surfaces.

Three variables control the display of solids for wireframe, hidden, and meshed representation. The *ISOLINES* variable controls the number of tessellation lines that are used to visually define cylindrical surfaces for wireframes. The *DISPSILH* variable can be toggled on or off to display silhouette lines for wireframe displays. *FACETRES* is the variable that controls the density of the mesh apparent with *Hide*. The variables are accessed by typing the names at the Command: prompt.

ISOLINES

This variable sets the number of tessellation lines that appear on a curved surface when shown in wireframe representation. The default setting for *ISOLINES* is 4 (Fig. 32-1). A solid of extrusion shows fewer tessellation lines (the current *ISOLINES* setting less 4) to speed regeneration time.

A higher setting gives better visualization of the curved surfaces but takes more computing time (Fig. 32-2). After changing the *ISOLINES* setting, *Regen* the drawing to see the new display.

Figure 32-1 —————————— **Figure 32-2** ——————————

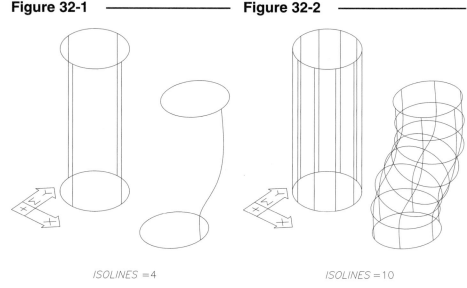

ISOLINES =4   ISOLINES =10

NOTE: In an isometric view attained by *3D Viewpoint*, the 4 lines (like Fig. 32-1) appear to overlap—they align when viewed from any perfect isometric angle.

DISPSILH

This variable can be turned on to display the limiting element contour lines, or silhouette, of curved shapes for a wireframe display (Fig. 32-3). The default setting is 0 (off). Since the silhouette lines are <u>viewpoint dependent</u>, significant computing time is taken to generate the display. You should <u>not</u> leave *DISPSILH* on 1 during construction.

Figure 32-3 ————

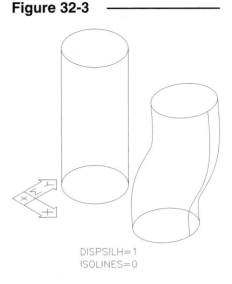

DISPSILH=1
ISOLINES=0

DISPSILH has a special function when used with *Hide*. When *DISPSILH* has the default setting of 0 and a *Hide* is performed, the solids appear opaque but display the mesh lines (Fig. 32-4). If *DISPSILH* is set to 1 before *Hide* is performed, the solids appear opaque but do <u>not</u> display the mesh lines (Fig. 32-5).

Figure 32-4 ———————— **Figure 32-5** ————————

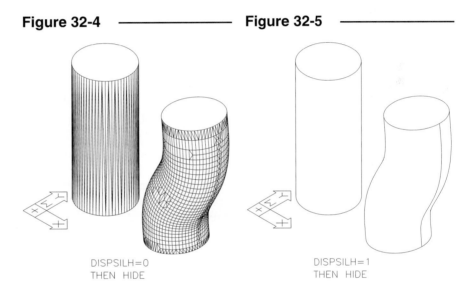

DISPSILH=0
THEN HIDE

DISPSILH=1
THEN HIDE

FACETRES

FACETRES controls the <u>density of the mesh</u> that is automatically created when a <u>Hide or a *Shade*</u> is performed. The default setting is .5, as shown in Figure 32-6. Decreasing the value produces a coarser mesh (Fig. 32-7), while increasing the value produces a finer mesh. The higher the value, the more computation time involved to generate the display or plot. The density of the mesh is actually a

Figure 32-6 ———————— **Figure 32-7** ————————

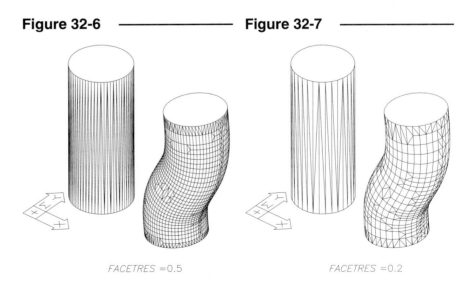

FACETRES =0.5

FACETRES =0.2

factor of both the *FACETRES* setting and the *VIEWRES* setting. Increasing *VIEWRES* also makes the mesh more dense. *FACETRES* can be set to any value between .01 and 10.

The *Preferences* dialog box can be used to change the settings for the *ISOLINES, DISPSILH,* and *FACET-RES* variables (Fig. 32-8). The upper-left corner of the *Performance* tab (*Solid model object display*) has two edit boxes and one checkbox that allow changing these variables as follows.

Figure 32-8 ——————————

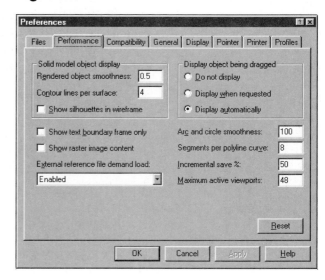

Rendered object smoothness	FACETRES
Contour lines per surface	ISOLINES
Show silhouettes in wireframe	DISPSILH

ANALYZING SOLID MODELS

Two commands in AutoCAD allow you to inquire about and analyze the solid geometry. *Massprop* calculates a variety of properties for the selected ACIS solid model. AutoCAD does the calculation and lists the information in screen or text window format. The data can be saved to a file for future exportation to a report document or analysis package. The *Interfere* command finds the interference of two or more solids and highlights the overlapping features so you can make necessary alterations.

MASSPROP

Pull-down Menu	COMMAND (TYPE)	ALIAS (TYPE)	Short-cut	Screen (side) Menu	Tablet Menu
Tools *Inquiry >* *Mass Properties*	*MASSPROP*	...	...	*TOOLS 1* *Massprop*	*U,7*

Since solid models define a complete description of the geometry, they are ideal for mass properties analysis. The *Massprop* command automatically computes a variety of mass properties.

Mass properties are useful for a variety of applications. The data generated by the *Massprop* command can be saved to an .MPR file for future exportation in order to develop bills of material, stress analysis, kinematics studies, and dynamics analysis.

Applying the *Massprop* command to a solid model produces a text screen displaying the following list of calculations (Fig. 32-9):

Figure 32-9 ——————————

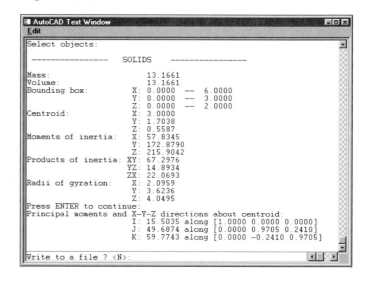

Mass	Mass is the quantity of matter that a solid contains. Mass is determined by density of the material and volume of the solid. Mass is not dependent on gravity and, therefore, different from but proportional to weight. Mass is also considered a measure of a solid's resistance to linear acceleration (overcoming inertia).
Volume	This value specifies the amount of space occupied by the solid.
Bounding Box	These lengths specify the extreme width, depth, and height of the selected solid.
Centroid	The centroid is the geometrical center of the solid. Assuming the solid is composed of material that is homogeneous (uniform density), the centroid is also considered the center of mass and center of gravity. Therefore, the solid can be balanced when supported only at this point.
Moments of Inertia	Moments convey how the mass is distributed around the X, Y, and Z axes of the current coordinate system. These values are a measure of a solid's resistance to <u>angular</u> acceleration (mass is a measure of a solid's resistance to <u>linear</u> acceleration). Moments of inertia are helpful for stress computations.
Products of Inertia	These values specify the solid's resistance to <u>angular</u> acceleration with respect to two axes at a time (XY, YZ, or ZX). Products of inertia are also useful for stress analysis.
Radii of Gyration	If the object were a concentrated solid mass without holes or other features, the radii of gyration represent these theoretical dimensions (radius about each axis) such that the same moments of inertia would be computed.
Principal Moments and X, Y, Z Directions	In structural mechanics, it is sometimes important to determine the orientation of the axes about which the moments of inertia are at a maximum. When the moments of inertia about centroidal axes reach a maximum, the products of inertia become zero. These particular axes are called the principal axes, and the corresponding moments of inertia with respect to these axes are the principal moments (about the centroid).

Notice the "Mass:" is reported as having the same value as "Volume." This is because AutoCAD Release 14 solids cannot have material characteristics assigned to them, so AutoCAD assumes a density value of 1. To calculate the mass of a selected solid in AutoCAD, use a reference guide (such as a machinist's handbook) to find the material density and multiply the value times the reported volume (mass = volume x density).

INTERFERE

Pull-down Menu	COMMAND (TYPE)	ALIAS (TYPE)	Short-cut	Screen (side) Menu	Tablet Menu
Draw *Solids >* *Interference*	*INTERFERE*	*INF*	...	DRAW 2 SOLIDS *Interfer*	...

In AutoCAD, unlike real life, it is possible to create two solids that occupy the same physical space. *Interfere* checks solids to determine whether or not they interfere (occupy the same space). If there is interference, *Interfere* reports the overlap and allows you to create a new solid from the interfering volume, if you desire. Normally, you specify two sets of solids for AutoCAD to check against each other:

```
Command: Interfere
Select the first set of solids...
Select objects: PICK
Select objects: Enter
1 solid selected.
Select the second set of solids...
Select objects: PICK
Select objects: Enter
1 solid selected.
Comparing 1 solid against 1 solid.
Interfering solids (first  set): 1
                (second set): 1
Interfering pairs:           1
Create interference solids? <N>: y
Command:
```

If you answer "yes" to the last prompt, a new solid is created equal to the exact size and volume of the interference. The original solids are not changed in any way. If no interference is found, AutoCAD reports "Solids do not interfere."

For example, consider the two solids shown in Figure 32-10. (The parts are displayed in wireframe representation.) The two shapes fit together as an assembly. The locating pin on the part on the right should fit in the hole in the left part.

Figure 32-10 —————————————

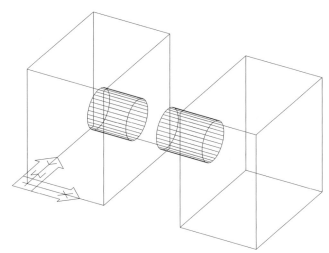

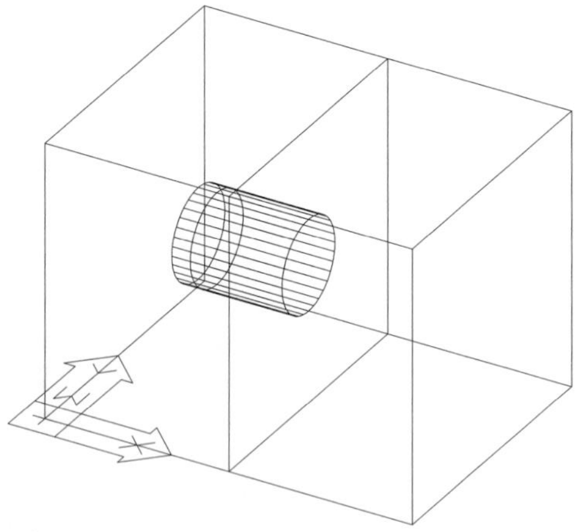

Sliding the parts together until the two vertical faces meet produces the assembly shown in Figure 32-11. There appears to be some inconsistency in the assembly of the hole and the pin. Either the pin extends beyond the hole (interference) or the hole is deeper than necessary (no interference). Using *Interfere*, you can find an overlap and create a solid is created by answering "yes" to "Create interference solids?" Use *Move* with the *Last* selection option to view and analyze the solid of interference.

Figure 32-11 —————————————

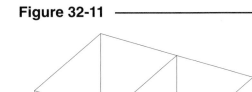

You can compare <u>more than two</u> solids against each other with *Interfere*. This is accomplished by selecting <u>all desired solids at the first prompt</u> and none at the second:

```
Select the first set of solids... PICK
Select the second set of solids... Enter
```

AutoCAD then compares all solids in the first set against each other. If more than one interference is found, AutoCAD highlights intersecting solids, one pair at a time.

CREATING SECTIONS FROM SOLIDS

Two AutoCAD commands are intended to create sections from solid models. *Section* is a drafting feature that creates a 2D "section view." The cross-section is determined by specifying a cutting plane. A cross-section view is automatically created based on the solid geometry that intersects the cutting plane. The original solid is not affected by the action of *Section*. *Slice* actually cuts the solid at the specified cutting plane. *Slice* therefore creates two solids from the original one and offers the possibility to retain both halves or only one. Many options are available for placement of the cutting plane.

SECTION

Pull-down Menu	COMMAND (TYPE)	ALIAS (TYPE)	Short-cut	Screen (side) Menu	Tablet Menu
Draw Solids > Section	*SECTION*	*SEC*	...	DRAW 2 SOLIDS *Section*	...

Section creates a 2D cross-section of a solid or set of solids. The cross-section created by *Solsect* is considered a traditional 2D section view. The cross-section is defined by a cutting plane, and the resulting section is determined by any solid material that passes through the cutting plane. The cutting plane can be specified by a variety of methods. The options for establishing the cutting plane are listed in the command prompt:

```
Command: section
Select objects: PICK
Select objects: Enter
Sectioning plane by Entity/Last/Zaxis/View/XY/YZ/ZX /<3points>:
```

For example, assume a cross-section is desired for the geometry shown in Figure 32-12. To create the section, you must define the cutting plane.

Figure 32-12 ─────────────────────

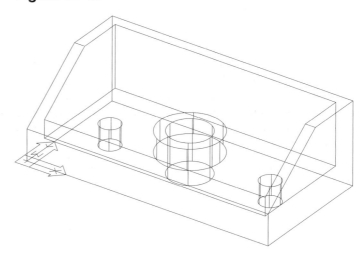

For this case, the *ZX* option is used and the requested point is defined to establish the position of the plane, as shown in Figure 32-13. The cross-section will be created on this plane.

Figure 32-13 ────────────

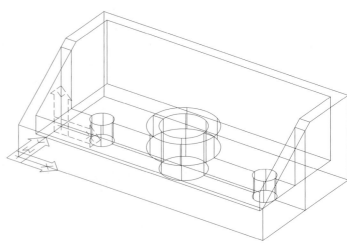

Once the cutting plane is established, a cross-section is automatically created by *Section*. The resulting geometry is a *Region* created on the <u>current</u> layer.

The resulting *Region* can be moved if needed as shown in Figure 32-14. If you want to use the *Region* to create a 2D section complete with hatch lines, several steps are required. First, the *Region* must be *Exploded* into individual *Regions* (4, in this case). Then each *Region* must be *Exploded* again to break the shapes into *Lines* that *Bhatch* can interpret as boundaries. You may need to create a *UCS* on the plane of the shapes before using *Bhatch*. Other *Lines* may be needed to make a complete section view.

Figure 32-14 ────────────

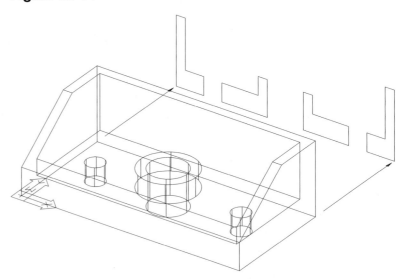

SLICE

Pull-down Menu	COMMAND (TYPE)	ALIAS (TYPE)	Short-cut	Screen (side) Menu	Tablet Menu
Draw *Solids >* *Slice*	*SLICE*	SL	...	DRAW 2 SOLIDS *Slice*	...

Slice creates a true solid section. *Slice* cuts an ACIS solid or set of solids on a specified cutting plane. The original solid is converted to two solids. You have the option to keep both halves or only the half that you specify. Examine the following command syntax:

    Command: slice
    Select objects: PICK
    Select objects: Enter
    Slicing plane by Object/Last/Zaxis/View/XY/YZ/ZX/<3points>: (option)
    Point on XY plane <0,0,0>: PICK
    Both sides/<Point on desired side of the plane>: PICK
    Command:

Entering *B* at the "Both sides/<Point on desired side of plane>:" prompt retains the solids on both sides of the cutting plane. Otherwise, you can pick a point on either side of the plane to specify which half to keep.

Figure 32-15 ────────────

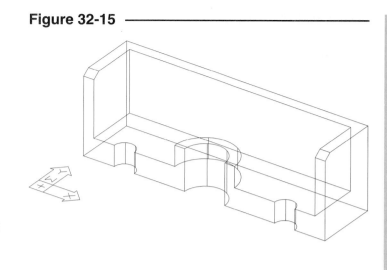

For example, using the solid model shown previously in Figure 32-12, you can use *Slice* to create a new sectioned solid shown here. The *ZX* method is used to define the cutting plane midway through the solid. Next, the new solid to retain was specified by PICKing a point on that geometry. The resulting sectioned solid is shown in Figure 32-15. Note that the solid on the near side of the cutting plane was not retained.

CONVERTING SOLIDS

AutoCAD Release 14 provides several utilities for converting solids to and from other file formats. Older AutoCAD solid models created with AME (Advanced Modeling Extension) can be converted to a Release 14 ACIS model with some success.

AMECONVERT

Pull-down Menu	COMMAND (TYPE)	ALIAS (TYPE)	Short-cut	Screen (side) Menu	Tablet Menu
...	*AMECONVERT*	...	...	...	...

Ameconvert is a conversion utility to convert older solid models (created with AME Release 2 or 2.1) to AutoCAD Release 14 ACIS solid models. The command is simple to use; however, the conversion may change the model slightly. If the selected solids to convert are not AME Release 2 or 2.1, AutoCAD ignores the request.

The newer Release 14 solid modeler, ACIS, creates solids of higher accuracy than the older AME modeler. Because of this new accuracy, objects that are converted may change in appearance and form. For example, two shapes that were originally considered sufficiently close and combined by Boolean operations in the old modeler may be interpreted as slightly offset in the ACIS modeler. This can occur for converted features such as fillets, chamfers, and through holes. Occasionally, a solid created in the older AME model is converted into two or more solids in Release 14 ACIS modeler.

Solids that are converted to Release 14 ACIS solids can be edited using Release 14 solids commands. Converted solids that do not convert as expected can usually be edited in Release 14 to produce an equivalent, but more accurate, solid.

CHAPTER EXERCISES

1. *Open* the **SADL-SL** drawing that you created in Chapter 31 Exercises. Calculate *Mass Properties* for the saddle. Write the report out to a file named **SADL-SL.MPR**. Use a text editor to examine the file.

2. *Open* the **SADL-SL** drawing again. Use *Slice* to cut the model in half longitudinally. Use an appropriate method to establish the "slicing plane" in order to achieve the resulting model, as shown in Figure 32-16. Use *SaveAs* and assign the name **SADL-CUT**.

Figure 32-16 ——————————————

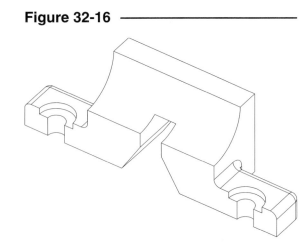

3. *Open* the **PULLY-SL** drawing that you created in Chapter 31 Exercises.

   Make a *New Layer* named **SECTION** and set it *Current*. Then use the *Section* command to create a full section "view" of the pulley. Establish a vertical cutting plane through the center of the model. Remove the section view object (*Region*) with the *Move* command, translating **100** units in the **X** direction. The model and the new section view should appear as in Figure 32-17 (*Hide* was performed on the pulley to enhance visualization). Complete the view by establishing a *UCS* at the section view, then adding the *Bhatch*, as shown in Figure 32-17. Finally, create the necessary *Lines* to complete the view. *SaveAs* **PULLY-SC**.

Figure 32-17 ——————————————

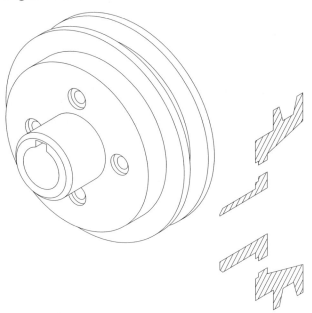

4. *Open* the **SADL-SL** drawing that you created in the Chapter 31 Exercises. Change the *ISOLINES* setting to display **10** tessellation lines. Use the *Hide* command to create a meshed hidden display. Change the *FACETRES* setting to display a coarser mesh and use *Hide* again. Make a plot of the model with the coarse mesh and with hidden lines (check the *Hide Lines* box in the *Plot Configuration* dialog box).

   Next, change the *FACETRES* setting to display a fine mesh. Use *Hide* to reveal the change. Make a plot of the model with *Hide Lines* checked to display the fine mesh. Then set the *DISPSILH* variable to **1** and make another plot with lines hidden. What is the difference in the last two plots? *Save* the **SADL-SL** file with the new settings.

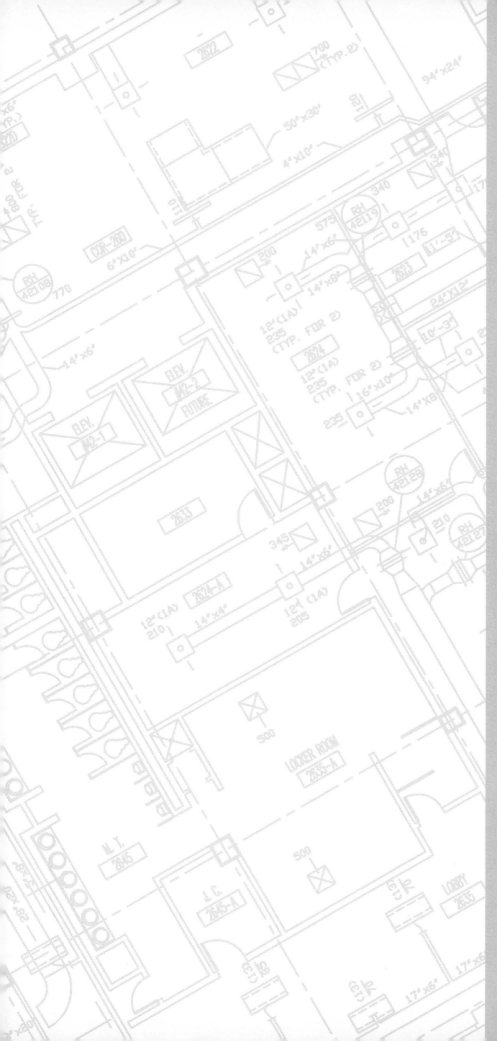

A, B, C, D

APPENDICES

Contents

APPENDIX A

System Variables

Variable	Characteristic	Description
ACADPREFIX	(Read-only) Type: String Not saved	Stores the directory path, if any, specified by the ACAD environment variable, with path separators appended if necessary.
ACADVER	(Read-only) Type: String Not saved	Stores the AutoCAD version number, which can have values like 14 or 14a. This variable differs from the DXF file $ACADVER header variable, which contains the drawing database level number.
ACISOUTVER	Type: Integer Saved in: Drawing Initial value: 16	Controls the ACIS version of SAT files created using the *ACISOUT* command. Currently, *ACISOUT* supports only a value of 16 (for ACIS version 1.6). Future releases of AutoCAD will support additional ACIS versions.
AFLAGS	Type: Integer Not saved Initial value: 0	Sets attribute flags for *ATTDEF* bit-code. It is the sum of the following: 0 No attribute mode selected 1 Invisible 2 Constant 4 Verify 8 Preset
ANGBASE	Type: Real Saved in: Drawing Initial value: 0.0000	Sets the base angle to 0 with respect to the current UCS.
ANGDIR	Type: Integer Saved in: Drawing Initial value: 0	Sets the positive angle direction from angle 0 with respect to the current UCS. 0 Counter-clockwise 1 Clockwise
APBOX	Type: Integer Saved in: Registry Initial value: 1	Turns the AutoSnap aperture box on or off. The aperture box is displayed in the center of the cursor when you snap to an object. This option is available only when Marker, Magnet, or Snaptip is selected. 0 Aperture box is not displayed 1 Aperture box is displayed
APERTURE	Type: Integer Saved in: Registry Initial value: 10	Sets *APERTURE* command object snap target height, in pixels.
AREA	(Read-only) Type: Real Not saved	Stores the last area computed by *AREA, LIST,* or *DBLIST.*
ATTDIA	Type: Integer Saved in: Drawing Initial value: 0	Controls whether *INSERT* uses a dialog box for attribute value entry. 0 Issues prompts on the command line 1 Uses a dialog box

R14

R14

Variable	Characteristic	Description
ATTMODE	Type: Integer Saved in: Drawing Initial value: 1	Controls display of attributes. 0 Off makes all attributes invisible 1 Normal retains current visibility of each attribute: visible attributes are displayed; invisible attributes are not 2 On makes all attributes visible
ATTREQ	Type: Integer Saved in: Drawing Initial value: 1	Determines whether *INSERT* uses default attribute settings during insertion of blocks. 0 Assumes the defaults for the values of all attributes 1 Turns on prompts or dialog box for attribute values, as specified at *ATTDIA*
AUDITCTL	Type: Integer Saved in: Registry Initial value: 0	Controls whether *AUDIT* creates an ADT file (audit report). 0 Prevents writing of ADT files 1 Writes ADT files
AUNITS	Type: Integer Saved in: Drawing Initial value: 0	Sets units for angles. 0 Decimal degrees 1 Degrees/minutes/seconds 2 Gradians 3 Radians 4 Surveyor's units
AUPREC	Type: Integer Saved in: Drawing Initial value: 0	Sets number of decimal places for angular units displayed on the Status line or when listing an object.
AUTOSNAP	Type: Integer Saved in: Registry Initial value: 7	Controls the display of the AutoSnap marker and Snaptip and turns the AutoSnap magnet on or off. It is the sum of the following bit values: 0 Turns off the marker, Snaptip, and magnet 1 Turns on the marker 2 Turns on the Snaptip 4 Turns on the magnet
BACKZ	(Read-only) Type: Real Saved in: Drawing	Stores the back clipping plane offset from the target plane for the current viewport, in drawing units. Meaningful only if the back clipping bit in VIEWMODE is on. The distance of the back clipping plane from the camera point can be found by subtracting *BACKZ* from the camera-to-target distance.
BLIPMODE	Type: Integer Saved in: Drawing Initial value: 0	Controls whether marker blips are visible. 0 Turns off marker blips 1 Turns on marker blips
CDATE	(Read-only) Type: Real Not saved	Sets calendar date and time.

R14

Variable	Characteristic	Description
CECOLOR	Type: String Saved in: Drawing Initial value: "BY-LAYER"	Sets the color of new objects.
CELTSCALE	Type: Real Saved in: Drawing Initial value: 1.0000	Sets the current object linetype scaling factor. This sets the linetype scaling for new objects relative to the *LTSCALE* setting. A line created with *CELTSCALE*=2 in a drawing with *LTSCALE* set to 0.5 would appear the same as a line created with *CELTSCALE*=1 in a drawing with *LTSCALE*=1.
CELTYPE	Type: String Saved in: Drawing Initial value: "BY-LAYER"	Sets the linetype of new objects.
CHAMFERA	Type: Real Saved in: Drawing Initial value: 0.5000	Sets the first chamfer distance.
CHAMFERB	Type: Real Saved in: Drawing Initial value: 0.5000	Sets the second chamfer distance.
CHAMFERC	Type: Real Saved in: Drawing Initial value: 1.0000	Sets the first chamfer length.
CHAMFERD	Type: Real Saved in: Drawing Initial value: 0.0000	Sets the first chamfer angle.
CHAMMODE	Type: Integer Not saved Initial value: 0	Sets the input method by which AutoCAD creates chamfers. 0 Requires two chamfer distances 1 Requires one chamfer distance and an angle
CIRCLERAD	Type: Real Not saved Initial value: 0.0000	Sets the default circle radius. A zero sets no default.
CLAYER	Type: String Saved in: Drawing Initial value: 0	Sets the current layer.
CMDACTIVE	(Read-only) Type: Integer Not saved	Stores the bit code that indicates whether an ordinary command, transparent command, script, or dialog box is active. It is the sum of the following: 1 Ordinary command is active 2 Ordinary command and a transparent command are active 4 Script is active 8 Dialog box is active 16 AutoLISP or *DDE* command is active

R13

R13

Variable	Characteristic	Description
CMDDIA	Type: Integer Saved in: Registry Initial value: 1	Controls whether dialog boxes are turned on for *PLOT* and external database commands. 0 Turns off dialog boxes 1 Turns on dialog boxes
CMDECHO	Type: Integer Not saved Initial value: 1	Controls whether AutoCAD echoes prompts and input during the AutoLISP **(command)** function. 0 Turns off echoing 1 turns on echoing
CMDNAMES	(Read-only) Type: String Not saved	Displays the name of the currently active command and transparent command. For example, *LINE'ZOOM* indicates that the *ZOOM* command is being used transparently during the *LINE* command. This variable is designed for use with programming interfaces such as AutoLISP, DIESEL, and ActiveX Automation.
CMLJUST	Type: Integer Saved in: Drawing Initial value: 0	Specifies multiline justification. 0 Top 1 Middle 2 Bottom
CMLSCALE	Type: Real Saved in: Drawing Initial value: 1.0000	Controls the overall width of a multiline. A scale factor of 2.0 produces a multiline that is twice as wide as the style definition. A zero scale factor collapses the multiline into a single line. A negative scale factor flips the order of the offset lines (that is, the smallest or most negative is placed on top when the multiline is drawn from left to right).
CMLSTYLE	Type: String Saved in: Drawing Initial value: "STANDARD"	Sets the multiline style that AutoCAD uses to draw the multiline.
COORDS	Type: Integer Saved in: Drawing Initial value: 1	Controls when coordinates are updated on the Status line. 0 Coordinate display is updated as you specify points with the pointing device 1 Display of absolute coordinates is continuously updated 2 Display of absolute coordinates is continuously updated, and distance and angle from last point are displayed when a distance or angle is requested.
CURSORSIZE	Type: Integer Saved in: Registry Initial value: 5	Determines the size of the cursor as a percentage of the screen size. Valid settings range from 1 to 100 percent. When set to 100, the cursor are full-screen and the ends of the cursor are never visible. When less than 100, the ends of the cursor may be visible when the cursor is moved to one edge of the screen.

R13

R14

Variable	Characteristic	Description
CVPORT	Type: Integer Saved in: Drawing Initial value: 2	Sets the identification number of the current viewport. You can change this value, thereby changing the current viewport, if the following conditions are met: • The identification number you specify is that of an active viewport. • A command in progress has not locked cursor movement to that viewport. • Tablet mode is off.
DATE	(Read-only) Type: Real Not saved	Stores the current date and time represented as a Julian date and fraction in a real number: <Julian date>.<Fraction> For example, on January 29, 1993, at 2:29:35 in the afternoon, the *DATE* variable would contain 2446460.603877364. Your computer clock provides the date and time. The time is represented as a fraction of a day. To compute differences in time, subtract the times returned by *DATE*. To extract the seconds since midnight from the value returned by *DATE*, use AutoLISP expressions: **(setq s (getvar "DATE"))** **(setq seconds (* 86400.0 – s (fix s)))** The *DATE* system variable returns a true Julian date only if the system clock is set to UTC/Zulu (Greenwich Mean Time). *TDCREATE* and *TDUPDATE* have the same format as *DATE*, but their values represent the creation time and last update time of the current drawing.
DBMOD	(Read-only) Type: Integer Not saved	Indicates the drawing modification status using bit code. It is the sum of the following: 1 Object database modified 4 Database variable modified 8 Window modified 16 View modified
DCTCUST	Type: String Saved in: Registry Initial value: ""	Displays the path and file name of the current custom spelling dictionary.
DCTMAIN	Type: String Saved in: Registry Initial value: *varies by country*	Displays the file name of the current main spelling dictionary. The full path is not shown, because this file is expected to reside in the *support* directory. You can specify a default main spelling dictionary using *SETVAR*. When prompted for a new value for *DCTMAIN*, you can enter one of the keywords below. *Keyword Language Name* enu American English ena Australian English

Variable	Characteristic	Description
DCTMAIN (continued)		
		ens British English (ise)
		enz British English (ize)
		ca Catalan
		cs Czech
		da Danish
		nl Dutch (primary)
		nls Dutch (secondary)
		fi Finnish
		fr French (unaccented capitals)
		fra French (accented capitals)
		de German (Scharfes s)
		ded German (Dopple s)
		it Italian
		no Norwegian (Bokmal)
		non Norwegian (Nynorsk)
		pt Portuguese (Iberian)
		ptb Portuguese (Brazilian)
		ru Russian (infrequent io)
		rui Russian (frequent io)
		es Spanish (unaccented capitals)
		esa Spanish (accented capitals)
		sv Swedish
DELOBJ	Type: Integer Saved in: Drawing Initial value: 1	Controls whether objects used to create other objects are retained or deleted from the drawing database. 0 Objects are retained 1 Objects are deleted
DEMANDLOAD	Type: Integer Saved in: Registry Initial value: 3	Specifies if and when AutoCAD demand loads a third-party application if a drawing contains custom objects created in that application. 0 Turns off demand loading. 1 Demand loads the source application when you open a drawing that contains custom objects. This setting does not demand load the application when you invoke one of the application's commands. 2 Demand loads the source application when you invoke one of the application's commands. This setting does not demand load the application when you open a drawing that contains custom objects. 3 Demand loads the source application when you open a drawing that contains custom objects or when you invoke one of the application's commands.
DIASTAT	(Read-only) Type: Integer Not saved	Stores the exit method of the most recently used dialog box. 0 Cancel 1 OK

R13

R14

Variable	Characteristic	Description
DIMADEC	Type: Integer Saved in: Drawing Initial value: -1	Controls the number of places of precision displayed for angular dimension text. The default value (-1) uses the *DIMDEC* system variable setting to generate places of precision. When *DIMADEC* is set to a value from 0 to 8, angular dimensions display precisions that differ from their linear dimension values. -1 Angular dimension is drawn using the number of decimal places corresponding to the *DIMDEC* setting 0-8 Angular dimension is drawn using the number of decimal places corresponding to the *DIMADEC* setting
DIMALT	Type: Switch Saved in: Drawing Initial value: Off	Controls use of alternate units in dimensions. See also *DIMALTD, DIMALTF, DIMALTZ, DIMALTTZ, DIMALTTD,* and *DIMAPOST.* Off Disables alternate units On Enables alternate units
DIMALTD	Type: Integer Saved in: Drawing Initial value: 2	Controls the number of decimal places in alternate units.
DIMALTF	Type: Real Saved in: Drawing Initial value: 25.4000	Controls scale factor in alternate units. If *DIMALT* is turned on, *DIMALTF* multiplies linear dimensions by a factor to produce a value in an alternate system of measurement. The initial value represents the number of millimeters in an inch.
DIMALTTD	Type: Integer Saved in: Drawing Initial value: 2	Sets the number of decimal places for the tolerance values in the alternate units of a dimension. *DIMALTTD* sets this value when you enter it on the command line or set it under Alternate Units in the *DDIM* Annotation dialog box.
DIMALTTZ	Type: Integer Saved in: Drawing Initial value: 0	Toggles suppression of zeros in tolerance values. 0 Suppresses zero feet and precisely zero inches. 1 Includes zero feet and precisely zero inches. 2 Includes zero feet and suppresses zero inches. 3 Includes zero inches and suppresses zero feet. To the preceding values, add: 4 Suppresses leading zeros. 8 Suppresses trailing zeros. *DIMALTTZ* sets this value when you enter it on the command line or set it under Alternate Units in the Annotation dialog box.
DIMALTU	Type: Integer Saved in: Drawing Initial value: 2	Sets the units format for alternate units of all dimension style family members except angular. 1 Scientific 2 Decimal

R14

R13

Variable	Characteristic	Description
DIMALTU (continued)		3 Engineering 4 Architectural (stacked) 5 Fractional (stacked) 6 Architectural (stacked) 7 Fractional 8 Windows Desktop (decimal format using Control Panel settings for decimal separator and number group systems) *DIMALTU* sets this value when you enter it on the command line or set it under Alternate Units in the Annotation dialog box.
DIMALTZ	Type: Integer Saved in: Drawing Initial value: 0	Controls the suppression of zeros for alternate unit dimension values. *DIMALTZ* stores this value when you enter it on the command line or set it under Alternate Units in the Annotation dialog box. *DIMALTZ* values 0-3 affect feet-and-inch dimensions only. 0 Suppresses zero feet and precisely zero inches 1 Includes zero feet and precisely zero inches 2 Includes zero feet and suppresses zero inches 3 Includes zero inches and suppresses zero feet 4 Suppresses leading zeroes in decimal dimensions (for example, 0.5000 becomes .5000) 8 Suppresses trailing zeroes in decimal dimensions (for example, 12.5000 becomes 12.5) 12 Suppresses both leading and trailing zeroes (for example, 0.5000 becomes .5)
DIMAPOST	Type: String Saved in: Drawing Initial value: ""	Specifies a text prefix or suffix (or both) to the alternate dimension measurement for all types of dimensions except angular. For instance, if the current units are Architectural, *DIMALT* is on, *DIMALTF* is 25.4 (the number of millimeters per inch), *DIMALTD* is 2, and *DIMAPOST* is set to "mm," a distance of 10 units would be displayed as 10"[254.00mm]. To turn off an established prefix or suffix (or both), set it to a single period (.).
DIMASO	Type: Switch Saved in: Drawing Initial value: On	Controls the associative of dimension objects. Off Creates no association between the various elements of the dimension. The lines, arcs, arrowheads, and text of a dimension are drawn as separate objects. On Creates an association between the elements of the dimension. The elements are formed into a single object. If the definition point on the object moves, the dimension value is updated. The *DIMASO* value is not stored in a dimension style.

Variable	Characteristic	Description
DIMASZ	Type: Real Saved in: Drawing Initial value: 0.1800	Controls the size of dimension line and leader line arrowheads. Also controls the size of hook lines. Multiples of the arrowhead size determine whether dimension lines and text should fit between the extension lines. *DIMASZ* is also used to scale arrowhead blocks if set by *DIMBLK*. *DIMASZ* has no effect when *DIMTSZ* is other than zero.
DIMAUNIT	Type: Integer Saved in: drawing Initial value: 0	Sets the angle format for angular dimension. 0 Decimal degrees 1 Degrees/minutes/seconds 2 Gradians 3 Radians 4 Surveyor's units *DIMAUNIT* sets this value when entered on the command line or set from the Primary Units area of the Annotation dialog box.
DIMBLK	Type: String Saved in: Drawing Initial value: ""	Sets the name of a block to be drawn instead of the normal arrowhead at the ends of dimension lines or leader lines. To turn off an established block name, set it to a single period (.).
DIMBLK1	Type: String Saved in: Drawing Initial value: ""	If *DIMSAH* is on, specifies a user-defined arrowhead block for the first end of the dimension line. This variable contains the name of a previously defined block. To turn off an established block name, set it to a single period (.).
DIMBLK2	Type: String Saved in: Drawing Initial value: ""	If *DIMSAH* is on, specifies a user-defined arrowhead block for the second end of the dimension line. This variable contains the name of a previously defined block. To turn off an established block name, set it to a single period (.).
DIMCEN	Type: Real Saved in: Drawing Initial value: 0.0900	Controls drawing of circle or arc center marks and centerlines by *DIMCENTER, DIMDIAMETER,* and *DIMRADIUS.* 0 No center marks or lines are drawn <0 Centerlines are drawn >0 Center marks are drawn The absolute value specifies the size of the mark portion of the centerline. *DIMDIAMETER* and *DIMRADIUS* draw the center mark or line only if the dimension line is placed outside the circle or arc.

Variable	Characteristic	Description
DIMCLRD	Type: Integer Saved in: Drawing Initial value: 0	Assigns colors to dimension lines, arrowheads, and dimension leader lines. Also controls the color of leader lines created with *LEADER*. The color can be any valid color number. Integer equivalents for *BYBLOCK* and *BYLAYER* are 0 and 256, respectively.
DIMCLRE	Type: Integer Saved in: Drawing Initial value: 0	Assigns colors to extension lines of dimensions. The color can be any valid color number. See *DIMCLRD*.
DIMCLRT	Type: Integer Saved in: Drawing Initial value: 0	Assigns colors to dimension text. The color can be any valid color. See *DIMCLRD*.
DIMDEC	Type: Integer Saved in: Drawing Initial value: 4	Sets the number of decimal places displayed for the primary units of a dimension. When the dimension family member is Parent, the precision is based on the units or angle format you have selected. *DIMDEC* stores this value when you enter it on the command line or set it under Primary Units in the Annotation dialog box.
DIMDLE	Type: Real Saved in: Drawing Initial value: 0.0000	Sets the distance the dimension line extends beyond the extension line when oblique strokes are drawn instead of arrowheads.
DIMDLI	Type: Real Saved in: Drawing Initial value: 0.3800	Controls the spacing of the dimension lines in baseline dimensions. Each dimension line is offset from the previous line by this amount, if necessary, to avoid drawing over it. Changes made with *DIMDLI* are not applied to existing dimensions.
DIMEXE	Type: Real Saved in: Drawing Initial value: 0.1800	Specifies how far to extend the extension line beyond the dimension line.
DIMEXO	Type: Real Saved in: Drawing Initial value: 0.0625	Specifies how far extension lines are offset from origin points. If you point directly at the corners of an object to be dimensioned, the extension lines do not touch the object.
DIMFIT	Type: Integer Saved in: Drawing Initial value: 3	Controls the placement of text and arrowheads inside or outside extension lines based on the available space between the extension lines. When space is available, AutoCAD always places text and arrowheads between the extension lines. Otherwise, the *DIMFIT* setting affects the placement of text and arrowheads as follows: 0 Text and arrows 1 Text only 2 Arrows only 3 Best fit

R13

Variable	Characteristic	Description
DIMFIT (continued)		4 Leader 5 No leader *DIMFIT* sets this value when you enter it on the command line or set it under Fit in the Format dialog box.
DIMGAP	Type: Real Saved in: Drawing Initial value: 0.0900	Sets the distance around the dimension text when the dimension line breaks to accommodate dimension text. Also sets the gap between annotation and a hook line created with *LEADER*. A negative *DIMGAP* value creates basic dimensioning—dimension text with a box around it. AutoCAD also uses *DIMGAP* as the minimum length for pieces of the dimension line. When calculating the default position for the dimension text, it positions the text inside the extension lines only if doing so breaks the dimension lines into two segments at least as long as *DIMGAP*. Text placed above or below the dimension line is moved inside only if there is room for the arrowheads, dimension text, and a margin between them at least as large as *DIMGAP*: 2 * (*DIMASZ* + *DIMGAP*). *DIMGAP* also sets the gap between a tolerance symbol and its feature control frame.
DIMJUST	Type: Integer Saved in: Drawing Initial value: 0	Controls the horizontal position of dimension text. 0 Centered along the dimension line between the extension lines 1 Next to the first extension line 2 Next to the second extension line 3 Above and aligned with the first extension line 4 Above and aligned with the second extension line
DIMLFAC	Type: Real Saved in: Drawing Initial value: 1.0000	Sets a global scale factor for linear dimensioning measurements. All linear distances measured by dimensioning (including radii, diameters, and coordinates) are multiplied by the *DIMLFAC* setting before being converted to dimension text. *DIMLFAC* has no effect on angular dimensions, and it is not applied to the values held in *DIMTM*, *DIMTP*, or *DIMRND*. If you are creating a dimension in paper space and *DIMLFAC* is not zero, AutoCAD multiplies the distance measured by the absolute value of *DIMLFAC*. In model space, negative values for *DIMLFAC* are ignored, and the value 1.0 is used instead.

R13

Variable	Characteristic	Description
DIMLFAC (continued)		AutoCAD computes a value for *DIMLFAC* if you try to change *DIMLFAC* from the Dim prompt while in paper space and you select the *Viewport* option. Dim: **dimlfac** Current value <1.0000> New value (Viewport): **v** Select viewport to set scale: AutoCAD calculates the scaling of model space to paper space and assigns the negative of this value to *DIMLFAC*.
DIMLIM	Type: Switch Saved in: Drawing Initial value: Off	Generates dimension limits as the default text. Setting *DIMLIM* on forces *DIMTOL* to be off. Off Dimension limits are not generated as default text On Dimension limits are generated as default text
DIMPOST	Type: String Saved in: Drawing Initial value: ""	Specifies a text prefix or suffix (or both) to the dimension measurement. For example, to establish a suffix for millimeters, set *DIMPOST* to mm; a distance of 19.2 units would be displayed as 19.2mm. If tolerances are turned on, the suffix is applied to the tolerances as well as to the main dimension. Use <> to indicate placement of the text in relation to the dimension value. For example, enter **<>mm** to display a 5.0 millimeter radial dimension as "5.0mm." If you entered **mm<>**, the dimension would be displayed as "mm5.0." Use the <> mechanism for angular dimensions.
DIMRND	Type: Real Saved in: Drawing Initial value: 0.0000	Rounds all dimensioning distances to the specified value. For instance, if *DIMRND* is set to 0.25, all distances round to the nearest 0.25 unit. If you set *DIMRND* to 1.0, all distances round to the nearest integer. Note that the number of digits edited after the decimal point depends on the precision set by *DIMDEC*. *DIMRND* does not apply to angular dimensions.
DIMSAH	Type: Switch Saved in: Drawing Initial value: Off	Controls use of user-defined arrowhead blocks at the ends of the dimension line. Off Normal arrowheads or user-defined arrowhead blocks set by *DIMBLK* are used On User-defined arrowhead blocks are used *DIMBLK1* and *DIMBLK2* specify different user-defined arrowhead blocks for each end of the dimension line.

Variable	Characteristic	Description
DIMSCALE	Type: Real Saved in: Drawing Initial value: 1.0000	Sets the overall scale factor applied to dimensioning variables that specify sizes, distances, or offsets. Also affects the scale of leader objects created with *LEADER*. 0.0 AutoCAD computes a reasonable default value based on the scaling between the current model space viewport and paper space. If you are in paper space, or in model space and not using the paper space feature, the scale factor is 1.0. >0 AutoCAD computes a scale factor that leads text sizes, arrowhead sizes, and other scaled distances to plot at their face value. *DIMSCALE* does not affect tolerances or measured lengths, coordinates, or angles.
DIMSD1	Type: Switch Saved in: Drawing Initial value: Off	Controls suppression of the first dimension line. When turned on, suppresses the display of the dimension line and arrowhead between the first extension line and the text.
DIMSD2	Type: Switch Saved in: Drawing Initial value: Off	Controls suppression of the second dimension line. When turned on, suppresses the display of the dimension line and arrowhead between the second extension line and the text.
DIMSE1	Type: Switch Saved in: Drawing Initial value: Off	Suppresses display of the first extension line. Off Extension line is not suppressed On Extension line is suppressed
DIMSE2	Type: Switch Saved in: Drawing Initial value: Off	Suppresses display of the second extension line. Off Extension line is not suppressed On Extension line is suppressed
DIMSHO	Type: Switch Saved in: Drawing Initial value: On	Controls redefinition of dimension objects while dragging. Associative dimensions recompute dynamically as they are dragged. On some computers, dynamic dragging can be very slow, so you can set *DIMSHO* to off to drag the original image instead. The *DIMSHO* value is not stored in a dimension style.
DIMSOXD	Type: Switch Saved in: Drawing Initial value: Off	Suppresses drawing of dimension lines outside the extension lines. Off Dimensions lines are not suppressed On Dimension lines are suppressed If the dimension lines would be outside the extension lines and *DIMTIX* is on, setting *DIMSOXD* to on suppresses the dimension line. If *DIMTIX* is off, *DIMSOXD* has not effect.
DIMSTYLE	(Read-only) Type: String Saved in: Drawing	Sets the current dimension style by name. To change the dimension style, use *DDIM* or *DIMSTYLE*.

Variable	Characteristic	Description
DIMTAD	Type: Integer Saved in: Drawing Initial value: 0	Controls the vertical position of text in relation to the dimension line. 0 Centers the dimension text between the extension lines. 1 Places the dimension text above the dimension line except when the dimension line is not horizontal and text inside the extension lines is forced horizontal (*DIMTIH* = 1). The distance from the dimension line to the baseline of the lowest line of text is the current *DIMGAP* value. 2 Places the dimension text on the side of the dimension line farthest away from the defining points. 3 Places the dimension text to conform to Japanese Industrial Standards (JIS).
DIMTDEC	Type: Integer Saved in: Drawing Initial value: 4	Sets the number of decimal places to display in tolerance values for the primary units in a dimension. *DIMTDEC* stores this value when you enter it on the command line or set it under Primary Units in the Annotation dialog box.
DIMTFAC	Type: Real Saved in: Drawing Initial value: 1.0000	Specifies the height of tolerance text and the relative size of stacked fractions. Use *DIMTFAC* for plus and minus tolerance strings when *DIMTOL* is on and *DIMTM* is not equal to *DIMTP* or when *DIMLIM* is on.
DIMTIH	Type: Switch Saved in: Drawing Initial value: On	Controls the position of dimension text inside the extension lines for all dimension types except ordinate. Off Aligns text with the dimension line On Draws text horizontally
DIMTIX	Type: Switch Saved in: Drawing Initial value: Off	Draws text between extension lines. Off The result varies with the type of dimension. For linear and angular dimensions, AutoCAD places text inside the extension lines if there is sufficient room. For radius and diameter dimensions that don't fit inside the circle or arc, *DIMTIX* has no effect and always forces the text outside the circle or arc. On Draws dimension text between the extension lines even if AutoCAD would ordinarily place it outside those lines.
DIMTM	Type: Real Saved in: Drawing Initial value: 0.0000	When *DIMTOL* or *DIMLIM* is on, *DIMTM* sets the minimum (or lower) tolerance limit for dimension text. AutoCAD accepts signed values for *DIMTM*. If *DIMTOL* is on and *DIMTP* and *DIMTM* are set to the same value, AutoCAD draws a tolerance value.

R13

Variable	Characteristic	Description
DIMTM (continued)		If *DIMTM* and *DIMTP* values differ, the upper tolerance is drawn above the lower, and a plus sign is added to the *DIMTP* value if it is positive. For *DIMTM*, AutoCAD uses the negative of the value you enter (adding a minus sign if you specify a positive number and a plus sign if you specify a negative number). No sign is added to a value of zero.
DIMTOFL	Type: Switch Saved in: Drawing Initial value: Off	Controls whether a dimension line is drawn between the extension lines even when the text is placed outside. For radius and diameter dimensions (when *DIMTIX* is off), it draws a dimension line and arrowheads inside the circle or arc and places the text and leader outside. Off Does not draw dimensions lines between the measured points when arrowheads are placed outside the measured points On Draws dimension lines between the measured points even when arrowheads are placed outside the measured points
DIMTOH	Type: Switch Saved in: Drawing Initial value: On	When turned on, controls the position of dimension text outside the extension lines. Off Aligns text with the dimension line On Draws text horizontally
DIMTOL	Type: Switch Saved in: Drawing Initial value: Off	Appends tolerances to dimension text. Setting *DIMTOL* on forces *DIMLIM* off.
DIMTOLJ	Type: Integer Saved in: Drawing Initial value: 1	Sets the vertical justification for tolerance values relative to the nominal dimension text. 0 Bottom 1 Middle 2 Top
DIMTP	Type: Real Saved in: Drawing Initial value: 0.0000	When *DIMTOL* or *DIMLIM* is on, sets the maximum (or upper) tolerance limit for dimension text. AutoCAD accepts signed values for *DIMTP*. If *DIMTOL* is on and *DIMTP* and *DIMTM* are set to the same value, AutoCAD draws a tolerance value. If *DIMTM* and *DIMTP* values differ, the upper tolerance is drawn above the lower and a plus sign is added to the *DIMTP* value if it is positive.
DIMTSZ	Type: Real Saved in: Drawing Initial value: 0.0000	Specifies the size of oblique strokes drawn instead of arrowheads for linear, radius, and diameter dimensioning. 0 Draws arrowheads >0 Draws oblique strokes instead of arrowheads The size of the oblique strokes is determined by this value multiplied by the *DIMSCALE* value.

Variable	Characteristic	Description
DIMTVP	Type: Real Saved in: Drawing Initial value: 0.0000	Controls the vertical position of dimension text above or below the dimension line. AutoCAD uses the *DIMTVP* value when *DIMTAD* is off. The magnitude of the vertical offset of text is the product of the text height and *DIMTVP*. Setting *DIMTVP* to 1.0 is equivalent to setting *DIMTAD* to on. AutoCAD splits the dimension line to accommodate the text only if the absolute value of *DIMTVP* is less than 0.7.
DIMTXSTY	Type: String Saved in: Drawing Initial value: "STANDARD"	Specifies the text style of the dimension.
DIMTXT	Type: Real Saved in: Drawing Initial value: 0.1800	Specifies the height of dimension text, unless the current text style has a fixed height.
DIMTZIN	Type: Integer Saved in: Drawing Initial value: 0	Controls the suppression of zeros in tolerance values. *DIMTZIN* stores this value when you enter it on the command line or set it under Primary Units in the Annotation dialog box. *DIMTZIN* values 0-3 affect feet-and-inch dimensions only. 0 Suppresses zero feet and precisely zero inches 1 Includes zero feet and precisely zero inches 2 Includes zero feet and suppresses zero inches 3 Includes zero inches and suppresses zero feet 4 Suppresses leading zeroes in decimal dimensions (for example, 0.5000 becomes .5000) 8 Suppresses trailing zeroes in decimal dimensions (for example, 12.5000 becomes 12.5) 12 Suppresses both leading and trailing zeroes (for example, 0.5000 becomes .5)
DIMUNIT	Type: Integer Saved in: Drawing Initial value: 2	Sets the units format for all dimension style family members except angular. 1 Scientific 2 Decimal 3 Engineering 4 Architectural (stacked) 5 Fractional (stacked) 6 Architectural 7 Fractional 8 Windows Desktop (decimal format using Control Panel settings for decimal separator and number grouping symbols)
DIMUPT	Type: Switch Saved in: Drawing Initial value: Off	Controls options for user-positioned text. Off Cursor controls only the dimension line location On Cursor controls both the text position and the dimension line location

Variable	Characteristic	Description
DIMZIN	Type: Integer Saved in: Drawing Initial value: 0	Controls the suppression of zeroes in the primary unit value. *DIMZIN* stores this value when you enter it on the command line or set it under Primary Units in the Annotation dialog box. *DIMZIN* values 0-3 affect feet-and-inch dimensions only. 0 Suppresses zero feet and precisely zero inches 1 Includes zero feet and precisely zero inches 2 Includes zero feet and suppresses zero inches 3 Includes zero inches and suppresses zero feet 4 Suppresses leading zeroes in decimal dimensions (for example, 0.5000 becomes .5000) 8 Suppresses trailing zeroes in decimal dimensions (for example, 12.5000 becomes 12.5) 12 Suppresses both leading and trailing zeroes (for example, 0.5000 becomes .5) *DIMZIN* also affects real-to-string conversions performed by the AutoLISP **(rtos)** and **(angtos)** functions.
DISPSILH	Type: Integer Saved in: Drawing Initial value: 0	Controls display of silhouette curves of solid objects in wireframe mode. Also controls whether mesh is drawn or suppressed when a solid object is hidden. 0 Off 1 On
DISTANCE	(Read-only) Type: Real Not saved	Stores the distance computed by *DIST*.
DONUTID	Type: Real Not saved Initial value: 0.5000	Sets the default for the inside diameter of a donut.
DONUTOD	Type: Real Not saved Initial value: 1.0000	Sets the default for the outside diameter of a donut. It must be nonzero. If *DONUTID* is larger than *DONUTOD*, the two values are swapped by the next command.
DRAGMODE	Type: Integer Saved in: Drawing Initial value: 2	Controls display of objects being dragged. 0 Does not display an outline of the object as you drag it 1 Displays the outline of the object as you drag it only if you enter **drag** on the command line after selecting the object to drag 2 Auto; always displays an outline of the object as you drag it
DRAGP1	Type: Integer Saved in: Registry Initial value: 10	Sets regen-drag input sampling rate.
DRAGP2	Type: Integer Saved in: Registry Initial value: 25	Sets fast-drag input sampling rate.

R13

Variable	Characteristic	Description
DWGCODEPAGE	(Read-only) Type: String Saved in: Drawing	Stores the same value as *SYSCODEPAGE* (for compatibility).
DWGNAME	(Read-only) Type: String Not saved	Stores the drawing name as entered by the user. If the drawing has not been named yet, *DWGNAME* defaults to "Drawing.dwg." If the user specified a drive/directory prefix, it is stored in the *DWGPREFIX* variable.
DWGPREFIX	(Read-only) Type: String Not saved	Stores the drive/directory prefix for the drawing.
DWGTITLED	(Read-only) Type: Integer Not saved	Indicates whether the current drawing has been named. 0 Drawing has not been named 1 Drawing has been named
EDGEMODE	Type: Integer Saved in: Registry Initial value: 0	Controls how *TRIM* and *EXTEND* determine cutting and boundary edges. 0 Uses the selected edge without an extension 1 Extends or trims the selected object to an imaginary extension of the cutting or boundary edge Line, arc, elliptical arc, ray, and polyline are objects eligible for natural extension. The natural extension of a line or ray is an unbounded line (xline); an arc is a circle; and an elliptical arc is an ellipse. A polyline is broken down into its line and arc components, which are extended to their natural boundaries.
ELEVATION	Type: Real Saved in: Drawing Initial value: 0.0000	Stores the current elevation relative to the current UCS for the current space.
EXPERT	Type: Integer Not saved Initial value: 0	Controls whether certain prompts are issued. 0 Issues all prompts normally. 1 Suppresses "About to regen, proceed?" and "Really want to turn the current layer off?" 2 Suppresses the preceding prompts and "Block already defined. Redefine it?" (*BLOCK*) and "A drawing with this name already exists. Overwriteit?" (*SAVE* or *WBLOCK*). 3 Suppresses the preceding prompts and those issued by *LINETYPE* if you try to load a linetype that's already loaded or create a new linetype in a file that already defines it. 4 Suppresses the preceding prompts and those issued by *UCS Save* and *VPORTS Save* if the name you supply already exists.

R13

Variable	Characteristic	Description
EXPERT (continued)		
		5 Suppresses the preceding prompts and those issued by the *DIMSTYLE Save* option and *DIMOVERRIDE* if the dimension style name you supply already exists (the entries are rede fined).
		When a prompt is suppressed by *EXPERT*, the operation in question is performed as though you entered **y** at the prompt. The setting of *EXPERT* can affect scripts, menu macros, AutoLISP, and the command functions.
EXPLMODE	Type: Integer Not saved Initial value: 1	Controls whether *EXPLODE* supports non-uniformly scaled (NUS) blocks. 0 Does not explode NUS blocks 1 Explodes NUS blocks
EXTMAX	(Read-only) Type: 3D Point Saved in: Drawing	Stores the upper-right point of the drawing extents. Expands outward as new objects are drawn, shrinks only with *ZOOM All* or *ZOOM Extents*. Reported in World coordinates for the current space.
EXTMIN	(Read-only) Type: 3D Point Saved in: Drawing	Stores the lower-left point of the drawing extents. Expands outward as new objects are drawn, shrinks only with *ZOOM All* or *ZOOM Extents*. Reported in World coordinates for the current space.
FACETRES	Type: Real Saved in: Drawing Initial value: 0.5	Further adjusts the smoothness of shaded and rendered objects and objects with hidden lines removed. Valid values are from 0.01 to 10.0.
FILEDIA	Type: Integer Saved in: Registry Initial value: 1	Suppresses display of the file dialog boxes. 0 Dialog boxes are not displayed. You can still request a file dialog box to appear by entering a tilde (~) in response to the command's prompt. The same is true for AutoLISP and ADS functions. 1 File dialog boxes are displayed. However, if a script or AutoLISP/ADS program is active, AutoCAD displays an ordinary prompt.
FILLETRAD	Type: Real Saved in: Drawing Initial value: 0.5000	Stores the current fillet radius.
FILLMODE	Type: Integer Saved in: Drawing Initial value: 1	Specifies whether multilines, traces, solids, solid-fill hatches, and wide polylines are filled in. 0 Objects are not filled in 1 Objects are filled
FONTALT	Type: String Saved in: Registry Initial value: "simple.shx"	Specifies the alternate font to be used when the specified font file cannot be located. If an alternate font is not specified, AutoCAD displays the Alternate Font dialog box. The dialog box is displayed in the following cases:

R13

R13

R13

Variable	Characteristic	Description
FONTALT (continued)		1. A Release 13 drawing is opened; *FONTALT* is not set or not found; and a TrueType, SHX, or PostScript font is not found for a defined text style.
		2. A Release 14 drawing is opened, *FONTALT* is not set or not found, and a SHX or PostScript font is not found for a defined text style. For missing TrueType fonts in Release 14 drawings, AutoCAD automatically substitutes the closest TrueType font available.
		3. The Browse button is pressed in the Preferences dialog box when you specify an alternate font.
		AutoCAD validates the alternate font specified for *FONTALT*. If the font name or font file name is not found, the message "Font not found" is displayed. Enter either a TrueType font name (for example, Times New Roman Bold) or a TrueType file name (for example *timebd.ttf*). When a TrueType file name is entered for *FONTALT*, AutoCAD returns the font name in place of the file name) if the font is registered with the operating system.
FONTMAP	Type: String Saved in: Registry Initial value: "acad.fmp"	Specifies the font mapping file to be used. A font mapping file contains one font mapping per line; the original font used in the drawing and the font to be substituted for it are separated by a semicolon (;). For example, to substitute the Times TrueType font for the Romans font, the line in the mapping file would read as follows: **romanc.shx;times.ttf** If *FONTMAP* does not point to a font mapping file, if the FMP file is not found, or if the font file name specified in the FMP file is not found, AutoCAD uses the font defined in the style. If the font in the style is not found, AutoCAD substitutes the font according to substitution rules.
FRONTZ	(Read-only) Type: Real Saved in: Drawing	Stores the front clipping plane offset from the target plane for the current viewport, in drawing units. Meaningful only if the front clipping bit in *VIEWMODE* is on and the front-clip-not-at-eye bit is also on. The distance of the front clipping plane from the camera point is found by subtracting *FRONTZ* from the camera-to-target distance.
GRIDMODE	Type: Integer Saved in: Drawing Initial value: 0	Specifies whether the grid is turned on or off. 0 Turns the grid off 1 Turns the grid on

Variable	Characteristic	Description
GRIDUNIT	Type: 2D point Saved in: Drawing Initial value: 0.5000, 0.5000	Specifies the grid spacing (*X* and *Y*) for the current viewport.
GRIPBLOCK	Type: Integer Saved in: Registry Initial value: 0	Controls the assignment of grips in blocks. 0 Assigns a grip only to the insertion point of the block 1 Assigns grips to objects within the block
GRIPCOLOR	Type: Integer Saved in: Registry Initial value: 5	Controls the color of nonselected grips (drawn as box outlines). The valid range is 1-255.
GRIPHOT	Type: Integer Saved in: Registry Initial value: 1	Controls the color of selected grips (drawn as filled boxes). The valid range is 1-255.
GRIPS	Type: Integer Saved in: Registry Initial value: 1	Controls use of selection set grips for the Stretch, Move, Rotate, Scale, and Mirror grip modes. 0 Turns off grips 1 Turns on grips To adjust the size of the grips and the effective selection area used by the cursor when you snap to a grip, use the *GRIPSIZE* system variable.
GRIPSIZE	Type: Integer Saved in: Registry Initial value: 3	Sets the size of the box drawn to display the grip, in pixels. The valid range is 1-255.
HANDLES	(Read-only) Type: Integer Saved in: Drawing Initial value: On	Reports whether object handles can be accessed by applications. Off Handles are not available On Handles are available
HIGHLIGHT	Type: Integer Not saved Initial value: 1	Controls object highlighting; does not affect objects selected with grips. 0 Turns off object selection highlighting 1 Turns on object selection highlighting
HPANG	Type: Real Not saved Initial value: 0	Specifies the hatch pattern angle.
HPBOUND	Type: Integer Saved in: Registry Initial value: 1	Controls the object type created by *BHATCH* and *BOUNDARY*. 0 Creates a region 1 Creates a polyline
HPDOUBLE	Type: Integer Not saved Initial value: 0	Specifies hatch pattern doubling for user-defined patterns. 0 Turns off hatch pattern doubling 1 Turns on hatch pattern doubling
HPNAME	Type: String Not saved Initial value: "ANSI31"	Sets a default hatch pattern name of up to 34 characters, no spaces allowed. Returns "" if there is no default. Enter a period (.) to set no default.

R13

Variable	Characteristic	Description
HPSCALE	Type: Real Not saved Initial value: 1.0000	Specifies the hatch pattern scale factor; must be nonzero.
HPSPACE	Type: Real Not saved Initial value: 1.0000	Specifies the hatch pattern line spacing for user-defined simple patterns; must be nonzero.
INDEXCTL	Type: Integer Saved in: Drawing Initial value: 0	Controls whether layer and spatial indexes are created and saved in drawing files. 0 No indexes are created 1 Layer index is created 2 Spatial index is created 3 Layer and spatial indexes are created
INETLOCATION	Type: Real Saved in: Registry Initial value: "www.autodesk. com/acaduser"	Stores the Internet location used by *BROWSER*.
INSBASE	Type: 3D point Saved in: Drawing Initial value: 0.0000, 0.0000,0.0000	Stores insertion basepoint set by *BASE*, expressed as a UCS coordinate for the current space.
INSNAME	Type: String Not saved Initial value: ""	Sets a default block name for *DDINSERT* or *INSERT*. The name must conform to symbol naming conventions. Returns "" if no default is set. Enter a period (.) to set no default.
ISAVEBAK	Type: Integer Saved in: Registry Initial value: 1	Improves the speed of incremental saves, especially for large drawings. *ISAVEBAK* controls the creation of a backup file (BAK). In Windows, copying the file data to create a BAK file for large drawings takes a major portion of the incremental save time. 0 No BAK file is created (even for a full save) 1 A BAK file is created **WARNING:** In some cases (such as a power failure in the middle of a save), it's possible that drawing data can be lost.
ISAVEPERCENT	Type: Integer Saved in: Registry Initial value: 50	Determines the amount of wasted space tolerated in a drawing file. The value of *ISAVEPERCENT* is an integer between 0 and 100. The default value of 50 means that the estimate of wasted space within the file does not exceed 50% of the total file size. Wasted space is eliminated by periodic full saves. When the estimate exceeds 50%, the next save will be a full save. This resets the wasted space estimate to 0. If *ISAVEPERCENT* is set to 0, every save is a full save.

R14

R13

Variable	Characteristic	Description
ISOLINES	Type: Integer Saved in: Drawing Initial value: 4	Specifies the number of isolines per surface on objects. Valid integer values are from 0 to 2047.
LASTANGLE	(Read-only) Type: Real Not saved	Stores the end angle of the last arc entered relative to the *XY* plane of the current UCS for the current space.
LASTPOINT	Type: 3Dpoint Saved in: Drawing Initial value: 0.0000, 0.0000,0.0000	Stores the last point entered, expressed as a UCS coordinate for the current space; referenced by the at symbol (@) during keyboard entry.
LASTPROMPT	(Read-only) Type: String Not saved Initial value: ""	Stores the last string echoed to the command line. This string is identical to the last line seen at the command line and includes any user input.
LENSLENGTH	(Read-only) Type: Real Saved in: Drawing	Stores the length of the lens (in millimeters) used in perspective viewing for the current viewport.
LIMCHECK	Type: Integer Saved in: Drawing Initial value: 0	Controls creation of objects outside the drawing limits. 0 Objects can be created outside the limits 1 Objects cannot be created outside the limits
LIMMAX	Type: 2D point Saved in: Drawing Initial value: 12.0000,9.0000	Stores the upper-right drawing limits for the current space, expressed as World coordinate.
LIMMIN	Type: 2D point Saved in: Drawing Initial value: 0.0000, 0.0000	Stores the lower-left drawing limits for the current space, expressed as a World coordinate.
LISPINIT	Type: Integer Saved in: Registry Initial value: 1	Specifies whether AutoLISP-defined functions and variables are preserved when you open a new drawing or whether they are valid in the current drawing session only. 0 AutoLISP functions and variables are preserved from drawing to drawing 1 AutoLISP functions and variables are valid in current drawing only (Release 13 behavior)
LOCALE	(Read-only) Type: String Not saved Initial value: "en"	Displays the ISO language code of the current AutoCAD version you are running.
LOGFILEMODE	Type: Integer Saved in: Registry Initial value: 0	Specifies whether the contents of the text window are written to a log file. 0 Log file is not maintained 1 Log file is maintained

R13

R14

R14

R13

R14

Variable	Characteristic	Description
LOGFILENAME	Type: String Saved in: Registry Initial value: "C:\ACADR14\ acad.log"	Specifies the path for the log file. The log file is created if you select Maintain a Log File on the General tab in the Preferences dialog box. **NOTE:** The initial value varies depending on where you installed AutoCAD.
LOGINNAME	(Read-only) Type: String Not saved	Displays the user's name as configured or as input when AutoCAD is loaded. The maximum length for a login name is 30 characters.
LTSCALE	Type: Real Saved in: Drawing Initial value: 1.0000	Sets the global linetype scale factor (cannot equal zero).
LUNITS	Type: Integer Saved in: Drawing Initial value: 2	Sets linear units. 1 Scientific 2 Decimal 3 Engineering 4 Architectural 5 Fractional
LUPREC	Type: Integer Saved in: Drawing Initial value: 4	Sets the number of decimal places displayed for linear units in commands (except dimensioning commands), variables, and output. This includes coordinate display, *LIST, DIST, ID, AREA,* and *DDMODIFY.*
MAXACTVP	Type: Integer Saved in: Drawing Initial value: 48	Sets the maximum number of viewports that can be active at one time.
MAXOBJMEM	Type: Integer Not saved Initial value: 0	Controls the object pager and specifies how much virtual memory AutoCAD allows the drawing to use before it starts paging the drawing out to disk into the object pager's swap files. The default value of 0 turns off the object pager. The object pager is also turned off when this variable is set to a negative value or the value 2,147,483,647. When *MAXOBJMEM* is set to any other value, the object pager is turned on and the specified value is used as the upper limit for the object pager's virtual memory. The *ACADMAXOBJMEM* environment variable also controls the object pager. **WARNING:** If you reboot your system without exiting AutoCAD, these swap files will not be deleted, so you should delete them. Do not delete them from within AutoCAD.
MAXSORT	Type: Integer Saved in: Registry Initial value: 200	Sets the maximum number of symbol names or block names sorted by listing commands. If the total number of items exceeds this value, no items are sorted.

Variable	Characteristic	Description
MEASUREMENT	Type: Integer Saved in: Drawing Initial value: 0	Sets drawing units as English or metric. Specifically, *MEASUREMENT* controls which hatch pattern and linetype file an existing drawing uses when it is opened. 0 English; AutoCAD uses the hatch pattern file and linetype file designated by the ANSIHatch and ANSILinetype registry settings 1 Metric; AutoCAD uses the hatch pattern file and linetype file designated by the ISOHatch and ISOLinetype registry settings The drawing units for new drawings are controlled by the *MEASUREINIT* registry variable (*MEASUREINIT* uses the same values as *MEASUREMENT*). The *MEASUREMENT* setting of a drawing always overrides the *MEASUREINIT* registry setting.
MENUCTL	Type: Integer Saved in: Registry Initial value: 1	Controls the page switching of the screen menu. 0 Screen menu does not switch pages in response to keyboard command entry 1 Screen menu does switch pages in response to keyboard command entry
MENUECHO	Type: Integer Not saved Initial value: 0	Sets menu echo and prompt control bits. It is the sum of the following: 1 Suppresses echo of menu items (^P in a menu item toggles echoing) 2 Suppresses display of system prompts during menu 4 Disables ^P toggle of menu echoing 8 Displays input/output strings; debugging aid for DIESEL macros
MENUNAME	(Read-only) Type: String Saved in: Application header	Stores the *MENUGROUP* name. If the current primary menu has no *MENUGROUP* name, the menu file includes the path if the file location is not specified in AutoCAD's environment setting.
MIRRTEXT	Type: Integer Saved in: Drawing Initial value: 1	Controls how *MIRROR* reflects text. 0 Retains text direction 1 Mirrors the text
MODEMACRO	Type: String Not saved Initial value: ""	Displays a text string on the Status line, such as the name of the current drawing, time/date stamp, or special modes. Use *MODEMACRO* to display a string of text, or use special text strings written in the DIESEL macro language to have AutoCAD evaluate the macro from time to time and base the Status line on user-selected conditions.
MTEXTED	Type: String Saved in: Registry Initial value: "Internal"	Sets the name of the program to use for editing multiline text objects. The default editor is the multiline text editor. If you enter a period (.), the system's default text editor is used.

R14

R13

Variable	Characteristic	Description
OFFSETDIST	Type: Real Not saved Initial value: 1.0000	Sets the default offset distance. <0 Offsets an object through a specified point >0 Sets the default offset distance
OLEHIDE	Type: Integer Saved in: Registry Initial value: 0	Controls the display of OLE objects in AutoCAD. 0 All OLE objects are visible 1 OLE objects are visible in paper space only 2 OLE objects are visible in model space only 3 No OLE objects are visible *OLEHIDE* affects both screen display and printing.
ORTHOMODE	Type: Integer Saved in: Drawing Initial value: 0	Constrains cursor movement to the perpendicular. When *ORTHOMODE* is turned on, the cursor can move only horizontally or vertically relative to the UCS and the current grid rotation angle. 0 Turns off Ortho mode 1 Turns on Ortho mode
OSMODE	Type: Integer Saved in: Drawing Initial value: 0	Sets running object snap modes using the following bit-codes. 0 *None* 1 *Endpoint* 2 *Midpoint* 4 *Center* 8 *Node* 16 *Quadrant* 32 *Intersection* 64 *Insertion* 128 *Perpendicular* 256 *Tangent* 512 *Nearest* 1024 *Quick* 2048 *Apparent Intersection* To specify more than one object snap, enter the sum of their values. For example, entering **3** specifies the *Endpoint* (bit code 1) and *Midpoint* (bit code 2) object snaps. Entering **4095** specifies all object snaps.
OSNAPCOORD	Type: Integer Saved in: Registry Initial value: 2	Controls whether coordinates entered on the command line override running object snaps. 0 Running object snap settings override keyboard coordinate entry 1 Keyboard entry overrides object snap settings 2 Keyboard entry overrides object snap settings except in scripts
PDMODE	Type: Integer Saved in: Drawing Initial value: 0	Controls how point objects are displayed. For information about values to enter, see *POINT*.

R14

R14

R13

Variable	Characteristic	Description
PDSIZE	Type: Real Saved in: Drawing Initial value: 0.0000	Sets the display size for point objects. 0 Creates a point at 5% of the graphics area height >0 Specifies an absolute size <0 Specifies a percentage of the viewport size
PELLIPSE	Type: Integer Saved in: Drawing Initial value: 0	Controls the ellipse type created with *ELLIPSE*. 0 Creates a true ellipse object 1 Creates a polyline representation of an ellipse
PERIMETER	(Read-only) Type: Real Not saved	Stores the last perimeter value computed by *AREA, LIST,* or *DBLIST.*
PFACEVMAX	(Read-only) Type: Integer Not saved	Sets the maximum number of vertices per face.
PICKADD	Type: Integer Saved in: Registry Initial value: 1	Controls additive selection of objects. 0 Turns off *PICKADD*. The objects most recently selected, either individually or by windowing, become the selection set. Previously selected objects are removed from the selection set. Add more objects to the selection set by holding down Shift while selecting. 1 Turns on *PICKADD*. Each object selected, either individually or by windowing, is added to the current selection set. To remove objects from the set, hold down Shift while selecting.
PICKAUTO	Type: Integer Saved in: Registry Initial value: 1	Controls automatic windowing at the Select Objects prompt. 0 Turns off *PICKAUTO* 1 Draws a selection window (for either Window or Crossing selection) automatically at the Select Objects prompt
PICKBOX	Type: Integer Saved in: Registry Initial value: 3	Sets object selection target height, in pixels.
PICKDRAG	Type: Integer Saved in: Registry Initial value: 0	Controls the method of drawing a selection window. 0 Draws the selection window using two points. Click the pointing device at one corner and then at the other corner. 1 Draws the selection window using dragging. Click at one corner, hold down the pick button on the pointing device, drag, and release the button at the other corner.
PICKFIRST	Type: Integer Saved in: Registry Initial value: 1	Controls whether you select objects before (noun-verb selection) or after you issue a command. 0 Turns off *PICKFIRST* 1 Turns on *PICKFIRST*

Variable	Characteristic	Description
PICKSTYLE	Type: Integer Saved in: Drawing Initial value: 1	Controls use of group selection and associative hatch selection. 0 No group selection or associative hatch selection 1 Group selection 2 Associative hatch selection 3 Group selection and associative hatch selection
PLATFORM	(Read-only) Type: String Not saved	Indicates which platform of AutoCAD is in use. One of the following strings may appear: "Microsoft Windows NT Version 3.51 (x86)" "Microsoft Windows NT Version 4.00 (x86)" "Microsoft Windows Version 4.00 (x86)"
PLINEGEN	Type: Integer Saved in: Drawing Initial value: 0	Sets how linetype patterns are generated around the vertices of a two-dimensional polyline. Does not apply to polylines with tapered segments. 0 Polylines are generated to start and end with a dash at each vertex 1 Generates the linetype in a continuous pattern around the vertices of the polyline
PLINETYPE	Type: Integer Saved in: Registry Initial value: 2	Specifies whether AutoCAD uses optimized 2D polylines. *PLINETYPE* controls both the creation of new polylines with the *PLINE* command and the conversion of existing polylines in drawings from previous releases. 0 Polylines in older drawings are not converted on open; *PLINE* creates old-format polylines 1 Polylines in older drawings are not converted on open; *PLINE* creates optimized polylines 2 Polylines in older drawings are converted on open; *PLINE* creates optimized polylines For more information on the two formats, see *CONVERT*. *PLINETYPE* also controls the polyline type created with the following commands: *BOUNDARY* (when object type is set to Polyline), *DONUT, ELLIPSE* (when *PELLIPSE* is set to 1), *PEDIT* (when selecting a line or arc), *POLYGON*, and *SKETCH* (when *SKPOLY* is set to 1).
PLINEWID	Type: Real Saved in: Drawing Initial value: 0.0000	Stores the default polyline width.
PLOTID	Type: String Saved in: Registry Initial value: ""	Changes the default plotter, based on its assigned description, and retains the text string of the current plotter description. Change to another configured plotter by entering its full or partial description. For example, if you have a plotter named "Draft Plotter," enter **draft**. Applications can use *PLOTTER* to step through the available plotter descriptions retained by *PLOTID* and thus control the default plotter.

R13

R14

Variable	Characteristic	Description
PLOTROTMODE	Type: Integer Saved in: Registry Initial value: 1	Controls the orientation of plots. 0 Rotates the effective plotting area so that the corner with the Rotation icon aligns with the paper at the lower-left for 0, top-left for 90, top-right for 180, and lower-right for 270 1 Aligns the lower-left corner of the effective plotting area with the lower-left corner of the paper
PLOTTER	Type: Integer Saved in: Registry Initial value: 0	Changes the default plotter, based on its assigned integer, and retains an integer number that AutoCAD assigns for each configured plotter. This number can be in the range of 0 up to the number of configured plotters. You may configure up to 29 plotters. Change to another configured plotter by entering its valid, assigned number. For example, if you configured four plotters, the valid numbers are 0 through 3. **NOTE:** AutoCAD does not permanently assign a number to a given plotter. If you delete a plotter configuration, AutoCAD assigns new numbers to each of the remaining configured plotters. AutoCAD also updates the value of *PLOTTER*.
POLYSIDES	Type: Integer Not saved Initial value: 4	Sets the default number of sides for *POLYGON*. The range is 3-1024.
POPUPS	(Read-only) Type: Integer Not saved	Displays the status of the currently configured display driver. 0 Does not support dialog boxes, the menu bar, pull-down menus, and icon menus 1 Supports these features
PROJECTNAME	Type: String Saved in: Drawing Initial value: ""	Assigns a project name to the current drawing. The project name points to a section in the registry which can contain one or more search paths for each project name defined. Used when an xref or image is not found in its original hard-coded path. Project names and their search directories are created through the Files tab of the Preferences dialog box. Project names make it easier for users to manage xrefs and images when drawings are exchanged between customers, or if users have different drive mappings to the same location on a server. If the xref or image is not found at the hard-coded path, the project paths associated with the project name are searched. The search order is hard-coded path, then project name search path, then AutoCAD search path.

R13

R14

Variable	Characteristic	Description
PROJMODE	Type: Integer Saved in: Registry Initial value: 1	Sets the current Projection mode for trimming or extending. 0 True 3D mode (no projection) 1 Project to the *XY* plane of the current UCS 2 Project to the current view plane
PROXYGRAPHICS	Type: Integer Saved in: Drawing Initial value: 1	Specifies whether images of proxy objects are saved in the drawing. 0 Image is not saved with the drawing; a bounding box is displayed instead 1 Image is saved with the drawing
PROXYNOTICE	Type: Integer Saved in: Registry Initial value: 1	Displays a notice when a proxy is created. A proxy is created when you open a drawing containing custom objects created by an application that is not present. A proxy is also created when you issue a command that unloads a custom object's parent application. 0 No proxy warning is displayed 1 Proxy warning is displayed
PROXYSHOW	Type: Integer Saved in: Registry Initial value: 1	Controls the display of proxy objects in a drawing. 0 Proxy objects are not displayed 1 Graphic images are displayed for all proxy objects 2 Only the bounding box is displayed for all proxy objects
PSLTSCALE	Type: Integer Saved in: Drawing Initial value: 1	Controls paper space linetype scaling. 0 No special linetype scaling. Linetype dash lengths are based on the drawing units of the space (model or paper) in which the objects were created, scaled by the global *LTSCALE* factor. 1 Viewport scaling governs linetype scaling. If *TILEMODE* is set to 0, dash lengths are based on paper space drawing units, even for objects in model space. In this mode, viewports can have varying magnifications, yet display linetypes identically. For a specific linetype, the dash lengths of a line in a viewport are the same as the dash lengths of a line in paper space. You can still control the dash lengths with *LTSCALE*.
PSPROLOG	Type: String Saved in: Registry Initial value: ""	Assigns a name for a prolog section to be read from the *acad.psf* file when you are using *PSOUT*.
PSQUALITY	Type: Integer Saved in: Registry Initial value: 75	Controls the rendering quality of PostScript images and whether they are drawn as filled objects or as outlines. 0 Turns off PostScript image generation <0 Sets the number of pixels per AutoCAD drawing unit for the PostScript resolution

Variable	Characteristic	Description
PSQUALITY (continued)		>0 Sets the number of pixels per drawing unit but uses the absolute value; causes AutoCAD to show the PostScript paths as outlines and does not fill them
QTEXTMODE	Type: Integer Saved in: Drawing Initial value: 0	Controls how text is displayed. 0 Turns off Quick Text mode; displays characters 1 Turns on Quick Text mode; displays a box in place of text
RASTERPREVIEW	Type: Integer Saved in: Registry Initial value: 1	Controls whether BMP preview images are saved with the drawing. 0 No preview image is created 1 Preview image created
REGENMODE	Type: Integer Saved in: Drawing Initial value: 1	Controls automatic regeneration of the drawing. 0 Turns off *REGENAUTO* 1 Turns on *REGENAUTO*
RE-INIT	Type: Integer Not saved Initial value: 0	Reinitializes the digitizer, digitizer port, and *acad.pgp* file using the following bit-codes: 1 Digitizer I/O port reinitialization 4 Digitizer reinitialization 16 PGP file reinitialization (reload) To specify more than one reinitialization, enter the sum of their values, for example, **5** to specify both digitizer port (1) and digitizer reinitialization (4).
RTDISPLAY	Type: Integer Saved in: Registry Initial value: 1	Controls the display of raster images during real-time zoom or pan. 0 Displays raster image content 1 Displays raster image outline only *RTDISPLAY* is saved in the current profile.
SAVEFILE	(Read-only) Type: String Saved in: Registry Initial value: "auto.sv$"	Stores current auto-save file name.
SAVENAME	(Read-only) Type: String Not saved Initial value: ""	Stores the file name and directory path of the current drawing once you save it.
SAVETIME	Type: Integer Saved in: Registry Initial value: 120	Sets the automatic save interval, in minutes. 0 Turns off automatic saving >0 Automatically saves the drawing at intervals specified by the nonzero integer The *SAVETIME* timer starts as soon as you make a change to a drawing. It is reset and restarted by a manual *SAVE*, *SAVEAS*, or *QSAVE*. The current drawing is saved to *auto.sv$*.

R13

R14

Variable	Characteristic	Description
SCREENBOXES	(Read-only) Type: Integer Saved in: Registry	Stores the number of boxes in the screen menu area of the graphics area. If the screen menu is turned off, *SCREENBOXES* is zero. On platforms that permit the AutoCAD graphics area to be resized or the screen menu to be reconfigured during an editing session, the value of this variable might change during the editing session.
SCREENMODE	(Read-only) Type: Integer Saved in: Registry	Stores a bit-code indicating the graphics/text state of the AutoCAD display. It is the sum of the following bit values: 0 Text screen is displayed 1 Graphics area is displayed 2 Dual-screen display is configured
SCREENSIZE	(Read-only) Type: 2D point Not saved	Stores current viewport size in pixels (*X* and *Y*).
SHADEDGE	Type: Integer Saved in: Drawing Initial value: 3	Controls shading of edges in rendering. 0 Faces shaded, edges not highlighted 1 Faces shaded, edges drawn in background color 2 Faces not filled, edges in object color 3 Faces in object color, edges in background color
SHADEDIF	Type: Integer Saved in: Drawing Initial value: 70	Sets the ratio of diffuse reflective light to ambient light (in percentage of diffuse reflective light).
SHPNAME	Type: String Not saved Initial value: ""	Sets a default shape name; must conform to symbol naming conventions. If no default is set, it returns ""; enter a period (.) to set no default.
SKETCHINC	Type: Real Saved in: Drawing Initial value: 0.1000	Sets the record increment for *SKETCH*.
SKPOLY	Type: Integer Saved in: Registry Initial value: 0	Determines whether *SKETCH* generates lines or polylines.
SNAPANG	Type: Real Saved in: Drawing Initial value: 0	Sets the snap and grid rotation angle for the current viewport. The angle you specify is relative to the current UCS. **NOTE:** Changes to this variable are not reflected in the grid until the display is refreshed. AutoCAD does *not* automatically redraw when variables are changed.
SNAPBASE	Type: 2Dpoint Saved in: Drawing Initial value: 0.0000, 0.0000	Sets the snap and grid origin point for the current viewport relative to the current UCS. **NOTE:** Changes to this variable are not reflected in the grid until the display is refreshed. AutoCAD does *not* automatically redraw when variables are changed.

Variable	Characteristic	Description
SNAPISOPAIR	Type: Integer Saved in: Drawing Initial value: 0	Controls the isometric plane for the current viewport. 0 Left 1 Top 2 Right
SNAPMODE	Type: Integer Saved in: Drawing Initial value: 0	Turns Snap mode on and off. 0 Snap off 1 Snap on for the current viewport.
SNAPSTYL	Type: Integer Saved in: Drawing Initial value: 0	Sets snap style for the current viewport. 0 Standard 1 Isometric
SNAPUNIT	Type: 2D Saved in: Drawing Initial value: 0.5000, 0.5000	Sets the snap spacing for the current viewport. If the *SNAPSTYL* system variable is set to 1, AutoCAD automatically adjusts the *X* value of *SNAPUNIT* to accommodate the isometric snap. **NOTE:** Changes to this system variable are not reflected in the grid until the display is refreshed. AutoCAD does *not* automatically redraw when system variables are changed.
SORTENTS	Type: Integer Saved in: Drawing Initial value: 96	Controls *DDSELECT* object sort order operations. *SORTENTS* uses the following bit codes: 0 Disables *SORTENTS* 1 Sorts for object selection 2 Sorts for object snap 4 Sorts for redraws 8 Sorts for *MSLIDE* slide creation 16 Sorts for *REGEN*s 32 Sorts for plotting 64 Sorts for PostScript output To select more than one, enter the sum of their codes. For example, enter **3** to specify sorting for both object selection and object snap. The initial value of 96 enables sorting for plotting and PostScript output only. Setting additional sorting options can result in slower regeneration and redrawing times.
SPLFRAME	Type: Integer Saved in: Drawing Initial value: 0	Controls display of splines and spline-fit polylines. 0 Does not display the control polygon for splines and spline-fit polylines. Displays the fit surface of a polygon mesh, not the defining mesh. Does not display the invisible edges of 3D faces orpolyface meshes. 1 Displays the control polygon for splines and spline-fit polylines. Only the defining mesh of a surface-fit polygon mesh is displayed (not the fit surface). Invisible edges of 3D faces or polyface meshes are displayed.

Variable	Characteristic	Description
SPLINESEGS	Type: Integer Saved in: Drawing Initial value: 8	Sets the number of line segments to be generated for each spline-fit polyline generated by *PEDIT* Spline.
SPLINETYPE	Type: Integer Saved in: Drawing Initial value: 6	Sets the type of curve generated by *PEDIT* Spline. 5 Quadratic B-spline 6 Cubic B-spline
SURFTAB1	Type: Integer Saved in: Drawing Initial value: 6	Sets the number of tabulations to be generated for *RULESURF* and *TABSURF*. Also sets the mesh density in the *M* direction for *REVSURF* and *EDGESURF*.
SURFTAB2	Type: Integer Saved in: Drawing Initial value: 6	Sets the mesh density in the *N* direction for *REVSURF* and *EDGESURF*.
SURFTYPE	Type: Integer Saved in: Drawing Initial value: 6	Controls the type of surface fitting to be performed by *PEDIT* Smooth. 5 Quadratic B-spline surface 6 Cubic B-spline surface 8 Bezier surface
SURFU	Type: Integer Saved in: Drawing Initial value: 6	Sets the surface density in the *M* direction.
SURFV	Type: Integer Saved in: Drawing Initial value: 6	Sets the surface density in the *N* direction.
SYSCODEPAGE	(Read-only) Type: String Not saved	Indicates the system code page specified in the *acad.xmx* file. Codes are as follows: ascii dos860 dos932 iso8859-7 big5 dos861 gb2312 iso8859-8 dos437 dos863 iso8859-1 iso8859-9 dos850 dos864 iso8859-2 johab dos852 dos865 iso8859-3 ksc5601 dos855 dos866 iso8859-4 mac-roman dos857 dos869 iso8859-6
TABMODE	Type: Integer Not saved Initial value: 0	Controls use of the tablet. 0 Turns off Tablet mode 1 Turns on Tablet mode
TARGET	(Read-only) Type: 3D point Saved in: Drawing	Stores a location (as a UCS coordinate) of the target point for the current viewport.
TDCREATE	(Read-only) Type: Real Saved in: Drawing	Stores the time and date the drawing was created.
TDINDWG	(Read-only) Type: Real Saved in: Drawing	Stores the total editing time.

Variable	Characteristic	Description
TDUPDATE	(Read-only) Type: Real Saved in: Drawing	Stores the time and date of the last update/save.
TDUSRTIMER	(Read-only) Type: Real Saved in: Drawing	Stores the user-elapsed timer.
TEMPPREFIX	(Read-only) Type: String Not saved	Contains the directory name (if any) configured for placement of temporary files, with a path separator appended.
TEXTEVAL	Type: Integer Not saved Initial value: 0	Controls the method of evaluation of text strings. 0 All responses to prompts for text strings and attribute values are taken literally 1 Text starting with an opening parenthesis [(] or an exclamation mark (!) is evaluated as an AutoLISP expression, as for nontextual input **NOTE:** *DTEXT* takes all input literally regard less of the setting of *TEXTEVAL*.
TEXTFILL	Type: Integer Saved in: Registry Initial value: 1	Controls the filling of TrueType fonts while plotting, exporting with *PSOUT*, and rendering. 0 Outputs text as outlines 1 Outputs text as filled images.
TEXTQLTY	Type: Integer Saved in: Drawing Initial value: 50	Sets the resolution of TrueType fonts while plotting, exporting with *PSOUT*, and rendering. Values represent dots per inch. Lower values decrease resolution and increase plotting speed. Higher values increase resolution and decrease plotting speed.
TEXTSIZE	Type: Real Saved in: Drawing Initial value: 0.2000	Sets the default height for new text objects drawn with the current text style (has no effect if the style has a fixed height).
TEXTSTYLE	Type: String Saved in: Drawing Initial value: "STANDARD"	Sets the name of the current text style.
THICKNESS	Type: Real Saved in: Drawing Initial value: 1	Sets the current 3D solid thickness.
TILEMODE	Type: Integer Saved in: drawing Initial value: 1	Controls access to paper space and the behavior of viewports. 0 Turns on paper space and viewport object (uses *MVIEW*). AutoCAD clears the graphics area and prompts you to create one or more viewports. 1 Turns on Release 10 Compatibility mode (uses *VPORTS*). AutoCAD restores the most recently active tiled-viewport configuration. Paper space objects—including viewport objects—are not displayed, and *MVIEW, MSPACE, PSPACE,* and *VPLAYER* cannot be used.

R13

Variable	Characteristic	Description
TOOLTIPS	Type: Integer Saved in: Registry Initial value: 1	Controls the display of tooltips. 0 Turns off display of tooltips 1 Turns on display of tooltips
TRACEWID	Type: Real Saved in: Drawing Initial value: 0.0500	Sets the default trace width.
TREEDEPTH	Type: Integer Saved in: Drawing Initial value: 3020	Specifies the maximum depth, that is, the number of times the tree-structured spatial index may divide into branches. 0 Suppresses the spatial index entirely, eliminating the performance improvements it provides in working with large drawings. This setting assures that objects are always processed in database order, making it unnecessary ever to set the *SORTENTS* system variable. >0 Turns on *TREEDEPTH*. An integer of up to four digits is valid. The first two digits refer to model space, and the second two digits refer to paper space. <0 Treats model space objets as two-dimensional (Z coordinates are ignored), as is always the case with paper space objects. Such a setting is appropriate for 2D drawings and makes more efficient use of memory without loss of performance. **NOTE:** You cannot use *TREEDEPTH* transparently.
TREEMAX	Type: Integer Saved in: Registry Initial value: 10000000	Limits memory consumption during drawing regeneration by limiting the number of nodes in the spatial index (oct-tree). By imposing a fixed limit with *TREEMAX*, you can load drawings created on systems with more memory than your system and with a larger *TREEDEPTH* than your system can handle. These drawings, if left unchecked, have an oct-tree large enough to eventually consume more memory than is available to your computer. *TREEMAX* also provides a safeguard against experimentation with inappropriately high *TREEDEPTH* values. The initial default for *TREEMAX* is 10000000 (10 million), a value high enough to effectively disable *TREEMAX* as a control for *TREEDEPTH*. The value to which you should set *TREEMAX* depends on your system's available RAM. You get about 15,000 oct-tree nodes per megabyte of RAM.

Variable	Characteristic	Description
TREEMAX (continued)		If you want an oct-tree to use up to, but no more than, 2 megabytes of RAM, set *TREEMAX* to 30000 (2×15,000). If AutoCAD runs out of memory allocating oct-tree nodes, restart AutoCAD, set *TREEMAX* to a smaller number, and try loading the drawing again. AutoCAD might occasionally run into the limit you set with *TREEMAX*. Follow the resulting prompt instructions. Your ability to increase *TREEMAX* depends on your computer's available memory.
TRIMMODE	Type: Integer Saved in: Registry Initial value: 1	Controls whether AutoCAD trims selected edges for chamfers and fillets. 0 Leaves selected edges intact. 1 Trims selected edges to the endpoints of chamfer lines and fillet arcs
UCSFOLLOW	Type: Integer Saved in: Drawing Initial value: 0	Generates a plan view whenever you change from one UCS to another. You can set *UCSFOLLOW* separately for each viewport. If *UCSFOLLOW* is on for a particular viewport, AutoCAD generates a plan view in that viewport whenever you change coordinate systems. Once the new UCS has been established, you can use *VPOINT, DVIEW, PLAN,* or *VIEW* to change the view of the drawing. It will change to a plan view again the next time you change coordinate systems. 0 UCS does *not* affect the view 1 Any UCS change causes a change to plan view of the new UCS in the current viewport The setting of *UCSFOLLOW* is maintained separately for paper space and model space and can be accessed in either, but the setting is ignored while in paper space (it is always treated as if set to 0). Although you can define a non-World UCS in paper space, the view remains in plan view to the World Coordinate System.
UCSICON	Type: Integer Saved in: Drawing Initial value: 1	Displays the user coordinate system icon for the current viewport using bit code. It is the sum of the following: 0 No icon displayed 1 On; icon is displayed 2 Origin; if icon is displayed, the icon floats to the UCS origin if possible
UCSNAME	(Read-only) Type: String Saved in: Drawing	Stores the name of the current coordinate system for the current space. Returns a null string if the current UCS is unnamed.
UCSORG	(Read-only) Type: 3D point Saved in: Drawing	Stores the origin point of the current coordinate system for the current space. This value is always stored as a World coordinate.

R13

Variable	Characteristic	Description
UCSXDIR	(Read-only) Type: 3D point Saved in: Drawing	Stores the *X* direction of the current UCS for the current space.
UCSYDIR	(Read-only) Type: 3D point Saved in: Drawing	Stores the *Y* direction of the current UCS for the current space.
UNDOCTL	(Read-only) Type: Integer Not saved	Stores a bit code indicating the state of the *UNDO* feature. It is the sum of the following values: 0 *UNDO* is turned off 1 *UNDO* is turned on 2 Only one command can be undone 4 Turns on the Auto option 8 A group is currently active
UNDOMARKS	(Read-only) Type: Integer Not saved	Stores the number of marks that have been placed in the *UNDO* control stream by the Mark option. The Mark and Back options are not available if a group is currently active.
UNITMODE	Type: Integer Saved in: Drawing Initial value: 0	Controls the display format for units. 0 Display fractional, feet and inches, and surveyor's angles as previously set 1 Displays fractional, feet and inches, and surveyor's angles in input format
USERI1-5	Type: Integer Saved in: Drawing Initial value: 0	*USERI1, USERI2, USERI3, USERI4,* and *USERI5* are used for storage and retrieval of integer values.
USERR1-5	Type: Real Saved in: Drawing Initial value: 0.0000	*USERR1, USERR2, USERR3, USERR4,* and *USERR5* are used for storage and retrieval of real numbers.
USERS1-5	Type: String Not saved Initial value: ""	*USERS1, USERS2, USERS3, USERS4,* and *USERS5* are used for storage and retrieval of text string data.
VIEWCTR	(Read-only) Type: 3D point Saved in: Drawing	Stores the center of view in the current viewport, expressed as a UCS coordinate.
VIEWDIR	(Read-only) Type: 3D vector Saved in: Drawing	Stores the viewing direction in the current viewport expressed in UCS coordinates. This describes the camera point as a 3D offset from the target point.
VIEWMODE	(Read-only) Type: Integer Saved in: Drawing	Controls Viewing mode for the current viewport using bit code. The value is the sum of the following: 0 Turned off 1 Perspective view active 2 Front clipping on 4 Back clipping on 8 UCS Follow mode on

R14

Variable	Characteristic	Description
VIEWMODE (continued)		
		16 Front clip not at eye. If on, the front clip distance (*FRONTZ*) determines the front clipping plane. If off, *FRONTZ* is ignored, and the front clipping plane is set to pass through the camera point (vectors behind the camera are not displayed). This flag is ignored if the front clipping bit (2) is off.
VIEWSIZE	(Read-only) Type: Real Saved in: Drawing	Stores the height of the view in the current viewport, expressed in drawing units.
VIEWTWIST	(Read-only) Type: Real Saved in: Drawing	Stores the view twist angle for the current viewport.
VISRETAIN	Type: Integer Saved in: Drawing Initial value: 1	Controls visibility of layers in xref files. 0 The xref layer definition in the current drawing takes precedence over these settings: On/Off, Freeze/Thaw, color and linetype settings for xref-dependent layers. 1 On/Off, Freeze/Thaw, color, and linetype settings for xref-dependent layers in the current drawing take precedence over the xref layer definition.
VSMAX	(Read-only) Type: 3D point Saved in: Drawing	Stores the upper-right corner of the current viewport's virtual screen, expressed as a UCS coordinate.
VSMIN	(Read-only) Type: 3D point Saved in: Drawing	Stores the lower-left corner of the current viewport's virtual screen, expressed as a UCS coordinate.
WORLDUCS	(Read-only) Type: Integer Not saved	Indicates whether the UCS is the same as the World Coordinate System. 0 Current UCS is different from the World Coordinate System 1 Current UCS is the same as the World Coordinate System
WORLDVIEW	Type: Integer Saved in: Drawing Initial value: 1	Controls whether the UCS changes to the WCS during *DVIEW* or *VPOINT*. 0 Current UCS remains unchanged 1 Current UCS is changed to the WCS for the duration of *DVIEW* or *VPOINT; DVIEW* and *VPOINT* command input is relative to the current UCS
XCLIPFRAME	Type: Integer Saved in: Drawing Initial value: 0	Controls visibility of xref clipping boundaries. 0 Clipping boundary is not visible 1 Clipping boundary is visible

R14

Variable	Characteristic	Description
XLOADCTL	Type: Integer Saved in: Registry Initial value: 1	Turns xref demand loading on and off and controls whether it opens the original drawing or a copy. 0 Turns off demand loading; entire drawing is loaded 1 Turns on demand loading; reference file is kept open 2 Turns on demand loading; a copy of the reference file is opened When XLOADCTL is set to 2, the reference copy is stored in the AutoCAD temporary files directory (defined by *PREFERENCES*) or in a user-specified directory.
XLOADPATH	Type: String Saved in: Registry Initial value: ""	Creates a path for storing temporary copies of demand-loaded xref files. For more information, see *XLOADCTL*.
XREFCTL	Type: Integer Saved in: Registry Initial value: 0	Controls whether AutoCAD writes external reference log (XLG) files. 0 Xref log (XLG) files are not written 1 Xref log (XLG) files are written

Source: AutoCAD *Command Reference*

R14

APPENDIX B

AutoCAD Release 14 Command Alias List

Command	Alias	Command	Alias
3DARRAY	3A	DDVPOINT	VP
3DFACE	3F	DIMALIGNED	DAL
3DPOLY	3P	DIMALIGNED	DIMALI
ALIGN	AL	DIMANGULAR	DAN
APPLOAD	AP	DIMANGULAR	DIMANG
ARC	A	DIMBASELINE	DBA
AREA	AA	DIMBASELINE	DIMBASE
ARRAY	AR	DIMCENTER	DCE
ASEADMIN	AAD	DIMCONTINUE	DCO
ASEEXPORT	AEX	DIMCONTINUE	DIMCONT
ASELINKS	ALI	DIMDIAMETER	DDI
ASEROWS	ARO	DIMDIAMETER	DIMDIA
ASESELECT	ASE	DIMEDIT	DED
ASESQLED	ASQ	DIMEDIT	DIMED
ATTDEF	-AT	DIMLINEAR	DIMLIN
ATTEDIT	-ATE	DIMLINEAR	DLI
BHATCH	BH	DIMORDINATE	DIMORD
BHATCH	H	DIMORDINATE	DOR
BLOCK	-B	DIMOVERRIDE	DIMOVER
BMAKE	B	DIMOVERRIDE	DOV
-BOUNDARY	-BO	DIMRADIUS	DIMRAD
BOUNDARY	BO	DIMRADIUS	DRA
BREAK	BR	DIMSTYLE	DIMSTY
CHAMFER	CHA	DIMSTYLE	DST
CHANGE	-CH	DIMTEDIT	DIMTED
CIRCLE	C	DIST	DI
COPY	CO	DIVIDE	DIV
COPY	CP	DONUT	DO
DDATTDEF	AT	DRAWORDER	DR
DDATTE	ATE	DSVIEWER	AV
DDCHPROP	CH	DTEXT	DT
DDCOLOR	COL	DVIEW	DV
DDEDIT	ED	ELLIPSE	EL
DDGRIPS	GR	ERASE	E
DDIM	D	EXPLODE	X
DDINSERT	I	EXPORT	EXP
DDMODIFY	MO	EXTEND	EX
DDOSNAP	OS	EXTRUDE	EXT
DDRENAME	REN	FILLET	F
DDRMODES	RM	FILTER	FI
DDSELECT	SE	GROUP	-G
DDUCS	UC	GROUP	G
DDUCSP	UCP	HATCH	-H
DDUNITS	UN	HATCHEDIT	HE
DDVIEW	V	HIDE	HI

Command	Alias	Command	Alias
-IMAGE	-IM	REDRAW	R
IMAGE	IM	REDRAWALL	RA
IMAGEADJUST	IAD	REGEN	RE
IMAGEATTACH	IAT	REGENALL	REA
IMAGECLIP	ICL	REGION	REG
IMPORT	IMP	RENAME	-REN
INSERT	-I	RENDER	RR
INSERTOBJ	IO	REVOLVE	REV
INTERFERE	INF	ROTATE	RO
INTERSECT	IN	RPREF	RPR
-LAYER	-LA	SCALE	SC
LAYER	LA	SCRIPT	SCR
LEADER	LE	SECTION	SEC
LEADER	LEAD	SETVAR	SET
LENGTHEN	LEN	SHADE	SHA
LINE	L	SLICE	SL
-LINETYPE	-LT	SNAP	SN
LINETYPE	LT	SOLID	SO
LIST	LI	SPELL	SP
LIST	LS	SPLINE	SPL
LTSCALE	LTS	SPLINEDIT	SPE
MATCHPROP	MA	STRETCH	S
MEASURE	ME	STYLE	ST
MIRROR	MI	SUBTRACT	SU
MLINE	ML	TABLET	TA
MOVE	M	THICKNESS	TH
MSPACE	MS	TILEMODE	TI
-MTEXT	-T	TILEMODE	TM
MTEXT	MT	TOLERANCE	TOL
MTEXT	T	TOOLBAR	TO
MVIEW	MV	TORUS	TOR
OFFSET	O	TRIM	TR
-OSNAP	-OS	UNION	UNI
-PAN	-P	UNITS	-UN
PAN	P	VIEW	-V
PASTESPEC	PA	VPOINT	-VP
PEDIT	PE	WBLOCK	W
PLINE	PL	WEDGE	WE
PLOT	PRINT	XATTACH	XA
POINT	PO	-XBIND	-XB
POLYGON	POL	XBIND	XB
PREFERENCES	PR	XCLIP	XC
PREVIEW	PRE	XLINE	XL
PSPACE	PS	-XREF	-XR
PURGE	PU	XREF	XR
QUIT	EXIT	ZOOM	Z
RECTANGLE	REC		

R14

APPENDIX C

CAD References for *Fundamentals of Graphics Communication*, 2e, Bertoline, et al.

NA=Not Applicable—capabilities not provided in AutoCAD or not discussed.

APPENDIX D

CAD References for *Technical Graphics Communication*, 2e, Bertoline, et al.

NA=Not Applicable—capabilities not provided in AutoCAD or not discussed.

INDEX